Introduction to the Finite Element Method and Implementation with MATLAB®

Connecting theory with numerical techniques using MATLAB®, this practical textbook equips students with the tools required to solve finite element problems. This hands-on guide covers a wide range of engineering problems through nine well-structured chapters including solid mechanics, heat transfer, and fluid dynamics; equilibrium, steady state and transient; and 1-D, 2-D, and 3-D problems. Engineering problems are discussed using case-study examples, which are solved using a systematic approach, both by examining the steps manually and by implementing a complete MATLAB® code. This topical coverage is supplemented by discourse on meshing, with a detailed explanation and implementation of 2-D meshing algorithms. Introducing theory and numerical techniques alongside comprehensive examples, this text increases engagement and provides students with the confidence needed to implement their own computer codes to solve given problems.

Dr. Gang Li is a professor and D. W. Reynolds Emerging Scholar of Mechanical Engineering at Clemson University. He was an awardee of the National Science Foundation Early Career Award. He is an associate editor of the *Journal of Computational Electronics* and serves on the ASME Committee on Computing in Applied Mechanics. Dr. Li's scholarly articles on computational mechanics and finite element method frequently appear in leading journals.

An enormously accessible, didactic, and comprehensive text treating key engineering topics and providing the reader with the necessary elements of linear algebra, numerical methods, and meshing techniques, as well as numerous programming examples using MATLAB®. Professor Li's book can be used by teachers in the classroom for final-year undergraduate and graduate students, and by anyone else interested in learning the theory and computational implementation of the finite element method.

Gabriel Potirniche, University of Idaho

From one of the experts in the field, this book on the finite element method is a comprehensive and thorough guide for graduate and senior undergraduate students. The book is engaging not just in content but also in delivery. Its focus on step-by-step explanation and implementation is particularly useful for helping students to connect the theory and practice. The reusable MATLAB® functions and programs that are integrated with the theoretical content reinforce the important components of FEA and provide a unique learning experience. Detailed description of numerical analysis and meshing techniques is also a major plus since these topics are barely covered in existing FEA textbooks. This introductory FEA book is suitable for students of all engineering majors.

Narayana Aluru, University of Illinois

Introduction to the Finite Element Method and Implementation with MATLAB®

GANG LI
Clemson University, South Carolina

CAMBRIDGE
UNIVERSITY PRESS

University Printing House, Cambridge CB2 8BS, United Kingdom

One Liberty Plaza, 20th Floor, New York, NY 10006, USA

477 Williamstown Road, Port Melbourne, VIC 3207, Australia

314–321, 3rd Floor, Plot 3, Splendor Forum, Jasola District Centre, New Delhi – 110025, India

103 Penang Road, #05–06/07, Visioncrest Commercial, Singapore 238467

Cambridge University Press is part of the University of Cambridge.

It furthers the University's mission by disseminating knowledge in the pursuit of education, learning, and research at the highest international levels of excellence.

www.cambridge.org
Information on this title: www.cambridge.org/9781108471688
DOI: 10.1017/9781108559058

© Cambridge University Press 2021

This publication is in copyright. Subject to statutory exception and to the provisions of relevant collective licensing agreements, no reproduction of any part may take place without the written permission of Cambridge University Press.

First published 2021
Reprinted 2021

Printed in the United Kingdom by TJ Books Limited, Padstow Cornwall

A catalogue record for this publication is available from the British Library.

ISBN 978-1-108-47168-8 Hardback

Additional resources for this publication at www.cambridge.org/introtofem

Cambridge University Press has no responsibility for the persistence or accuracy of URLs for external or third-party internet websites referred to in this publication and does not guarantee that any content on such websites is, or will remain, accurate or appropriate.

To my family

Contents

	Preface	*page* xi
1	**Introduction**	1
	1.1 Finite Element Analysis and Its Procedure	1
	1.2 A Brief History of the Finite Element Method	7
	1.3 FEA Applications, Software, and Trend	9
	1.4 Major FEA Codes	13
	References	14
2	**Mathematical Preliminaries**	16
	2.1 Overview	16
	2.2 Vector and Matrix Algebra	16
	2.3 Vector Calculus	30
	2.4 Taylor Series Expansion	33
	2.5 Integration by Parts	34
	2.6 Divergence Theorem	36
	2.7 Fundamentals of Variational Calculus	36
	2.8 Summary	48
	2.9 Problems	48
	References	51
3	**Numerical Analysis Methods**	52
	3.1 Overview	52
	3.2 Numerical Interpolation and Approximation	52
	3.2.1 Lagrange Interpolation	53
	3.2.2 Hermite Interpolation	61
	3.2.3 Piecewise Polynomial Interpolation	64
	3.3 Numerical Differentiation	66
	3.4 Numerical Integration	69
	3.5 Systems of Linear Equations	76
	3.6 Summary	79
	3.7 Problems	80
	References	82

4 General Procedure of FEA for Linear Static Analysis: 1-D Problems — 84
- 4.1 Overview — 84
- 4.2 General Procedure of FEA for Linear Static Analysis — 84
- 4.3 1-D Elasticity — 86
  - 4.3.1 A One-Dimensional Axially Loaded Elastic Bar — 86
  - 4.3.2 Computer Implementation — 111
- 4.4 1-D Steady State Heat Transfer — 126
  - 4.4.1 Steady State Heat Transfer in a 1-D Rod — 126
  - 4.4.2 Computer Implementation — 136
- 4.5 1-D Fluid Flow — 139
  - 4.5.1 Steady State Mass Transport by 1-D Fluid Flow — 140
  - 4.5.2 Computer Implementation — 146
- 4.6 Summary — 149
- 4.7 Problems — 150
- *References* — 154

5 FEA for Multi-Dimensional Scalar Field Problems — 155
- 5.1 Overview — 155
- 5.2 2-D Steady State Heat Transfer — 155
  - 5.2.1 Steady State Heat Transfer in a Two-Dimensional Plate — 155
  - 5.2.2 Computer Implementation — 210
- 5.3 3-D Steady State Heat Transfer — 216
  - 5.3.1 3-D Heat Transfer in a Heat Sink — 217
  - 5.3.2 Computer Implementation — 235
- 5.4 Numerical Issues and Performance Considerations — 239
  - 5.4.1 Convergence Considerations — 240
  - 5.4.2 The Patch Test — 243
  - 5.4.3 Element Quality — 244
- 5.5 Further Considerations — 246
- 5.6 Summary — 247
- 5.7 Problems — 249
- *References* — 258

6 Mesh Generation — 259
- 6.1 Overview — 259
  - 6.1.1 Geometric Modeling of Objects — 260
  - 6.1.2 2-D Meshing Methods — 262
- 6.2 Modeling of 2-D Geometries Using Planar Straight Line Graph — 266
- 6.3 Delaunay Triangulation and Refinement Meshing — 268
  - 6.3.1 Delaunay Triangulation — 269
  - 6.3.2 Delaunay Refinement — 275
- 6.4 Computer Implementation — 282
  - 6.4.1 Data Structures — 283

		6.4.2 Auxiliary Geometry Operations	294
		6.4.3 Meshing Functions	299
	6.5	Summary	315
	6.6	Problems	315
	References		318

7 FEA for Multi-Dimensional Vector Field Problems — 319

	7.1	Overview	319
	7.2	2-D Elasticity	319
		7.2.1 Linear Elasticity: A Brief Review	320
		7.2.2 2-D Elasticity: Thin Plate Subject to External Loads	330
		7.2.3 Computer Implementation	360
	7.3	3-D Elasticity	363
		7.3.1 3-D Elastic Structure Subjected to External Loads	363
		7.3.2 Computer Implementation	373
	7.4	2-D Steady State Incompressible Viscous Flow	378
		7.4.1 2-D Steady State Cavity Driven Flow	378
		7.4.2 Computer Implementation	388
	7.5	Summary	394
	7.6	Problems	395
	References		404

8 Structural Elements — 405

	8.1	Overview	405
	8.2	Trusses	405
		8.2.1 3-D Truss System Subjected to External Loads	406
		8.2.2 Computer Implementation	416
	8.3	Beams and Space Frames	419
		8.3.1 3-D Space Frame Structure Subjected to External Loads	419
		8.3.2 Computer Implementation	435
	8.4	Plates	438
		8.4.1 Kirchhoff Plates Subjected to External Loads	438
		8.4.2 Computer Implementation	452
	8.5	Summary	456
	8.6	Problems	456
	References		462

9 FEA for Linear Time-Dependent Analysis — 463

	9.1	Overview	463
	9.2	2-D Transient Heat Transfer	463
		9.2.1 Transient Heat Transfer in a Square Plate	464
		9.2.2 Computer Implementation	474
	9.3	Elastodynamics	480

　　　　9.3.1 Vibration and Dynamic Response of Continuum
　　　　　　　Structures   480
　　　　9.3.2 Computer Implementation   494
9.4　Summary   499
9.5　Problems   500

Index   506

Preface

With unprecedented computing power, data transfer speed, and data processing capacity, today's engineering is experiencing a revolution of the digital age. What we are witnessing today is the explosive trend of digitization and computation of everything: from atoms to the Milky Way, from 3-D printing to brain engineering, from artificial intelligence to virtual reality ... Among other key technologies, physics based computing lies at the core of this revolution, and the Finite Element Method (FEM) is one of the most powerful enabling tools to do that. This is precisely why Finite Element Analysis (FEA) is one of the core courses taught in many Engineering departments, at undergraduate and/or graduate level. While there are many good introductory textbooks to the Finite Element Method, the 12 years I have spent teaching this subject have shown me that the conventional introductory approach covers much of the theory, but doesn't allow for much practical application. At the end of the semester, students could rarely implement a complete computer code to solve a given engineering problem. In addition to this, I found that students were able to learn the subject most effectively when they were required to implement their own finite element codes to solve given engineering problems. This observation gave me the motivation to write this introductory level textbook which emphasizes the connection from the mathematical foundation of the method, to the procedure of the numerical analysis, and then further to the implementation of a computer code. For the past several years I have been implementing this idea in my own teaching, and writing my own notes for my classes. The student learning outcomes and feedback were positive and encouraging. Now that the content of my teaching notes has become sufficiently mature, I feel that I am ready to transform the notes into a textbook.

This book is designed for senior undergraduate and first-year graduate students. As prerequisites, the students are expected to have:

1. A basic knowledge of linear algebra, ordinary and partial differential equations, and vector analysis.
2. A basic knowledge of solid mechanics, heat transfer, and fluid dynamics.
3. Basic skills with a programming language (MATLAB®, C, C++, Java, Python, Fortran, etc.)

The topics of the textbook include: an introduction to the mathematical foundation, solution procedure, and numerical implementation of the finite element method; applications to heat transfer, fluid flow, and structural analysis of solids; introduction to transient and dynamic analysis; analysis strategies using finite elements; introduction to solid modeling, meshing, data structures, and computer algorithms for the development of finite element computer codes in MATLAB.

The textbook aims to enable students to:

1 Gain an in-depth understanding of the mathematical formulation and the numerical implementation of the finite element method for solving engineering problems.
2 Apply numerical algorithms and techniques that are necessary for the numerical implementation of FEM.
3 Develop their own computer codes and obtain Finite Element (FE) solutions for a variety of engineering problems, including static/transient heat transfer, structural, elasticity, and fluid dynamics problems.

Some of the unique features of the book include the following:

- Implementation oriented step-by-step demonstration of the finite element procedure for solving linear solid mechanics, heat transfer, and fluid flow problems.
- Introduction of practical modeling and the numerical techniques required to implement a complete finite element program using MATLAB: data structures and storage, solid modeling, meshing, data visualization, solution of linear system of equations, numerical integration and differentiation, interpolation and approximations.
- The book gives the big picture at the beginning and enables the students to quickly understand the framework of the method. The students will be able to implement the method and use their own codes to solve 1-D elasticity problems in just three weeks. After that, more advanced numerical techniques and more complicated 2-D and 3-D problems can be learned by building upon this established understanding and implemented program.
- MATLAB codes are provided for all numerical examples and exercise problems. All the MATLAB codes can be accessed and downloaded from the book's online resource portal at www.cambridge.org/introtofem.
- Fifteen engineering problems of elasticity, heat transfer and fluid flow and mesh creation examples fully worked and described in step-by-step detail.
- Over 110 end-of-chapter problems for students to practice.

Pedagogically, I would suggest go through Chapter 2 quickly and only discuss the contents that senior undergraduate or first year graduate students may not be familiar with, such as vector and matrix norms, condition number, and variational calculus. After that, I would skip Chapter 3 and directly start the 1-D elasticity problem. The numerical analysis methods are presented in a separate chapter (Chapter 3) to avoid repetition. When the relevant numerical methods are needed in the discussion of the

FEA procedure for various types of physical problems, I will go back to Chapter 3 and discuss the content. This book is organized in a way that it tells a story for each type of engineering problem covered. At the beginning of each story an engineering problem is given. The story expands itself when we go on analyzing and solving the problem step by step. The story ends when the FE solution to the problem is obtained. The math and numerical methods are integrated parts of the storyline. Once the story is done, the students are expected to have gained intimate knowledge of all things related to the FEA of that type of problem. In fact, the students are expected to be able to work the problem manually and solve the problem on paper by using the FEM, with the help of MATLAB just for the algebraic calculations. With that, the computer implementation becomes natural and straightforward. As an added benefit, the students will be able to check their computer implementation by comparing intermediate results obtained from their MATLAB codes with their manual calculations. This dual-path learning process is carried out for types of the most problems covered in the book. The concepts and methods that are common for multiple types of problems are reinforced multiple times over the chapters to enable the students to better retain what they have learned. On the other hand, when we go through the steps of solving a different problem in a new chapter, the new or different content in the story becomes obvious and easy to grasp since the students have previous stories for comparison.

I would like to thank the students and colleagues who have provided their feedback, advice, and criticisms for the improvement of this book. I am also grateful to the editors and staff members at Cambridge University Press, especially Steve Elliot, Lisa Pinto, and Stefanie Seaton, for their help in publishing this manuscript. I would also like to thank Dale and Jackie Reynolds for their generous support through the Dale Reynolds 67 Emerging Faculty Scholar Endowment at Clemson University.

1 Introduction

1.1 Finite Element Analysis and Its Procedure

While there is no "standard" definition of finite element method (FEM) or finite element analysis (FEA), the general consensus in the literature is that finite element analysis is a general purpose computational technique for obtaining numerical solutions to mathematical models of physical problems. Many physical problems can be described using a (or a set of) mathematical equation(s).

Compared to other numerical methods, the most significant advantage of FEM is its versatility, which lies in two aspects: (1) it is versatile for arbitrary complex geometries, (2) it is versatile for a wide range of physical problems including structural mechanics, heat transfer, fluid dynamics, electromagnetics, and problems in many other areas. The versatility of the method can be attributed to a few characteristics of the method:

- It has a well-established mathematical foundation. The method itself is closely related to the fundamental problems of variational calculus, that is, finding a unique function that minimizes a functional that has this function as a variable. Since this fundamental problem is essentially the fundamental problem of many engineering problems, a method of finding the solution to this fundamental problem is essentially a method of solving all these engineering problems. Furthermore, its convergence properties and error bounds can be obtained by using functional analysis methods. For these reasons, the behavior and performance of finite element solutions are relatively well understood and the accuracy of the solutions is guaranteed by its convergence properties.
- The finite element discretization, approximation, and numerical integration schemes are simple yet powerful numerical schemes that work well on irregular geometries. The implementation of finite element models into computer codes is straightforward and can easily be modularized, making the software development tasks less complex.
- The compact support of the shape functions enables small interaction distances among the nodes, i.e., the nodes in an element are only affected by those in the neighboring elements. This property leads to very sparse coefficient matrices in the linear systems generated by the finite element discretization. In addition, in most cases, the coefficient matrices are symmetric due to the "bilinear" form in the finite element formulation. Linear systems with sparse and symmetric coefficient

matrices require small memory storage and can be solved efficiently by using special algorithms suitable for symmetric matrices.

Due to the simplicity and versatility of the FEM, since the first commercial code NASTRAN (short for "NASA structural analysis") released in 1969, FEA techniques and tools have been quickly adopted and widely used in many of the manufacturing and consumer goods industries. Among those, the automotive, aerospace and defense, and electrical and electronics industries are the ones whose product design and development rely heavily on FEA. In parallel, the FEA software and service sector has grown into a billion dollar industry itself. It has been forecast that the global FEA software market is to expand at a compound annual growth rate of 9.1% from 2016 to 2024 (Goldstein 2019). The market is anticipated to reach USD 2.6 billion by the end of 2024.

So what does FEA involve exactly? When we talk about FEA, depending on the context, we may actually refer to very different scenarios. There is a very large difference in the meaning of "FEA" when a design engineer is performing a finite element design analysis for an engineering design of a product (e.g. an auto part), and when a software development engineer is conducting finite element analysis for a class of engineering problems for the purpose of developing finite element models and implementing a computational solver as the final product of his/her line of work. Figure 1.1 illustrates this difference. In essence, the design engineer is a user of FEA software tools. A design engineer is only concerned with how good his/her design of a physical product or solution to a particular engineering problem is in terms of performance and cost. In comparison, the FE developer focuses on the finite element method itself, works on building, mathematically and numerically, and implementing FE model(s), and typically deals with not just one but a class of engineering problems. In other words, the FE developer develops finite element models for a class of engineering problems and implements them into a finite element software package for design engineers to use. From this point of view, this book is for "FE developers." We discuss the basic theories, principles, techniques, and methods that constitute the foundation and building blocks of the finite element method. The intended audience of this book are senior undergraduate and graduate students who would like to become FE developers or want to acquire an under-the-hood understanding of FEM, as well as junior FE development professionals who need a systematic overview of the method. This book does not discuss how to use commercial FE packages to perform FEA for various engineering problems. Instead, we focus on the mathematical formulation, computational modeling, and implementation aspects of the FEM for several classes of engineering problems, namely linear elasticity, heat transfer, and fluid dynamics problems.

Broadly speaking, a complete FEA procedure can be divided into (1) formulation and pre-processing, (2) solution, and (3) post-processing stages. Here we use an example to briefly illustrate, from an internal point of view, a typical FEA procedure for solving a linear elasticity problem. Figure 1.2 shows a thin plate with holes and notches subjected to uniformly distributed tensile loads at its left and right edges. Assuming the dimensions, material properties, and loads are given, and the structure

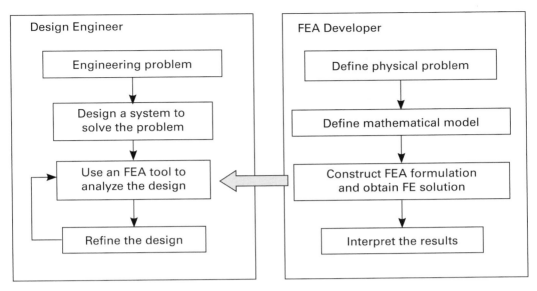

Figure 1.1 The meaning of FEA for a design engineer and an FE developer.

is under isothermal condition, the objective of FEA is to obtain the deformation and stress distribution of the plate.

Before an analysis strategy and modeling approach are determined, we first need to have a clear (and correct) definition of the physical system and problem. In this step, we would ask ourselves a set of questions:

- What type of system is this: a single or a multi-object system?
- What type of physical behavior/phenomenon are we investigating? Is it a mechanics, heat transfer, electromagnetics, acoustics, or multiphysics problem? Is it an equilibrium, steady state, or transient/dynamic problem? Is it a linear or nonlinear problem?
- What is the dimensionality of the problem: 1-D, 2-D or 3-D?
- What are the relevant physical properties of the object(s) in the system?

The answers to these questions give the definition of the physical system under consideration. In our example shown in Fig. 1.2, the system is a single structure system. We are interested in the equilibrium mechanical behavior of this structure and would like to obtain its deformation and stress profiles under the given loading condition. The structure can be treated as a two-dimensional problem due to its geometric feature (thin plate) and loading condition (in-plane loading). The plate is symmetric with respect to both x- and y-axes. If the given load is small, small displacement is expected and the theory of linear elasticity applies. The required material properties include Young's modulus and Poisson's ratio.

Once the physical problem is defined, the next step is defining a mathematical model to describe the physical behavior of the system. A well-defined mathematical model should faithfully represent the physics of the given problem and should contain

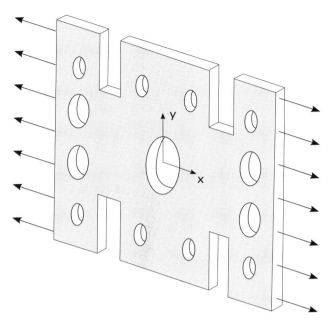

Figure 1.2 A thin plate is subjected to uniformly distributed loading on its left and right edges.

proper geometric information of the domain, loads, and boundary conditions. The mathematical equations (called governing equations) along with the loading and boundary conditions should be sufficient to guarantee a unique solution to the model.

For our example, due to the symmetry of the geometry and loads as shown in Fig. 1.2, the deformation of the plate is symmetric about the x- and y-axes. Therefore, it is only necessary to calculate the deformation and stress of a quarter of the plate and the rest are just the mirror images of the quarter. Figure 1.3 shows the computational domain (denoted as Ω) which is a quarter of the plate, geometric boundary (denoted as Γ), boundary conditions, and loads. Note that the roller-type boundary conditions at the bottom and left edges of the computational domain are the results of the symmetric displacement field along the x- and y-axes. That is, only x-displacement along the x-axis and only y-displacement along the y-axis are allowed if the displacement field is symmetric along the x- and y-axes. As the physical problem is defined as a linear static elasticity problem, the governing equation is the well-known differential equation of equilibrium

$$\nabla \cdot \sigma = \mathbf{b} \qquad in \; \Omega$$

where σ is the Cauchy stress tensor and $\mathbf{b}$ is the body force vector. The constitutive relation is the generalized Hooke's law

$$\sigma = \mathbf{C}\epsilon$$

where ϵ is the strain vector and $\mathbf{C}$ is the material stiffness matrix. The boundary conditions are

$u_x = 0$ along the left edge of the domain
$u_y = 0$ along the bottom edge of the domain
$\sigma_{xx} = P$ along the right edge of the domain.

The governing equations and boundary conditions altogether give a complete set of mathematical equations describing the mechanical behavior of the plate and having a unique solution.

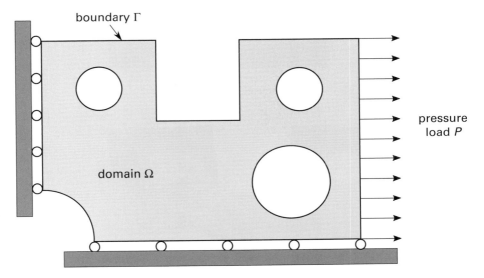

Figure 1.3 Domain of the computational model of the plate.

The steps described above are categorized to be in the pre-processing stage. Next, in the solution stage, we first convert the differential governing equations into one or several integral equations called weak form. The computational domain is discretized into a set of non-overlapping small regions of primitive shapes called elements. In each element, the element vertices and sometimes also a set of points on the element edges are set to be the "nodes" on which the unknown physical variable(s) of interest is (are) to be calculated. This geometric discretization process is called meshing. A mesh of the plate in our example is shown in Fig. 1.4. Then, the unknown physical variables are approximated over each element by using functions of simple forms. Such functions are called shape functions or basis functions. The approximation process is referred to as the finite element approximation. Substituting the finite element approximation into the weak form and discretizing the weak form according to the geometric discretization, the weak form can be converted into a set of algebraic equations that can be solved numerically by using a computer.

In the solution stage, the process of (1) converting the differential governing equation(s) into weak form, (2) discretization of the computational domain and the weak form, and (3) approximating the unknown variables over the elements is referred to as developing the finite element formulation of a mathematical model. Having obtained

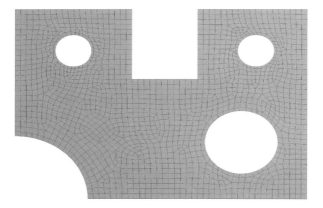

Figure 1.4 Meshing the computational model.

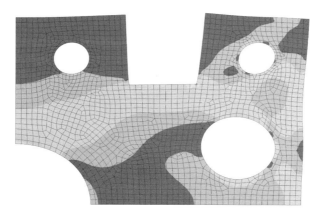

Figure 1.5 Von-Mises stress distribution in the deformed plate.

the discretized finite element formulation, by using numerical analysis methods and algorithms such as numerical integration and numerical solution of simultaneous algebraic equations, one can structure and code a program that uses the input information of geometry, physical properties, loads, and boundary conditions, and calculates the desired unknown physical variables over the nodes and elements. For our plate example, the primary unknown physical variables are the displacements. By solving the simultaneous algebraic equations obtained, the displacements of the nodes are calculated. Adding the displacements to the original positions of the nodes, the deformed shape of the plate is obtained, as shown in Fig. 1.5. After that, by using the now-known nodal displacements, strains can be obtained by calculating the derivatives of the displacements, and stresses can be calculated by using Hooke's law. The calculated strains and stresses can be visualized using contour plots. For example, the von-Mises stress distribution over the plate is visualized in Fig. 1.5. The calculation of the secondary physical quantities such as the strains and stresses in this example, and the visualization and interpretation of the numerical results all belong to the post-processing stage of FEA.

1.2 A Brief History of the Finite Element Method

The finite element method, whether in its early days or as in its current state, was not invented or developed by a single individual. The method was formed gradually by many pioneers from two sides: the mathematics side and the engineering side. The major developments before the FEM became a general purpose practical analysis tool are depicted in Fig. 1.6.

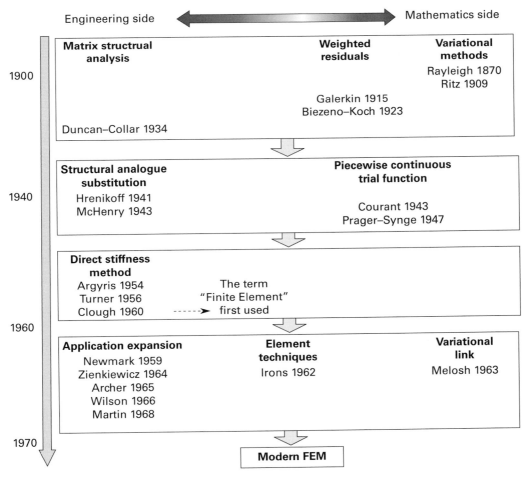

Figure 1.6 A brief history of the finite element method.

On the mathematics side, much earlier than the emergence of the concept of finite element, two classes of methods were developed for solving differential equations: the method of weighted residuals and the variational method. In the weighted residual method, the unknown solution is expressed as a linear combination of a set of "trial" functions. The coefficients of the trial functions are obtained by requiring a weighted integral of the residual to be zero. Depending on the choice of the weight function, the weighted residual method can be divided into several sub-categories: collocation, Galerkin, least squares, sub-domain methods, etc. As a kind of weighted residual method, the least squares method was originated by Gauss in 1795 for least squares estimation. The least squares method uses the derivatives of the residual with respect to the unknown coefficients of the trial functions as the weight functions. In comparison, Russian engineer Galerkin (1915) used the trial functions themselves as the weight function, leading to the Galerkin weighted residual method. The sub-domain weighted residual method was first employed by Biezeno and Koch (1923) for structural stability problems. A different type of method for solving differential equations is based on the variational principle. The variational method was first used by Rayleigh (1870) and later by Ritz (1909). While developed independently, there is a close relationship between the weighted residual method and the variational method. In fact, as will be shown later in this book, the Galerkin weighted residual method is equivalent to the variational method in many cases. Although the variation and weighted residual methods enabled approximated solutions to a variety of types of differential equations, their practical use was hindered by the requirement that the trial functions must span the entire domain and satisfy the boundary conditions. The application of the methods was difficult in the analysis of large structural systems.

On the engineering side, the pioneers were structural engineers. Starting from the early 1930s, there was a trend to represent large structures with complex geometry by using smaller and simpler structures connected to each other. A representative method is the matrix structural analysis approach developed by Duncan and Collar (1934). In the early 1940s, Hrennikoff (1941) and McHenry (1943), separately, developed lattice analogies to represent continuum structures using connected beams and bars. The lattice analogy only achieved limited success. It did not draw much attention and investment until the then called direct stiffness method (DSM) was generalized by Turner et al. in 1956 (Turner 1956). Also in the 1940s, Courant (1943) used piecewise linear interpolation over 2-D triangular elements as Rayleigh–Ritz trial functions. Unfortunately, his work was not followed up immediately as it was advertised as "generalized finite differences" in his article. In 1947, Prager and Synge (1947) also proposed a concept of regional discretization similar to Courant's work. During the period of 1952 to 1964 at Boeing, Turner oversaw the creation and expansion of the DSM, a procedure that constructs the stiffness matrices for beam, truss, and two-dimensional triangular and rectangular plane stress elements, and assembles them to obtain the global structure stiffness matrix. The success of the method with the use of digital computer in 1950s prompted further development of the element stiffness based methods and resource commitment from Boeing. Influenced by Turner's work, Argyris (1955) proposed matrix structural analysis methods using energy principles. In continuing Turner's work, Clough (1960) coined the term "Finite Element" in 1960.

The decade following 1960 was the golden age of FEM development. A series of developments in theories, techniques, and methods was completed during this time, greatly enhancing the applicability and performance of the finite element method. Melosh (1963) showed that the conforming displacement based models are a form of Rayleigh–Ritz method with the minimization of potential energy, and systematized the variational derivation of stiffness elements. This started the convergence of the Argyris' dual formulation of energy methods and the DSM of Turner, and also marked the beginning of the method to solve nonstructural applications. In terms of element techniques, Irons, Zienkiewicz, and others (Ergatoudis, Irons, and Zienkiewicz 1968) invented isoparametric elements, shape functions, and the patch test. For expanding the capability of the FEM for engineering applications, Newmark (1959) developed time integration schemes for structural dynamic analysis, and Archer (1965) developed the consistent-mass matrix. Zienkiewicz, Martin, and Wilison all made substantial contributions to expand the FEM's application to a wide range of engineering areas in the 1960s. By mid 1970s, the finite element method had largely evolved into its present form.

1.3 FEA Applications, Software, and Trend

Accompanied by the development of the finite element method itself, the first set of commercial FEA packages were developed in the late 1960s and early 1970s. The well-known ones included NASTRAN, ANSYS, and ABAQUS. The first commercial version of NASTRAN FEA software was released in 1969. The first commercial version of ANSYS software was labeled version 2.0 and released in 1971. The first ABAQUS version was released in 1979.

In its early days, FEA was mainly used in the aerospace industry and had a small footprint in civil engineering. In the 1970s, along with the advances in computer aided design and manufacturing (CAD/CAM) technology, especially the development of mathematical representations of arbitrary curved surfaces, the application of FEA quickly spread into new areas such as the automotive industry. However, before 1980, its application was still largely limited by the high cost of computers and their limited computing power. In 1964, the IBM 360 mainframe computer produced 1 MFLOPS (floating point operations per second) computing power with a price tag of $2.5~3M. In 1976, Cray-1, a supercomputer at that time, had 80 MFLOPS computing power at a cost of $5~8M. Due to this limitation, large FEA calculations could only be performed by large companies and government agencies such as national labs. Entering the 1980s, the computing power of the large computers grew very rapidly while the cost remained more or less the same. At the same time, much smaller workstations with microprocessors, such as Apollo Computer and Sun Microsystems, appeared. Such systems typically had MFLOPS performance for a cost of $15~100K. FEA became more affordable. In the early 1980s, FEM application became popular in the automotive industry.

Starting from the end of the 1980s, the development of smaller high performance workstations and personal computers (PCs) started to grow at a stunning speed. For example, in 1991, a business-class PC with Intel 486/33 microprocessor produces 30 MFLOPS computing performance, and costs around $4K if equipped with 4 MB of RAM, a 200 MB hard disk and 14 inches display. This trend of exponential growth of computing power known as Moore's law has continued to the present day. Today, the computing power of a 6-core Intel Core i7 8700 desktop processor's scientific computing power reaches as much as 70 GFLOPS. That's almost 900 times more powerful than Cray-1. The price of the Intel processor is under $400, compared to the $5~8M price tag for Cray-1 in 1976. The cost per GFLOPS is cut down further due to the rise of the GPUs (Graphic Processing Unit). For example, NVDIA's GeForce GTX 970 delivers 3,494 GFLOPS processing power for single precision floating point operations. The price of the unit is slightly above $300 in 2018, which makes the cost per GFLOPs less than $0.1. Figure 1.7 depicts the 10 orders of magnitude decrease of cost per GFLOPs over the last 60 years.

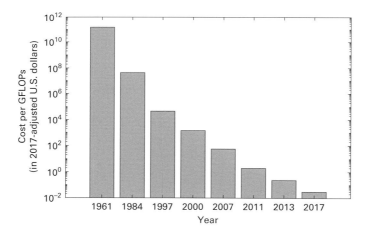

Figure 1.7 The decrease of cost per GFLOPs over the last 60 years.

Accompanied by the explosive growth of the computing power that became easily accessible to engineers and researchers, the modeling capabilities and solving power of FEM kept growing at a fast pace. The application of FEA quickly spread into disciplines and fields outside of structural mechanics. As early as 1960, FEM was applied to heat transfer, fluid mechanics, and then thermomechanical problems. For fluid dynamics problems, FE formulations were developed in parallel with the development of other numerical methods including finite difference and finite volume methods. In 1965, Gladwell (1965) first introduced FEM formulation for acoustics problems. In the early 1980s, the FEM was employed for solving electrostatic and electromagnetic problems. Then, the FEM was applied to solve quantum mechanical systems (Ram-Mohan, Saigal, Dossa, & Shertzer 1990), and systems in earth science, plasma physics, bio and chemical engineering, and more. In 1994, the first flight of Boeing 777 was tested entirely via ANSYS. FEA became the most versatile computational analysis tool across academia and industries.

Starting from the mid 1990s, the FEM has experienced several waves of new development due to the rapid advances in microelectronics, information technology, micro- and nano-technology, as well as parallel, distributed, and more recently cloud computing. Among others, four significant trends have pushed the boundary of the FEM:

- **Generalization of FEM.** Based on the fundamental principles of FEM, including the variational principles and method of weighted residuals, new methods and techniques were developed to improve the capability of the method for analyzing more complex problems, or to enhance the method's performance and reduce its limitations. Several major developments include:
  - *Discontinuous Galerkin method.* In view of the advantages of FEM and finite volume method (FVM), the Discontinuous Galerkin method (DG-EFM) was first proposed by Reed and Hill in 1973. Similar to FEM, the DG-EFM is also based on the weak formulation of the physical system. However, the Galerkin weighted residuals are set to vanish locally and the local domains are connected through numerical fluxes at their interfaces, similar to the interface fluxes in FVM. Due to its localization of the weak form, DG-FEM has a higher flexibility of local approximation and mesh adaptivity. The last decade has seen a large increase in the research activities of DG-EFM. It has been applied to many engineering problems including electromagnetics, structural mechanics, fluid dynamics, plasma physics, etc.
  - *Extended FEM (XFEM).* The extended FEM was introduced by Belytschko and Black in 1999. XFEM is an extension of the standard FEM by adding discontinuous functions and nodal enriched degrees of freedom in an element based on the concept of partition of unity. The method has been successfully applied to model the propagation of various discontinuities such as cracks and weak and material interfaces. XFEM has also been implemented in commercial codes like ABAQUS and Altair Radioss.
  - *Spectral element methods.* The spectral element method (SEM) is a variation of the standard FEM that uses high order spectral functions as the basis functions. The spectral element method was first introduced by Patera in 1984. Due to the high order orthogonal basis functions, the SEM has naturally diagonal mass matrices, exponential convergence characteristics, and built-in p-refinement capability. Due to these properties, the SEM has been used for applications requiring high accuracy and convergence rate. It is also advantageous to use spectral elements in domain decomposition and component mode synthesis calculations.
  - *Meshless methods.* Generating high quality mesh can be computationally time consuming, especially for complex geometries. In view of this difficulty, the

idea of making the entire finite element analysis meshless or mesh-free was very attractive. Since Belytscho's Element Free Galerkin method (1996), many variations of the meshless methods have been developed. Broadly defined, meshless methods contain two key steps: construction of meshless approximation functions or shape functions and their derivatives, and meshless discretization of the governing partial differential equations (PDEs). Least squares, kernel-based, and radial basis function approaches are three techniques that have gained considerable attention for construction of meshless shape functions. The meshless discretization of the partial differential equations can be categorized into three classes: cell integration, local point integration, and point collocation.

- **Parallel and cloud computing.** Over the past 20 years, parallel computing was popularized by ever faster networks, distributed systems, and multi-processor and multi-core processor computer architectures. Parallel application programming interface (API) standards such as openMP and MPI were established for shared memory and distributed systems, respectively, in the late 1990s. Following this trend, almost all major FEA codes have been parallelized for high performance computing (HPC) systems. The parallelization greatly increased the upper limit of the degrees of freedom (DOFs) for a simulation. For example, it was reported by a group of Japanese scientists in 2015 that they were able to perform earthquake simulations with 1.08 trillion DOFs on a HPC with 663,552 cores for 29.7 s per time step (Ichimura et al. 2015). More recently, cloud computing has become more popular, offering the opportunity of cloud based FEA via bridging the computing resources and the mass market. In 2013, SimScale released the world's first cloud-based 3-D simulation platform.

- **Mutliphysics simulation.** Multiphysics modeling and simulation of materials, structures, and systems involves coupling of multiple physical phenomena and obtaining a consistent solution among the associated physical energy domains. For FEA, multiphysics simulation is not a new development. During the early days of FEM, coupled analysis of thermomechanical and fluid-structure interaction problems were already investigated. However, there is a clear trend that we are facing more and more complex and challenging problems involving multiple interacting physical processes. For example, many micro- and nano-electromechanical systems (MEMS/NEMS) based ultra small and sensitive sensors are operated by using electrostatically driven mechanical structures. Batteries are typically governed by the coupling of chemical reactions, fluid dynamics, and ion transport. In the modeling of multiphysics systems, governing equations of the physical domains are solved either simultaneously or iteratively, and the physical variables that are shared by these physical domains are updated at every iteration and every time step to satisfy all the governing equations. While the major commercial FEA packages all have some level of multiphysics capability, COMSOL, first version released in 1998, is a package that focuses on multiphysics problems.

- **Multiscale modeling.** When modern science and technology continue to explore "the room at the bottom," the lower size limit of practical engineering components and systems, such as microelectronics circuits and ultra small sensors/actuators, has been pushed from submicron to nanometers. As the characteristic length of such systems scales down to several tens of nanometers, nanoscale effects, such as quantum effects, material defects, and surface effects become significant. Classical models based on continuum assumptions or the computational design tools that have been developed for microsystems and macrosystems may not be directly applicable for nanoscale systems. However, although the characteristic length of the nanoscale components is often a few nanometers, the entire system could still be of the order of micrometers and contain millions of atoms. In this case, atomistic simulation methods such as ab initio calculations, molecular dynamics (MD), and Monte Carlo (MC) simulations, that can be employed for an accurate analysis of systems comprising several hundreds to several tens of thousands of atoms, are computationally impractical for design and analysis. Since the mid 1990s, multiscale modeling has drawn much attention and many methods have been developed over the last 20 years. Broadly defined, there are three multiscale modeling strategies: direct coupling, top-down, and bottom-up approaches. Direct coupling methods typically decompose the physical domain into atomistic, continuum, and interface regions. Atomistic and continuum calculations are performed separately and the interface regions are used to exchange information between the atomistic and continuum regions. Top-down approaches, such as the quasicontinuum (QC) method, the bridging scale method, and the heterogeneous multiscale method, solve the continuum equations by extracting constitutive laws from the underlying atomistic descriptions. In contrast, bottom-up methods such as the coarse-grained molecular dynamics and multigrid bridging approaches coarse-grain the atoms of the system into macroatoms and the fundamental equations defined on the atoms are coarse-grained into equivalent macroscale equations. Although much progress has been made, multiscale modeling is still in its early stages and attracting much research interest.

1.4 Major FEA Codes

There are numerous FEA codes available today. Some of them have a long history and some are relatively new. In addition, quite a few of them are open source. Separating the commercial and open source ones, some of the major FEA codes are listed below.

Commercial FEA codes:

- Abaqus (Company: Dassault Systems)
- Adina (Company: Adina)
- ANSYS (Company: ANSYS)
- Autodesk Simulation (Company: Autodesk)
- Cero Simulate (Company: PTC)
- COMSOL (Company: COMSOL)
- HyperWorks (Company: Altair)

- LS-DYNA (Company: LSTC)
- NASTRAN (Company: MSC software, Siemens)
- NISA (Company: CISL)
- Marc (Company: MSC software).

Open source FEA codes:

- CalculiX: www.calculix.de/
- Code_Aster: www.code-aster.org/spip.php?rubrique2
- Elmer: www.csc.fi/web/elmer
- FEniCS: https://fenicsproject.org/
- FreeFEM++: www3.freefem.org/

References

Archer, J. S. (1965), "Consistent matrix formulations for structural analysis using finite-element techniques," *AIAA Journal* **3**(10), 1910–1918.

Argyris, J. H. (1955), "Energy theorems and structural analysis: a generalized discourse with applications on energy principles of structural analysis including the effects of temperature and non-linear stress-strain relations part i. general theory," *Aircraft Engineering and Aerospace Technology* **27**(2), 42–58.

Clough, R. W. (1960), "The finite element method in plane stress analysis," in *Proceedings of 2nd ASCE conference on electronic computation*, Pittsburgh Pa., Sept. 8 and 9, 1960.

Courant, R. (1943), "Variational methods for the solution of problems of equilibrium and vibrations," *Bulletin of the American Mathematical Society* **49**, 1–23.

Duncan, W. & Collar, A. (1934), "Lxxiv. a method for the solution of oscillation problems by matrices," *The London, Edinburgh, and Dublin Philosophical Magazine and Journal of Science* **17**(115), 865–909.

Ergatoudis, I., Irons, B. & Zienkiewicz, O. (1968), "Curved, isoparametric, quadrilateral elements for finite element analysis," *International Journal of Solids and Structures* **4**(1), 31–42.

Gladwell, G. (1965), "A finite element method for acoustics," in *Proceedings of fifth international conference on acoustics*, (Liege, 1965), paper L33.

Goldstein (2019), "Global Finite Element Analysis (FEA)," *Software Market Outlook 2016–2024*, Technical report, Goldstein Research.

Hrennikoff, A. (1941), "Solution of problems of elasticity by the framework method," *Journal of Applied Mechanics* **8**(4), 169–175.

Ichimura, T., Fujita, K., Quinay, P. E. B., Maddegedara, L., Hori, M., Tanaka, S., Shizawa, Y., Kobayashi, H. & Minami, K. (2015), "Implicit nonlinear wave simulation with 1.08 t dof and 0.270 t unstructured finite elements to enhance comprehensive earthquake simulation," in *High Performance Computing, Networking, Storage and Analysis, 2015 SC-International Conference*, IEEE, pp. 1–12.

McHenry, D. (1943), "A lattice analogy for the solution of stress problems," *Journal of the Institution of Civil Engineers* **21**(2), 59–82.

Melosh, R. J. (1963), "Basis for derivation of matrices for the direct stiffness method," *AIAA Journal* **1**(7), 1631–1637.

Newmark, N. M. (1959), "A method of computation for structural dynamics," *Journal of the Engineering Mechanics Division* **85**(3), 67–94.

Prager, W. & Synge, J. L. (1947), "Approximations in elasticity based on the concept of function space," *Quarterly of Applied Mathematics* **5**(3), 241–269.

Ram-Mohan, L. R., Saigal, S., Dossa, D. & Shertzer, J. (1990), "The finite-element method for energy eigenvalues of quantum mechanical systems," *Computers in Physics* **4**(1), 50–59.

Turner, M. (1956), "Stiffness and deflection analysis of complex structures," *Journal of the Aeronautical Sciences* **23**(9), 805–823.

2 Mathematical Preliminaries

2.1 Overview

Briefly stated, the finite element method can be understood as a numerical method of solving physical problems. It is supported by three pillars: mathematical theories, numerical methods, and physical principles. In this chapter, we discuss basic mathematical concepts and methods that will be used as tools in the development of finite element formulations and solution of finite element models. Basic knowledge of linear algebra, calculus of vectors and matrices, variational calculus, and integral equations is necessary in the derivation of finite element formulations. These fundamental mathematical concepts and methods are reviewed in this chapter as the building blocks for the content of this book. Numerical methods for numerical approximation, differentiation, integration, discretization, and solution of linear systems will be discussed in Chapter 3. The mathematical tools and numerical methods are then utilized along with relevant physical principles in the illustration of the FEA procedure for different types of physical problems in later chapters.

2.2 Vector and Matrix Algebra

In linear algebra, a vector is defined as an ordered list of numbers. In n-dimensional space, a vector contains n numbers listed in a column, denoted as

$$\mathbf{a} = \begin{Bmatrix} a_1 \\ a_2 \\ \vdots \\ a_n \end{Bmatrix} \tag{2.1}$$

and its transpose is given by

$$\mathbf{a}^T = \{a_1 \ a_2 \ \cdots \ a_n\} \tag{2.2}$$

where an element in the vector is denoted as a_i.

Scalar Product

The dot (or inner, scalar) product of two vectors $\mathbf{a}$ and $\mathbf{b}$ is defined as

$$\mathbf{a} \cdot \mathbf{b} = \mathbf{a}^T \mathbf{b} = \|\mathbf{a}\| \|\mathbf{b}\| \cos\theta \tag{2.3}$$

where $\|\mathbf{a}\|$ and $\|\mathbf{b}\|$ are the lengths of the two vectors, and θ is the angle between $\mathbf{a}$ and $\mathbf{b}$. In index notation

$$\mathbf{a} \cdot \mathbf{b} = a_1 b_1 + a_2 b_2 + \cdots + a_n b_n = \sum_{i=1}^{n} a_i b_i. \qquad (2.4)$$

There are several important properties of the scalar product:

1. The scalar product is commutative: $\mathbf{a} \cdot \mathbf{b} = \mathbf{b} \cdot \mathbf{a}$ or $\mathbf{a}^T \mathbf{b} = \mathbf{b}^T \mathbf{a}$.
2. The scalar product is distributive: $\mathbf{a} \cdot (\mathbf{b} + \mathbf{c}) = \mathbf{a} \cdot \mathbf{b} + \mathbf{a} \cdot \mathbf{c}$.
3. If $\mathbf{a}$ and $\mathbf{b}$ are perpendicular to each other, then $\mathbf{a} \cdot \mathbf{b} = 0$.
4. If $\mathbf{a}$ and $\mathbf{b}$ are parallel to each other, then $\mathbf{a} \cdot \mathbf{b} = \|\mathbf{a}\| \|\mathbf{b}\|$.

Vector Product

The cross (or outer, vector) product of two vectors $\mathbf{a}$ and $\mathbf{b}$ in three-dimensional space is defined as

$$\mathbf{a} \times \mathbf{b} = \begin{vmatrix} \mathbf{i} & \mathbf{j} & \mathbf{k} \\ a_1 & a_2 & a_3 \\ b_1 & b_2 & b_3 \end{vmatrix}$$

where $\mathbf{i}$, $\mathbf{j}$, and $\mathbf{k}$ are unit base vectors

$$\mathbf{i} = \begin{Bmatrix} 1 \\ 0 \\ 0 \end{Bmatrix} \qquad \mathbf{j} = \begin{Bmatrix} 0 \\ 1 \\ 0 \end{Bmatrix} \qquad \mathbf{k} = \begin{Bmatrix} 0 \\ 0 \\ 1 \end{Bmatrix}.$$

There are several important properties of the vector product:

1. The vector product is not commutative: $\mathbf{a} \times \mathbf{b} \neq \mathbf{b} \times \mathbf{a}$. Instead, $\mathbf{a} \times \mathbf{b} = -\mathbf{b} \times \mathbf{a}$
2. The vector product is distributive: $\mathbf{a} \times (\mathbf{b} + \mathbf{c}) = \mathbf{a} \times \mathbf{b} + \mathbf{a} \times \mathbf{c}$.
3. If $\mathbf{a}$ and $\mathbf{b}$ are parallel to each other, then $\mathbf{a} \times \mathbf{b} = 0$.

Example 2.1 Given $\mathbf{a} = \{2 \ 0 \ 0\}^T$ and $\mathbf{b} = \{2 \ 1 \ 0\}^T$, calculate $\mathbf{c} = \mathbf{a} \times \mathbf{b}$

Solution

$$\mathbf{c} = \mathbf{a} \times \mathbf{b} = \begin{vmatrix} \mathbf{i} & \mathbf{j} & \mathbf{k} \\ 2 & 0 & 0 \\ 2 & 1 & 0 \end{vmatrix} = 0\mathbf{i} + 0\mathbf{j} + 2\mathbf{k}. \qquad (2.5)$$

The geometric interpretation of the cross product is shown in Fig. 2.1. The cross product of $\mathbf{a}$ and $\mathbf{b}$ generates a new vector which is perpendicular to the plane formed by $\mathbf{a}$ and $\mathbf{b}$. The positive direction of $\mathbf{c}$ is determined by using the right-hand rule: going from $\mathbf{a}$ to $\mathbf{b}$, curl your right hand fingers into a half circle, your thumb now points in the positive direction of $\mathbf{c}$. The magnitude or length of $\mathbf{c}$ is given by

$$\|\mathbf{c}\| = \|\mathbf{a}\| \|\mathbf{b}\| \sin \theta \qquad (2.6)$$

which is twice the area of the triangle formed by **a** and **b**, as shown in Fig. 2.1. It is easily seen that in this example

$$\|\mathbf{c}\| = \|\mathbf{a}\|\|\mathbf{b}\|\sin\theta = 2 \times 1 = 2. \tag{2.7}$$

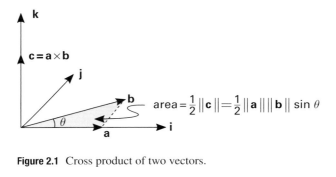

Figure 2.1 Cross product of two vectors.

Matrix in a Linear System

Considering a set of linear simultaneous equations

$$5x_1 - 4x_2 + x_3 = 0 \tag{2.8}$$
$$-4x_1 + 6x_2 - 4x_3 - 4x_4 = 1 \tag{2.9}$$
$$x_1 - 4x_2 + 6x_3 - 4x_4 = 0 \tag{2.10}$$
$$x_2 - 4x_3 + 5x_4 = 0. \tag{2.11}$$

Using matrix notation, the set of equations can be rewritten as

$$\begin{bmatrix} 5 & -4 & 1 & 0 \\ -4 & 6 & -4 & -4 \\ 1 & -4 & 6 & -4 \\ 0 & 1 & -4 & 5 \end{bmatrix} \begin{Bmatrix} x_1 \\ x_2 \\ x_3 \\ x_4 \end{Bmatrix} = \begin{Bmatrix} 0 \\ 1 \\ 0 \\ 0 \end{Bmatrix}. \tag{2.12}$$

In short form, we have

$$\mathbf{Ax} = \mathbf{b} \tag{2.13}$$

where **A** is a matrix, **x** and **b** are vectors. A matrix is defined as a table of numbers or alternatively a two-dimensional array of numbers, whose rows and columns can be regarded as horizontal and vertical vectors, respectively. A general form of a matrix **A** is given by

$$\mathbf{A} = [A_{ij}] = \begin{bmatrix} A_{11} & A_{12} & A_{13} & \cdots & A_{1n} \\ A_{21} & A_{22} & A_{23} & \cdots & A_{2n} \\ A_{31} & A_{32} & A_{33} & \cdots & A_{3n} \\ \vdots & \vdots & \vdots & \ddots & \vdots \\ A_{m1} & A_{m2} & A_{m3} & \cdots & A_{mn} \end{bmatrix} \tag{2.14}$$

where m and n are the number of rows and columns, respectively. $\mathbf{A}_{m \times n}$ has the order of $m \times n$. An element in the i-th row and j-th column of $\mathbf{A}$ is denoted as A_{ij}. If $m = n$, $\mathbf{A}$ is called a square matrix. If $n = 1$, $\mathbf{A}$ becomes a column vector $\mathbf{A}_{m \times 1}$. If $m = 1$, $\mathbf{A}$ becomes a row vector $\mathbf{A}_{1 \times n}$.

The transpose of $\mathbf{A}$ is written as $\mathbf{A}^T$. It is obtained by interchanging rows and columns of $\mathbf{A}^T$. Written in index form, $A^T_{ij} = A_{ji}$.

Example 2.2 Find the transpose of

$$\mathbf{A} = \begin{bmatrix} A_{11} & A_{12} & A_{13} \\ A_{21} & A_{22} & A_{23} \end{bmatrix}$$

Solution

$$\mathbf{A}^T = \begin{bmatrix} A_{11} & A_{21} \\ A_{12} & A_{22} \\ A_{13} & A_{23} \end{bmatrix}$$

Special Matrices

Several types of special matrices are of particular importance in the context of finite element analysis. They are summarized as follows.

Diagonal matrix. A diagonal matrix is a square matrix in which only the elements on the diagonal line running from the upper left corner to the lower right corner (i.e., the main diagonal) can be nonzero. Written in index form,

$$A_{ij} \begin{cases} \neq 0 & i = j \\ = 0 & i \neq j. \end{cases} \quad (2.15)$$

Identity matrix. An identity matrix is a diagonal matrix with all the main diagonal elements equal to one. Multiplying an $n \times n$ identity matrix by any $n \times 1$ vector does not change that vector.

$$\mathbf{I}_{n \times n} = \begin{bmatrix} 1 & 0 & \cdots & 0 \\ 0 & 1 & \cdots & 0 \\ \vdots & \vdots & \vdots & \vdots \\ 0 & 0 & \cdots & 1 \end{bmatrix}. \quad (2.16)$$

In index form

$$I_{ij} = \begin{cases} 1 & i = j \\ 0 & i \neq j. \end{cases} \quad (2.17)$$

Symmetric matrix. By definition, a symmetric matrix is a square matrix that is equal to its transpose.

$$\mathbf{A} = \mathbf{A}^T \quad \Rightarrow \quad \begin{cases} \mathbf{A} \text{ is square} \\ A_{ij} = A_{ji}. \end{cases} \quad (2.18)$$

Note that a diagonal matrix is symmetric.

Skew or anti-symmetric matrix. A skew-symmetric matrix is a square matrix whose transpose is equal to its negative

$$\mathbf{A}^T = -\mathbf{A} \quad \text{or} \quad A_{ji} = -A_{ij}. \quad (2.19)$$

For example, the following matrix is a skew-symmetric matrix

$$\mathbf{A} = \begin{bmatrix} 0 & -2 & 3 \\ 2 & 0 & 4 \\ -3 & -4 & 0 \end{bmatrix}.$$

Note that the main diagonal elements of a skew-symmetric matrix are all zeros, since they must be their own negative. In addition, it is shown below that any matrix can be written as the summation of a symmetric matrix and a skew-symmetric matrix.

$$\mathbf{A} = \frac{1}{2}\left(\mathbf{A} + \mathbf{A}^T\right) + \frac{1}{2}\left(\mathbf{A} - \mathbf{A}^T\right) \quad (2.20)$$

where the first term on the right hand side is a symmetric matrix and the second is a skew-symmetric matrix.

Triangular matrix. An upper triangular matrix is a square matrix in which all the elements below the main diagonal are zero. A lower triangular matrix is a square matrix in which all the elements above the main diagonal are zero. Figure 2.2 illustrates the configurations of triangular matrices. The shaded regions are nonzero regions.

For an upper (lower) triangular matrix, if the main diagonal elements are zeros, the matrix is called strictly upper (lower) triangular.

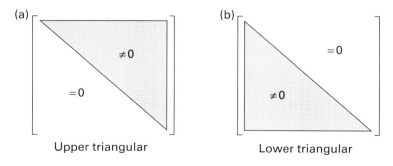

Figure 2.2 Triangular matrices: upper (a) and lower (b) triangular matrices.

2.2 Vector and Matrix Algebra

Band matrix and symmetric band matrix. A band matrix is a square matrix in which only a narrow (band) region around the main diagonal contains nonzero elements. As shown in Fig. 2.3, the shaded region contains nonzero elements.

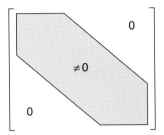

Figure 2.3 Band matrix.

It is clear that a symmetric band matrix is a band matrix with $\mathbf{A} = \mathbf{A}^T$. The mathematical definition is given by

$$A_{ij} = A_{ji} \tag{2.21}$$
$$A_{ij} = 0 \qquad \text{for } j > i + m_A \text{ or } j < i - m_A \tag{2.22}$$

where m_A is called the half-bandwidth. As shown in Fig. 2.4, $2m_A + 1 = 7$ is the bandwidth of $\mathbf{A}$. What if $m_A = 0$? The matrix becomes a diagonal matrix if $m_A = 0$.

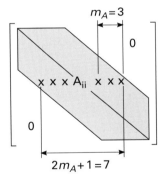

Figure 2.4 Band width of a band matrix.

Matrix Operations

Basic matrix operations include:

Addition

$$\mathbf{C} = \mathbf{A} + \mathbf{B} \qquad or \tag{2.23}$$
$$C_{ij} = A_{ij} + B_{ij} \tag{2.24}$$

Subtraction

$$\mathbf{C} = \mathbf{A} - \mathbf{B} \qquad or \tag{2.25}$$
$$C_{ij} = A_{ij} - B_{ij} \tag{2.26}$$

Multiplication

$$\mathbf{C}_{p \times q} = \mathbf{A}_{p \times n} \mathbf{B}_{n \times q} \qquad or \qquad (2.27)$$

$$C_{ij} = \sum_{r=1}^{n} A_{ir} B_{rj}. \qquad (2.28)$$

Note that matrix multiplication is:

- Distributive $\Rightarrow \mathbf{A}(\mathbf{B} + \mathbf{C}) = \mathbf{AB} + \mathbf{AC}$
- Associative $\Rightarrow \mathbf{A}(\mathbf{BC}) = (\mathbf{AB})\mathbf{C}$
- Not commutative $\Rightarrow \mathbf{AB} \neq \mathbf{BA}$.

Scalar Multiplication

$$\mathbf{C} = \beta \mathbf{A} \qquad \Rightarrow c_{ij} = \beta a_{ij} \qquad (2.29)$$

where β is a scalar.

Example 2.3 Compute $\mathbf{u}^T \mathbf{A} \mathbf{u}$ using the symmetry property of $\mathbf{A}$

$$\mathbf{A} = \begin{bmatrix} 3 & 2 & 1 \\ 2 & 4 & 2 \\ 1 & 2 & 6 \end{bmatrix} \qquad \mathbf{u} = \begin{Bmatrix} 1 \\ 2 \\ -1 \end{Bmatrix}$$

Solution
Let $\mathbf{A} = \mathbf{U} + \mathbf{D} + \mathbf{L}$, where

$$\mathbf{U} = \begin{bmatrix} 0 & 2 & 1 \\ 0 & 0 & 2 \\ 0 & 0 & 0 \end{bmatrix} \qquad \mathbf{D} = \begin{bmatrix} 3 & 0 & 0 \\ 0 & 4 & 0 \\ 0 & 0 & 6 \end{bmatrix}.$$

Since $\mathbf{A}$ is symmetric, $\mathbf{L} = \mathbf{U}^T$. Therefore,

$$\begin{aligned}
\mathbf{u}^T \mathbf{A} \mathbf{u} &= \mathbf{u}^T (\mathbf{U} + \mathbf{D} + \mathbf{L}) \mathbf{u} \\
&= \mathbf{u}^T \mathbf{U} \mathbf{u} + \mathbf{u}^T \mathbf{D} \mathbf{u} + \mathbf{u}^T \mathbf{U}^T \mathbf{u} \\
&= \mathbf{u}^T \mathbf{U} \mathbf{u} + \mathbf{u}^T \mathbf{D} \mathbf{u} + (\mathbf{u}^T \mathbf{U} \mathbf{u})^T \\
&= 2 \mathbf{u}^T \mathbf{U} \mathbf{u} + \mathbf{u}^T \mathbf{D} \mathbf{u} \\
&= 2(-1) + 25 \\
&= 23.
\end{aligned}$$

Determinant of a Matrix

The determinant of a matrix is a value computed from the elements of a square matrix. The determinant of a matrix **A** is denoted by $det\mathbf{A}$. A matrix is not invertible or singular if and only if $det\mathbf{A} = 0$.

The cofactor matrix **C** of a matrix **A** is defined by

$$C_{ij} = (-1)^{i+j} M_{ij} \tag{2.30}$$

where M_{ij} is the determinant of the smaller matrix obtained by eliminating the i-th row and j-th column of **A**. Then $det\mathbf{A}$ can be obtained by the Laplace (or cofactor) expansion along any row i or column j as

$$det\mathbf{A} = A_{i1}C_{i1} + A_{i2}C_{i2} + \cdots + A_{in}C_{in} \tag{2.31}$$

or

$$det\mathbf{A} = A_{1j}C_{1j} + A_{2j}C_{2j} + \cdots + A_{nj}C_{nj}. \tag{2.32}$$

In the case of a 2×2 matrix the determinant is obtained as

$$det \begin{bmatrix} a & b \\ c & d \end{bmatrix} = a \times d - b \times c. \tag{2.33}$$

Example 2.4 Compute the determinant of

$$\begin{bmatrix} 1 & 2 & 2 \\ -3 & 0 & -1 \\ 3 & -1 & 6 \end{bmatrix}. \tag{2.34}$$

Solution
Using Laplace expansion along the first row, we have

$$det \begin{bmatrix} 1 & 2 & 2 \\ -3 & 0 & -1 \\ 3 & -1 & 6 \end{bmatrix} = 1 \times \begin{vmatrix} 0 & -1 \\ -1 & 6 \end{vmatrix} + (-2) \times \begin{vmatrix} -3 & -1 \\ 3 & 6 \end{vmatrix} + 2 \times \begin{vmatrix} -3 & 0 \\ 3 & -1 \end{vmatrix}$$

$$= 1 \times (-1) + (-2) \times (-15) + 2 \times 3 = 35.$$

Matrix Inversion

The inverse of a matrix **A** is denoted as $\mathbf{A}^{-1}$. If the inverse of **A** exists, then **A** is non-singular and

$$\mathbf{A}\mathbf{A}^{-1} = \mathbf{A}^{-1}\mathbf{A} = \mathbf{I}. \tag{2.35}$$

If **A** is singular, then the determinant of **A** is zero, $det\mathbf{A} = 0$, and $\mathbf{A}^{-1}$ does not exist. If **A** is non-singular, the inverse of **A** can be determined by

$$\mathbf{A}^{-1} = \frac{1}{det\mathbf{A}} \mathbf{C}^T \tag{2.36}$$

where $det\mathbf{A}$ is the determinant of matrix $\mathbf{A}$ and $\mathbf{C}$ is the cofactor matrix of $\mathbf{A}$.

Example 2.5 Compute the inversion of

$$\begin{bmatrix} 1 & -1 & 0 \\ -1 & 2 & -1 \\ 0 & -1 & 2 \end{bmatrix}. \qquad (2.37)$$

Solution

$$\begin{bmatrix} 1 & -1 & 0 \\ -1 & 2 & -1 \\ 0 & -1 & 2 \end{bmatrix}^{-1} = \frac{1}{3-2-0} \begin{bmatrix} 3 & 2 & 1 \\ 2 & 2 & 1 \\ 1 & 1 & 1 \end{bmatrix}^{T} = \begin{bmatrix} 3 & 2 & 1 \\ 2 & 2 & 1 \\ 1 & 1 & 1 \end{bmatrix}. \qquad (2.38)$$

Checking the correctness of the solution:

$$\begin{bmatrix} 1 & -1 & 0 \\ -1 & 2 & -1 \\ 0 & -1 & 2 \end{bmatrix} \begin{bmatrix} 3 & 2 & 1 \\ 2 & 2 & 1 \\ 1 & 1 & 1 \end{bmatrix} = \begin{bmatrix} 1 & 0 & 0 \\ 0 & 1 & 0 \\ 0 & 0 & 1 \end{bmatrix}. \qquad (2.39)$$

Another sufficient and necessary condition for a matrix $\mathbf{A}_{n \times n}$ to be singular is that its row or column vectors are linear dependent. That is, there exists a non-trivial combination of the row (or column) vectors of $\mathbf{A}$, $\mathbf{a}_1, \mathbf{a}_2, \ldots, \mathbf{a}_n$, such that

$$c_1 \mathbf{a}_1 + c_2 \mathbf{a}_2 + \ldots + c_n \mathbf{a}_n = 0 \qquad (2.40)$$

where the coefficients $c_1, c_2, \ldots, c_n$ are not all zero (i.e. non-trivial). Some examples of singular matrices that can be easily tested against Eq. (2.40) are shown below.

$$\mathbf{A} = \begin{bmatrix} 1 & 2 & 3 \\ 2 & 4 & 6 \\ 0 & 3 & 1 \end{bmatrix}, \quad \mathbf{A} = \begin{bmatrix} 1 & 2 & 3 \\ 0 & 0 & 0 \\ -3 & 3 & 1 \end{bmatrix}, \quad \mathbf{A} = \begin{bmatrix} 1 & 2 & 0 \\ 9 & 0 & 0 \\ -5 & 3 & 0 \end{bmatrix}.$$
$$(2.41)$$

Matrix Eigenvalues and Eigenvectors

A scalar number λ and a vector $\mathbf{x}$ are called an eigenvalue and an eigenvector of a square matrix $\mathbf{A}$, respectively, if they satisfy the equation

$$\mathbf{A}\mathbf{x} = \lambda \mathbf{x}. \qquad (2.42)$$

The geometric interpretation is that the vector $\mathbf{x}$ is transformed by $\mathbf{A}$ into a vector going in the same direction as $\mathbf{x}$ (i.e. overlaying $\mathbf{x}$) but stretched by a factor of λ. For example, imagine we have a set of unit vectors spreading uniformly around the origin point $(0,0)$ to represent vectors of all orientations in the 2-D Cartesian coordinate system. Then we multiply a matrix

$$\mathbf{A} = \begin{bmatrix} 1.5 & 0.75 \\ 0.75 & 1.5 \end{bmatrix} \qquad (2.43)$$

2.2 Vector and Matrix Algebra

by the set of vectors as shown in Fig. 2.5. The matrix transforms the original vectors into a set of new vectors. Typically the resultant vectors have different directions and magnitudes from the original vectors. However, if a resultant vector has the same direction as the original vector, then the original unit vector is an eigenvector of the matrix, and the magnitude ratio of the resultant and original vectors is the eigenvalue associated with the eigenvector.

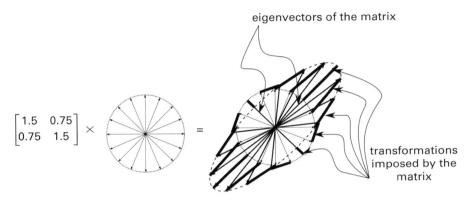

Figure 2.5 Multiplying a matrix by a set of unit vectors in the 2-D Cartesian coordinate system.

Moving the right hand side of Eq. (2.42) to the left hand side, we have

$$\mathbf{A}\mathbf{x} - \lambda\mathbf{x} = \mathbf{0} \qquad (2.44)$$

$$\Rightarrow (\mathbf{A} - \lambda\mathbf{I})\mathbf{x} = \mathbf{0}. \qquad (2.45)$$

It is easy to see that $\mathbf{x} = \mathbf{0}$ is a solution to Eq. (2.45). This solution is called a trivial solution of the problem. For the homogeneous linear system Eq. (2.45) to have a non-trivial solution, the determinant of the coefficient matrix $\mathbf{A} - \lambda\mathbf{I}$ must be zero. That is,

$$|\mathbf{A} - \lambda\mathbf{I}| = 0. \qquad (2.46)$$

If the matrix $\mathbf{A}$ is 3×3, then Eq. (2.46) becomes

$$\begin{vmatrix} A_{11} - \lambda & A_{12} & A_{13} \\ A_{21} & A_{22} - \lambda & A_{23} \\ A_{31} & A_{32} & A_{33} - \lambda \end{vmatrix} = 0. \qquad (2.47)$$

Writing out the determinant given in Eq. (2.47) results in a cubic equation in λ. This cubic equation is called the characteristic equation of $\mathbf{A}$. The cubic characteristic equation has three roots which are the eigenvalues of $\mathbf{A}$.

Example 2.6 Find the eigenvalues and eigenvectors of the following matrix

$$\begin{bmatrix} 2 & 3 & 0 \\ 1 & 5 & 2 \\ 0 & 0 & 7 \end{bmatrix}. \qquad (2.48)$$

Solution

The determinant of $(\mathbf{A} - \lambda\mathbf{I}) = 0$ gives the characteristic equation:

$$|\mathbf{A} - \lambda\mathbf{I}| = \begin{vmatrix} 2-\lambda & 3 & 0 \\ 1 & 5-\lambda & 2 \\ 0 & 0 & 7-\lambda \end{vmatrix} = (2-\lambda)(5-\lambda)(7-\lambda) - 3(7-\lambda) = 0$$

$$\Rightarrow (7-\lambda)\left(\lambda^2 - 7\lambda + 7\right) = 0. \tag{2.49}$$

The three roots of the cubic characteristic equation, Eq. (2.49), are $\lambda_1 = 7$, $\lambda_2 = \frac{7+\sqrt{21}}{2}$, and $\lambda_3 = \frac{7-\sqrt{21}}{2}$.

For $\lambda_1 = 7$, the homogeneous equations $\mathbf{Ax} - \lambda\mathbf{x} = \mathbf{0}$ becomes

$$\begin{bmatrix} -5 & 3 & 0 \\ 1 & -2 & 2 \\ 0 & 0 & 0 \end{bmatrix} \begin{Bmatrix} x_1 \\ x_2 \\ x_3 \end{Bmatrix} = \begin{Bmatrix} 0 \\ 0 \\ 0 \end{Bmatrix}. \tag{2.50}$$

By following the standard row reduction procedure of the augmented matrix

$$\begin{bmatrix} -5 & 3 & 0 & 0 \\ 1 & -2 & 2 & 0 \\ 0 & 0 & 0 & 0 \end{bmatrix} \tag{2.51}$$

we have the row-reduced echelon form of the linear system Eq. (2.50)

$$\begin{bmatrix} 1 & -3/5 & 0 & 0 \\ 0 & 1 & -10/7 & 0 \\ 0 & 0 & 0 & 0 \end{bmatrix}. \tag{2.52}$$

The row-reduced echelon form Eq. (2.52) shows that x_3 is a free variable, and $x_2 = 10/7 x_3$ and $x_1 = 3/5 x_2 = 6/7 x_3$. Hence the general solution is

$$\mathbf{x} = x_3 \begin{Bmatrix} 6/7 \\ 10/7 \\ 1 \end{Bmatrix}. \tag{2.53}$$

Since any nonzero scalar multiplied by an eigenvector of a matrix is also an eigenvector with the same eigenvalue, typically we compute the eigenvectors with unit length. For the eigenvector given in Eq. (2.53), unit length requires $x_1^2 + x_2^2 + x_3^2 = 1$. Therefore, after normalization,

$$\mathbf{x} = \begin{Bmatrix} 0.4411 \\ 0.7352 \\ 0.5147 \end{Bmatrix}. \tag{2.54}$$

By following the same procedure, the eigenvectors corresponding to λ_2 and λ_3 are obtained as $\mathbf{x} = \{-0.6205 \ -0.7842 \ 0\}^T$ and $\mathbf{x} = \{-0.9669 \ 0.2250 \ 0\}^T$, respectively.

The eigenvalues and eigenvectors have the following important properties:

- If **A** is symmetric and all its elements are real, the eigenvalues and their corresponding eigenvectors are all real.
- If **A** is a positive definite symmetric real matrix, its eigenvalues are all positive.
- **A** has one or more zero eigenvalues when it is singular.

Integration of a Matrix

In some cases, the elements of a matrix **A** can be functions. For example,

$$\mathbf{A} = \begin{bmatrix} x & x+2 \\ 2 & x^2 \end{bmatrix}.$$

An integral of **A** is obtained by integrating each element of the matrix. For example,

$$\int_0^1 \mathbf{A}\,dx = \begin{bmatrix} \int_0^1 x\,dx & \int_0^1 (x+2)\,dx \\ \int_0^1 2\,dx & \int_0^1 x^2\,dx \end{bmatrix} = \begin{bmatrix} \frac{1}{2} & \frac{5}{2} \\ 2 & \frac{1}{3} \end{bmatrix}.$$

Vector Norms

In linear algebra, a norm is a measure representing some sort of length or size of a vector. For this reason, a norm of a vector is a non-negative number and is equal to zero only if the vector is the zero vector. The norms of a vector are defined in general through the *p*-norm expression which is defined as follows.

p-Norm of a Vector $\mathbf{x}_{n\times 1}$

$$\|\mathbf{x}\|_p = \left(\sum_{i=1}^n |x_i|^p\right)^{1/p} \tag{2.55}$$

where $p \in \{1, 2, 3, \ldots\}$ As special cases of the general *p*-norm, one can define 1-norm, 2-norm, ..., ∞-norm:

- 1-norm: $\|\mathbf{x}\|_1 = \sum_{i=1}^n |x_i|$.

  For example, if $\mathbf{x} = \{x_1 \; x_2\}^T$, then $\|\mathbf{x}\|_1 = |x_1| + |x_2|$. Here $\|\mathbf{x}\|_1$ is the distance between the tail and the head of **x** as measured in "city blocks."

- 2-norm: $\|\mathbf{x}\|_2 = \left(\sum_{i=1}^n |x_i|^2\right)^{1/2}$

  $\|\mathbf{x}\|_2$ represents the length or magnitude of **x**, often just denoted as $\|\mathbf{x}\|$.

- ∞-norm: $\|\mathbf{x}\|_\infty = \max_i |x_i|$

  $\|\mathbf{x}\|_\infty$ represents the maximum magnitude of **x**'s elements.

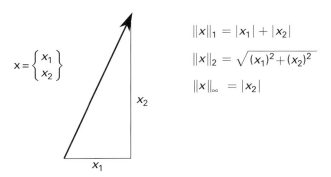

Figure 2.6 Geometric interpretation of vector norms.

The geometric interpretation is illustrated in Fig. 2.6. The most commonly used norm is the 2-norm which is also called the L^2-norm of the vector.

Example 2.7 Calculate the 1-, 2-, 3-, 4-, and ∞-norm of vector $\mathbf{A} = \{3\ 4\}^T$.

Solution
1-norm= $|3| + |4| = 7$

2-norm= $\sqrt{|3|^2 + |4|^2} = 5$

3-norm= $\left(|3|^3 + |4|^3\right)^{1/3} = 4.498$

4-norm= $\left(|3|^4 + |4|^4\right)^{1/4} = 4.285$

∞-norm= $max(|3|, |4|) = 4$

Matrix Norms

The *p*-norm of a matrix $\mathbf{A}$ is defined as

$$\|\mathbf{A}\|_p = \max_{\mathbf{x} \neq 0} \frac{\|\mathbf{A}\mathbf{x}\|_p}{\|\mathbf{x}\|_p}. \tag{2.56}$$

Note that $\mathbf{A}\mathbf{x}$ is a vector, call it $\mathbf{b}$. The matrix $\mathbf{A}$ changes the vector $\mathbf{x}$ into another vector $\mathbf{b}$, therefore, the matrix represents a transformation operation on the vector $\mathbf{x}$, i.e.,

$$\|\mathbf{A}\|_p = \max_{\mathbf{x} \neq 0} \frac{\|\mathbf{A}\mathbf{x}\|_p}{\|\mathbf{x}\|_p} = \max_{\mathbf{x} \neq 0} \frac{\|\mathbf{b}\|_p}{\|\mathbf{x}\|_p}. \tag{2.57}$$

Since vector norms can be viewed as some sort of "size" of the vector, a norm of $\mathbf{A}$ can be viewed as the maximum ratio of stretching the matrix can do to all the vectors except $\mathbf{0}$.

2.2 Vector and Matrix Algebra

- 1-norm: $\|\mathbf{A}\|_1 = \max_j \sum_{i=1}^{n} |a_{ij}|$
  Maximum absolute column sum of the matrix.
- 2-norm: $\|\mathbf{A}\|_2 = \left(\max_{eigen\ value} (\mathbf{A}^T \mathbf{A}) \right)^{1/2}$
- ∞-norm: $\|\mathbf{A}\|_\infty = \max_i \sum_{j=1}^{n} |a_{ij}|$
  Maximum absolute row sum of the matrix.

Condition Number of a Matrix

The condition number of a matrix is defined as

$$cond(\mathbf{A}) = \|\mathbf{A}\| \cdot \|\mathbf{A}^{-1}\|. \tag{2.58}$$

Note that $\|\mathbf{A}\|$ represents the maximum stretching the matrix can do to any nonzero vector, and $\|\mathbf{A}^{-1}\|$ represents the maximum ratio of compression the matrix can do to any nonzero vector.

When the 2-norm is used, it can be shown that the condition number of a $n \times n$ matrix $\mathbf{A}$ can be determined by (Haftka 1990)

$$cond(\mathbf{A}) = \frac{\lambda_n}{\lambda_1} \tag{2.59}$$

where λ_1 and λ_n are the first and n-th eigenvalue of $\mathbf{A}$, respectively. If $\mathbf{A}$ is singular, $cond(\mathbf{A}) = \infty$. $\mathbf{A}$ is said to be ill-conditioned if $cond(\mathbf{A})$ is large. The accuracy of a computer solution to the linear system $\mathbf{Ax} = \mathbf{b}$ is given by (Heath 1997)

$$\frac{\|\Delta \mathbf{x}\|}{\|\mathbf{x}\|} \approx cond(\mathbf{A}) \cdot \epsilon_{machine} \tag{2.60}$$

where $\Delta \mathbf{x} = \mathbf{x}_{solution} - \mathbf{x}_{exact}$ is the error of the computer solution compared to the exact solution, and $\epsilon_{machine}$ is the machine precision. For single precision floating point numbers, $\epsilon_{machine} \approx 10^{-7}$. For double precision floating point numbers, $\epsilon_{machine} \approx 10^{-16}$.

Example 2.8 An entry in the computer solution vector $\mathbf{x}$ of $\mathbf{Ax} = \mathbf{b}$ is shown below

$$\mathbf{x} = \left\{ \begin{array}{c} \vdots \\ 0.1234567890123456 \\ \vdots \end{array} \right\}. \tag{2.61}$$

The equations are solved by using double precision numbers and the condition number of the matrix $\mathbf{A}$ is 10^5. Estimate the accuracy of the result.

Solution

Since double precision numbers are used, $\epsilon_{machine} = 10^{-16}$, and

$$\frac{\|\Delta \mathbf{x}\|}{\|\mathbf{x}\|} \approx cond(\mathbf{A}) \cdot \epsilon_{machine} = 10^5 \times 10^{-16} = 10^{-11}. \tag{2.62}$$

Assuming the number 0.1234567890123456 is a representative of the **x** vector, then

$$\frac{error}{0.1234567890123456} \approx 10^{-11} \tag{2.63}$$

$$error \approx 10^{-12}. \tag{2.64}$$

The number of significant digits after the decimal point is 11. In other words, the result is accurate to the 11th digit after the decimal point.

In the example shown above, if the condition number of the matrix **A** is 10^{15}, then by following the calculations illustrated in the example, we find that the result is only accurate to the first digit after the decimal point. In this case, the result is not accurate (10% error) and the matrix **A** is ill-conditioned (condition number is 10^{15}).

Another situation is that when there is a perturbation in **A**, denoted as $\Delta \mathbf{A}$, the condition number of the matrix provides an upper bound on the variation of the solution **x** due to $\Delta \mathbf{A}$, i.e.,

$$\frac{\|\Delta \mathbf{x}\|}{\|\mathbf{x} + \Delta \mathbf{x}\|} \leq cond(\mathbf{A}) \frac{\|\Delta \mathbf{A}\|}{\|\mathbf{A}\|}. \tag{2.65}$$

Note that the upper bound represents the worst-case scenario. While it is possible to find a $\Delta \mathbf{A}$ such that Eq. (2.65) becomes an equality, the actual $\Delta \mathbf{x}$ is typically much smaller then the upper bound.

2.3 Vector Calculus

In this section, the basic rules of vector differentiation that will be used in the derivation of the finite element formulations are described and illustrated.

Gradient of a Scalar Field

A scalar function (or "field") $u(x, y, z)$ is a function that gives a scalar value for a given spatial point (x, y, z). For example, density, temperature, and pressure of a given material occupying a spatial volume are all scalar functions. The gradient of u, however, is a vector field which is defined as

2.3 Vector Calculus

$$\nabla u = \begin{Bmatrix} \dfrac{\partial u}{\partial x} \\ \dfrac{\partial u}{\partial y} \\ \dfrac{\partial u}{\partial z} \end{Bmatrix}. \tag{2.66}$$

A vector field is a field that gives a set of scalar values (called components) at each spatial point (x, y, z) in a given volume. For example, displacement, velocity, and acceleration of a given material occupying a spatial volume are all vector fields since they are all vectors with x, y, and z components at each spatial point in the volume.

Note that the gradient operator ∇ is a vector operator that performs a directional differentiaton operation on a given function, i.e.,

$$\nabla = \begin{Bmatrix} \dfrac{\partial}{\partial x} \\ \dfrac{\partial}{\partial y} \\ \dfrac{\partial}{\partial z} \end{Bmatrix}. \tag{2.67}$$

Gradient of a Vector Field

The gradient of a vector field $\mathbf{v} = \{u_1 \ u_2 \ u_3\}^T$ gives a matrix of derivatives, which is given by

$$\nabla \mathbf{u} = \begin{bmatrix} \dfrac{\partial u_1}{\partial x} & \dfrac{\partial u_1}{\partial y} & \dfrac{\partial u_1}{\partial z} \\ \dfrac{\partial u_2}{\partial x} & \dfrac{\partial u_2}{\partial y} & \dfrac{\partial u_2}{\partial z} \\ \dfrac{\partial u_3}{\partial x} & \dfrac{\partial u_3}{\partial y} & \dfrac{\partial u_3}{\partial z} \end{bmatrix}. \tag{2.68}$$

Divergence of a Vector Field

The divergence of a vector field is the dot/scalar product of the gradient operator and the vector field, i.e.,

$$\nabla \cdot \mathbf{u} = \begin{Bmatrix} \dfrac{\partial}{\partial x} \\ \dfrac{\partial}{\partial y} \\ \dfrac{\partial}{\partial z} \end{Bmatrix} \cdot \begin{Bmatrix} u_1 \\ u_2 \\ u_3 \end{Bmatrix} = \dfrac{\partial u_1}{\partial x} + \dfrac{\partial u_2}{\partial y} + \dfrac{\partial u_3}{\partial z}. \tag{2.69}$$

Curl of a Vector Field

The curl of a vector field, $\nabla \times \mathbf{u}$, is the vector product of the gradient operator and the vector field, which describes the infinitesimal rotation of the vector field:

$$\nabla \times \mathbf{u} = \begin{vmatrix} \mathbf{i} & \mathbf{j} & \mathbf{k} \\ \dfrac{\partial}{\partial x} & \dfrac{\partial}{\partial y} & \dfrac{\partial}{\partial z} \\ u_1 & u_2 & u_3 \end{vmatrix} \quad (2.70)$$

$$= \left(\frac{\partial u_3}{\partial y} - \frac{\partial u_2}{\partial z}\right)\mathbf{i} + \left(\frac{\partial u_1}{\partial z} - \frac{\partial u_3}{\partial x}\right)\mathbf{j} + \left(\frac{\partial u_2}{\partial x} - \frac{\partial u_1}{\partial y}\right)\mathbf{k}. \quad (2.71)$$

Or written in vector form

$$\nabla \times \mathbf{u} = \left\{ \begin{array}{c} \dfrac{\partial u_3}{\partial y} - \dfrac{\partial u_2}{\partial z} \\ \dfrac{\partial u_1}{\partial z} - \dfrac{\partial u_3}{\partial x} \\ \dfrac{\partial u_2}{\partial x} - \dfrac{\partial u_1}{\partial y} \end{array} \right\}. \quad (2.72)$$

Laplacian of a Scalar Field

The Laplacian operator, ∇^2, is equivalent to divergence of gradient, that is

$$\nabla^2 u = \nabla \cdot (\nabla u) = \frac{\partial^2 u}{\partial x^2} + \frac{\partial^2 u}{\partial y^2} + \frac{\partial^2 u}{\partial z^2}. \quad (2.73)$$

Laplacian of a Vector Field

$$\nabla^2 \mathbf{u} = \left\{ \begin{array}{c} \nabla^2 u_1 \\ \nabla^2 u_2 \\ \nabla^2 u_3 \end{array} \right\} = \left\{ \begin{array}{c} \dfrac{\partial^2 u_1}{\partial x^2} + \dfrac{\partial^2 u_1}{\partial y^2} + \dfrac{\partial^2 u_1}{\partial z^2} \\ \dfrac{\partial^2 u_2}{\partial x^2} + \dfrac{\partial^2 u_2}{\partial y^2} + \dfrac{\partial^2 u_2}{\partial z^2} \\ \dfrac{\partial^2 u_3}{\partial x^2} + \dfrac{\partial^2 u_3}{\partial y^2} + \dfrac{\partial^2 u_3}{\partial z^2} \end{array} \right\}. \quad (2.74)$$

Some Useful Identities

$$\nabla(uv) = u\nabla v + v\nabla u \quad (2.75)$$

$$\nabla^2(uv) = v\nabla^2 u + 2\nabla u \cdot \nabla v + u\nabla^2 v \quad (2.76)$$

$$\nabla \cdot (a\mathbf{u}) = a\nabla \cdot \mathbf{u} + \mathbf{u} \cdot \nabla a \quad (2.77)$$

$$\nabla \times (a\mathbf{u}) = a\nabla \times \mathbf{u} - \mathbf{u} \times \nabla a \quad (2.78)$$

$$\nabla \cdot \nabla u = \nabla^2 u \quad (2.79)$$

$$\nabla \cdot (\mathbf{u} \times \mathbf{v}) = \mathbf{v} \cdot (\nabla \times \mathbf{u}) - \mathbf{u} \cdot (\nabla \times \mathbf{v}) \quad (2.80)$$

$$\nabla \cdot (\nabla \times \mathbf{u}) = 0 \quad (2.81)$$

$$\nabla \times \nabla u = \mathbf{0}. \quad (2.82)$$

Example 2.9 Prove that $\nabla(uv) = v\nabla u + u\nabla v$

Solution

$$\nabla(uv) = \begin{Bmatrix} \dfrac{\partial(uv)}{\partial x} \\ \dfrac{\partial(uv)}{\partial y} \\ \dfrac{\partial(uv)}{\partial z} \end{Bmatrix} = \begin{Bmatrix} \dfrac{v\partial u + u\partial v}{\partial x} \\ \dfrac{v\partial u + u\partial v}{\partial y} \\ \dfrac{v\partial u + u\partial v}{\partial z} \end{Bmatrix} = \begin{Bmatrix} \dfrac{v\partial u}{\partial x} \\ \dfrac{v\partial u}{\partial y} \\ \dfrac{v\partial u}{\partial z} \end{Bmatrix} + \begin{Bmatrix} \dfrac{u\partial v}{\partial x} \\ \dfrac{u\partial v}{\partial y} \\ \dfrac{u\partial v}{\partial z} \end{Bmatrix}$$

$$= v\nabla u + u\nabla v.$$

Example 2.10 Prove that $\nabla \times \nabla u = \mathbf{0}$

Solution

$$\nabla \times \nabla u = \begin{vmatrix} \mathbf{i} & \mathbf{j} & \mathbf{k} \\ \dfrac{\partial}{\partial x} & \dfrac{\partial}{\partial y} & \dfrac{\partial}{\partial z} \\ \dfrac{\partial u}{\partial x} & \dfrac{\partial u}{\partial y} & \dfrac{\partial u}{\partial z} \end{vmatrix}$$

$$= \left(\dfrac{\partial^2 u}{\partial y \partial z} - \dfrac{\partial^2 u}{\partial y \partial z}\right)\mathbf{i} + \left(\dfrac{\partial^2 u}{\partial x \partial z} - \dfrac{\partial^2 u}{\partial x \partial z}\right)\mathbf{j} + \left(\dfrac{\partial^2 u}{\partial x \partial y} - \dfrac{\partial^2 u}{\partial x \partial y}\right)\mathbf{k} = \mathbf{0}.$$

2.4 Taylor Series Expansion

Taylor series expansion is a series expansion of a smooth function in the vicinity of a point. The series is infinite and contains terms that are calculated using the values of the function's derivatives at the point.

Functions of One Variable

$$f(x + \Delta x) = f(x) + \frac{df(x)}{dx}\Delta x + \frac{1}{2}\frac{d^2 f(x)}{dx^2}(\Delta x)^2 + \frac{1}{3!}\frac{d^3 f(x)}{dx^3}(\Delta x)^3 + R \quad (2.83)$$

where R is the remainder (containing terms with order > 3 in Eq. (2.83)).

Functions of Two Variables

$$f(x+\Delta x, y+\Delta y) = f(x,y) + \frac{\partial f(x,y)}{\partial x}\Delta x + \frac{\partial f(x,y)}{\partial y}\Delta y$$
$$+ \frac{1}{2}\left(\frac{\partial^2 f(x,y)}{\partial x^2}(\Delta x)^2 + 2\frac{\partial^2 f(x,y)}{\partial x \partial y}(\Delta x)(\Delta y) + \frac{\partial^2 f(x,y)}{\partial y^2}(\Delta y)^2\right) + R. \tag{2.84}$$

Functions of m Variables

$$f(x_1+\Delta x_1, x_2+\Delta x_2, \ldots, x_m+\Delta x_m) = f(x_1, x_2, \ldots, x_m)$$
$$+ \sum_{k=1}^{n} \frac{1}{k!}\left((\Delta x_1)\frac{\partial}{\partial x_1} + (\Delta x_2)\frac{\partial}{\partial x_2} + \ldots + (\Delta x_m)\frac{\partial}{\partial x_m}\right)^k f(x_1, x_2, \ldots, x_m) + R. \tag{2.85}$$

where n is the expansion order.

Generally speaking, if the function's derivatives are finite and the distance from the point (i.e. $\Delta x, \Delta y, \Delta z$) is small, as the order of Δx, Δy, and Δz becomes higher, the value of the higher order terms becomes smaller, and thus the contribution from the higher order terms to the Taylor series expansion becomes insignificant. In this case, the higher order terms can be removed from the series and the value of the function can be approximated by the first few terms of the series. This technique is the basis of many numerical approximation and discretization methods, as will be discussed in the later chapters.

2.5 Integration by Parts

In the derivation of finite element equations, integrals of the following form are often encountered,

$$\int_a^b u(x)v'(x)dx \tag{2.86}$$

where the prime "$'$" denotes the derivative of the function $v(x)$ with respect to x. As will be discussed later, sometimes it is desirable to transfer the derivative from $v(x)$ to $u(x)$ in the integral. This can be achieved by using a technique called integration by parts. The integration by parts formula is given by

$$\int_a^b u(x)v'(x)dx = u(x)v(x)\Big|_a^b - \int_a^b u'(x)v(x)dx. \tag{2.87}$$

Note that the derivative is moved from $v(x)$ to $u(x)$ in the integral on the right hand side. The proof of Eq. (2.87) is given below:

$$\int_a^b u(x)v'(x)dx + \int_a^b u'(x)v(x)dx$$
$$= \int_a^b \left[u(x)v'(x) + u'(x)v(x)\right] dx$$
$$= \int_a^b [u(x)v(x)]' dx$$
$$= [u(x)v(x)]\big|_a^b.$$

Example 2.11 Evaluate the integral $\int_0^2 xe^{2x}dx$

Solution
Let $x = u(x)$ and $e^{2x} = v'(x)$, then $v(x) = \frac{1}{2}e^{2x}$

By using Eq. (2.87), we have

$$\int_0^2 xe^{2x}dx = \left(x\frac{1}{2}e^{2x}\right)\bigg|_0^2 - \int_0^2 \frac{1}{2}e^{2x}dx$$
$$= e^4 - \frac{1}{4}e^{2x}\bigg|_0^2$$
$$= \frac{3}{4}e^4 + \frac{1}{4}.$$

Example 2.12 Evaluate the integral $\int x^2 \sin(2x)dx$

Solution
Let $x^2 = u(x)$ and $\sin(2x) = v'(x)$, then $v(x) = -\frac{1}{2}\cos(2x)$

By using Eq. (2.87), we have

$$\int x^2 \sin(2x)dx = -x^2 \frac{1}{2}\cos(2x) + \int x\cos(2x)dx.$$

To evaluate the integral on the right hand side, we need to do integration by parts again. For the integral, let $x = u(x)$ and $\cos(2x) = v'(x)$, then $v(x) = \frac{1}{2}\sin(2x)$.

We have

$$\int x^2 \sin(2x)dx = -\frac{1}{2}x^2 \cos(2x) + \int x\cos(2x)dx$$
$$= -\frac{1}{2}x^2 \cos(2x) + \frac{1}{2}x\sin(2x) - \frac{1}{2}\int \sin(2x)dx$$
$$= -\frac{1}{2}x^2 \cos(2x) + \frac{1}{2}x\sin(2x) + \frac{1}{4}\cos(2x) + c.$$

Note that c is an arbitrary constant.

2.6 Divergence Theorem

The divergence theorem, also known as Gauss' theorem, converts a domain integral into a surface integral, or vice versa. The divergence theorem is given by

$$\int_{\Omega} (\nabla \cdot \mathbf{u}) d\Omega = \int_{\Gamma} \mathbf{u} \cdot \mathbf{n} d\Gamma \qquad (2.88)$$

where Ω denotes the volume, Γ denotes the surface, and $\mathbf{n}$ is the outward unit normal vector on the surface, as shown in Fig. 2.7. Note that $\mathbf{u}$ is a vector field. The divergence theorem can be used to derive a formula which is useful in obtaining finite element equations for 2-D and 3-D problems. The formula is called Green's formula which is given by

$$\int_{\Omega} f(\nabla \cdot \mathbf{u}) d\Omega = \int_{\Gamma} f \mathbf{u} \cdot \mathbf{n} d\Gamma - \int_{\Omega} \nabla f \cdot \mathbf{u} d\Omega. \qquad (2.89)$$

The proof of Green's formula is left to the reader as an exercise.

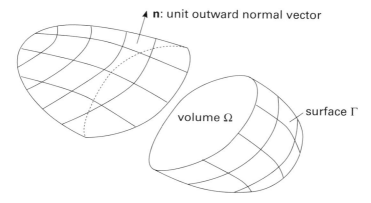

Figure 2.7 Divergence theorem.

2.7 Fundamentals of Variational Calculus

In this section, fundamentals of variational calculus are introduced as a mathematical preparation for the derivation of finite element weak form. For that purpose, the content of this section is limited to the fundamental and inverse problems of variational calculus. For more detailed discussion on this subject, the reader is referred to (Gelfand, Silverman et al. 2000).

The Fundamental Problem of Variational Calculus

The fundamental problem of the calculus of variations is to find a function $u(x)$ such that

$$\Pi = \int_a^b F(x, u, u') dx \qquad (2.90)$$

reaches an extreme value and $u(x)$ satisfies the boundary conditions:

$$u(a) = u_1 \tag{2.91}$$
$$u(b) = u_2. \tag{2.92}$$

In the above equations, $u' = \frac{du}{dx}$ and Π is called a functional. Since $u(x)$ is a function of x, different expressions of $u(x)$ would give a different value of Π. Therefore, Π is a function of function $u(x)$. By definition, a functional is a function of function(s) that gives a specific value for a given function.

Example 2.13 Find the shortest curve joining two points $P_1(0,0)$ and $P_2(1,1)$, see Fig. 2.8, where the curve length is given by

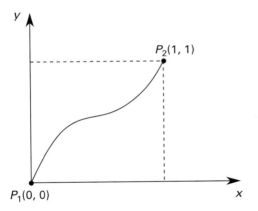

Figure 2.8 A curve joining two points $P_1(0,0)$ and $P_2(1,1)$.

$$\Pi = \int_{P_1}^{P_2} dS = \int_0^1 \sqrt{1+(y')^2}\,dx \tag{2.93}$$

and the boundary conditions are given by

$$y(0) = 0 \tag{2.94}$$
$$y(1) = 1. \tag{2.95}$$

Solution
It is shown in Eqs. (2.93–2.95) that the function of the curve should satisfy certain conditions: (1) $y(x)$ should be differentiable, and (2) $y(x)$ should satisfy the boundary conditions. It is clear that many curves can satisfy these conditions. Any of these curves is called an admissible function. By definition, an admissible function is a function $y(x)$ that satisfies certain conditions. These conditions are usually differentiability conditions (e.g., $y(x)$ is differentiable) and boundary conditions (e.g. $y(0) = 0$ and $y(1) = 1$).

If the integral Π reaches an extreme value then its first variation vanishes:

$$\delta \Pi = 0 \qquad \text{(Stationary condition)} \qquad (2.96)$$

where δ is the variational operator. For a given admissible function $u(x)$, δu denotes the variation of $u(x)$. As shown in Fig. 2.9, the variation of $u(x)$, δu, is defined as

$$\delta u = \hat{u}(x) - u(x) \qquad (2.97)$$

where $\hat{u}(x)$ is an arbitrary (but different) admissible function in the vicinity of $u(x)$.

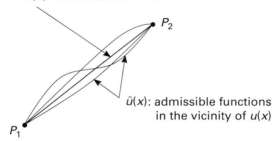

Figure 2.9 Admissible functions.

To better understand the above problem, it is necessary to clarify several concepts:

1. Π is called a functional, i.e., a function of functions.
2. It is required that $u(x)$ have a continuous derivative $u'(x)$.
3. Determining the stationary value of a functional Π is similar to the problem in the calculus where an extreme value (or a point of inflexion) of a single function is sought. In other words, finding a function $u(x)$ such that functional Π reaches an extremum is similar to finding a value of the variable x such that function $f(x)$ reaches an extremum.
4. It is an optimization problem.

Several important properties of the variational operator δ are listed below:

$$\frac{d}{dx}(\delta u) = \delta\left(\frac{du}{dx}\right) \qquad (2.98)$$

$$\delta(u'^2) = 2u'\delta u' \qquad (2.99)$$

$$\delta(u+v) = \delta u + \delta v \qquad (2.100)$$

$$\delta\left(\int u\,dx\right) = \int \delta u\,dx. \qquad (2.101)$$

Now look at $F(x, u, u')$ with a variation in u and u', δu and $\delta(u')$, respectively. The Taylor's expansion of F about u and u' can written as

2.7 Fundamentals of Variational Calculus

$$\hat{F} = F(x, u + \delta u, u' + \delta(u')) = F(x, u, u') + \left(\frac{\partial F}{\partial u} \delta u + \frac{\partial F}{\partial u'} \delta(u') \right)$$
$$+ \frac{1}{2} \left(\frac{\partial^2 F}{\partial u^2} (\delta u)^2 + 2 \frac{\partial^2 F}{\partial u \partial u'} \delta u \delta(u') + \frac{\partial^2 F}{\partial u'^2} [\delta(u')]^2 \right) + higher\ order\ terms.$$
(2.102)

The first variation of F is defined by

$$\delta F = \hat{F} - F = \left(\frac{\partial F}{\partial u} \delta u + \frac{\partial F}{\partial u'} \delta(u') \right) \quad (neglect\ higher\ order\ terms). \quad (2.103)$$

The first variation of the functional Π is given by:

$$\delta \Pi = \delta \int_a^b F dx \tag{2.104}$$

$$= \int_a^b \delta F dx \tag{2.105}$$

$$= \int_a^b \left(\frac{\partial F}{\partial u} \delta u + \frac{\partial F}{\partial u'} \delta(u') \right) dx \tag{2.106}$$

$$= \int_a^b \frac{\partial F}{\partial u} \delta u dx + \int_a^b \frac{\partial F}{\partial u'} \delta(u') dx. \tag{2.107}$$

Note that $\delta(u') = (\delta u)'$, so Eq. (2.107) can be rewritten as

$$\delta \Pi = \int_a^b \frac{\partial F}{\partial u} \delta u dx + \int_a^b \frac{\partial F}{\partial u'} (\delta u)' dx. \tag{2.108}$$

By using integration by parts for the second term on the right hand side of Eq. (2.108), we obtain

$$\delta \Pi = \int_a^b \frac{\partial F}{\partial u} \delta u dx + \left. \frac{\partial F}{\partial u'} \delta u \right|_a^b - \int_a^b \frac{d}{dx} \left(\frac{\partial F}{\partial u'} \right) \delta u dx. \tag{2.109}$$

Since $\delta u = 0$ at $x = a$ and $x = b$

$$\delta \Pi = \int_a^b \frac{\partial F}{\partial u} \delta u dx - \int_a^b \frac{d}{dx} \left(\frac{\partial F}{\partial u'} \right) \delta u dx \tag{2.110}$$

$$= \int_a^b \left[\frac{\partial F}{\partial u} - \frac{d}{dx} \left(\frac{\partial F}{\partial u'} \right) \right] \delta u dx. \tag{2.111}$$

Since δu is arbitrary within (a, b), $\delta \Pi = 0$ gives

$$\frac{\partial F}{\partial u} - \frac{d}{dx} \left(\frac{\partial F}{\partial u'} \right) = 0. \tag{2.112}$$

Equation (2.112) is called the Euler equation. The Euler equation is typically a partial differential equation (PDE) that should be satisfied point-wise in the domain. Due to the requirement of the point-wise satisfaction of the Euler equation, it is also called the strong form of the governing equation. In contrast, the integral form, which represents the vanishing of the first variation of the functional, is called the weak form. As shown in Fig. 2.10, given a functional statement, both the weak form and the strong form can be obtained straightforwardly.

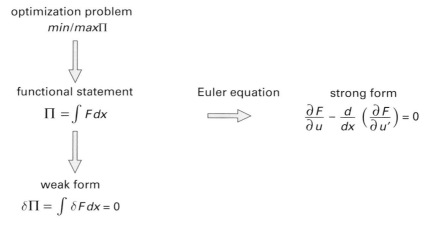

Figure 2.10 Weak and strong form of the governing equations.

Example 2.14 Find the shortest curve joining two points $P_1(0,0)$ and $P_2(1,1)$.

Solution

The curve length functional is given by

$$\Pi = \int F(x,y,y')dx = \int_0^1 \sqrt{1+(y')^2}dx \qquad (2.113)$$

i.e.

$$F(x,y,y') = \sqrt{1+(y')^2}. \qquad (2.114)$$

The Euler equation is then

$$\frac{\partial F}{\partial u} - \frac{d}{dx}\left(\frac{\partial F}{\partial u'}\right) = 0$$

$$\Rightarrow 0 - \frac{d}{dx}\left(\frac{y'}{\sqrt{1+(y')^2}}\right) = 0$$

$$\Rightarrow \frac{y'}{\sqrt{1+(y')^2}} = const$$

$$\Rightarrow y' = const$$

$$\Rightarrow y = c_1 x + c_2. \qquad (2.115)$$

Applying boundary conditions $y(0) = 0$ and $y(1) = 1$, we obtain

$$c_1 \cdot 0 + c_2 = 0 \qquad (2.116)$$

$$c_1 \cdot 1 + c_2 = 1 \qquad (2.117)$$

which gives $c_1 = 1$ and $c_2 = 0$. Therefore, the shortest curve joining two points $P_1(0, 0)$ and $P_2(1, 1)$ is $y = x$.

Bilinear Functionals

In the previous section, the functional Π is a function of a single variable function $u(x)$ and its first derivative, i.e.,

$$\Pi(u) = \int F(x, u, u') dx. \tag{2.118}$$

A functional can contain multiple variable functions. For example,

$$\Pi(u, v) = \int F(x, u, u', v, v') dx. \tag{2.119}$$

The functional given by Eq. (2.119) is called a two variable functional. A functional is said to be bilinear if it is linear in both u and v. In other words, if u and v can be expressed as

$$u = \alpha u_1 + \beta u_2 \tag{2.120}$$
$$v = \alpha v_1 + \beta v_2 \tag{2.121}$$

then

$$\Pi(\alpha u_1 + \beta u_2, v) = \alpha \Pi(u_1, v) + \beta \Pi(u_2, v) \tag{2.122}$$
$$\Pi(u, \alpha v_1 + \beta v_2) = \alpha \Pi(u, v_1) + \beta \Pi(u, v_2) \tag{2.123}$$

where α and β are real numbers, u_1, u_2, v_1, and v_2 are functions. If a two variable functional is bilinear, we denote it as $B(u, v)$. A bilinear functional $B(u, v)$ is said to be symmetric if

$$B(u, v) = B(v, u). \tag{2.124}$$

The following are a few examples of symmetric bilinear functionals

$$B(u, v) = \int uv \, dx \tag{2.125}$$

$$B(u, v) = \int k(x) \frac{du}{dx} \frac{dv}{dx} dx \tag{2.126}$$

$$B(u, v) = \int \left[p(x) \frac{d^2u}{dx^2} \frac{d^2v}{dx^2} + q(x) uv \right] dx. \tag{2.127}$$

The Inverse Problem of the Calculus of Variations

The inverse problem of the calculus of variations is the task of constructing a functional $\Pi(u)$ whose Euler equation is precisely a given differential equation. As shown in Fig. 2.11, a differential equation (i.e., the Euler equation) is defined in the domain Ω as

$$\mathscr{L} u = f \quad \text{in } \Omega, \tag{2.128}$$

with the boundary conditions:

$$u = \bar{u} \quad \text{on } \Gamma_u \tag{2.129}$$

$$\frac{\partial u}{\partial \mathbf{n}} = \bar{q} \quad \text{on } \Gamma_q \tag{2.130}$$

where $\mathscr{L}$ denotes the differential operator, Γ_u and Γ_q are the portions of the surface of the domain where u and $\frac{\partial u}{\partial \mathbf{n}}$ boundary conditions are prescribed, respectively. Note that $\Gamma_u \cup \Gamma_q = \Gamma$. The boundary condition given in Eq. (2.129) is called essential or Dirichlet (type) boundary condition, and that in Eq. (2.130) is called natural or Neumann (type) boundary condition. Since the right hand side of the boundary conditions is not zero, the boundary conditions are called non-homogeneous.

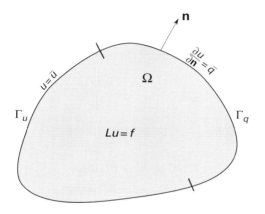

Ω: domain

$\Gamma = \Gamma_u \cup \Gamma_q$: domain boundary

Figure 2.11 An arbitrary computational domain.

An important theorem that can be employed to obtain the functional $\Pi(u)$ from the differential equation is stated as follows.

Theorem 1:

For differential equations (PDE/ODE) of the form

$$\mathscr{L}u = f, \quad \text{in } \Omega \tag{2.131}$$

with boundary conditions

$$u = \bar{u} \quad \text{on } \Gamma_u \tag{2.132}$$

$$\frac{\partial u}{\partial \mathbf{n}} = \bar{q} \quad \text{on } \Gamma_q \tag{2.133}$$

if the following conditions are satisfied:

1. the differential operator $\mathscr{L}$ is symmetric and positive
2. Eq. (2.131) has a solution

3 for an arbitrary function $w(x)$ with $w(x) = 0$ on Γ_u, the integral

$$\int_\Omega w\mathscr{L}u d\Omega \tag{2.134}$$

can be converted into a symmetric bilinear form with boundary terms

$$\int_\Omega w\mathscr{L}u d\Omega = B(w,u) + bt(w) \tag{2.135}$$

where $bt(w)$ denotes the boundary terms resulting from the conversion, then the solution u assigns a minimum value to the functional

with
$$\Pi = \frac{1}{2}B(u,u) - l(u) \tag{2.136}$$

$$l(u) = \int_\Omega uf d\Omega - bt(u) \tag{2.137}$$

where $bt(u)$ is the same boundary term shown in Eq. (2.135) except that w is replaced by u. The stationary condition of the functional can be obtained by invoking $\delta\Pi = 0$ which gives

$$\delta\Pi = B(\delta u, u) - l(\delta u) = 0. \tag{2.138}$$

The solution u to Eq. (2.138) minimizes the functional Π.

Here, we omit the proof of the above theorem. We say the differential operator $\mathscr{L}$ is symmetric if, for any two smooth functions u and v defined over the domain with $u = v = 0$ on the boundary,

$$\int_\Omega v\mathscr{L}u d\Omega = \int_\Omega u\mathscr{L}v d\Omega. \tag{2.139}$$

The differential operator $\mathscr{L}$ is positive if

$$\int_\Omega u\mathscr{L}u d\Omega \geq 0 \qquad (= 0 \text{ iff } u = 0). \tag{2.140}$$

Example 2.15 In $[a,b]$, we have

$$\mathscr{L}u = -\frac{d^2u}{dx^2} + q(x)u = f(x) \qquad q(x) \geq 0 \tag{2.141}$$

$$u(a) = u(b) = 0. \tag{2.142}$$

Obtain the functional $\Pi(u)$ whose Euler equation is Eq. (2.141).

Solution
The linear differential operator $\mathscr{L}$ is

$$\mathscr{L} = -\frac{d^2}{dx^2} + q(x).$$

We first verify that $\mathscr{L}$ is symmetric:

$$\int_a^b v\mathscr{L}u\,d\Omega = \int_a^b v\left(-\frac{d^2u}{dx^2} + q(x)u\right)dx$$

$$= \int_a^b -v\frac{d^2u}{dx^2}dx + \int_a^b vq(x)u\,dx$$

$$= -v\frac{du}{dx}\bigg|_a^b + \int_a^b \frac{dv}{dx}\frac{du}{dx}dx + \int_a^b vq(x)u\,dx$$

$$= -v\frac{du}{dx}\bigg|_a^b + u\frac{dv}{dx}\bigg|_a^b - \int_a^b u\frac{d^2v}{dx^2}dx + \int_a^b vq(x)u\,dx$$

$$= \int_a^b u\left(-\frac{d^2v}{dx^2} + q(x)v\right)dx - \underbrace{v\frac{du}{dx}\bigg|_a^b}_{=0} + \underbrace{u\frac{dv}{dx}\bigg|_a^b}_{=0}$$

$$= \int_\Omega u\mathscr{L}v\,d\Omega.$$

Therefore, $\mathscr{L}$ is symmetric.
Next we verify that $\mathscr{L}$ is positive:

$$\int_a^b u\mathscr{L}u\,d\Omega = \int_a^b u\left(-\frac{d^2u}{dx^2} + q(x)u\right)dx$$

$$= \int_a^b -u\frac{d^2u}{dx^2}dx + \int_a^b uq(x)u\,dx$$

$$= -\underbrace{u\frac{du}{dx}\bigg|_a^b}_{=0} + \int_a^b \frac{du}{dx}\frac{du}{dx}dx + \int_a^b uq(x)u\,dx$$

$$= \int_a^b \left(\frac{du}{dx}\right)^2 dx + \int_a^b q(x)u^2\,dx \geq 0.$$

Therefore, $\mathscr{L}$ is positive.
By using integration by parts, the integral $\int_a^b w\mathscr{L}u\,d\Omega$ can be converted to a symmetric bilinear form as

$$\int_a^b w\mathscr{L}u\,d\Omega = \int_a^b w\left(-\frac{d^2u}{dx^2} + q(x)u\right)dx$$

$$= -\int_a^b w\frac{d^2u}{dx^2}dx + \int_a^b q(x)wu\,dx$$

$$= \int_a^b \frac{du}{dx}\frac{dw}{dx}dx + \int_a^b q(x)wu\,dx - \underbrace{w\frac{du}{dx}\bigg|_a^b}_{=0}$$

$$= \int_a^b \left(\frac{du}{dx}\frac{dw}{dx} + q(x)wu\right)dx$$

$$= B(w,u).$$

2.7 Fundamentals of Variational Calculus

It is obvious that $B(w, v)$ is symmetric. Note that $bt(w)$ is zero in this case. According to Theorem 1, the functional of this problem is

$$\Pi = \frac{1}{2}B(u, u) - l(u)$$

$$= \frac{1}{2}\int_a^b \left(\frac{du}{dx}\frac{du}{dx} + q(x)u \cdot u\right) dx - \int_a^b ufdx$$

$$= \int_a^b \frac{1}{2}\left(\frac{du}{dx}\right)^2 dx + \int_a^b \frac{1}{2}q(x)u^2 dx - \int_a^b ufdx.$$

Invoking $\delta\Pi = 0$, we have

$$B(\delta u, u) - l(\delta u) = 0$$

that is,

$$\int_a^b \left(\frac{d\delta u}{dx}\frac{du}{dx} + q(x)u\delta u\right) dx - \int_a^b \delta u f dx = 0. \quad (2.143)$$

The solution to Eq. (2.143) minimizes the functional Π. Equation (2.143) is called the variational form or the weak form of the problem given by Eqs. (2.141, 2.142). In finite element terminology, the set of Eqs. (2.141, 2.142) is called the strong form. The variational form, Eq. (2.143), is also called the weak form. The adjectives "strong" and "weak" represent the continuity requirement of the function $u(x)$ in the equations. In Eq. (2.141), it is required that at least the first derivative of $u(x)$, du/dx, is continuous so that d^2u/dx^2 is meaningful everywhere in the domain. However, the highest derivative is du/dx in Eq. (2.143), which can be discontinuous at a finite number of points for the integration. That is, only the continuity of $u(x)$ is required. Therefore, in the weak form, the continuity requirement of $u(x)$ is weakened.

Next we verify the Euler equation:
In Eq. (2.143),

$$F = \frac{1}{2}\left(\frac{du}{dx}\right)^2 + \frac{1}{2}q(x)u^2 - uf.$$

Therefore,

$$\frac{\partial F}{\partial u} - \frac{d}{dx}\left(\frac{\partial F}{\partial u'}\right) = 0$$

$$\Rightarrow qu - f - \frac{d}{dx}\left(\frac{du}{dx}\right) = 0$$

$$\Rightarrow -\frac{d^2u}{dx^2} + qu = f.$$

Example 2.16 Obtain the functional for the following governing equation and boundary conditions defined in $[a, b]$

$$-\frac{d^2u}{dx^2} + q(x)u = f(x) \qquad q(x) > 0 \qquad (2.144)$$

$$u(a) = \bar{u} \qquad (2.145)$$

$$\frac{du}{dx}(b) = \bar{q}. \qquad (2.146)$$

Solution

The symmetry and positivity of the differential operator $\mathscr{L}$ has been proven in the previous example. By using integration by parts, the integral $\int_\Omega w\mathscr{L}u d\Omega$ can be converted to a symmetric bilinear form with a boundary term

$$\int_\Omega w\mathscr{L}u d\Omega = \int_a^b w\left(-\frac{d^2u}{dx^2} + q(x)u\right) dx$$

$$= -\int_a^b w\frac{d^2u}{dx^2}dx + \int_a^b q(x)wu dx$$

$$= \int_a^b \frac{du}{dx}\frac{dw}{dx}dx + \int_a^b q(x)wu dx - w(b)\overset{=\bar{q}}{\frac{du}{dx}(b)} + \overset{=0}{w(a)}\frac{du}{dx}(a)$$

$$= \int_a^b \left(\frac{du}{dx}\frac{dw}{dx} + q(x)wu\right) dx - w(b)\bar{q}$$

$$= B(w, u) + bt(w).$$

where $bt(w) = -w(b)\bar{q}$. This example is different from the previous example as the natural boundary condition is non-homogeneous. For this reason, the boundary term $bt(w)$ is nonzero. According to Theorem 1, the functional of this problem is

$$\Pi = \frac{1}{2}B(u, u) - l(u)$$

$$= \int_a^b \frac{1}{2}\left(\frac{du}{dx}\right)^2 dx + \int_a^b \frac{1}{2}q(x)u^2 dx - \int_a^b uf dx - u(b)\bar{q}.$$

We obtain the weak (or variational) form by invoking $\delta\Pi = 0$,

$$\int_a^b \frac{d\delta u}{dx}\frac{du}{dx}dx + \int_a^b \delta u q(x)u dx - \int_a^b \delta u f dx - \delta u(b)\bar{q} = 0.$$

Example 2.17 Obtain the functional for the following governing equation and boundary conditions defined as

$$-\nabla \cdot (k\nabla u) + qu = f \qquad q > 0 \qquad in \; \Omega \qquad (2.147)$$

$$u = \bar{u} \quad \text{on } \Gamma_u \tag{2.148}$$

$$\frac{\partial u}{\partial \mathbf{n}} = \bar{q} \quad \text{on } \Gamma_q \tag{2.149}$$

where $\mathbf{n} = \{n_x \; n_y \; n_z\}^T$ and,

$$\frac{\partial u}{\partial \mathbf{n}} = \frac{\partial u}{\partial x}n_x + \frac{\partial u}{\partial y}n_y + \frac{\partial u}{\partial z}n_z = \nabla u \cdot \mathbf{n} \tag{2.150}$$

and Γ_u and Γ_q are the portions of the surface of the domain where the essential and natural boundary conditions are prescribed, respectively.

Solution

The differential operator of Eq. (2.147) is

$$\mathscr{L} = -\nabla \cdot (k\nabla) + q.$$

The proof of symmetry and positivity of the differential operator $\mathscr{L}$ is left to the reader as an exercise. By using Green's formula, the integral $\int_\Omega w \mathscr{L} u \, d\Omega$ can be converted to a symmetric bilinear form with a boundary term

$$\begin{aligned}
\int_\Omega w \mathscr{L} u \, d\Omega &= \int_\Omega w \left[-\nabla \cdot (k\nabla u) + qu \right] d\Omega \\
&= -\int_\Omega w \left[\nabla \cdot (k\nabla u) \right] d\Omega + \int_\Omega qwu \, d\Omega \\
&= \int_\Omega \nabla w \cdot (k\nabla u) d\Omega + \int_\Omega qwu \, d\Omega - \int_\Gamma w(k\nabla u) \cdot \mathbf{n} \, d\Gamma \\
&= \int_\Omega \left[(\nabla w)^T k \nabla u + qwu \right] d\Omega - \int_\Gamma wk \frac{\partial u}{\partial \mathbf{n}} d\Gamma \\
&= \int_\Omega \left[(\nabla w)^T k \nabla u + qwu \right] d\Omega - \int_{\Gamma_q} wk\bar{q} \, d\Gamma \\
&= B(w, u) + bt(w).
\end{aligned}$$

Clearly, $B(w, u)$ is a symmetric bilinear functional. According to Theorem 1, the functional of this problem is

$$\Pi = \frac{1}{2} B(u, u) - l(u) = \int_\Omega \frac{1}{2} \left[(\nabla u)^T k \nabla u + qu^2 \right] d\Omega - \int_\Omega uf \, d\Omega - \int_{\Gamma_q} uk\bar{q} \, d\Gamma.$$

Invoking $\delta \Pi = 0$, we obtain the weak form

$$\int_\Omega (\nabla \delta u)^T k \nabla u \, d\Omega + \int_\Omega \delta u q u \, d\Omega - \int_\Omega \delta u f \, d\Omega - \int_{\Gamma_q} \delta u k \bar{q} \, d\Gamma = 0.$$

From the basic theory of variational calculus, it is shown that finding the optimal function in an admissible function space is achievable once the objective functional is constructed. Therefore, finding the functional is critical. In the following several important comments about the functional are provided:

1. For general differential equations with non-homogeneous boundary conditions, it could be quite involved to construct the functional.
2. Not every differential equation has a corresponding functional.
3. Once a functional has been established for a certain class of problems, it can be applied to all the problems in that class and therefore provides a general analysis tool.
4. For many problems, the functional has clear physical meaning (e.g., energy) and can be obtained straightforwardly.

2.8 Summary

In this chapter, a set of basic mathematical concepts and techniques are reviewed. Although the reader should already be familiar with most of the concepts, a quick review is necessary because good understanding of these concepts is essential for the further development of the numerical analysis methods, finite element formulations, and solution techniques. The section on variational calculus could be new to the reader. While the variational calculus section in this book is designed to be self-contained, further reading on this topic is recommended to the interested reader.

Upon completion of this chapter, you should be able to:
- perform vector and matrix calculations
- understand the basic vector calculus concepts and methods, and know how to use them to prove the identities listed in Section 2.3
- understand the concept of matrix condition number and its implication in the solution of a linear system
- understand the calculus theorems that are important to FEA: Taylor's series expansion, integration by parts, and divergence theorem
- have basic understanding of the fundamental problem of variational calculus
- understand how the weak form is derived from variational calculus and what the physical meaning of the variational principles discussed in this chapter is.

2.9 Problems

2.1 Assuming that f and g are scalar fields, and $\mathbf{v}$ is a vector field, show that
(a) $\nabla \cdot (f\mathbf{v}) = f\nabla \cdot \mathbf{v} + \mathbf{v} \cdot \nabla f$
(b) $\nabla \cdot (f\nabla g) = f\nabla^2 g + \nabla f \cdot \nabla g$
(c) $\nabla \cdot (f\nabla g) - \nabla \cdot (g\nabla f) = f\nabla^2 g - g\nabla^2 f$
(d) $\nabla \times (\nabla f) = 0$.

2.2 In MATLAB, create a 12×8 matrix of random numbers (use "rand"). Move through the matrix, element by element, and set any value that is less than 0.3 to 0 and any value that is greater than (or equal to) 0.3 to 1. (Note: if you are not familiar with MATLAB, the following website contains a comprehensive user's guide: www.mathworks.com/access/helpdesk/help/techdoc/.)

2.3 In MATLAB, enter the three matrices

$$A = \begin{bmatrix} 1 & \frac{1}{2} & \frac{1}{3} & \frac{1}{4} \\ \frac{1}{2} & \frac{1}{3} & \frac{1}{4} & \frac{1}{5} \\ \frac{1}{3} & \frac{1}{4} & \frac{1}{5} & \frac{1}{6} \\ \frac{1}{4} & \frac{1}{5} & \frac{1}{6} & \frac{1}{7} \end{bmatrix} \quad B = \begin{bmatrix} 1.0000 & 0.5000 & 0.3333 & 0.2500 \\ 0.5000 & 0.3333 & 0.2500 & 0.2000 \\ 0.3333 & 0.2500 & 0.2000 & 0.1667 \\ 0.2500 & 0.2000 & 0.1667 & 0.1429 \end{bmatrix}$$

$$C = \begin{bmatrix} 16 & -120 & 240 & -140 \\ -120 & 1200 & -2700 & 1680 \\ 240 & -2700 & 6480 & -4200 \\ -140 & 1680 & -4200 & 2800 \end{bmatrix}$$

(a) Calculate $A - B$. Use the "format long" command to show the result. Return to the standard short format with "format short."

(b) Calculate AC and BC. Define what is meant by the inverse of a square matrix. What is the inverse of the matrix A? Or of the matrix C?

2.4 Write the following system of linear equations in the form $Ax = B$ and use the MATLAB command "$A \backslash B$" to solve the system.

$$3x + 3y + 4z = 2$$
$$x + y + 4z = 2$$
$$2x + 5y + 4z = 3.$$

2.5 In MATLAB, enter the matrices

$$A_{11} = \begin{bmatrix} 1 & 0 \\ 0 & -1 \end{bmatrix} \quad A_{12} = zeros(2, 2) \quad A_{22} = \begin{bmatrix} 0 & 1 \\ 1 & 0 \end{bmatrix}$$

(a) Form the 4×4 matrix A using the following MATLAB construction:
"$A = [A_{11} \; A_{12}; \; A_{12} \; A_{22}]$."

(b) Find the smallest value of integer n greater than 1 such that $A^n = A$.

2.6 Let

$$A = \begin{bmatrix} 1 & 4 \\ 2 & -1 \end{bmatrix}.$$

Use the MATLAB determinant command "det" to compute "$det(2 * eye(2) - A)$." Find a positive integer value of t such that $det(tI - A) = 0$.

2.7 Prove Green's formula, Eq. (2.89).

2.8 Using the divergence theorem, show that the volume, V, enclosed by the surface Γ can be obtained by evaluating the following surface integral:

$$V = \frac{1}{3} \int_\Gamma \mathbf{n} \cdot \mathbf{x} \, d\Gamma$$

where $\mathbf{x}$ is the position vector from the origin of a Cartesian coordinate system to a point on the surface, and $\mathbf{n}$ is the unit outward normal vector at that point.

2.9 A function u is defined in the domain (shaded) as shown in Fig. 2.12.

$$u = \sqrt{x^2 + (y-3)^2}.$$

Write a MATLAB program to draw a contour plot of u in the domain, as shown in Fig. 2.13. Use a mesh grid of 201×201 and at least 10 contour lines. The color outside the domain should be white. (Hint: useful MATLAB functions: meshgrid, colormap, contourf.)

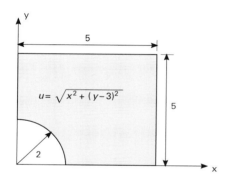

Figure 2.12 Function defined in a domain.

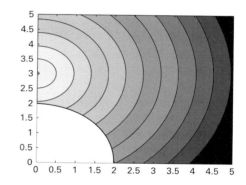

Figure 2.13 Contour plot.

2.10 What is the Euler equation for the following functional?

$$\int_a^b f(x,y)\sqrt{1+(y')^2}\,dx.$$

2.11 Find function $y(x)$ to make the following functional stationary.

$$\int_a^b \frac{(y')^2}{x^3}\,dx.$$

2.12 Find a functional Π such that minimizing the functional is equivalent to satisfying the following differential equation

$$\frac{\partial^2 u}{\partial x^2} + \frac{\partial^2 u}{\partial y^2} - Q(x,y) = 0,$$

and the boundary conditions

$$\frac{\partial u}{\partial x}n_x + \frac{\partial u}{\partial y}n_y + \bar{q} = 0, \qquad \text{along boundary } S_1,$$

$$u = \bar{u}, \qquad \text{along the rest of the boundary,}$$

where $\bar{q}$ and $\bar{u}$ are constants, n_x and n_y are the x- and y-components of the outward normal vector.

References

Gelfand, I. M., Silverman, R. A. et al. (2000), *Calculus of variations*, Courier Corporation.
Haftka, R. T. (1990), "Stiffness-matrix condition number and shape sensitivity errors," *AIAA Journal* **28**(7), 1322–1324.
Heath, M. T. (1997), *Scientific computing: an introductory survey*, McGraw-Hill.

3 Numerical Analysis Methods

3.1 Overview

In this chapter, we discuss several topics of numerical analysis. The topics include numerical interpolation, differentiation, integration, and solution of linear systems of equations. These numerical techniques are essential in the finite element method. Functions and subroutines performing these numerical calculations will be used repeatedly in the implementation of FE codes for engineering analysis. While numerical analysis is a broad area, this chapter only covers the methods and techniques that are used in the development and solution of finite element formulations and models in this book. More detailed and in-depth discussions of numerical analysis methods can be found in the reference books listed at the end of the chapter. It should be noted that MATLAB has many built-in functions for performing numerical analysis tasks. However, this book is intended to reveal the underlying theories and techniques of the numerical methods. Therefore, we explain various numerical methods from the basics and create the MATLAB functions from scratch.

3.2 Numerical Interpolation and Approximation

When we want to use a mathematical function to represent or fit to a discrete set of data points that have some kind of relationship between their values and their positions in space, there are typically two choices: interpolation or approximation. By definition, interpolation is the construction of a function that will precisely go through the data points, as shown in Fig. 3.1(a). The function is called interpolation function. In comparison, approximation is to find a function that follows the data points by certain rule(s). For example: finding a polynomial function of certain order that is the closest (i.e., with minimal total distances) to a set of given data points. In such cases, the approximation function does not necessarily go through all or any of the data points, as shown in Fig. 3.1(b).

In the context of finite element analysis, polynomial interpolation is dominantly employed in the numerical representation of the unknown function(s) over the elements. Therefore, we confine ourselves to interpolation methods here.

3.2 Numerical Interpolation and Approximation

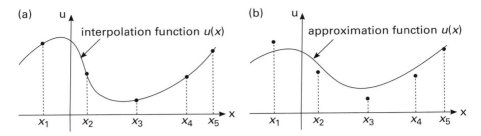

Figure 3.1 Interpolation vs approximation.

3.2.1 Lagrange Interpolation

As shown in Fig. 3.1(a), a numerical interpolation is the process of constructing a mathematical function, called the interpolation function, that goes through the data points. There are, however, multiple choices for the type of the interpolation function. Common types include polynomial, trigonometric, exponential, and rational functions. The most popular and very effective type is the polynomial interpolation function.

A polynomial interpolation function of x can be written as

$$u(x) = a_1 \cdot 1 + a_2 x + a_3 x^2 + a_4 x^3 + \cdots \tag{3.1}$$

where $1, x, x^2, \ldots$ are referred to as monomial basis functions, and $a_1, a_2, a_3, \ldots$, are the unknown coefficients. To find the coefficients, we need a set of prescribed conditions: the polynomial function $u(x)$ equals $u_1, u_2, u_3, \ldots$ at $x_1, x_2, x_3, \ldots$, respectively. These prescribed conditions give rise to a set of simultaneous equations:

$$u(x_1) = a_1 \cdot 1 + a_2 x_1 + a_3 x_1^2 + a_4 x_1^3 + \cdots = u_1$$
$$u(x_2) = a_1 \cdot 1 + a_2 x_2 + a_3 x_2^2 + a_4 x_2^3 + \cdots = u_2$$
$$u(x_3) = a_1 \cdot 1 + a_2 x_3 + a_3 x_3^2 + a_4 x_3^3 + \cdots = u_3.$$
$$\cdots \tag{3.2}$$

Equation (3.2) is called a linear system as the equations only contain linear functions of the unknowns. Assuming the data points are all distinct in x, for Eq. (3.2) to have a unique solution of the coefficients $a_1, a_2, a_3, \ldots$, the number of equations (i.e. the number of data points) must be equal to the number of coefficients. Therefore, it is clear that if there are n data points, the interpolation function should have n monomial basis coefficients and should be of order $n - 1$, that is

$$u(x) = a_1 \cdot 1 + a_2 x + a_3 x^2 + a_4 x^3 + \cdots a_n x^{n-1}. \tag{3.3}$$

Written in matrix form, we have

$$\begin{bmatrix} 1 & x_1 & x_1^2 & \cdots & x_1^{n-1} \\ 1 & x_2 & x_2^2 & \cdots & x_2^{n-1} \\ 1 & x_3 & x_3^2 & \cdots & x_3^{n-1} \\ \vdots & \vdots & \vdots & & \vdots \\ 1 & x_n & x_n^2 & \cdots & x_n^{n-1} \end{bmatrix} \begin{Bmatrix} a_1 \\ a_2 \\ a_3 \\ \vdots \\ a_n \end{Bmatrix} = \begin{Bmatrix} u_1 \\ u_2 \\ u_3 \\ \vdots \\ u_n \end{Bmatrix}. \qquad (3.4)$$

By solving Eq. (3.4), the coefficients $a_1, a_2, a_3, \ldots, a_n$ can be obtained. The polynomial interpolation function is uniquely determined.

Example 3.1 Construct a polynomial interpolation function for the two data points shown in Fig. 3.2.

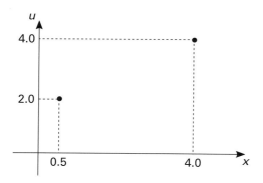

Figure 3.2 Find a polynomial function interpolating two data points.

Solution
Writing

$$u(0.5) = u_1 = 2.0 \qquad (3.5)$$
$$u(4.0) = u_2 = 4.0 \qquad (3.6)$$

gives two equations, which means we are able to determine two coefficients. Therefore, we let $u(x)$ have the form of

$$u(x) = a_1 + a_2 x \qquad (3.7)$$

which is a linear function. Substituting Eq. (3.7) into Eqs. (3.5, 3.6), the coefficients can be determined by solving two equations

$$a_1 + a_2 \times 0.5 = 2.0 \qquad (3.8)$$
$$a_1 + a_2 \times 4.0 = 4.0. \qquad (3.9)$$

In matrix form, we have

$$\begin{bmatrix} 1 & 0.5 \\ 1 & 4.0 \end{bmatrix} \begin{Bmatrix} a_1 \\ a_2 \end{Bmatrix} = \begin{Bmatrix} 2.0 \\ 4.0 \end{Bmatrix}. \quad (3.10)$$

Solving the linear system, we obtain $a_1 = 1.7143$ and $a_2 = 0.5714$. The linear interpolation function is then $u(x) = 1.7143 + 0.5714x$, as shown in Fig. 3.3.

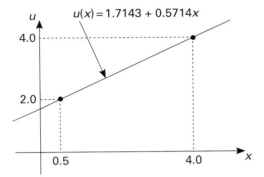

Figure 3.3 The linear interpolation function.

The example above shows that the number of data points determines the order of interpolation function. In the following example, a polynomial interpolation function is constructed over three data points.

Example 3.2 Construct a polynomial interpolation function for the three data points shown in Fig. 3.4.

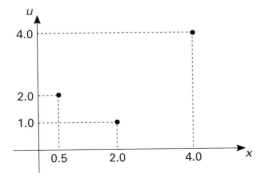

Figure 3.4 Three data points for use with Example 3.2.

Solution

We have three data points specified so three polynomial coefficients can be determined. Therefore the interpolating polynomial has the form

$$u(x) = a_1 + a_2 x + a_3 x^2 \tag{3.11}$$

and the coefficients can be obtained by solving

$$\begin{bmatrix} 1 & x_1 & x_1^2 \\ 1 & x_2 & x_2^2 \\ 1 & x_3 & x_3^2 \end{bmatrix} \begin{Bmatrix} a_1 \\ a_2 \\ a_3 \end{Bmatrix} = \begin{Bmatrix} u_1 \\ u_2 \\ u_3 \end{Bmatrix}$$

$$\Rightarrow \begin{bmatrix} 1 & 0.5 & 0.5^2 \\ 1 & 2.0 & 2.0^2 \\ 1 & 4.0 & 4.0^2 \end{bmatrix} \begin{Bmatrix} a_1 \\ a_2 \\ a_3 \end{Bmatrix} = \begin{Bmatrix} 2.0 \\ 1.0 \\ 4.0 \end{Bmatrix}. \tag{3.12}$$

The solution of the linear system Eq. (3.12) gives

$$a_1 = 2.9524 \qquad a_2 = -2.2143 \qquad a_3 = 0.6190.$$

Therefore, the quadratic interpolation function is
$u(x) = 2.9524 - 2.2143x + 0.6190x^2$, as shown in Fig. 3.5.

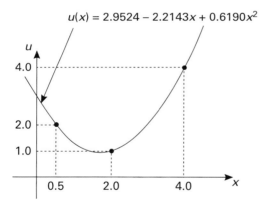

Figure 3.5 Quadratic interpolation function.

For n data points, the n coefficients can be obtained by solving the linear system given in Eq. (3.4). The coefficient matrix in Eq. (3.4) is called the Vandermonde matrix. To calculate the coefficients of the polynomial, the inverse of the Vandermonde matrix must be computed. When n increases, the computational cost of calculating the inverse increases and the Vandermonde matrix becomes ill-conditioned quickly. Therefore, in practice, the interpolation polynomial is not obtained by inverting the Vandermonde matrix. Instead, the polynomial interpolation function is obtained by using another approach: Lagrange interpolation. In 1-D space, the Lagrange interpolation of the unknown polynomial going through n data points is constructed by

3.2 Numerical Interpolation and Approximation

$$u(x) = \sum_{i=1}^{n} N_i(x) u_i \qquad (3.13)$$

where n is the number of data points over an interval, u_i is the value of the data point i, and $N_i(x)$ is the i-th Lagrange basis function given by

$$N_i(x) = \frac{\Pi_{k=1, k \neq i}^{n}(x - x_k)}{\Pi_{k=1, k \neq i}^{n}(x_i - x_k)} \qquad i = 1, 2, \ldots, n. \qquad (3.14)$$

Equation (3.13) states that the interpolating polynomial of n distinct data points can be expressed as a linear combination of the n Lagrange basis functions with the values of the data points as their coefficients. As we will see in later chapters, in finite element analysis, $N_i(x)$, $i = 1, 2, \ldots, n$ are also called the shape functions. A MATLAB code for computing the 1-D Lagrange basis functions is shown below.

```
% Input :   data_points_x : a n x 1 matrix which stores the x− coordinates
%           of the data points to be interpolated .
% Input :   x_vector : a np x 1 array storing the x− coordinates of a set
%           of points where the Lagrange basis function is evaluated at
% Output: N, a np x n matrix storing N(i,j) which represents Ni(x_j)
function [N]=CompLagBasis1D(data_points_x, x_vector)
np=size( data_points_x ,1);      % number of input points
n=size(x_vector ,1);              % number of x points
N=ones(np,n);                     % N matrix stores result
for i=1:np                        % loop over i: i−th basis function
  for j=1:n                       % loop over j: j−th output location
    for k=1:np;                   % loop over k: compute the product
      if i~=k                     % skip when x_i = x_k
        N(i,j)=N(i,j)*(x_vector(j)−data_points_x(k,1)) /...
            ( data_points_x (i ,1)−data_points_x(k,1));
      end
    end
  end
end
```

Note that in the MATLAB code, a set of x-coordinates should be given in the vector "x-vector" as the input. This is because the code can only return the values of the Lagrange basis functions at discrete locations. In numerical analysis, a computer code does not solve or compute the functions analytically, but does that numerically at a set of discrete points. Therefore, coordinates of discrete locations must be specified for the MATLAB code to calculate the value of the functions at these locations.

From Eq. (3.14), it is easy to see that for two data points, $n = 2$, the shape functions are linear:

$$N_1(x) = \frac{(x - x_2)}{(x_1 - x_2)} \qquad N_2(x) = \frac{(x - x_1)}{(x_2 - x_1)}. \qquad (3.15)$$

The linear shape functions are plotted in Fig. 3.6 below. It is clear that the linear shape functions are straight lines going either from 0 to 1 or from 1 to 0 within the interval of

$[x_1, x_2]$. Another characteristic of the Lagrange basis functions is that a basis function is equal to 0 at all but one data point. For the linear (two data points) case, as shown in the figure, $N_1(x_1) = 1$, $N_1(x_2) = 0$ and $N_2(x_1) = 0$, $N_2(x_2) = 1$. Conventionally, the index of the basis functions is set such that $N_i(x_i) = 1$ and $N_i(x_j) = 0$, $i \neq j$.

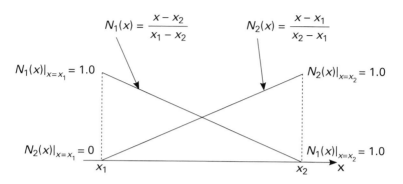

Figure 3.6 Linear Lagrange basis (or shape) functions.

Example 3.3 Construct the polynomial interpolation function for the two data points shown in Fig. 3.2 in Example 3.1.

Solution
Following Eq. (3.15) and Eq. (3.13), the linear interpolation function is given by

$$u(x) = \sum_{i=1}^{2} N_i(x) u_i$$

$$= \frac{(x - x_2)}{(x_1 - x_2)} u_1 + \frac{(x - x_1)}{(x_2 - x_1)} u_2$$

$$= \frac{(x - 4)}{(0.5 - 4.0)} 2.0 + \frac{(x - 0.5)}{(4.0 - 0.5)} 4.0$$

$$= \frac{4}{7} x + \frac{12}{7}$$

$$= 0.5714x + 1.7143. \tag{3.16}$$

The solution is the same as that obtained in Example 3.1. However, by using the Lagrange interpolation, there is no need to solve a linear system of equations.

For $n = 3$, the Lagrange interpolation basis functions are quadratic. From Eq. (3.14), we obtain

$$N_1(x) = \frac{(x - x_2)(x - x_3)}{(x_1 - x_2)(x_1 - x_3)} \tag{3.17}$$

$$N_2(x) = \frac{(x-x_1)(x-x_3)}{(x_2-x_1)(x_2-x_3)} \qquad (3.18)$$

$$N_3(x) = \frac{(x-x_1)(x-x_2)}{(x_3-x_1)(x_3-x_2)}. \qquad (3.19)$$

The shape functions are shown in Fig. 3.7.

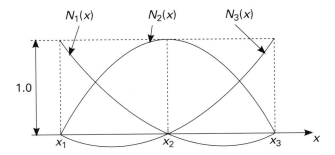

Figure 3.7 Quadratic Lagrange interpolation basis functions.

Example 3.4 Construct the polynomial interpolation function for the three data points shown in Fig. 3.4 in Example 3.2.

Solution
Following Eqs. (3.17-3.19) and Eq. (3.13), the quadratic interpolation function is given by

$$u(x) = \sum_{i=1}^{3} N_i(x)u_i$$

$$= \frac{(x-x_2)(x-x_3)}{(x_1-x_2)(x_1-x_3)}u_1 + \frac{(x-x_1)(x-x_3)}{(x_2-x_1)(x_2-x_3)}u_2 + \frac{(x-x_1)(x-x_2)}{(x_3-x_1)(x_3-x_2)}u_3$$

$$= \frac{(x-2)(x-4)}{(0.5-2)(0.5-4)}2 + \frac{(x-0.5)(x-4)}{(2-0.5)(2-4)}1 + \frac{(x-0.5)(x-2)}{(4-0.5)(4-2)}4$$

$$= 2.9524 - 2.2143x + 0.6190x^2. \qquad (3.20)$$

The solution is the same as that obtained in Example 3.2.

The following MATLAB code computes the Lagrange interpolation function going through a set of data points.

```
1  % Input : data_points : a n x 2 matrix . The first column stores the
2  %                       x- coordinates of the data points to be interpolated , the
3  %                       second column stores the function values at the data points .
4  % Input : x_vector : np x 1 array storing the x- coordinates of a set of
5  %                    points where the interpolation function is calculated .
```

```
6  % Output: u, n x 1 array  storing  the  value  of  the  interpolating  function
7  %                at the  points  specified  by x_vector.
8  function [u]=compLagIntp1D(data_points, x_vector)
9  np=size( data_points ,1);
10 n=size( x_vector ,1);
11 N=CompLagBasis1D(data_points(:,1), x_vector);
12 u=zeros(n,1);
13 for j=1:n
14   for i=1:np
15     u(j,1)=u(j,1)+N(i,j)*data_points (i,2);   %u(x_j)=sum(N_i( x_j )*u_i )
16   end
17 end
```

Using the MATLAB functions, we revisit Example 3.2. The following code computes and plots the Lagrange interpolation function over the interval of [0, 5]. The result is shown in Fig. 3.8.

```
1  dp (:,1) =[0.5 2 4]';              % x− coordinates of input data points
2  dp (:,2) =[2 1 4]';                % data values of input data points
3  x_vector =[0:0.05:5]';             % output points
4  u=CompLagIntp1D(dp, x_vector);     % compute the interpolation function
5                                     % at the output points
6  figure (1);                        % launch figure 1
7  plot( x_vector ,u,'k−','LineWidth',2);  % plot the interpolation function
8                                     % at the output points
9  hold on;                           % hold on
10 plot(dp (:,1) ,dp (:,2) ,'ko','LineWidth',5);  % plot the input data points
11 set(gca,' fontsize ',16);           % set font size of the figure
12 xlabel('x',' fontsize ',18);        % label x−axis
13 ylabel('u',' fontsize ',18);        % label y−axis
```

The 1-D Lagrange interpolation can be extended to fit a grid of data points in multiple dimensions. For example, the 2-D Lagrange interpolation basis functions over m-by-n data points are simply the products of 1-D Lagrange interpolation basis functions of orders m and n in x- and y-directions, respectively, i.e.,

$$N_{(i-1)\cdot n+j}(x,y) = \frac{\Pi_{p=1,p\neq i}^{m}(x-x_p)}{\Pi_{p=1,p\neq i}^{m}(x_i-x_p)} \frac{\Pi_{q=1,q\neq j}^{n}(y-y_q)}{\Pi_{q=1,q\neq j}^{n}(y_j-y_q)}$$

$$i = 1, 2, \ldots, m; j = 1, 2, \ldots, n. \qquad (3.21)$$

For example, the 2-D Lagrange interpolation basis functions for a 2-by-2 data point grid are obtained as

$$N_1(x,y) = \frac{(x-x_2)}{(x_1-x_2)}\frac{(y-y_2)}{(y_1-y_2)} \qquad N_2(x,y) = \frac{(x-x_2)}{(x_1-x_2)}\frac{(y-y_1)}{(y_2-y_1)}$$

$$N_3(x,y) = \frac{(x-x_1)}{(x_2-x_1)}\frac{(y-y_2)}{(y_1-y_2)} \qquad N_4(x,y) = \frac{(x-x_1)}{(x_2-x_1)}\frac{(y-y_1)}{(y_2-y_1)} \qquad (3.22)$$

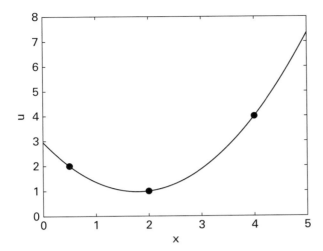

Figure 3.8 MATLAB result: quadratic Lagrange interpolation of three data points.

Likewise, in 3-D, the Lagrange interpolation basis functions are given by

$$N_{(i-1)\cdot n\cdot o+(j-1)\cdot o+k}(x, y, z)$$

$$= \frac{\Pi_{p=1, p\neq i}^{m}(x - x_p) \; \Pi_{q=1, q\neq j}^{n}(y - y_q) \; \Pi_{r=1, r\neq k}^{o}(z - z_r)}{\Pi_{p=1, p\neq i}^{m}(x_i - x_p) \; \Pi_{q=1, q\neq j}^{n}(y_j - y_q) \; \Pi_{r=1, r\neq k}^{o}(z_k - z_r)}$$

$$i = 1, 2, \ldots, m; j = 1, 2, \ldots, n; k = 1, 2, \ldots, o. \qquad (3.23)$$

High Order Polynomial Interpolation

While polynomial interpolation is effective in many cases, there are problems on how accurately a polynomial can represent the underlying data trend, especially for high order polynomials. A well-known example is the Runge phenomenon, as shown in Fig. 3.9. The figure shows the results of Lagrange interpolation for the function given by

$$f(x) = \frac{1}{1 + 25x^2}. \qquad (3.24)$$

The actual function is plotted using a solid line for comparison. The Lagrange interpolation is carried out by using 3, 7, and 11 equispaced data points and the resultant 2nd, 6th, and 10th order polynomial interpolation functions are plotted in the figure, respectively. It is shown that large oscillation occurs when using polynomials of high order. This example demonstrates that increasing the order of interpolating polynomials does not always improve accuracy.

3.2.2 Hermite Interpolation

In the Lagrange polynomial interpolation, the fitting polynomial is determined by letting the polynomial go through the data points. That is, the value of the polynomial

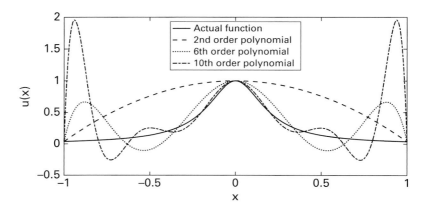

Figure 3.9 Runge phenomenon.

is prescribed at specified locations. In finite element analysis, sometimes an interpolating polynomial is sought with not only its value but also its derivative(s) given at specified locations. The procedure of developing such a polynomial is called Hermite interpolation. Taking a 1-D polynomial with its value and first derivative given at specified locations, the mathematical problem is given by

$$u(x_1) = a_1 \cdot 1 + a_2 x_1 + a_3 x_1^2 + a_4 x_1^3 + \cdots = u_1$$
$$\frac{du}{dx}(x_1) = a_2 + 2a_3 x_1 + 3a_4 x_1^2 + \cdots = p_1$$
$$u(x_2) = a_1 \cdot 1 + a_2 x_2 + a_3 x_2^2 + a_4 x_2^3 + \cdots = u_2$$
$$\frac{du}{dx}(x_2) = a_2 + 2a_3 x_2 + 3a_4 x_2^2 + \cdots = p_2$$
$$\cdots \tag{3.25}$$

where u_i and p_i are the value and first derivative of the polynomial at point i, respectively. It can be observed from Eq. (3.25) that two conditions (value and the first derivative) are specified at each data point. That is, each data point in Hermite interpolation gives two equations, which enables the determination of two unknown coefficients of the polynomial. Therefore, for n data points, a polynomial of order $2n - 1$ (having $2n$ coefficients) can be determined.

Example 3.5 Construct a Hermite polynomial interpolation function for the two data points shown in Fig. 3.10.

Solution
The two data points in the figure give rise to four equations for the Hermite interpolation function to have the values and derivatives at the points. Since four equations can determine four unknown coefficients, the Hermite polynomial is determined to be cubic. By using Eq. (3.25), a linear system of the unknown coefficients can be generated as

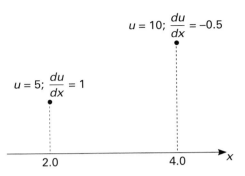

Figure 3.10 Hermite interpolation example problem.

$$\begin{bmatrix} 1 & 2 & 2^2 & 2^3 \\ 0 & 2 & 2\cdot 2 & 3\cdot 2^2 \\ 1 & 4 & 4^2 & 4^3 \\ 0 & 4 & 2\cdot 4 & 3\cdot 4^2 \end{bmatrix} \begin{Bmatrix} a_1 \\ a_2 \\ a_3 \\ a_4 \end{Bmatrix} = \begin{Bmatrix} 5 \\ 1 \\ 10 \\ -0.5 \end{Bmatrix}. \qquad (3.26)$$

Solving the linear system gives the solution $\{30\ -32\ 12.75\ -1.5\}^T$. Therefore, the cubic Hermite polynomial is $30 - 32x + 12.75x^2 - 1.5x^3$. The MATLAB plot of the function is shown in Fig. 3.11.

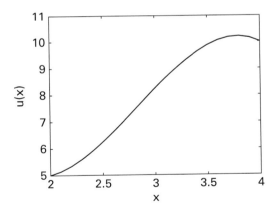

Figure 3.11 Example problem: MATLAB plot of Hermite interpolation.

The method of determining the Hermite interpolation function shown above is the method of undetermined coefficients, which requires the solutions of a linear system. Similar to the Lagrange interpolation, there are other methods available to calculate the Hermite interpolation polynomial. Two popular methods include the modified Lagrange basis method (Ciarlet and Raviart 1972) and the divided difference method (Hakopian 1982). The details of these methods are beyond the scope of this book. Here, without the proof, we simply provide a convenient result from the

Table 3.1 Data points for piecewise polynomial interpolation

x	0	1/3	2/3	1	4/3	5/3	2
u	5.1765	86.7692	11.5113	16.0000	−0.6819	−3.8745	−4.8552

literature: given an interval $[x_0, x_1]$ and value of $u(x)$ and $u' = du(x)/dx$ at x_0 and x_1, the cubic Hermite polynomial can be written as

$$u(x) = \left(1 + 2\frac{x - x_0}{x_1 - x_0}\right)\left(\frac{x_1 - x}{x_1 - x_0}\right)^2 u(x_0) + (x - x_0)\left(\frac{x_1 - x}{x_1 - x_0}\right)^2 u'(x_0)$$

$$+ \left(1 + 2\frac{x_1 - x}{x_1 - x_0}\right)\left(\frac{x_0 - x}{x_0 - x_1}\right)^2 u(x_1) + (x - x_1)\left(\frac{x_0 - x}{x_0 - x_1}\right)^2 u'(x_1). \tag{3.27}$$

3.2.3 Piecewise Polynomial Interpolation

Figure 3.9 shows that fitting a single polynomial to a large number of data points requires the order of the polynomial to be high, which will likely cause oscillating behavior in the interpolation function. To overcome this problem, the piecewise polynomial interpolation approach divides the interpolation domain into smaller parts. This process is called partition of the domain. Each resultant part then contains a small number of data points which can be interpolated using low-degree polynomials. The advantage of piecewise polynomial interpolation is that a large number of data points can be fitted with low-degree polynomials.

Consider a set of data points (x_1, u_1), (x_2, u_2), ..., (x_n, u_n) with $x_1 < x_2 < x_3 ... < x_n$, the partitioning process gives a set of small intervals called subintervals, and different interpolation polynomials are computed to fit the data points in different subintervals. The simplest piecewise polynomial interpolation is the 1-D linear piecewise interpolation which uses straight lines to connect successive data points. Figure 3.12 shows an example of the linear piecewise interpolation in comparison with the single polynomial interpolation. We aim to interpolate a MATLAB built-in function "humps(x)" given by

$$humps(x) = \frac{1}{(x - 0.3)^2 + 0.01} + \frac{1}{(x - 0.9)^2 + 0.04} - 6. \tag{3.28}$$

The set of data points used for interpolation is given in Table 3.1.

It is shown in Fig. 3.12 that, although its fitting quality is not ideal, the linear piecewise interpolation already describes the underlying function much better than the single 6th order polynomial interpolation in this case.

A natural choice following the linear piecewise interpolation is the quadratic piecewise interpolation. For the example problem shown in Fig. 3.12, the domain is divided into three parts: [0, 2/3], [2/3, 4/3], and [4/3, 2], each containing three data points: {(0, 5.1765), (1/3, 86.7692), (2/3, 11.5113)}, {(2/3, 11.5113), (1, 16),

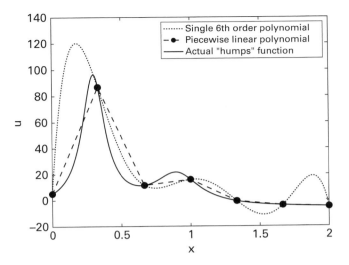

Figure 3.12 Piecewise linear interpolation of a set of data points.

$(4/3, -0.6819)\}$, and $\{(4/3, -0.6819), (5/3, -3.8745), (2, -4.8552)\}$, respectively. By using the code shown in the Lagrange interpolation section, one can construct the quadratic piecewise interpolation as shown in Fig. 3.13.

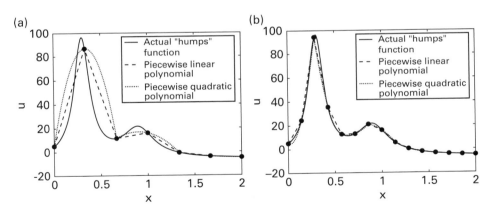

Figure 3.13 Linear and quadratic piecewise interpolation of the "humps" function.

Figure 3.13(a) shows that the quadratic piecewise interpolation gives a set of more curvy interpolants going through the data points. However, it is visually no better than the linear interpolants in fitting the underlying humps function. While it is true that when the data points are sparsely distributed in comparison to the geometric features of the underlying function, the quadratic piecewise interpolation typically gives more accurate fitting when the data point intervals are reduced, as shown in Fig. 3.13(b).

In addition to the simple piecewise polynomial interpolation we discussed above, there are other effective methods to construct the piecewise interpolants, as by definition there is a large degree of freedom in how to choose the partitioning scheme and continuity condition for the polynomial interpolants. For example, the cubic Hermite interpolation and cubic spline interpolation are among the popular piecewise polynomial interpolation methods (De Boor 1978).

3.3 Numerical Differentiation

For a sufficiently smooth function, the derivative of the function can be approximated numerically by using the Taylor series expansion. The method is referred to as the finite difference approximation (Mitchell & Griffiths 1980). For a 1-D function, recall the Taylor series expansion

$$f(x + \Delta x) = f(x) + \frac{df(x)}{dx}\Delta x + \frac{1}{2}\frac{d^2f(x)}{dx^2}(\Delta x)^2 + \frac{1}{3!}\frac{d^3f(x)}{dx^3}(\Delta x)^3 + \cdots \quad (3.29)$$

where ... represents the remainder containing terms with order of Δx higher than 3. Rearranging the terms, one can write the first derivative of the function as

$$\frac{df(x)}{dx} = \frac{f(x + \Delta x) - f(x)}{\Delta x} - \frac{1}{2}\frac{d^2f(x)}{dx^2}(\Delta x) - \frac{1}{3!}\frac{d^3f(x)}{dx^3}(\Delta x)^2 + \cdots \quad (3.30)$$

Assuming Δx is a small number, then the first derivative can be approximated as

$$\frac{df(x)}{dx} \approx \frac{f(x + \Delta x) - f(x)}{\Delta x}. \quad (3.31)$$

This approximated first derivative of $f(x)$ is first-order accurate since the truncated remainder terms are dominated by $\frac{1}{2}\frac{d^2f(x)}{dx^2}(\Delta x)$ which is first-order in Δx and denoted by $O(\Delta x)$. Figure 3.14 shows that the approximated first derivative given in Eq. (3.31) is simply the slope of the straight dashed line connecting $f(x)$ and $f(x + \Delta x)$. Equation (3.31) is called the forward difference formula.

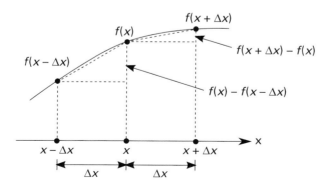

Figure 3.14 Finite difference approximation of derivatives of function $f(x)$.

If we use the Taylor series expansion to write out the function value at $x - \Delta x$, i.e.,

$$f(x - \Delta x) = f(x) - \frac{df(x)}{dx}\Delta x + \frac{1}{2}\frac{d^2f(x)}{dx^2}(\Delta x)^2 - \frac{1}{3!}\frac{d^3f(x)}{dx^3}(\Delta x)^3 + \cdots \quad (3.32)$$

Then by rearranging the terms in Eq. (3.32) and truncating the terms of second and higher orders of Δx, the first derivative of $f(x)$ can also be approximated as

$$\frac{df(x)}{dx} \approx \frac{f(x) - f(x - \Delta x)}{\Delta x}. \quad (3.33)$$

From Fig. 3.14, it is clear that this approximation is the slope of the straight dashed line connecting $f(x - \Delta x)$ and $f(x)$. Equation (3.33) is called the backward difference formula.

Compared to the forward and backward difference schemes, a more accurate approximation can be constructed by subtracting Eq. (3.32) from Eq. (3.29), i.e.,

$$f(x + \Delta x) - f(x - \Delta x) = 2\frac{df(x)}{dx}\Delta x + \frac{2}{3!}\frac{d^3f(x)}{dx^3}(\Delta x)^3 + \cdots \quad (3.34)$$

We have

$$\frac{df(x)}{dx} = \frac{f(x + \Delta x) - f(x - \Delta x)}{2\Delta x} - \frac{1}{6}\frac{d^3f(x)}{dx^3}(\Delta x)^2 + \cdots \quad (3.35)$$

Truncating the terms $\frac{1}{6}\frac{d^3f(x)}{dx^3}(\Delta x)^2 + \cdots$, we obtain the approximation scheme called the central difference

$$\frac{df(x)}{dx} = \frac{f(x + \Delta x) - f(x - \Delta x)}{2\Delta x}. \quad (3.36)$$

Note that the dominant term in the truncated remainder terms are second order in Δx. Therefore, the central difference formula is second-order accurate.

Example 3.6 Write a MATLAB code and calculate the first derivative of $x^3 - x^2$ at $x = 1$ by using the forward, backward, and central difference schemes and show the variation of the results as Δx reduces from 1.0 to 10^{-3}.

Solution
MATLAB code for computing the derivative:

```
x0=1.0;                  % the point where the derivative is calculated
dx =10.^[-3:.1:0];       % delta x values varying from 1 to 10^-3

for k=1:length(dx)
    % forward difference scheme
    dudx_f(k) = ((x0+dx(k))^3-(x0+dx(k))^2 -x0^3 + x0^2)/dx(k);
    % backward difference scheme
    dudx_b(k) = (x0^3-x0^2-(x0-dx(k))^3 +(x0-dx(k))^2)/dx(k);
    % central difference scheme
    dudx_c(k) = ((x0+dx(k))^3-(x0+dx(k))^2 ...
```

```
11                    -(x0-dx(k))^3+(x0-dx(k))^2)/(2*dx(k));
12 end
13
14 %- plot the results
15 figure(1);
16 semilogx(dx, dudx_f,'k--','LineWidth',2);
17 hold on
18 semilogx(dx, dudx_b,'k:','LineWidth',2);
19 semilogx(dx, dudx_c,'k','LineWidth',2);
20 set(gca,' fontsize ',16);
21 xlabel('x',' fontsize ',18);
22 ylabel('du/dx',' fontsize ',18);
23 legend('Forward difference','Backward diffference',...
24        'Central difference ','location','northwest');
```

The results are shown in Fig. 3.15.

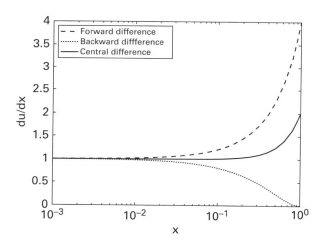

Figure 3.15 Finite difference approximation of the first derivative of function $f(x)$ at $x = 1$. The exact solution is 1.0.

Finally, adding Eq. (3.32) and Eq. (3.29) leads to

$$f(x + \Delta x) + f(x - \Delta x) = 2f(x) + \frac{d^2 f(x)}{dx^2}(\Delta x)^2 + O\left((\Delta x)^4\right) \quad (3.37)$$

Thus the second derivative of $f(x)$ can be written as

$$\begin{aligned} \frac{d^2 f(x)}{dx^2} &= \frac{f(x + \Delta x) - 2f(x) + f(x - \Delta x)}{(\Delta x)^2} + O\left((\Delta x)^2\right) \\ &\approx \frac{f(x + \Delta x) - 2f(x) + f(x - \Delta x)}{(\Delta x)^2}. \end{aligned} \quad (3.38)$$

Equation (3.38) is the central difference formula for the second derivative of $f(x)$. It is shown in Eq. (3.38) that the truncated remainder terms are second order in Δx. Hence Eq. (3.38) is second-order accurate.

For 2-D functions, we can use Taylor series expansions of functions of multiple variables, as described in Section 2.4, to derive the finite difference formulas for the derivatives of the functions. Without going through the straightforward derivations, the central difference formulas of the first and second derivatives of 2-D functions are given below. Given a 2-D function $f(x, y)$ we have

$$\frac{\partial f(x,y)}{\partial x} = \frac{f(x+\Delta x, y) - f(x-\Delta x, y)}{2\Delta x} \tag{3.39}$$

$$\frac{\partial f(x,y)}{\partial y} = \frac{f(x, y+\Delta y) - f(x, y-\Delta y)}{2\Delta y} \tag{3.40}$$

and

$$\frac{\partial^2 f(x,y)}{\partial x^2} = \frac{f(x+\Delta x, y) - 2f(x,y) + f(x-\Delta x, y)}{(\Delta x)^2} \tag{3.41}$$

$$\frac{\partial^2 f(x,y)}{\partial y^2} = \frac{f(x, y+\Delta y) - 2f(x,y) + f(x, y-\Delta y)}{(\Delta y)^2} \tag{3.42}$$

$$\frac{\partial^2 f(x,y)}{\partial x \partial y}$$
$$= \frac{f(x+\Delta x, y+\Delta y) + f(x-\Delta x, y-\Delta y) - f(x-\Delta x, y+\Delta y) - f(x+\Delta x, y-\Delta y)}{4\Delta x \Delta y}. \tag{3.43}$$

where the function is evaluated at the grid points shown in Fig. 3.16. The finite difference approximations of 3-D function derivatives can be obtained from a 3-D grid of points by using the same approach employed for calculating the derivatives of 2-D functions. The tedious expressions are not displayed here.

3.4 Numerical Integration

The numerical evaluation (approximately) of definite integrals is called numerical integration or numerical quadrature (Davis and Rabinowitz 2007). In general, a numerical quadrature formula is given by

$$\int_{x_1}^{x_2} f(x)dx \approx \sum_{i=1}^{ng} w_i f(x_i) \tag{3.44}$$

where ng is the number of quadrature points, w_i, $i = 1, 2, \ldots, ng$, are the weights, and x_i, $i = 1, 2, \ldots, ng$, are the quadrature points or nodes or abscissas. The formula given in Eq. (3.44) is called a quadrature rule.

Typically, the weights and quadrature points in a quadrature rule are determined by approximating the actual function $f(x)$ by a certain simple function

Numerical Analysis Methods

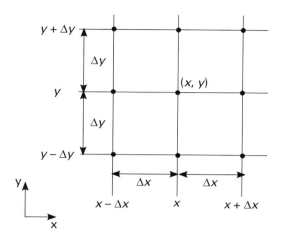

Figure 3.16 Finite difference grid for calculating first and second derivatives of function $f(x, y)$.

(usually polynomials) and then requiring this approximating simple function to be integrated exactly over the interval by the quadrature rule. The following example illustrates this procedure.

Example 3.7 A function $f(x)$ is shown in Fig. 3.17. The function is to be integrated over the interval $[x_1, x_2]$, i.e., we seek to compute $\int_{x_1}^{x_2} f(x)dx$.

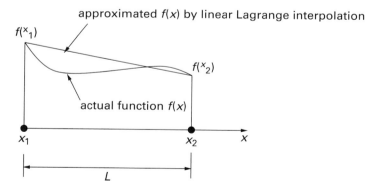

Figure 3.17 Numerical integration error with linear approximation.

Solution
The function is approximated by using the linear Lagrange interpolation as shown in Fig. 3.17, i.e.,

$$f(x) \approx \frac{x - x_2}{x_1 - x_2} f(x_1) + \frac{x - x_1}{x_2 - x_1} f(x_2). \tag{3.45}$$

Substituting Eq. (3.45) into the integral, we obtain

$$\int_{x_1}^{x_2} f(x)dx \approx \int_{x_1}^{x_2} \left[\frac{x-x_2}{x_1-x_2}f(x_1) + \frac{x-x_1}{x_2-x_1}f(x_2)\right]dx$$

$$= \left(\int_{x_1}^{x_2} \frac{x-x_2}{x_1-x_2}dx\right)f(x_1) + \left(\int_{x_1}^{x_2} \frac{x-x_1}{x_2-x_1}dx\right)f(x_2)$$

$$= \frac{x_2-x_1}{2}f(x_1) + \frac{x_2-x_1}{2}f(x_2)$$

$$= w_1 f(x_1) + w_2 f(x_2). \tag{3.46}$$

Equation (3.46) is a quadrature rule called the Trapezoid rule, where $w_1 = w_2 = \frac{x_2-x_1}{2} = \frac{L}{2}$ are the weights, and x_1 and x_2 are the quadrature points of the rule. Note that if the Lagrange interpolation is used, the two endpoints x_1 and x_2 of the interval are the two quadrature points. The geometric interpretation of the Trapezoid rule is shown in Fig. 3.18. The Trapezoid rule can be regarded as the summation of the areas of two rectangles as given in Eq. (3.46). The first rectangle has a width of w_1 and a height of $f(x_1)$ and the second is $w_2 \times f(x_2)$. The summation of the shaded rectangular areas is the approximation of $\int_{x_1}^{x_2} f(x)dx$ which by definition is the area covered by the actual function $f(x)$ within the interval of $[x_1, x_2]$. It is easy to see that if $f(x)$ is linear, this approximation is exact. Therefore, the Trapezoid rule is exact for linear functions.

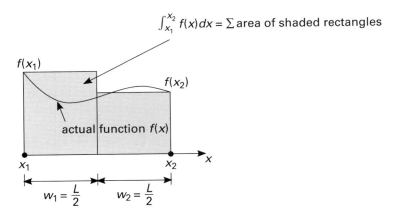

Figure 3.18 Geometric interpretation of Trapezoid quadrature rule.

If a quadratic Lagrange interpolation is used to approximate the function $f(x)$, as shown in Fig. 3.19, a different quadrature rule can be derived.

$$\int_{x_1}^{x_2} f(x)dx$$

$$\approx \int_{x_1}^{x_2} \left[\frac{(x-x_2)(x-x_3)}{(x_1-x_2)(x_1-x_3)}f(x_1) + \frac{(x-x_1)(x-x_3)}{(x_2-x_1)(x_2-x_3)}f(x_2) \right.$$

$$\left. + \frac{(x-x_1)(x-x_2)}{(x_3-x_1)(x_3-x_2)}f(x_3) \right] dx$$

$$= \frac{x_2-x_1}{6}f(x_1) + \frac{x_2-x_1}{6}f(x_2) + \frac{4(x_2-x_1)}{6}f(x_3)$$

$$= \frac{L}{6}f(x_1) + \frac{L}{6}f(x_2) + \frac{4L}{6}f(x_3)$$

$$= w_1 f(x_1) + w_2 f(x_2) + w_3 f(x_3). \tag{3.47}$$

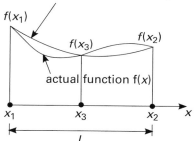

Figure 3.19 Numerical integration with quadratic approximation.

Equation (3.47) is a quadrature rule called Simpson's rule. As in the Trapezoid rule case, Simpson's rule integration can be regarded as approximating the area covered by $f(x)$ between x_1 and x_2 by using the summation of the areas of three rectangles, as shown in Fig. 3.20.

By using the same procedure, one can obtain the quadrature rules based on cubic, quartic Lagrange interpolations and so on. Quadrature rules that are obtained by using the Lagrange interpolations over equally spaced quadrature points in the interval are called the family of Newton–Cotes quadrature rules. In Newton–Cotes quadrature rules, the 2-point Trapezoid rule can integrate linear functions exactly (called polynomial degree 1). The 3-point Simpson's rule can integrate quadratic functions exactly (polynomial degree 2) since it is based on the quadratic Lagrange interpolation. However, due to the symmetry of the abscissas, Simpson's rule also integrates cubic functions exactly. Therefore, it is in fact of polynomial degree 3. In general, an odd-order Newton–Cotes rule gives an extra degree beyond that of the polynomial interpolant on which it is based, i.e.,

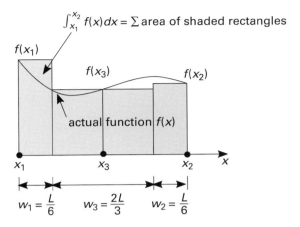

Figure 3.20 Geometric interpretation of Simpson's quadrature rule.

$$\text{polynomial degree of } n\text{-point Newton–Cotes rule} = \begin{cases} n-1 & \text{if } n \text{ is even} \\ n & \text{if } n \text{ is odd} \end{cases} \quad (3.48)$$

Another class of quadrature rules is called Gaussian quadrature or simply Gauss quadrature (we will call it Gauss quadrature in this book). The key idea of the Gauss quadrature rules is that both the weights and the locations of the quadrature points are determined such that the maximum polynomial degree of the quadrature rule is achieved. For example, to obtain the 2-point Gauss quadrature rule, we need to determine the two quadrature points (also called Gauss points) and the two weights. In Gauss quadrature, the locations of the Gauss points are typically not at the ends of the interval and are not equally spaced. For this reason, to distinguish the Gauss points from the endpoints or element nodes, $x_1, x_2, \ldots$, we denote the Gauss points as x_{g1}, $x_{g2}, \ldots$, and we denote their corresponding weights as $w_1, w_2, \ldots$, respectively. The four unknowns (two Gauss point locations and two weights) allow four constraint conditions to be specified for obtaining a unique solution for the Gauss quadrature rule. For this reason, we require the Gauss quadrature rule to integrate $f(x) = 1, x, x^2, x^3$ exactly (then any linear combination of $1, x, x^2, x^3$ can be integrated exactly), i.e.,

$$w_1 \times 1 + w_2 \times 1 = \int_{x_1}^{x_2} 1 dx$$
$$w_1 \times x_{g1} + w_2 \times x_{g2} = \int_{x_1}^{x_2} x dx$$
$$w_1 \times x_{g1}^2 + w_2 \times x_{g2}^2 = \int_{x_1}^{x_2} x^2 dx$$
$$w_1 \times x_{g1}^3 + w_2 \times x_{g2}^3 = \int_{x_1}^{x_2} x^3 dx. \quad (3.49)$$

Given the interval $[x_1, x_2]$, Eq. (3.49) can be solved. Note that equation system, Eq. (3.49), is nonlinear and needs a nonlinear solver to calculate the unknowns. In

Table 3.2 1-D Gauss quadrature rules for the interval $[-1, 1]$

Gauss points	Gauss point locations	Weights	Polynomial degree
1	$x_{g1} = 0$	$w_1 = 2$	1
2	$x_{g1} = -\frac{1}{\sqrt{3}}$ $x_{g2} = \frac{1}{\sqrt{3}}$	$w_1 = w_2 = 1.0$	3
3	$x_{g1} = -\sqrt{\frac{3}{5}}$ $x_{g2} = 0$ $x_{g3} = \sqrt{\frac{3}{5}}$	$w_1 = \frac{5}{9}$ $w_2 = \frac{8}{9}$ $w_3 = \frac{5}{9}$	5
4	$x_{g1} = -\sqrt{\frac{3 + 2\sqrt{\frac{6}{5}}}{7}}$ $x_{g2} = -\sqrt{\frac{3 - 2\sqrt{\frac{6}{5}}}{7}}$ $x_{g3} = \sqrt{\frac{3 - 2\sqrt{\frac{6}{5}}}{7}}$ $x_{g4} = \sqrt{\frac{3 + 2\sqrt{\frac{6}{5}}}{7}}$	$w_1 = \frac{18 - \sqrt{30}}{36}$ $w_2 = \frac{18 + \sqrt{30}}{36}$ $w_3 = \frac{18 + \sqrt{30}}{36}$ $w_4 = \frac{18 - \sqrt{30}}{36}$	7

general, for an n-point Gauss quadrature, $2n$ equations can be written in determining the n Gauss points and n weights. The $2n$ equations allow us to enforce the exact integration of polynomials up to the order of $2n - 1$. Therefore, an n-point Gauss quadrature is of polynomial order $2n - 1$. Since the equations are nonlinear, obtaining the weights and locations is not trivial. Tables containing pre-calculated values are often used. The 1-, 2-, 3-, and 4-point Gauss quadrature rules for the interval $[-1, 1]$ are listed in Table 3.2. For integrals with an arbitrary interval $[a, b]$, the Gauss points and weights can be obtained by scaling the values given in Table 3.2. The scaled quadrature rules are summarized in Table 3.3. A relative integration error is defined to measure the accuracy of a numerical quadrature:

$$\text{relative integration error} = \left| \frac{\text{computed value} - \text{exact value}}{\text{exact value}} \right| \quad (3.50)$$

3.4 Numerical Integration

Table 3.3 1-D Gauss quadrature rules for an interval [a, b]

Gauss points	Gauss point locations	Weights	Polynomial degree
1	$x_{g1} = x_0$	$w_1 = L$	1
2	$x_{g1} = x_0 - \frac{1}{\sqrt{3}}\left(\frac{L}{2}\right)$ $x_{g2} = x_0 + \frac{1}{\sqrt{3}}\left(\frac{L}{2}\right)$	$w_1 = w_2 = \frac{L}{2}$	3
3	$x_{g1} = x_0 - \frac{\sqrt{3}}{\sqrt{5}}\left(\frac{L}{2}\right)$ $x_{g2} = x_0$ $x_{g3} = x_0 + \frac{\sqrt{3}}{\sqrt{5}}\left(\frac{L}{2}\right)$	$w_1 = \frac{5}{9}\left(\frac{L}{2}\right)$ $w_2 = \frac{8}{9}\left(\frac{L}{2}\right)$ $w_3 = \frac{5}{9}\left(\frac{L}{2}\right)$	5
4	$x_{g1} = x_0 - \left(\frac{L}{2}\right)\sqrt{\frac{3+2\sqrt{\frac{6}{5}}}{7}}$ $x_{g2} = x_0 - \left(\frac{L}{2}\right)\sqrt{\frac{3-2\sqrt{\frac{6}{5}}}{7}}$ $x_{g3} = x_0 + \left(\frac{L}{2}\right)\sqrt{\frac{3-2\sqrt{\frac{6}{5}}}{7}}$ $x_{g4} = x_0 + \left(\frac{L}{2}\right)\sqrt{\frac{3+2\sqrt{\frac{6}{5}}}{7}}$	$w_1 = \frac{18-\sqrt{30}}{36}\left(\frac{L}{2}\right)$ $w_2 = \frac{18+\sqrt{30}}{36}\left(\frac{L}{2}\right)$ $w_3 = \frac{18+\sqrt{30}}{36}\left(\frac{L}{2}\right)$ $w_4 = \frac{18-\sqrt{30}}{36}\left(\frac{L}{2}\right)$	7

* where $L = b - a$ is the length of the interval and $x_0 = (a+b)/2$ is the center point of the interval [a, b].

Example 3.8 Use the 2-point Gauss quadrature rule to evaluate the integral

$$\int_0^1 \frac{x^3}{2} dx \tag{3.51}$$

Solution
The 2-point Gauss quadrature gives $x_{g1} = 0.5 - \frac{1}{\sqrt{3}}\frac{1}{2}$, $x_{g2} = 0.5 + \frac{1}{\sqrt{3}}\frac{1}{2}$, $w_1 = 0.5$ and $w_2 = 0.5$. We obtain

$$\int_0^1 \frac{x^3}{2} dx \approx \frac{1}{2}\left(0.5 - \frac{1}{\sqrt{3}}\frac{1}{2}\right)^3 \frac{1}{2} + \frac{1}{2}\left(0.5 + \frac{1}{\sqrt{3}}\frac{1}{2}\right)^3 \frac{1}{2} = \frac{1}{8}. \tag{3.52}$$

The exact value of the integral is 1/8. In this case the Gauss quadrature gives the exact result. This is due to the fact that the 2-point Gauss quadrature integrates any polynomial up to the order of 3 exactly.

Compared to the Newton–Cotes quadrature, the Gauss quadrature can achieve a higher polynomial degree with the same number of quadrature points. In the computer implementation of FEA, Gauss quadrature has become the preferred method for numerical integration.

A MATLAB function for obtaining the Gauss points and weights for the interval $[-1, 1]$ is given as follows

```
% Get Gauss points and weights for the given number of Gauss points
% in the interval of [-1,1]
% Input : n_guass_points : number of Gauss points
% Output : gauss_points : vectors store the locations of Gauss points
% Output : gauss_weights : the corresponding weights of the Gauss points
function [ gauss_points , gauss_weights]=Get1DGauss(n_gauss_points)
% set up empty return variables
gauss_points =zeros(n_gauss_points ,1) ;
gauss_weights=zeros(n_gauss_points ,1) ;
% switch block , assign Gauss points and weights for
% the given input parameter
switch n_gauss_points
   case 1                                      % 1 Gauss point
      gauss_points=0;
      gauss_weights=2.0;
   case 2                                      % 2 Gauss points
      gauss_points=[-1.0/sqrt(3) 1.0/ sqrt(3)]';
      gauss_weights=[1 1]';
   case 3                                      % 3 Gauss points
      gauss_points=[-sqrt(3/5) 0 sqrt (3/5) ]';
      gauss_weights=[5/9 8/9 5/9]';
   case 4                                      % 4 Gauss points
      gauss_points=[-0.861136 -0.339981 0.339981 0.861136]';
      gauss_weights=[0.34785 0.652145 0.652145 0.34785]';
   otherwise                                   % max 4 Gauss points here
      fprintf ('Error calling Get1DGauss\n') % add more if needed
end
```

3.5 Systems of Linear Equations

Systems of linear algebraic equations arise in almost all engineering, applied mathematics, and scientific computation problems. In finite element analysis, linear systems are discretized representations of governing equations, or the results of linear approximation of nonlinear equations. Therefore, efficient and accurate solution of linear systems is critical for many numerical methods to effectively solve a wide variety of practical computational problems.

In matrix-vector form, a system of linear equations can be written as

$$\mathbf{Ax} = \mathbf{b} \tag{3.53}$$

where $\mathbf{A}_{m\times n}$ is the coefficient matrix, $\mathbf{x}_{n\times 1}$ is the unknown vector to be computed, $\mathbf{b}_{m\times 1}$ is the right hand side vector. Depending on the characteristics of the coefficient matrix and the right hand side vector, the linear system can be categorized into several types. It is easy to see that, in the linear system, the number of rows represents the number of equations, and the number of columns is equal to the number of unknowns. When $m \neq n$, the matrix $\mathbf{A}$ is rectangular. If $m > n$, then the number of equations is larger than the number of unknowns. The system is said to be over-determined. If the equations in an over-determined system are independent from each other, then the system has no solution. If $m < n$, there are more unknowns than equations. It is called an under-determined system. An under-determined linear system has either no solution or infinitely many solutions depending on the type of $\mathbf{A}$ and $\mathbf{b}$. If $\mathbf{b} = \mathbf{0}$, then the system is called a homogeneous system. A homogeneous under-determined linear system always has infinitely many non-trivial solutions. If $\mathbf{b} \neq \mathbf{0}$, the system is non-homogeneous. In this case, if the rows of $\mathbf{A}$ are independent from each other, then the system has infinitely many solutions, otherwise, it has no solution. For example, the following system

$$\begin{bmatrix} 1 & 2 & 3 \\ 2 & 4 & 6 \end{bmatrix} \begin{Bmatrix} u_1 \\ u_2 \\ u_3 \end{Bmatrix} = \begin{Bmatrix} 0 \\ 1 \end{Bmatrix} \tag{3.54}$$

has no solution, and

$$\begin{bmatrix} 1 & 2 & 3 \\ 2 & 0 & -5 \end{bmatrix} \begin{Bmatrix} u_1 \\ u_2 \\ u_3 \end{Bmatrix} = \begin{Bmatrix} 0 \\ 1 \end{Bmatrix} \tag{3.55}$$

has infinitely many solutions.

In finite element analysis, the situation of $m \neq n$ rarely occurs. Instead, the matrix $\mathbf{A}$ is typically square, i.e., $m = n$. In this case, we also have homogeneous and non-homogeneous square linear systems. If the system is non-homogeneous, the system has a unique solution if the matrix $\mathbf{A}$ is non-singular. $\mathbf{A}_{n\times n}$ is said to be singular if it has any of the following equivalent properties:

1 The inverse of $\mathbf{A}$ does not exist.
2 The determinant of $\mathbf{A}$ is zero: $det(\mathbf{A}) = 0$.
3 The row vectors or column vectors of $\mathbf{A}$ are linearly dependent.
4 The rank of $\mathbf{A}$ is less than n.
5 There is a non-trivial solution of $\mathbf{x}$ for $\mathbf{Ax} = \mathbf{0}$.

If any one of the above is not true, then all of them are not true. The matrix is then said to be non-singular and the non-homogeneous linear system $\mathbf{A}_{n\times n}\mathbf{x}_{n\times 1} = \mathbf{b}_{n\times 1}$ has a unique solution. From the 5th property listed above, it is also concluded that, when $\mathbf{b} = \mathbf{0}$, the homogeneous system $\mathbf{Ax} = \mathbf{0}$ has a non-trivial solution if and only if $\mathbf{A}$ is singular.

Table 3.4 Numerical methods for solving linear system of equations

Direct methods	Iterative methods
• Gaussian elimination	• Gauss-Seidel
• Gauss-Jordan elimination	• Successive over relaxation (SOR)
• LU decomposition	• Conjugate gradient method
	• Generalized minimal residual method

To solve a linear system, the most direct and obvious way is to calculate

$$\mathbf{x} = \mathbf{A}^{-1}\mathbf{b} \qquad (3.56)$$

where $\mathbf{A}^{-1}$ is the inverse of $\mathbf{A}$. However, it is computationally very expensive to explicitly calculate the inverse of $\mathbf{A}$. The computational cost of a method is referred to as the complexity in numerical analysis. The general strategy is to transform the system into an equivalent problem that is easier to solve. This is important since solving linear systems is one of the essential computing routines which is often performed repeatedly in an engineering analysis. Therefore, the efficiency of solving linear systems is directly related to the efficiency of the entire analysis.

There are many methods available for the solution of a system of linear equations. Again, the ranking of the methods depends on the characteristics of $\mathbf{A}$. For example, whether $\mathbf{A}$ is symmetric, band, triangular, positive definite, or sparse determines which method should be used. Broadly defined, the solution methods can be categorized into direct (Davis 2006) and iterative methods (Saad 2003). Table 3.4 lists a few widely used numerical methods.

Generally speaking, iterative methods are more suitable for linear systems with a large sparse coefficient matrix $\mathbf{A}$. It requires less memory storage since it solves from one or a few equations at a time and reuses the memory from the last iteration. In comparison, direct methods are generally more robust. In addition, a major drawback of the iterative methods is that after solving $\mathbf{A}\mathbf{x} = \mathbf{b}_1$, one must start over again from the beginning in order to solve $\mathbf{A}\mathbf{x} = \mathbf{b}_2$.

MATLAB uses UMFPACK as the default solver for its built-in routines "\" (backslash), "lu," and "chol," etc. UMFPACK uses the Unsymmetric MultiFrontal method and direct sparse LU factorization to solve general sparse systems. The solution syntax is illustrated in the following example.

Example 3.9 Solve the linear system using MATLAB

$$\begin{bmatrix} 3 & 4 & 6 & 0 \\ 0 & 5 & 0 & 0 \\ 0 & 0 & 7 & -1 \\ 0 & 0 & 0 & -3 \end{bmatrix} \begin{Bmatrix} u_1 \\ u_2 \\ u_3 \\ u_4 \end{Bmatrix} = \begin{Bmatrix} -2 \\ 1 \\ -2 \\ 4 \end{Bmatrix}. \qquad (3.57)$$

Solution

The MATLAB code solving the equations as both a dense and a sparse system is given below

```
1  A=[3 4 6 0; 0 5 0 0; 0 0 7 -1; 0 0 0 -3];  % A matrix  (dense)
2  b=[-2 1 -2 4]';                             % b vector  (dense)
3  x=A\b                                        % solve  Ax=b
4  AS=sparse(4,4);                              % A matrix  (sparse)
5  AS(1,1)=3; AS(1,2)=4; AS(1,3)=6;             % set the entries
6  AS(2,2)=5; AS(3,3)=7; AS(3,4)=-1; AS(4,4)=-3; % of A matrix
7  bs=sparse([-2 1 -2 4]');                     % b vector  (sparse)
8  x=AS\bs                                      % solve  Ax=b
```

MATLAB output results are

```
      x=                    x=
      0.0190         (1,1)   0.0190
      0.2000         (2,1)   0.2000
     -0.4762         (3,1)  -0.4762
     -1.3333         (4,1)  -1.3333.
```

The left column is the result obtained by using the dense matrix and right is from the sparse version.

3.6 Summary

Upon completion of this chapter, you should be able to:

- understand the difference between the Lagrange, Hermite, and piecewise polynomial interpolations; know how to implement Lagrange interpolation for a given set of data points in MATLAB
- understand the concept of numerical differentiation by using Taylor series expansion; understand the difference between central difference, forward difference, and backward difference; know how to obtain the derivatives of 1-D and 2-D functions by using the numerical differentiation schemes
- understand the geometric interpretation of numerical integration; understand the difference between the numerical integration methods: Trapezoid, Simposon's, Newton–Cotes rules, and Gauss quadrature; know how to implement the integration rules using MATLAB to integrate 1-D functions
- understand the various scenarios of a linear system of simultaneous equations; know how to use MATLAB to solve a linear system.

3.7 Problems

3.1 An experiment has produced the following data:

x	0	0.5	1.0	6.0	7.0	9.0
u	0.0	1.6	2.0	2.0	1.5	0.0

We wish to interpolate the data with a smooth curve in the hope of obtaining reasonable values of u for values of x between the points at which measurements were taken.

(a) Using any method you like, determine the polynomial of degree 5 that interpolates the given data, and make a smooth plot of it over the range $0 \leq x \leq 9$.

(b) Similarly, determine a piecewise quadratic interpolation that interpolates the given data, and make a plot of it over the same range.

(c) Which interpolant seems to give more reasonable values between the given data points? Can you explain why each curve behaves the way it does?

(d) Might piecewise linear interpolation be a better choice for these particular data? Why?

3.2 Interpolating the data points

x	0	1	4	9	16	25	36	49	64
u	0	1	2	3	4	5	6	7	8

should give an approximation to the square root function.

(a) Compute the polynomial of degree 8 that interpolates these nine data points. Plot the resulting polynomial as well as the corresponding values given by the MATLAB built-in "sqrt" function over the interval $[0, 64]$.

(b) Use piecewise cubic Hermite interpolation functions to interpolate the same data points and again plot the resulting curve along with the built-in "sqrt" function.

(c) Which of the two interpolations is more accurate over most of the domain?

(d) Which of the two interpolations is more accurate between 0 and 1?

Note: use the value and the first derivative of the square root function at each data point for the Hermite interpolation.

3.3 The gamma function is defined by

$$\Gamma(t) = \int_0^\infty x^{t-1} e^{-x} dx, \qquad t > 0. \tag{3.58}$$

For an integer argument n, the gamma function has the value

$$\Gamma(n) = (n-1)! \tag{3.59}$$

so interpolating the data points

x	1	2	3	4	5
u	1	1	2	6	24

should yield an approximation to the gamma function over the given range.
(a) Compute the polynomial of degree 4 that interpolates these five data points. Plot the resulting polynomial as well as the corresponding values given by the built-in gamma function over the interval $[1, 5]$.
(b) Use piecewise linear interpolation functions to interpolate the same data and again plot the resulting curve along with the built-in "gamma" function.
(c) Which of the two interpolants is more accurate over most of the domain?

3.4 Suppose that the quadrature rule

$$\int_a^b f(x)dx \approx \sum_{i=1}^n w_i f(x_i) \qquad (3.60)$$

is exact for all constant functions. What does this imply about the weights w_i or the nodes x_i?

3.5 What is the polynomial degree of each of the following types of numerical quadrature rules?
(a) An n-point Newton–Cotes rule, where n is odd.
(b) An n-point Newton–Cotes rule, where n is even.
(c) An n-point Gaussian quadrature rule.
(d) What accounts for the difference between the answers to parts (a) and (b)?
(e) What accounts for the difference between the answers to parts (b) and (c)?

3.6 How might one use a standard one-dimensional quadrature routine to compute the value of a double integral over a rectangular region?

3.7 Compute the approximate value of the integral $\int_0^1 x^3 dx$, first by the Trapezoid rule and then by Simpson's rule.
(a) Estimate the errors of the two results.
(b) Would you expect the latter to be exact for this problem? Why?

3.8 Since

$$\int_0^1 \frac{4}{1+x^2} dx = \pi \qquad (3.61)$$

one can compute an approximate value for π using numerical integration of the given function. First, divide the integration interval $[0, 1]$ into a set of equal length sub-intervals of length h. Then use the trapezoid and Simpson quadrature rules to compute the integral over each sub-interval. Finally, sum the quadrature results over the sub-intervals to obtain the approximate value of π for various h. Try to characterize the error as a function of h for each rule, and also compare the accuracy of the rules with each other (based on the known value of π). Is there any point beyond which decreasing h yields no further improvement? Why?

3.9 Using any method you choose, evaluate the double integral

$$\iint e^{xy} dx dy \tag{3.62}$$

over each of the following regions:

(a) The unit square, i.e., $0 \leq x \leq 1, 0 \leq y \leq 1$.
(b) The quarter of the unit disc lying in the first quadrant, i.e., $x^2 + y^2 \leq 1, x \geq 0, y \geq 0$.

3.10 Given a sufficiently smooth function $f : R \to R$, use Taylor series to derive a second-order accurate, one-sided difference approximation to $f'(x)$ in terms of the values of $f(x), f(x + \Delta x)$, and $f(x + 2\Delta x)$.

3.11 In this exercise we will experiment with numerical differentiation using the data below.

x	0	1	2	3	4
u	1.0	2.7	5.8	6.6	7.5

Compute the derivative of the original data and also experiment with randomly perturbing the u values to determine the sensitivity of the resulting derivative estimates. For each method, comment on both the reasonableness of the derivative estimates and their sensitivity to perturbations. Note that the data are monotonically increasing, so one might expect the derivative always to be positive.

(a) Fit a polynomial of degree 4 to the data using the Lagrange interpolation, then differentiate the resulting polynomial and evaluate the derivative at each of the given x values.
(b) Interpolate the data with piecewise quadratic interpolation functions, differentiate the resulting piecewise quadratic polynomials, and evaluate the derivative at each of the given x values.
(c) Compare the results obtained from both approaches and comment on the accuracy and efficiency of the methods.

References

Ciarlet, P. G. & Raviart, P. (1972), "General Lagrange and Hermite interpolation in rn with applications to finite element methods," *Archive for Rational Mechanics and Analysis* **46**(3), 177–199.

Davis, P. J. & Rabinowitz, P. (2007), *Methods of numerical integration*, Courier Corporation.

References

Davis, T. A. (2006), *Direct methods for sparse linear systems*, Vol. 2, Siam.

De Boor, C. (1978), *A practical guide to splines*, Vol. 27, Springer-Verlag.

Hakopian, H. A. (1982), "Multivariate divided differences and multivariate interpolation of Lagrange and Hermite type," *Journal of Approximation Theory* **34**(3), 286–305.

Mitchell, A. R. & Griffiths, D. F. (1980), *The finite difference method in partial differential equations*, John Wiley.

Saad, Y. (2003), *Iterative methods for sparse linear systems*, Vol. 82, Siam.

4 General Procedure of FEA for Linear Static Analysis: 1-D Problems

4.1 Overview

In this chapter, we start the illustration of general procedure of FEA for linear static analysis through a step-by-step solution of a 1-D elasticity problem. We introduce the important concepts and numerical techniques that make up the finite element method as we go through the steps. The example problem is set up such that a MATLAB-aided hand calculation can be carried out through the steps. The reader is expected to work out the solution by following the steps and verify the results with the ones presented in the text. This approach ensures a thorough understanding of the FEA procedure and relevant concepts and techniques. Then, computer implementation of the procedure is discussed and a MATLAB code for solving the 1-D elasticity problem is provided. Thus, the 1-D problem is solved twice: first by hand calculation and then by creating a computer program. By completing two calculations for the same problem, we demonstrate how the understanding of the method obtained from the hand calculation is reflected in the computer implementation, and how the step-by-step calculation can be automated by a computer code. The chapter then goes on and demonstrates that other types of physical problems (heat transfer and fluid flow) can be solved by using the same FEA procedure. While the steps are described in less detail compared to the 1-D elasticity example, the demonstration of the FEA procedure for these problems aims to reinforce the understanding of the method and related numerical techniques.

4.2 General Procedure of FEA for Linear Static Analysis

In essence, finite element analysis is a procedure of converting the mathematical model of a physical problem into a set of computer solvable algebraic equations and obtain the numerical solution of the problem at a set of discrete locations, and also at discrete instances of time in a transient analysis. The general steps of FEA for a static or quasi-static problem are summarized in Table 4.1. In this chapter, the general steps are illustrated over several example (case study) problems of one-dimensional (1-D) continuum systems, including an elasticity problem, a heat transfer problem, and an advection–diffusion problem. It is demonstrated that these problems of different physical phenomena can be solved by following the same procedure of FEA. The

4.2 General Procedure of FEA for Linear Static Analysis

Table 4.1 General procedure of FEA for linear static analysis

No	Step	Description
1	Define the physical problem	Determine the system and physical behavior of interest, dimensionality, and computational domain. Answer questions such as: Does this system contain a single or multi objects? Is it a 1-D, 2-D, or 3-D problem? Is this a solid mechanics, heat transfer, or fluid dynamics problem? ...
2	Define the mathematical model	Use mathematical and physics principles to define governing equations (differential equations, functional descriptions, or algebraic equations).
3	Derive the weak form	Use mathematical tools to obtain the weak form equation(s).
4	Discretize the computational domain	Discretize (mesh) the computational domain.
5	Approximate the unknown physical quantities over the elements	Finite element approximation using shape functions.
6	Discretize the weak form equation over the elements	Compute the element matrices and vectors.
7	Assemble the global system equations	Assemble the element matrices and vectors into the global linear system.
8	Apply essential boundary conditions	Enforce the essential boundary conditions on the global linear system.
9	Solve the global system equations	Use a linear solver to solve the global system and obtain the solution of the unknown vector.
10	Perform post-processing	Perform visualization, analysis of the solution, and calculation of other physical quantities from the solution.

finite element method's versatility in solving different types of engineering problems is one of the major advantages it has over other numerical methods.

As shown in Table 4.1, there are ten steps in the general procedure. These steps can be categorized into three parts: (1) formulation and pre-processing, (2) solution, and (3) post-processing. The formulation and pre-processing part includes the first four steps involving problem definition, mathematical modeling, computational domain definition, and geometric discretization (meshing). The solution part contains Steps 5–9 in which the discretization of the mathematical model and numerical solution are performed. The post-processing part is the last step of the analysis carrying out data visualization and result interpretation. While the steps are for linear static analysis of continuum structures and systems, they largely remain the same for nonlinear and transient/dynamic analysis. The main difference in the analysis procedure for those problems lies in the numerical solutions of nonlinear system equations and/or time-dependent calculations. In addition, for simple discrete systems such as rigid bodies connected by springs, or truss systems, Steps 3–5 shown in Table 4.1 can simply be skipped.

In this chapter, the case study problems are first solved and demonstrated step by step through manual calculations. The derivation of the FE formulations is carried out by using the mathematical tools described in Chapter 2. The numerical approximations and calculations are performed by using the numerical analysis tools described in Chapter 3. After solving the problems manually, the implementation of the analysis steps in MATLAB is explained. The flow chart for the computer programs, data structures for data storage and retrieving, functions and subroutines of calculating shape functions, numerical integrations, and visualization are discussed. Finally, a complete set of MATLAB codes that solve the case study problems are presented.

4.3 1-D Elasticity

In this section, we go through the general procedure of FEA by solving a 1-D axially loaded elastic bar problem. The analysis steps are described in great detail. The finite element formulations are obtained by using multiple approaches. The mathematical derivations and numerical calculations are presented in a way that the reader can readily follow the steps and manually work out the solution. The objective of this section is to enable the reader to work the process manually, walk through the procedure step by step, and gain the most detailed knowledge on the theories and techniques involved in the solution of the problem.

4.3.1 A One-Dimensional Axially Loaded Elastic Bar

We consider a solid elastic bar with linearly varying cross sectional area, fixed at its right end and subjected to the axial loads as shown in Fig. 4.1. The bar is made of linear elastic material with Young's modulus, $E = 35 \times 10^3$ Ksi. The left end of the bar is subjected to a compressive pressure of 100 Ksi. We seek to find the displacement and stress distribution in the elastic bar. In the following, we solve the problem by using the FEA steps listed in Table 4.1.

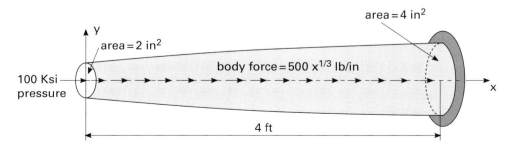

Figure 4.1 A 1-D axially loaded elastic bar.

Step 1: Define the Physical Problem
The system contains one object: the elastic bar. Since we seek to find the displacement and stress distribution in the bar, and the bar is elastic, it is a problem of linear elasticity. The bar is axially loaded and the cross section is small compared to the length of the bar. From the theory of elasticity, on a given cross section in the bar, the deformation and the stresses are close to uniform. Therefore, it is reasonable to assume that the displacement and stress of the bar only vary along the axis of the bar, i.e., the mechanical response of the bar under the applied loads is only a function of x, not a function of y and z. Based on this analysis, the problem can be reduced to a 1-D problem. The domain of the 1-D bar is $0 \leq x \leq 4$ ft. As we seek to find the deformation of the bar when it is in equilibrium, this in fact is a quasi-static problem.

Step 2: Define the Mathematical Model
To obtain the mathematical description of the problem, the relation between the applied loads and the deformation of an elastic body needs to be established. There are two fundamental laws that can be used to establish such relation: equations of equilibrium and Hooke's law.

Taking an infinitesimal segment of the elastic bar along the x-axis, a free body diagram can be drawn and the forces acting on the segment are shown in Fig 4.2.

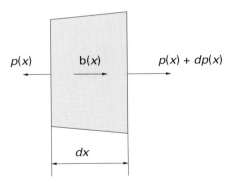

Figure 4.2 Free body diagram of an infinitesimal segment of the elastic bar.

In the figure, $p(x)$ and $p(x)+dp(x)$ are the internal forces acting on the left and right surfaces of the free body, respectively. The balance of the forces acting on the free body gives the equation of equilibrium in the x-direction, i.e.,

$$-p(x) + b(x)dx + p(x) + dp(x) = 0 \qquad (4.1)$$

$$\Rightarrow \frac{dp(x)}{dx} + b(x) = 0 \qquad (4.2)$$

where $b(x)$ is the axial body force per unit length.

The normal stress σ is defined by

$$\sigma(x) = \frac{p(x)}{A(x)} \tag{4.3}$$

where $A(x)$ is the cross sectional area of the infinitesimal segment. Substituting Eq. (4.3) into Eq. (4.2), we obtain

$$\frac{d(A(x)\sigma(x))}{dx} + b(x) = 0. \tag{4.4}$$

Hooke's law is given by

$$\sigma(x) = E\epsilon(x) \tag{4.5}$$

where E is the Young's modulus or the modulus of elasticity and $\epsilon(x)$ is the normal strain in the x-direction which is defined as

$$\epsilon(x) = \frac{du}{dx}. \tag{4.6}$$

Substituting Eqs. (4.6, 4.5) into Eq. (4.4), we obtain

$$\frac{d}{dx}\left(A(x)E(x)\frac{du}{dx}\right) + b(x) = 0. \tag{4.7}$$

Therefore, by using the definition of stress and strain, and Hooke's material law, the equation of equilibrium is written in terms of the displacement in the bar. Equation (4.7) is the governing differential equation of the elastic bar. Other than the differential equation, it is known that a pressure of 100 Ksi is applied at the left end, and the right end of the bar is fixed. These conditions can be written as

$$\sigma(x=0) = E\epsilon(x=0) = E\left.\frac{du}{dx}\right|_{x=0} = -100 \text{ Ksi} = -\bar{t} \tag{4.8}$$

$$u(x = L) = 0 \tag{4.9}$$

where $\bar{t}$ is the surface traction. The negative sign is denoting a compressive stress. Equations (4.8, 4.9) are called the boundary conditions. The boundary conditions combined with the governing differential equation is called the strong form (equations).

From the results of variational calculus shown in Chapter 2, Example 2.17, the functional statement of the strong form given in Eqs. (4.7–4.9) can be obtained as

$$\Pi = \frac{1}{2}\int_0^{48in} AE\left(\frac{du}{dx}\right)^2 dx - \int_0^{48in} b(x)u\,dx - uA\bar{t}\big|_{x=0}. \tag{4.10}$$

Note that, the cross sectional area A and Young's modulus E can be functions of x too. In this example, A is a function of x and E is constant. To simplify the notation, we omit (x) for the cross sectional area A.

Step 3: Derive the Weak Form

In the following, we obtain the weak form by using four different approaches, and show that they all lead to the same weak form. We demonstrate they are equivalent methods with different points of view to the problem.

Variational Principles

In Step 2, the governing differential equation and the functional statement of the 1-D elastic bar problem were obtained. In this step, an optimization criterion can be applied to the functional statement. From variational calculus, the functional reaches its extremum when its first variation vanishes, i.e.,

$$\delta \Pi = 0 \tag{4.11}$$

$$\Rightarrow \int_0^{48in} \left(\frac{d\delta u}{dx}\right) AE \left(\frac{du}{dx}\right) dx - \int_0^{48in} \delta u b(x) dx - \delta u A \bar{t} \Big|_{x=0} = 0. \tag{4.12}$$

Equation (4.12) is called the weak form (equation) for the 1-D elastic bar problem. Compared to the strong form, Eqs. (4.7–4.9), the weak form is "weak" in the sense that the continuity requirement of the displacement u is weakened in Eq. (4.12). From Eqs. (4.7–4.9), the first derivative of displacement u needs to be continuous and the second derivative of u needs to exist everywhere along the length of the bar so that the differential equation holds at any location in the bar. However, in Eq. (4.12), for the integral equation to be valid (or integrable), the displacement u and its variation δu need to be continuous, and their first derivatives are allowed to have jump discontinuity at a finite number of locations. The continuity requirement of the unknown functions is thus weakened. More discussion on the continuity requirement of the functions in the weak form will come later.

Minimizing the Potential Energy

The weak form given in Eq. (4.12) can also be obtained from an energy point of view. For problems of elasticity, equilibrium is reached when the total potential energy of the system is minimized. The potential energy is defined by

$$\text{potential energy} = \text{internal energy} - \text{external energy}. \tag{4.13}$$

In our case, the internal energy is given by the strain energy of the bar:

$$\text{internal energy} = \text{strain energy}$$

$$= \int_0^{48in} \frac{1}{2} \sigma \epsilon A \, dx$$

$$= \int_0^{48in} \frac{1}{A} E \epsilon^2 \, dx$$

$$= \int_0^{48in} \frac{1}{2} AE \left(\frac{du}{dx}\right)^2 dx \tag{4.14}$$

and the external energy is given by

$$\text{external energy} = \text{work done by external forces}$$

$$= \int_0^{48in} u b(x) dx + u A \bar{t} \Big|_{x=0}. \tag{4.15}$$

Therefore, the potential energy can be written as

$$\text{potential energy} = \frac{1}{2}\int_0^{48in} AE\left(\frac{du}{dx}\right)^2 dx - \int_0^{48in} ub(x)dx - uA\bar{t}\Big|_{x=0}. \quad (4.16)$$

Equation (4.16) is exactly the functional we have obtained from the variational calculus. This shows that the functional of the system is in fact the total potential energy of the system. The Theorem of Minimum Potential Energy states (Sokolnikoff, Specht et al. 1956): "Of all the displacements satisfying compatibility and the prescribed boundary conditions, those that satisfy the equilibrium equations make the potential energy a minimum." By minimizing the potential energy, i.e., enforcing the first variation of the potential energy to be zero,

$$\delta \text{ (potential energy)} = 0 \quad (4.17)$$

the weak form shown in Eq. (4.12) can be obtained.

Principle of Virtual Work

The principle of virtual work states that, when the system is in equilibrium, given a small perturbative virtual displacement δu (i.e., the variation of u in variational calculus), the work done by the internal stress (internal virtual work) is equal to the work done by the external forces (external virtual work).

The internal virtual work is given by

$$\text{internal virtual work} = \int_0^{48in} \delta\epsilon\sigma A dx = \int_0^{48in} \left(\frac{d\delta u}{dx}\right) AE \left(\frac{du}{dx}\right) dx \quad (4.18)$$

where $\delta\epsilon$ is the virtual strain due to the virtual displacement. Note that the internal stress is a negative "internal force." For example, for an infinitesimal element subjected to a tensile stress, the "internal force" exerted by the element is inward. The external virtual work includes the virtual work done by the body force and the virtual work done by the surface traction at $x = 0$, i.e.,

$$\text{external virtual work} = \int_0^{48in} \delta ub(x)dx + \delta uA\bar{t}\Big|_{x=0}. \quad (4.19)$$

The equality of the internal and external virtual work gives the weak form

$$\int_0^{48in} \left(\frac{d\delta u}{dx}\right) AE \left(\frac{du}{dx}\right) dx = \int_0^{48in} \delta ub(x)dx + \delta uA\bar{t}\Big|_{x=0}. \quad (4.20)$$

Galerkin Weighted Residual Method

In the general method of weight residual, we define an arbitrary continuous weight function $w(x)$ with

$$w(x) = 0 \quad \text{on } \Gamma_u \quad (4.21)$$

where Γ_u is the portion of the boundary where the essential boundary condition is specified. In the 1-D elastic bar problem, Γ_u is the right end of the bar $x = 48$ in.

4.3 1-D Elasticity

By multiplying the weight function $w(x)$ to both sides of the governing differential equation and integrating the resultant products over the domain, we obtain

$$\int_0^{48in} w \left[\frac{d}{dx}\left(AE\frac{du}{dx}\right) + b(x) \right] dx = \int_0^{48in} w \cdot 0 \, dx$$

$$\Rightarrow \int_0^{48in} w \left[\frac{d}{dx}\left(AE\frac{du}{dx}\right) + b(x) \right] dx = 0. \tag{4.22}$$

Equation (4.22) can be further written as

$$\int_0^{48in} w \left[\frac{d}{dx}\left(AE\frac{du}{dx}\right) \right] dx + \int_0^{48in} wb(x) dx = 0. \tag{4.23}$$

By using integration by parts for the first integral, Eq. (4.23) can be rewritten as

$$wAE\frac{du}{dx}\bigg|_0^{48in} - \int_0^{48in} \left(\frac{dw}{dx}\right) AE \left(\frac{du}{dx}\right) dx + \int_0^{48in} wb(x) dx = 0. \tag{4.24}$$

Since $w(x = 48 \text{ in}) = 0$ by definition, Eq. (4.24) can be further rewritten as

$$-wAE\frac{du}{dx}\bigg|_{x=0} - \int_0^{48in} \left(\frac{dw}{dx}\right) AE \left(\frac{du}{dx}\right) dx + \int_0^{48in} wb(x) dx = 0$$

$$\Rightarrow w A \bar{t}\big|_{x=0} - \int_0^{48in} \left(\frac{dw}{dx}\right) AE \left(\frac{du}{dx}\right) dx + \int_0^{48in} wb(x) dx = 0. \tag{4.25}$$

If we take the variation of u (or a virtual displacement) to be the weight function w, we obtain

$$\delta u A \bar{t}\big|_{x=0} - \int_0^{48in} \left(\frac{d\delta u}{dx}\right) AE \left(\frac{du}{dx}\right) dx + \int_0^{48in} \delta u b(x) dx = 0. \tag{4.26}$$

Equation (4.26) is the weak form of the 1-D elastic problem.

Finally, the following comments are provided as a summary for the step of obtaining the weak form:

1. In a weighted residual method, the weight function $w(x)$ is often referred to as the test function and the unknown displacement function $u(x)$ is often called the trial solution or trial function.
2. In Eq. (4.26), the variation of u, δu, can be considered as a special kind of weight function in Eq. (4.25). When δu is used as the weight function and assuming that we will approximate u and δu the same way in later steps, this method of weighted residual becomes the Galerkin weighted residual method.
3. There is no variation of the displacement on the boundary where u is specified, i.e., $\delta u = 0$ on the boundary with essential boundary conditions (Γ_u). Therefore, δu satisfies the requirements of the weight function.

4 We have shown in this section that one can use different methods/approaches to obtain the weak form. The map in Fig. 4.3 summarizes the paths to reach the weak form for elasticity problems.

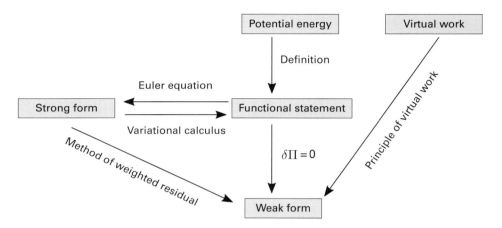

Figure 4.3 Approaches to obtain the weak form.

Step 4: Discretize the Computational Domain

In this step, we break up the 1-D computational domain ($x = [0, L]$) into a set of segments. Each segment is called an element. The element set is called a mesh. The process of discretizing the domain into elements is called meshing. For example, a finite element mesh of the 1-D elastic bar problem is shown in Fig. 4.4. The 1-D domain is discretized into five equal length elements. Filled circles are placed at the ends of the elements. The circles are called nodes. As shown in the figure, each element has two nodes, and in the interior of the domain, each node is shared by two elements. The numbers below the nodes are used to index the nodes. The circled numbers are the element numbers.

For 1-D domains, the meshing process is straightforward. One can easily create the mesh by using simple algorithms. However, the meshing process becomes more involved for 2-D and 3-D domains with complex geometry. Meshing of multi-dimensional domains is discussed in Chapter 6.

Step 5: Approximate the Unknown over the Elements

Assuming the displacements at the nodes $1, 2, \ldots, 6$ are $u_1, u_2, \ldots, u_6$, the task of this step is to approximate the unknown displacement field within each element. The displacements $u_1, u_2, \ldots, u_6$ are called nodal displacements. The standard approach in the finite element analysis is to approximate the unknown quantities within each element using polynomials. The polynomials are constructed such that they go through (interpolate) the nodal displacement values at the nodes. The method of constructing

4.3 1-D Elasticity

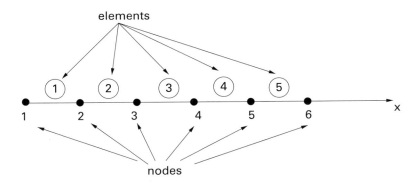

Figure 4.4 The 1-D domain is discretized into five equal length elements.

the interpolating polynomials is the Lagrange interpolation method, as described in Chapter 3.

Figure 4.5 show a linear interpolation within an element having two nodes. Since the displacement approximation in the element shown in Fig. 4.5 is linear, the element is called a linear element. If we put another node in the middle of the element, we have a three node quadratic element as shown in Fig. 4.6. Note that in the figure we use an empty circle to represent the interior node.

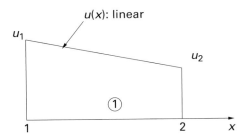

Figure 4.5 Approximate $u(x)$ as a linear function.

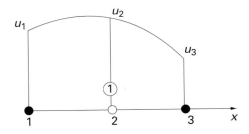

Figure 4.6 Approximate $u(x)$ as a quadratic function.

As discussed in Chapter 3, in the Lagrange interpolation the unknown displacement polynomial in an element is approximated by

$$u(x) = \sum_{i=1}^{n} N_i(x) u_i \qquad (4.27)$$

where n is the number of nodes in the element, u_i is the nodal displacement, and $N_i(x)$ is the i-th Lagrange basis function given by

$$N_j(x) = \frac{\Pi_{k=1, k \neq j}^{n}(x - x_k)}{\Pi_{k=1, k \neq j}^{n}(x_j - x_k)} \qquad j = 1, 2, \ldots, n. \qquad (4.28)$$

In finite element analysis, $N_i(x)$, $i = 1, 2, \ldots, n$, are also called the shape functions. For 2-node linear elements, $n = 2$, the shape functions are linear.

$$N_1(x) = \frac{(x - x_2)}{(x_1 - x_2)} \qquad N_2(x) = \frac{(x - x_1)}{(x_2 - x_1)}. \tag{4.29}$$

The shape functions of the 1-D linear element are shown in Fig. 4.7.

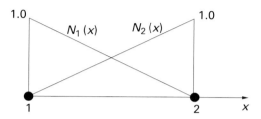

Figure 4.7 Linear shape functions.

For 3-node quadratic elements, $n = 3$, the shape functions are quadratic.

$$N_1(x) = \frac{(x - x_2)(x - x_3)}{(x_1 - x_2)(x_1 - x_3)} \tag{4.30}$$

$$N_2(x) = \frac{(x - x_1)(x - x_3)}{(x_2 - x_1)(x_2 - x_3)} \tag{4.31}$$

$$N_3(x) = \frac{(x - x_1)(x - x_2)}{(x_3 - x_1)(x_3 - x_2)}. \tag{4.32}$$

The shape functions are shown in Fig. 4.8.

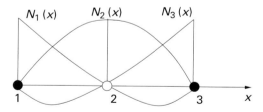

Figure 4.8 Quadratic shape functions.

In FEA, it important to understand several important properties of $N_i(x)$ which are listed below:

1 A shape function has a value of either 1 or 0 at a node:

$$N_i(x_j) = \begin{cases} 1 & i = j \\ 0 & i \neq j \end{cases} \tag{4.33}$$

2. The sum of all shape functions at any point equals 1:

$$\sum_{i=1}^{n} N_i(x) = 1 \qquad (4.34)$$

and

$$\sum_{i=1}^{n} N_i(x) x_i = x. \qquad (4.35)$$

3. C^n notation:
   A function is said to be of class C^n (in a specified domain) if the function and its first n derivatives are continuous (in the domain). A function of class C^n must also be of class $C^{n-1}, C^{n-2}, \ldots, C^0$ but it may or may not be $C^{n+1}, C^{n+2}, \ldots$. From this definition, the polynomial shape functions are C^∞ continuous within the elements. Therefore, the polynomial approximation of the unknown displacement u is C^∞ continuous within the elements. However, the polynomial approximation of u is only C^0 continuous across the elements (i.e., its first derivative is piecewise continuous) as shown in Fig. 4.9. Therefore, the polynomial finite element approximation is also called piecewise polynomial interpolation.

linear elements: C^0 continuous

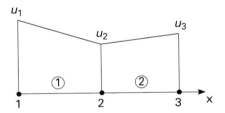

quadratic elements: C^0 continuous

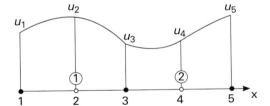

Figure 4.9 C^0 continuous shape functions.

Back to the 1-D elastic bar example, the mesh consists of five equal length linear elements. The linear elements and their shape functions are shown in Fig. 4.10.

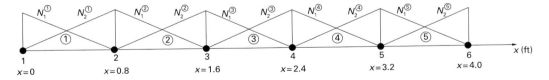

Figure 4.10 Shape functions of the 1-D elastic bar problem.

We have already obtained the linear shape functions which are given by

$$N_1(x) = \frac{(x - x_2)}{(x_1 - x_2)} \qquad N_2(x) = \frac{(x - x_1)}{(x_2 - x_1)}. \qquad (4.36)$$

General Procedure of FEA: 1-D Problems

Denoting the length of each element as L, the shape functions can be rewritten as

$$N_1(x) = \frac{(x_2 - x)}{L} \qquad N_2(x) = \frac{(x - x_1)}{L}. \qquad (4.37)$$

The shape functions and their first derivatives of the first element are then obtained as

$$N_1^{(1)}(x) = \frac{0.8 - x}{0.8} \qquad \frac{dN_1^{(1)}(x)}{dx} = -\frac{1}{0.8} \qquad (4.38)$$

$$N_2^{(1)}(x) = \frac{x - 0}{0.8} \qquad \frac{dN_2^{(1)}(x)}{dx} = \frac{1}{0.8} \qquad (4.39)$$

where the superscript ① denotes the first element. Note that the unit of the first derivatives of the shape functions in Eqs. (4.38–4.39) is 1/in. The shape functions and their first derivatives of the elements 2, 3, 4, and 5 are given by

$$N_1^{(2)}(x) = \frac{1.6 - x}{0.8} \qquad \frac{dN_1^{(2)}(x)}{dx} = -\frac{1}{0.8} \qquad (4.40)$$

$$N_2^{(2)}(x) = \frac{x - 0.8}{0.8} \qquad \frac{dN_2^{(2)}(x)}{dx} = \frac{1}{0.8} \qquad (4.41)$$

$$N_1^{(3)}(x) = \frac{2.4 - x}{0.8} \qquad \frac{dN_1^{(3)}(x)}{dx} = -\frac{1}{0.8} \qquad (4.42)$$

$$N_2^{(3)}(x) = \frac{x - 1.6}{0.8} \qquad \frac{dN_2^{(3)}(x)}{dx} = \frac{1}{0.8} \qquad (4.43)$$

$$N_1^{(4)}(x) = \frac{3.2 - x}{0.8} \qquad \frac{dN_1^{(4)}(x)}{dx} = -\frac{1}{0.8} \qquad (4.44)$$

$$N_2^{(4)}(x) = \frac{x - 2.4}{0.8} \qquad \frac{dN_2^{(4)}(x)}{dx} = \frac{1}{0.8} \qquad (4.45)$$

$$N_1^{(5)}(x) = \frac{4.0 - x}{0.8} \qquad \frac{dN_1^{(5)}(x)}{dx} = -\frac{1}{0.8} \qquad (4.46)$$

$$N_2^{(5)}(x) = \frac{x - 3.2}{0.8} \qquad \frac{dN_2^{(5)}(x)}{dx} = \frac{1}{0.8}. \qquad (4.47)$$

Several observations can be made from the above shape functions written out for the five linear elements:

1. The form and the shape of the shape functions are the same for all the elements, only the positions are shifted.
2. The first derivatives of the shape functions are constant.

The behavior of the shape functions of the elements are due to the equal length of the elements. If the element lengths were different, the slope (i.e., first derivative) of the shape functions would be different.

Step 6: Discretize the Weak Form over the Elements

The weak form of the 1-D elastic bar problem is given by Eq. (4.12) (or Eqs. (4.20), (4.26)),

$$\int_0^{4 \text{ ft}} \left(\frac{d\delta u}{dx}\right) AE \left(\frac{du}{dx}\right) dx - \int_0^{4 \text{ ft}} \delta ub(x)dx - \delta u A\bar{t}\Big|_{x=0} = 0. \qquad (4.48)$$

It is clear that the integrals are defined over the entire domain. They are called global integrals. In Step 4, the 1-D geometry is discretized or meshed into five elements. Using the geometric discretization, in this step, the global integrals are discretized into element integrals over the mesh. First, we put all the terms that do not contain the unknown u to the right hand side:

$$\int_0^{4 \text{ ft}} \left(\frac{d\delta u}{dx}\right) AE \left(\frac{du}{dx}\right) dx = \int_0^{4 \text{ ft}} \delta ub(x)dx + \delta u A\bar{t}\Big|_{x=0}. \qquad (4.49)$$

Next we break the global integrals in the weak form into element integrals. The first term on the left hand side can be rewritten as

$$\int_0^{4 \text{ ft}} \left(\frac{d\delta u}{dx}\right) AE \left(\frac{du}{dx}\right) dx = \int_0^{0.8 \text{ ft}} \left(\frac{d\delta u}{dx}\right) AE \left(\frac{du}{dx}\right) dx$$

$$+ \int_{0.8 \text{ ft}}^{1.6 \text{ ft}} \left(\frac{d\delta u}{dx}\right) AE \left(\frac{du}{dx}\right) dx$$

$$+ \int_{1.6 \text{ ft}}^{2.4 \text{ ft}} \left(\frac{d\delta u}{dx}\right) AE \left(\frac{du}{dx}\right) dx$$

$$+ \int_{2.4 \text{ ft}}^{3.2 \text{ ft}} \left(\frac{d\delta u}{dx}\right) AE \left(\frac{du}{dx}\right) dx$$

$$+ \int_{3.2 \text{ ft}}^{4.0 \text{ ft}} \left(\frac{d\delta u}{dx}\right) AE \left(\frac{du}{dx}\right) dx. \qquad (4.50)$$

The first term on the right hand side:

$$\int_0^{4 \text{ ft}} \delta ub(x)dx = \int_0^{0.8 \text{ ft}} \delta ub(x)dx + \int_{0.8 \text{ ft}}^{1.6 \text{ ft}} \delta ub(x)dx + \int_{1.6 \text{ ft}}^{2.4 \text{ ft}} \delta ub(x)dx$$

$$+ \int_{2.4 \text{ ft}}^{3.2 \text{ ft}} \delta ub(x)dx + \int_{3.2 \text{ ft}}^{4.0 \text{ ft}} \delta ub(x)dx. \qquad (4.51)$$

Note that the second term on the right hand side is associated with element ① only.

General Procedure of FEA: 1-D Problems

The unknown function displacement u is then replaced by using the finite element approximation:

$$u(x) = \sum_{i=1}^{n} N_i(x) u_i \tag{4.52}$$

and the derivative of u is replaced by

$$\frac{du(x)}{dx} = \sum_{i=1}^{n} \frac{dN_i(x)}{dx} u_i. \tag{4.53}$$

Assuming linear elements are used, for the first element, we have:

$$u^{①}(x) = \sum_{i=1}^{2} N_i^{①}(x) u_i^{①}$$
$$= N_1^{①}(x) u_1^{①} + N_2^{①}(x) u_2^{①} \tag{4.54}$$

and

$$\frac{du^{①}(x)}{dx} = \sum_{i=1}^{2} \frac{dN_i^{①}(x)}{dx} u_i^{①}$$
$$= \frac{dN_1^{①}(x)}{dx} u_1^{①} + \frac{dN_2^{①}(x)}{dx} u_2^{①}$$
$$= N_1'^{①}(x) u_1^{①} + N_2'^{①}(x) u_2^{①}.$$

Substituting the finite element approximation, the first element integral in the first term on the left hand side (Eq. (4.50)) can be rewritten as:

$$\int_0^{0.8 \text{ ft}} \left(\frac{d\delta u}{dx}\right) AE \left(\frac{du}{dx}\right) dx \approx \int_0^{0.8 \text{ ft}} \left(\frac{d\delta u}{dx}\right) AE \left[N_1'^{①}(x) N_2'^{①}(x)\right] \begin{Bmatrix} u_1^{①} \\ u_2^{①} \end{Bmatrix} dx. \tag{4.55}$$

Using the same approximation for δu (the Galerkin method), we obtain:

$$\delta u^{①}(x) = \sum_{i=1}^{2} N_i^{①}(x) \delta u_i^{①} = N_1^{①}(x) \delta u_1^{①} + N_2^{①}(x) \delta u_2^{①} \tag{4.56}$$

and

$$\frac{d\delta u^{①}(x)}{dx} = \sum_{i=1}^{2} \frac{dN_i^{①}(x)}{dx} \delta u_i^{①} = N_1'^{①}(x) \delta u_1^{①} + N_2'^{①}(x) \delta u_2^{①}. \tag{4.57}$$

Substituting the above approximation into Eq. (4.55), we obtain:

$$\int_0^{0.8 \text{ ft}} \left(\frac{d\delta u}{dx}\right) AE \left(\frac{du}{dx}\right) dx \approx$$

$$\int_0^{0.8 \text{ ft}} \left[\delta u_1^{①} \; \delta u_2^{①}\right] \begin{Bmatrix} N_1'^{①}(x) \\ N_2'^{①}(x) \end{Bmatrix} AE \left[N_1'^{①}(x) N_2'^{①}(x)\right] \begin{Bmatrix} u_1^{①} \\ u_2^{①} \end{Bmatrix} dx. \tag{4.58}$$

4.3 1-D Elasticity

Taking all the nodal displacements and the displacement variations (they are not functions of x) out of the integral, we have:

$$\int_0^{0.8 \text{ ft}} \begin{bmatrix} \delta u_1^{\circled{1}} & \delta u_2^{\circled{1}} \end{bmatrix} \begin{Bmatrix} N_1'^{\circled{1}}(x) \\ N_2'^{\circled{1}}(x) \end{Bmatrix} AE \begin{bmatrix} N_1'^{\circled{1}}(x) N_2'^{\circled{1}}(x) \end{bmatrix} \begin{Bmatrix} u_1^{\circled{1}} \\ u_2^{\circled{1}} \end{Bmatrix} dx$$

$$= \begin{bmatrix} \delta u_1^{\circled{1}} & \delta u_2^{\circled{1}} \end{bmatrix} \left(\int_0^{0.8 \text{ ft}} \begin{Bmatrix} N_1'^{\circled{1}}(x) \\ N_2'^{\circled{1}}(x) \end{Bmatrix} AE \begin{bmatrix} N_1'^{\circled{1}}(x) N_2'^{\circled{1}}(x) \end{bmatrix} dx \right) \begin{Bmatrix} u_1^{\circled{1}} \\ u_2^{\circled{1}} \end{Bmatrix}$$

$$= \begin{bmatrix} \delta u_1^{\circled{1}} & \delta u_2^{\circled{1}} \end{bmatrix} \left(\int_0^{0.8 \text{ t}} AE \begin{bmatrix} \left(N_1'^{\circled{1}}(x)\right)^2 & N_1'^{\circled{1}}(x)N_2'^{\circled{1}}(x) \\ N_2'^{\circled{1}}(x)N_1'^{\circled{1}}(x) & \left(N_2'^{\circled{1}}(x)\right)^2 \end{bmatrix} dx \right) \begin{Bmatrix} u_1^{\circled{1}} \\ u_2^{\circled{1}} \end{Bmatrix}$$

$$= \begin{bmatrix} \delta u_1^{\circled{1}} & \delta u_2^{\circled{1}} \end{bmatrix} \begin{bmatrix} \int_0^{0.8 \text{ft}} AE \left(N_1'^{\circled{1}}(x)\right)^2 dx & \int_0^{0.8 \text{ ft}} AE \left(N_1'^{\circled{1}}(x) N_2'^{\circled{1}}(x)\right) dx \\ \int_0^{0.8 \text{ft}} AE \left(N_1'^{\circled{1}}(x) N_2'^{\circled{1}}(x)\right) dx & \int_0^{0.8 \text{ft}} AE \left(N_2'^{\circled{1}}(x)\right)^2 dx \end{bmatrix} \begin{Bmatrix} u_1^{\circled{1}} \\ u_2^{\circled{1}} \end{Bmatrix}$$

$$= \begin{bmatrix} \delta u_1^{\circled{1}} & \delta u_2^{\circled{1}} \end{bmatrix} \begin{bmatrix} k_{11}^{\circled{1}} & k_{12}^{\circled{1}} \\ k_{21}^{\circled{1}} & k_{22}^{\circled{1}} \end{bmatrix} \begin{Bmatrix} u_1^{\circled{1}} \\ u_2^{\circled{1}} \end{Bmatrix}$$

$$= \begin{bmatrix} \delta u_1^{\circled{1}} & \delta u_2^{\circled{1}} \end{bmatrix} \mathbf{k}^{\circled{1}} \begin{Bmatrix} u_1^{\circled{1}} \\ u_2^{\circled{1}} \end{Bmatrix}. \tag{4.59}$$

where $\mathbf{k}^{\circled{1}}$ denotes the element (stiffness) matrix of element $\circled{1}$. Substituting the shape functions and the material and geometric properties into the equation above, the integrals can be evaluated individually, i.e.,

$$k_{11}^{\circled{1}} = \int_0^{0.8 \text{ ft}} AE \left(N_1'^{\circled{1}}(x)\right)^2 dx$$

$$= \int_0^{9.6 \text{ in}} \left(\frac{1}{24}x + 2\right) \text{in}^2 \times \left(35 \times 10^6 \text{ lb/in}^2\right) \left(\frac{1}{9.6 \text{ in}}\right)^2 dx$$

$$= 8.02 \times 10^6 \text{ lb/in}. \tag{4.60}$$

$$k_{12}^{\circled{1}} = \int_0^{0.8 \text{ ft}} AE \left(N_1'^{\circled{1}}(x) N_2'^{\circled{1}}(x)\right) dx$$

$$= \int_0^{9.6 \text{ in}} \left(\frac{1}{24}x + 2\right) \text{in}^2 \times \left(35 \times 10^6 \text{ lb/in}^2\right) \left(\frac{1}{9.6 \text{ in}}\right)\left(-\frac{1}{9.6 \text{ in}}\right) dx$$

$$= -8.02 \times 10^6 \text{ lb/in}. \tag{4.61}$$

$$k_{21}^{\circled{1}} = k_{12}^{\circled{1}} \tag{4.62}$$

General Procedure of FEA: 1-D Problems

and

$$k_{22}^{①} = \int_0^{0.8\text{ ft}} AE\left(N_2'^{①}(x)\right)^2 dx$$

$$= \int_0^{9.6\text{ in}} \left(\frac{1}{24}x + 2\right) in^2 \times \left(35 \times 10^6 \text{ lb/in}^2\right) \left(\frac{1}{9.6\text{ in}}\right)^2 dx$$

$$= 8.02 \times 10^6 \text{ lb/in}. \tag{4.63}$$

Therefore,

$$\mathbf{k}^{①} = \begin{bmatrix} k_{11}^{①} & k_{12}^{①} \\ k_{21}^{①} & k_{22}^{①} \end{bmatrix} = \begin{bmatrix} 8.02 & -8.02 \\ -8.02 & 8.02 \end{bmatrix} \times 10^6 \text{ lb/in}. \tag{4.64}$$

Having completed the substitution of the finite element approximation and the manual calculation, the first element integral is now represented by a product of the element vectors and element matrix:

$$\int_0^{0.8\text{ ft}} \left(\frac{d\delta u}{dx}\right) AE \left(\frac{du}{dx}\right) dx \approx \begin{bmatrix} \delta u_1^{①} & \delta u_2^{①} \end{bmatrix} \begin{bmatrix} k_{11}^{①} & k_{12}^{①} \\ k_{21}^{①} & k_{22}^{①} \end{bmatrix} \begin{Bmatrix} u_1^{①} \\ u_2^{①} \end{Bmatrix}$$

$$= \begin{bmatrix} \delta u_1^{①} & \delta u_2^{①} \end{bmatrix} \begin{bmatrix} 8.02 & -8.02 \\ -8.02 & 8.02 \end{bmatrix} \times 10^6 \begin{Bmatrix} u_1^{①} \\ u_2^{①} \end{Bmatrix}. \tag{4.65}$$

Note that the unit lb/in of the element matrix is omitted for brevity. Similarly, the second element integral on the left hand side can be calculated as:

$$\int_{0.8\text{ ft}}^{1.6\text{ ft}} \left(\frac{d\delta u}{dx}\right) AE \left(\frac{du}{dx}\right) dx$$

$$\approx \begin{bmatrix} \delta u_1^{②} & \delta u_2^{②} \end{bmatrix} \begin{bmatrix} \int_{0.8\text{ ft}}^{1.6\text{ ft}} AE\left(N_1'^{②}(x)\right)^2 dx & \int_{0.8\text{ ft}}^{1.6\text{ ft}} AE\left(N_1'^{②}(x)N_2'^{②}(x)\right) dx \\ \int_{0.8\text{ ft}}^{1.6\text{ ft}} AE\left(N_1'^{②}(x)N_2'^{②}(x)\right) dx & \int_{0.8\text{ ft}}^{1.6\text{ ft}} AE\left(N_2'^{②}(x)\right)^2 dx \end{bmatrix} \begin{Bmatrix} u_1^{②} \\ u_2^{②} \end{Bmatrix}$$

$$= \begin{bmatrix} \delta u_1^{②} & \delta u_2^{②} \end{bmatrix} \begin{bmatrix} k_{11}^{②} & k_{12}^{②} \\ k_{21}^{②} & k_{22}^{②} \end{bmatrix} \begin{Bmatrix} u_1^{②} \\ u_2^{②} \end{Bmatrix}$$

$$= \begin{bmatrix} \delta u_1^{②} & \delta u_2^{②} \end{bmatrix} \begin{bmatrix} 9.48 & -9.48 \\ -9.48 & 9.48 \end{bmatrix} \times 10^6 \begin{Bmatrix} u_1^{②} \\ u_2^{②} \end{Bmatrix}. \tag{4.66}$$

By using the same procedure, the third, fourth, and fifth element integrals can be calculated as:

$$\int_{1.6\,\text{ft}}^{2.4\,\text{ft}} \left(\frac{d\delta u}{dx}\right) AE \left(\frac{du}{dx}\right) dx \approx \begin{bmatrix} \delta u_1^{(3)} & \delta u_2^{(3)} \end{bmatrix} \begin{bmatrix} k_{11}^{(3)} & k_{12}^{(3)} \\ k_{21}^{(3)} & k_{22}^{(3)} \end{bmatrix} \begin{Bmatrix} u_1^{(3)} \\ u_2^{(3)} \end{Bmatrix}$$

$$= \begin{bmatrix} \delta u_1^{(3)} & \delta u_2^{(3)} \end{bmatrix} \begin{bmatrix} 1.094 & -1.094 \\ -1.094 & 1.094 \end{bmatrix} \times 10^7 \begin{Bmatrix} u_1^{(3)} \\ u_2^{(3)} \end{Bmatrix}. \tag{4.67}$$

$$\int_{2.4\,\text{ft}}^{3.2\,\text{ft}} \left(\frac{d\delta u}{dx}\right) AE \left(\frac{du}{dx}\right) dx \approx \begin{bmatrix} \delta u_1^{(4)} & \delta u_2^{(4)} \end{bmatrix} \begin{bmatrix} k_{11}^{(4)} & k_{12}^{(4)} \\ k_{21}^{(4)} & k_{22}^{(4)} \end{bmatrix} \begin{Bmatrix} u_1^{(4)} \\ u_2^{(4)} \end{Bmatrix}$$

$$= \begin{bmatrix} \delta u_1^{(4)} & \delta u_2^{(4)} \end{bmatrix} \begin{bmatrix} 1.24 & -1.24 \\ -1.24 & 1.24 \end{bmatrix} \times 10^7 \begin{Bmatrix} u_1^{(4)} \\ u_2^{(4)} \end{Bmatrix}. \tag{4.68}$$

$$\int_{3.2\,\text{ft}}^{4.0\,\text{ft}} \left(\frac{d\delta u}{dx}\right) AE \left(\frac{du}{dx}\right) dx \approx \begin{bmatrix} \delta u_1^{(5)} & \delta u_2^{(5)} \end{bmatrix} \begin{bmatrix} k_{11}^{(5)} & k_{12}^{(5)} \\ k_{21}^{(5)} & k_{22}^{(5)} \end{bmatrix} \begin{Bmatrix} u_1^{(5)} \\ u_2^{(5)} \end{Bmatrix}$$

$$= \begin{bmatrix} \delta u_1^{(5)} & \delta u_2^{(5)} \end{bmatrix} \begin{bmatrix} 1.385 & -1.385 \\ -1.385 & 1.385 \end{bmatrix} \times 10^7 \begin{Bmatrix} u_1^{(5)} \\ u_2^{(5)} \end{Bmatrix}. \tag{4.69}$$

Combining the results obtained above, we obtain:

$$\int_0^{4\,\text{ft}} \left(\frac{d\delta u}{dx}\right) AE \left(\frac{du}{dx}\right) dx$$

$$\approx \begin{bmatrix} \delta u_1^{(1)} & \delta u_2^{(1)} \end{bmatrix} \begin{bmatrix} k_{11}^{(1)} & k_{12}^{(1)} \\ k_{21}^{(1)} & k_{22}^{(1)} \end{bmatrix} \begin{Bmatrix} u_1^{(1)} \\ u_2^{(1)} \end{Bmatrix} + \begin{bmatrix} \delta u_1^{(2)} & \delta u_2^{(2)} \end{bmatrix} \begin{bmatrix} k_{11}^{(2)} & k_{12}^{(2)} \\ k_{21}^{(2)} & k_{22}^{(2)} \end{bmatrix} \begin{Bmatrix} u_1^{(2)} \\ u_2^{(2)} \end{Bmatrix}$$

$$+ \begin{bmatrix} \delta u_1^{(3)} & \delta u_2^{(3)} \end{bmatrix} \begin{bmatrix} k_{11}^{(3)} & k_{12}^{(3)} \\ k_{21}^{(3)} & k_{22}^{(3)} \end{bmatrix} \begin{Bmatrix} u_1^{(3)} \\ u_2^{(3)} \end{Bmatrix} + \begin{bmatrix} \delta u_1^{(4)} & \delta u_2^{(4)} \end{bmatrix} \begin{bmatrix} k_{11}^{(4)} & k_{12}^{(4)} \\ k_{21}^{(4)} & k_{22}^{(4)} \end{bmatrix} \begin{Bmatrix} u_1^{(4)} \\ u_2^{(4)} \end{Bmatrix}$$

$$+ \begin{bmatrix} \delta u_1^{(5)} & \delta u_2^{(5)} \end{bmatrix} \begin{bmatrix} k_{11}^{(5)} & k_{12}^{(5)} \\ k_{21}^{(5)} & k_{22}^{(5)} \end{bmatrix} \begin{Bmatrix} u_1^{(5)} \\ u_2^{(5)} \end{Bmatrix}. \tag{4.70}$$

Next we substitute the finite element approximation for the displacement variation into the integral on the right hand side of the weak form, i.e.,

General Procedure of FEA: 1-D Problems

$$\int_0^{4\text{ ft}} \delta ub(x)dx$$

$$= \int_0^{0.8\text{ ft}} \delta ub(x)dx + \int_{0.8\text{ ft}}^{1.6\text{ ft}} \delta ub(x)dx + \int_{1.6\text{ ft}}^{2.4\text{ ft}} \delta ub(x)dx + \int_{2.4\text{ ft}}^{3.2\text{ ft}} \delta ub(x)dx$$

$$+ \int_{3.2\text{ ft}}^{4.0\text{ ft}} \delta ub(x)dx$$

$$\approx \int_0^{0.8\text{ ft}} \begin{bmatrix} \delta u_1^{(1)} & \delta u_2^{(1)} \end{bmatrix} \begin{Bmatrix} N_1^{(1)}(x) \\ N_2^{(1)}(x) \end{Bmatrix} b(x)dx + \int_{0.8in}^{1.6\text{ ft}} \begin{bmatrix} \delta u_1^{(2)} & \delta u_2^{(2)} \end{bmatrix} \begin{Bmatrix} N_1^{(2)}(x) \\ N_2^{(2)}(x) \end{Bmatrix} b(x)dx$$

$$+ \int_{1.6\text{ ft}}^{2.4\text{ ft}} \begin{bmatrix} \delta u_1^{(3)} & \delta u_2^{(3)} \end{bmatrix} \begin{Bmatrix} N_1^{(3)}(x) \\ N_2^{(3)}(x) \end{Bmatrix} b(x)dx + \int_{2.4\text{ ft}}^{3.2\text{ ft}} \begin{bmatrix} \delta u_1^{(4)} & \delta u_2^{(4)} \end{bmatrix} \begin{Bmatrix} N_1^{(4)}(x) \\ N_2^{(4)}(x) \end{Bmatrix} b(x)dx$$

$$+ \int_{3.2\text{ ft}}^{4.0\text{ ft}} \begin{bmatrix} \delta u_1^{(5)} & \delta u_2^{(5)} \end{bmatrix} \begin{Bmatrix} N_1^{(5)}(x) \\ N_2^{(5)}(x) \end{Bmatrix} b(x)dx$$

$$= \begin{bmatrix} \delta u_1^{(1)} & \delta u_2^{(1)} \end{bmatrix} \begin{Bmatrix} \int_0^{0.8\text{ ft}} N_1^{(1)}(x)b(x)dx \\ \int_0^{0.8\text{ ft}} N_2^{(1)}(x)b(x)dx \end{Bmatrix} + \begin{bmatrix} \delta u_1^{(2)} & \delta u_2^{(2)} \end{bmatrix} \begin{Bmatrix} \int_{0.8in}^{1.6\text{ ft}} N_1^{(2)}(x)b(x)dx \\ \int_{0.8in}^{1.6\text{ ft}} N_2^{(2)}(x)b(x)dx \end{Bmatrix}$$

$$+ \begin{bmatrix} \delta u_1^{(3)} & \delta u_2^{(3)} \end{bmatrix} \begin{Bmatrix} \int_{1.6\text{ ft}}^{2.4\text{ ft}} N_1^{(3)}(x)b(x)dx \\ \int_{1.6\text{ ft}}^{2.4\text{ ft}} N_2^{(3)}(x)b(x)dx \end{Bmatrix} + \begin{bmatrix} \delta u_1^{(4)} & \delta u_2^{(4)} \end{bmatrix} \begin{Bmatrix} \int_{2.4\text{ ft}}^{3.2\text{ ft}} N_1^{(4)}(x)b(x)dx \\ \int_{2.4\text{ ft}}^{3.2\text{ ft}} N_2^{(4)}(x)b(x)dx \end{Bmatrix}$$

$$+ \begin{bmatrix} \delta u_1^{(5)} & \delta u_2^{(5)} \end{bmatrix} \begin{Bmatrix} \int_{3.2\text{ ft}}^{4.0\text{ ft}} N_1^{(5)}(x)b(x)dx \\ \int_{3.2\text{ ft}}^{4.0\text{ ft}} N_2^{(5)}(x)b(x)dx \end{Bmatrix}$$

$$= \begin{bmatrix} \delta u_1^{(1)} & \delta u_2^{(1)} \end{bmatrix} \begin{Bmatrix} f_1^{(1)} \\ f_2^{(1)} \end{Bmatrix} + \begin{bmatrix} \delta u_1^{(2)} & \delta u_2^{(2)} \end{bmatrix} \begin{Bmatrix} f_1^{(2)} \\ f_2^{(2)} \end{Bmatrix} + \begin{bmatrix} \delta u_1^{(3)} & \delta u_2^{(3)} \end{bmatrix} \begin{Bmatrix} f_1^{(3)} \\ f_2^{(3)} \end{Bmatrix}$$

$$+ \begin{bmatrix} \delta u_1^{(4)} & \delta u_2^{(4)} \end{bmatrix} \begin{Bmatrix} f_1^{(4)} \\ f_2^{(4)} \end{Bmatrix} + \begin{bmatrix} \delta u_1^{(5)} & \delta u_2^{(5)} \end{bmatrix} \begin{Bmatrix} f_1^{(5)} \\ f_2^{(5)} \end{Bmatrix}. \tag{4.71}$$

The vectors containing $f_1^{(1)}, f_2^{(1)}, \ldots$, are referred to as the element force vectors. Substituting the shape functions and the body force, the force vectors in Eq. (4.71) can be calculated, i.e.,

$$\int_0^{4\text{ ft}} \delta u b(x) dx \approx \begin{bmatrix} \delta u_1^{(1)} & \delta u_2^{(1)} \end{bmatrix} \begin{Bmatrix} 3279.1 \text{ lb} \\ 4372.1 \text{ lb} \end{Bmatrix} + \begin{bmatrix} \delta u_1^{(2)} & \delta u_2^{(2)} \end{bmatrix} \begin{Bmatrix} 5595.2 \text{ lb} \\ 6033.3 \text{ lb} \end{Bmatrix}$$

$$+ \begin{bmatrix} \delta u_1^{(3)} & \delta u_2^{(3)} \end{bmatrix} \begin{Bmatrix} 6757.9 \text{ lb} \\ 7067.0 \text{ lb} \end{Bmatrix} + \begin{bmatrix} \delta u_1^{(4)} & \delta u_2^{(4)} \end{bmatrix} \begin{Bmatrix} 7615.4 \text{ lb} \\ 7861.8 \text{ lb} \end{Bmatrix}$$

$$+ \begin{bmatrix} \delta u_1^{(5)} & \delta u_2^{(5)} \end{bmatrix} \begin{Bmatrix} 8313.2 \text{ lb} \\ 8521.4 \text{ lb} \end{Bmatrix}. \tag{4.72}$$

The last term in the weak form, Eq. (4.49), containing the natural boundary condition, is only associated with node 1 in the first element, i.e.,

$$\delta u A \bar{t}\big|_{x=0} = \delta u_1^{(1)} A_{(x=0)} \bar{t}, \tag{4.73}$$

Equation (4.73) can be rewritten in element vector form as:

$$\begin{bmatrix} \delta u_1^{(1)} & \delta u_2^{(1)} \end{bmatrix} \begin{Bmatrix} A_{(x=0)} \bar{t} \\ 0 \end{Bmatrix}. \tag{4.74}$$

Combining Eqs. (4.70, 4.71, 4.74), the original weak form can be expressed as:

$$\int_0^{4\text{ ft}} \left(\frac{d\delta u}{dx}\right) AE \left(\frac{du}{dx}\right) dx = \int_0^{4\text{ ft}} \delta u b(x) dx + \delta u A \bar{t}\big|_{x=0}$$

$$\Rightarrow \begin{bmatrix} \delta u_1^{(1)} & \delta u_2^{(1)} \end{bmatrix} \begin{bmatrix} k_{11}^{(1)} & k_{12}^{(1)} \\ k_{21}^{(1)} & k_{22}^{(1)} \end{bmatrix} \begin{Bmatrix} u_1^{(1)} \\ u_2^{(1)} \end{Bmatrix} + \begin{bmatrix} \delta u_1^{(2)} & \delta u_2^{(2)} \end{bmatrix} \begin{bmatrix} k_{11}^{(2)} & k_{12}^{(2)} \\ k_{21}^{(2)} & k_{22}^{(2)} \end{bmatrix} \begin{Bmatrix} u_1^{(2)} \\ u_2^{(2)} \end{Bmatrix}$$

$$+ \begin{bmatrix} \delta u_1^{(3)} & \delta u_2^{(3)} \end{bmatrix} \begin{bmatrix} k_{11}^{(3)} & k_{12}^{(3)} \\ k_{21}^{(3)} & k_{22}^{(3)} \end{bmatrix} \begin{Bmatrix} u_1^{(3)} \\ u_2^{(3)} \end{Bmatrix} + \begin{bmatrix} \delta u_1^{(4)} & \delta u_2^{(4)} \end{bmatrix} \begin{bmatrix} k_{11}^{(4)} & k_{12}^{(4)} \\ k_{21}^{(4)} & k_{22}^{(4)} \end{bmatrix} \begin{Bmatrix} u_1^{(4)} \\ u_2^{(4)} \end{Bmatrix}$$

$$+ \begin{bmatrix} \delta u_1^{(5)} & \delta u_2^{(5)} \end{bmatrix} \begin{bmatrix} k_{11}^{(5)} & k_{12}^{(5)} \\ k_{21}^{(5)} & k_{22}^{(5)} \end{bmatrix} \begin{Bmatrix} u_1^{(5)} \\ u_2^{(5)} \end{Bmatrix}$$

$$= \begin{bmatrix} \delta u_1^{(1)} & \delta u_2^{(1)} \end{bmatrix} \begin{Bmatrix} f_1^{(1)} \\ f_2^{(1)} \end{Bmatrix} + \begin{bmatrix} \delta u_1^{(2)} & \delta u_2^{(2)} \end{bmatrix} \begin{Bmatrix} f_1^{(2)} \\ f_2^{(2)} \end{Bmatrix} + \begin{bmatrix} \delta u_1^{(3)} & \delta u_2^{(3)} \end{bmatrix} \begin{Bmatrix} f_1^{(3)} \\ f_2^{(3)} \end{Bmatrix}$$

$$+ \begin{bmatrix} \delta u_1^{(4)} & \delta u_2^{(4)} \end{bmatrix} \begin{Bmatrix} f_1^{(4)} \\ f_2^{(4)} \end{Bmatrix} + \begin{bmatrix} \delta u_1^{(5)} & \delta u_2^{(5)} \end{bmatrix} \begin{Bmatrix} f_1^{(5)} \\ f_2^{(5)} \end{Bmatrix}$$

$$+ \begin{bmatrix} \delta u_1^{(1)} & \delta u_2^{(1)} \end{bmatrix} \begin{Bmatrix} A_{(x=0)} \bar{t} \\ 0 \end{Bmatrix}. \tag{4.75}$$

Equation (4.75) is the discretized version of the original weak form. The element integrals are expressed in terms of the nodal displacements and their variations, and the element stiffness matrices and force vectors. The element stiffness matrices and force vectors are calculated in this step. The unknown nodal displacements are to be calculated in Steps 7–9.

Step 7: Assemble the Global System Equations

Before we proceed to assemble the global system equations, it is necessary to introduce the concepts of the element (or local) node index and the global node index. As shown in Fig. 4.11, the node indices 1, 2, ...,6 are the global node indices in which each node has a unique index and each index denotes a unique node. When working on an individual element, nodes are also denoted by node indices. In this case, the element node indices are defined just for the element. For example, all the linear elements shown in the figure have a node 1 and a node 2. However, the node 2 of element 2 is the global node 3 and the node 2 of element 3 is the global node 4. The mapping between the global node indices and the element node indices is called the connectivity information.

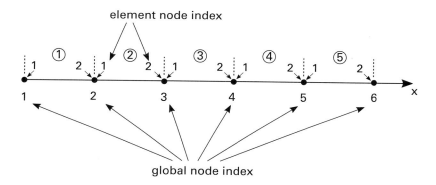

Figure 4.11 Local and global node indices.

For the mesh shown in Fig. 4.11, the mapping between the local element nodal displacements and the global nodal displacements gives:

$$u_1^{(1)} \to u_1; \quad u_2^{(1)} \to u_2; \quad u_1^{(2)} \to u_2; \quad u_2^{(2)} \to u_3; \quad \ldots \tag{4.76}$$

and

$$\delta u_1^{(1)} \to \delta u_1; \quad \delta u_2^{(1)} \to \delta u_2; \quad \delta u_1^{(2)} \to \delta u_2; \quad \delta u_2^{(2)} \to \delta u_3; \quad \ldots \tag{4.77}$$

Replacing the element nodal displacements and their variations by the global nodal displacements and their variations in the discretized weak form given in Eq. (4.73), we obtain:

4.3 1-D Elasticity

$$[\delta u_1 \; \delta u_2] \begin{bmatrix} k_{11}^{(1)} & k_{12}^{(1)} \\ k_{21}^{(1)} & k_{22}^{(1)} \end{bmatrix} \begin{Bmatrix} u_1 \\ u_2 \end{Bmatrix} + [\delta u_2 \; \delta u_3] \begin{bmatrix} k_{11}^{(2)} & k_{12}^{(2)} \\ k_{21}^{(2)} & k_{22}^{(2)} \end{bmatrix} \begin{Bmatrix} u_2 \\ u_3 \end{Bmatrix}$$

$$+ [\delta u_3 \; \delta u_4] \begin{bmatrix} k_{11}^{(3)} & k_{12}^{(3)} \\ k_{21}^{(3)} & k_{22}^{(3)} \end{bmatrix} \begin{Bmatrix} u_3 \\ u_4 \end{Bmatrix} + [\delta u_4 \; \delta u_5] \begin{bmatrix} k_{11}^{(4)} & k_{12}^{(4)} \\ k_{21}^{(4)} & k_{22}^{(4)} \end{bmatrix} \begin{Bmatrix} u_4 \\ u_5 \end{Bmatrix}$$

$$+ [\delta u_5 \; \delta u_6] \begin{bmatrix} k_{11}^{(5)} & k_{12}^{(5)} \\ k_{21}^{(5)} & k_{22}^{(5)} \end{bmatrix} \begin{Bmatrix} u_5 \\ u_6 \end{Bmatrix}$$

$$= [\delta u_1 \; \delta u_2] \begin{Bmatrix} f_1^{(1)} \\ f_2^{(1)} \end{Bmatrix} + [\delta u_2 \; \delta u_3] \begin{Bmatrix} f_1^{(2)} \\ f_2^{(2)} \end{Bmatrix} + [\delta u_3 \; \delta u_4] \begin{Bmatrix} f_1^{(3)} \\ f_2^{(3)} \end{Bmatrix}$$

$$+ [\delta u_4 \; \delta u_5] \begin{Bmatrix} f_1^{(4)} \\ f_2^{(4)} \end{Bmatrix} + [\delta u_5 \; \delta u_6] \begin{Bmatrix} f_1^{(5)} \\ f_2^{(5)} \end{Bmatrix}$$

$$+ [\delta u_1 \; \delta u_2] \begin{Bmatrix} A_{(x=0)} \bar{t} \\ 0 \end{Bmatrix}. \tag{4.78}$$

The terms in the equation above can not be added together without multiplying out the vectors and matrices, due to the fact that different nodes are involved in each element. To resolve this problem, we expand all the terms to a global setting as follows:

$$[\delta u_1 \; \delta u_2 \; \delta u_3 \; \delta u_4 \; \delta u_5 \; \delta u_6] \begin{bmatrix} k_{11}^{(1)} & k_{12}^{(1)} & 0 & 0 & 0 & 0 \\ k_{21}^{(1)} & k_{22}^{(1)} & 0 & 0 & 0 & 0 \\ 0 & 0 & 0 & 0 & 0 & 0 \\ 0 & 0 & 0 & 0 & 0 & 0 \\ 0 & 0 & 0 & 0 & 0 & 0 \\ 0 & 0 & 0 & 0 & 0 & 0 \end{bmatrix} \begin{Bmatrix} u_1 \\ u_2 \\ u_3 \\ u_4 \\ u_5 \\ u_6 \end{Bmatrix}$$

$$+ [\delta u_1 \; \delta u_2 \; \delta u_3 \; \delta u_4 \; \delta u_5 \; \delta u_6] \begin{bmatrix} 0 & 0 & 0 & 0 & 0 & 0 \\ 0 & k_{11}^{(2)} & k_{12}^{(2)} & 0 & 0 & 0 \\ 0 & k_{21}^{(2)} & k_{22}^{(2)} & 0 & 0 & 0 \\ 0 & 0 & 0 & 0 & 0 & 0 \\ 0 & 0 & 0 & 0 & 0 & 0 \\ 0 & 0 & 0 & 0 & 0 & 0 \end{bmatrix} \begin{Bmatrix} u_1 \\ u_2 \\ u_3 \\ u_4 \\ u_5 \\ u_6 \end{Bmatrix}$$

$$+ [\delta u_1 \; \delta u_2 \; \delta u_3 \; \delta u_4 \; \delta u_5 \; \delta u_6] \begin{bmatrix} 0 & 0 & 0 & 0 & 0 & 0 \\ 0 & 0 & 0 & 0 & 0 & 0 \\ 0 & 0 & k_{11}^{(3)} & k_{12}^{(3)} & 0 & 0 \\ 0 & 0 & k_{21}^{(3)} & k_{22}^{(3)} & 0 & 0 \\ 0 & 0 & 0 & 0 & 0 & 0 \\ 0 & 0 & 0 & 0 & 0 & 0 \end{bmatrix} \begin{Bmatrix} u_1 \\ u_2 \\ u_3 \\ u_4 \\ u_5 \\ u_6 \end{Bmatrix}$$

General Procedure of FEA: 1-D Problems

$$+ \begin{bmatrix} \delta u_1 & \delta u_2 & \delta u_3 & \delta u_4 & \delta u_5 & \delta u_6 \end{bmatrix} \begin{bmatrix} 0 & 0 & 0 & 0 & 0 & 0 \\ 0 & 0 & 0 & 0 & 0 & 0 \\ 0 & 0 & 0 & 0 & 0 & 0 \\ 0 & 0 & 0 & k_{11}^{(4)} & k_{12}^{(4)} & 0 \\ 0 & 0 & 0 & k_{21}^{(4)} & k_{22}^{(4)} & 0 \\ 0 & 0 & 0 & 0 & 0 & 0 \end{bmatrix} \begin{Bmatrix} u_1 \\ u_2 \\ u_3 \\ u_4 \\ u_5 \\ u_6 \end{Bmatrix}$$

$$+ \begin{bmatrix} \delta u_1 & \delta u_2 & \delta u_3 & \delta u_4 & \delta u_5 & \delta u_6 \end{bmatrix} \begin{bmatrix} 0 & 0 & 0 & 0 & 0 & 0 \\ 0 & 0 & 0 & 0 & 0 & 0 \\ 0 & 0 & 0 & 0 & 0 & 0 \\ 0 & 0 & 0 & 0 & 0 & 0 \\ 0 & 0 & 0 & 0 & k_{11}^{(5)} & k_{12}^{(5)} \\ 0 & 0 & 0 & 0 & k_{21}^{(5)} & k_{22}^{(5)} \end{bmatrix} \begin{Bmatrix} u_1 \\ u_2 \\ u_3 \\ u_4 \\ u_5 \\ u_6 \end{Bmatrix}$$

$$= \begin{bmatrix} \delta u_1 & \delta u_2 & \delta u_3 & \delta u_4 & \delta u_5 & \delta u_6 \end{bmatrix} \begin{Bmatrix} f_1^{(1)} \\ f_2^{(1)} \\ 0 \\ 0 \\ 0 \\ 0 \end{Bmatrix} + \begin{bmatrix} \delta u_1 & \delta u_2 & \delta u_3 & \delta u_4 & \delta u_5 & \delta u_6 \end{bmatrix} \begin{Bmatrix} 0 \\ f_1^{(2)} \\ f_2^{(2)} \\ 0 \\ 0 \\ 0 \end{Bmatrix}$$

$$+ \begin{bmatrix} \delta u_1 & \delta u_2 & \delta u_3 & \delta u_4 & \delta u_5 & \delta u_6 \end{bmatrix} \begin{Bmatrix} 0 \\ 0 \\ f_1^{(3)} \\ f_2^{(3)} \\ 0 \\ 0 \end{Bmatrix} + \begin{bmatrix} \delta u_1 & \delta u_2 & \delta u_3 & \delta u_4 & \delta u_5 & \delta u_6 \end{bmatrix} \begin{Bmatrix} 0 \\ 0 \\ 0 \\ f_1^{(4)} \\ f_2^{(4)} \\ 0 \end{Bmatrix}$$

$$+ \begin{bmatrix} \delta u_1 & \delta u_2 & \delta u_3 & \delta u_4 & \delta u_5 & \delta u_6 \end{bmatrix} \begin{Bmatrix} 0 \\ 0 \\ 0 \\ 0 \\ f_1^{(5)} \\ f_2^{(5)} \end{Bmatrix} + \begin{bmatrix} \delta u_1 & \delta u_2 & \delta u_3 & \delta u_4 & \delta u_5 & \delta u_6 \end{bmatrix} \begin{Bmatrix} A_{(x=0)} \bar{t} \\ 0 \\ 0 \\ 0 \\ 0 \\ 0 \end{Bmatrix}.$$

(4.79)

It is observed that in Eq. (4.79), the arbitrary displacement variations can be canceled from both sides of the equation. The remaining terms can be added together straightforwardly as:

$$\begin{bmatrix} k_{11}^{(1)} & k_{12}^{(1)} & 0 & 0 & 0 & 0 \\ k_{21}^{(1)} & k_{22}^{(1)}+k_{11}^{(2)} & k_{12}^{(2)} & 0 & 0 & 0 \\ 0 & k_{21}^{(2)} & k_{22}^{(2)}+k_{11}^{(3)} & k_{12}^{(3)} & 0 & 0 \\ 0 & 0 & k_{21}^{(3)} & k_{22}^{(3)}+k_{11}^{(4)} & k_{12}^{(4)} & 0 \\ 0 & 0 & 0 & k_{21}^{(4)} & k_{22}^{(4)}+k_{11}^{(5)} & k_{12}^{(5)} \\ 0 & 0 & 0 & 0 & k_{21}^{(5)} & k_{22}^{(5)} \end{bmatrix} \begin{Bmatrix} u_1 \\ u_2 \\ u_3 \\ u_4 \\ u_5 \\ u_6 \end{Bmatrix} = \begin{Bmatrix} f_1^{(1)} + A_{(x=0)}\bar{t} \\ f_2^{(1)} + f_1^{(2)} \\ f_2^{(2)} + f_1^{(3)} \\ f_2^{(3)} + f_1^{(4)} \\ f_2^{(4)} + f_1^{(5)} \\ f_2^{(5)} \end{Bmatrix}.$$

(4.80)

Equation (4.80) is the assembled global system equation. The construction of the global matrix and the global vector is called the assembly process. Substituting the value of the calculated integrals, we obtain:

$$\begin{bmatrix} 8.02 & -8.02 & 0 & 0 & 0 & 0 \\ -8.02 & 17.5 & -9.48 & 0 & 0 & 0 \\ 0 & -9.48 & 20.42 & -10.94 & 0 & 0 \\ 0 & 0 & -10.94 & 23.34 & -12.4 & 0 \\ 0 & 0 & 0 & -12.4 & 26.25 & -13.85 \\ 0 & 0 & 0 & 0 & -13.85 & 13.85 \end{bmatrix} \times 10^3 \begin{Bmatrix} u_1 \\ u_2 \\ u_3 \\ u_4 \\ u_5 \\ u_6 \end{Bmatrix} = \begin{Bmatrix} 203.279 \\ 9.967 \\ 12.791 \\ 14.682 \\ 16.175 \\ 8.521 \end{Bmatrix}.$$

(4.81)

Note that, in Eqs. (4.80, 4.81), the natural (stress) boundary condition is already included in the global vector. However, the essential (displacement) boundary condition has not been included. Essential boundary conditions and concentrated loads are applied in Step 8.

Step 8: Apply the Essential Boundary Conditions

In our example problem, no concentrated nodal forces are applied. The application of concentrated nodal force will be discussed in later chapters. To apply the essential boundary conditions (i.e. the displacement boundary condition in this example), several approaches have been developed. In this section, we introduce two popular methods: the direct substitution method and the penalty method. Another widely used method, the method of Lagrange multipliers, will be discussed in later chapters.

The first method to be discussed is the direct substitution method. In this method, the prescribed essential boundary conditions are substituted into the global system equations obtained in Step 7 (Eq. (4.81)). After the substitution, the columns corresponding to the nodes with essential boundary conditions are multiplied by the prescribed nodal values and moved to the right hand side of the system equations, and the rows corresponding to these nodes are then deleted. In the 1-D elastic bar problem, the essential boundary condition is given by:

$$u(x)|_{x=4 \text{ ft}} = 0 \quad \Rightarrow \quad u_6 = 0. \quad (4.82)$$

General Procedure of FEA: 1-D Problems

Substituting Eq. (4.82) into Eq. (4.81), we obtain:

$$\begin{bmatrix} 802e3 & -8.02e3 & 0 & 0 & 0 & 0 \\ -8.02e3 & 17.5e3 & -9.48e3 & 0 & 0 & 0 \\ 0 & -9.48e3 & 20.42e3 & -10.94e3 & 0 & 0 \\ 0 & 0 & -10.94e3 & 23.34e3 & -12.4e3 & 0 \\ 0 & 0 & 0 & -12.4e3 & 26.25e3 & -13.85e3 \\ 0 & 0 & 0 & 0 & 0 & 1 \end{bmatrix} \begin{Bmatrix} u_1 \\ u_2 \\ u_3 \\ u_4 \\ u_5 \\ u_6 \end{Bmatrix}$$

$$= \begin{Bmatrix} 203.2791 \\ 9.9673 \\ 12.7912 \\ 14.6824 \\ 16.1750 \\ 0 \end{Bmatrix}. \tag{4.83}$$

Equation (4.83) can be rewritten as:

$$\begin{bmatrix} 8.02e3 & -8.02e3 & 0 & 0 & 0 \\ -8.02e3 & 17.5e3 & -9.48e3 & 0 & 0 \\ 0 & -9.48e3 & 20.42e3 & -10.94e3 & 0 \\ 0 & 0 & -10.94e3 & 23.34e3 & -12.4e3 \\ 0 & 0 & 0 & -12.4e3 & 26.25e3 \\ 0 & 0 & 0 & 0 & 0 \end{bmatrix} \begin{Bmatrix} u_1 \\ u_2 \\ u_3 \\ u_4 \\ u_5 \end{Bmatrix}$$

$$+ \begin{Bmatrix} 0 \\ 0 \\ 0 \\ 0 \\ -13.85e3 \\ 1 \end{Bmatrix} u_6 = \begin{Bmatrix} 203.2791 \\ 9.9673 \\ 12.7912 \\ 14.6824 \\ 16.1750 \\ 0 \end{Bmatrix} \tag{4.84}$$

$$\Rightarrow \begin{bmatrix} 8.02e3 & -8.02e3 & 0 & 0 & 0 \\ -8.02e3 & 17.5e3 & -9.48e3 & 0 & 0 \\ 0 & -9.48e3 & 20.42e3 & -10.94e3 & 0 \\ 0 & 0 & -10.94e3 & 23.34e3 & -12.4e3 \\ 0 & 0 & 0 & -12.4e3 & 26.25e3 \\ 0 & 0 & 0 & 0 & 0 \end{bmatrix} \begin{Bmatrix} u_1 \\ u_2 \\ u_3 \\ u_4 \\ u_5 \end{Bmatrix}$$

$$= \begin{Bmatrix} 203.2791 \\ 9.9673 \\ 12.7912 \\ 14.6824 \\ 16.1750 \\ 0 \end{Bmatrix} - \begin{Bmatrix} 0 \\ 0 \\ 0 \\ 0 \\ -13.85e3 \\ 1 \end{Bmatrix} 0 \qquad (4.85)$$

$$\Rightarrow \begin{bmatrix} 8.02e3 & -8.02e3 & 0 & 0 & 0 \\ -8.02e3 & 17.5e3 & -9.48e3 & 0 & 0 \\ 0 & -9.48e3 & 20.42e3 & -10.94e3 & 0 \\ 0 & 0 & -10.94e3 & 23.34e3 & -12.4e3 \\ 0 & 0 & 0 & -12.4e3 & 26.25e3 \\ 0 & 0 & 0 & 0 & 0 \end{bmatrix} \begin{Bmatrix} u_1 \\ u_2 \\ u_3 \\ u_4 \\ u_5 \end{Bmatrix} = \begin{Bmatrix} 203.2791 \\ 9.9673 \\ 12.7912 \\ 14.6824 \\ 16.1750 \\ 0 \end{Bmatrix}$$

$$\Rightarrow \begin{bmatrix} 8.02e3 & -8.02e3 & 0 & 0 & 0 \\ -8.02e3 & 17.5e3 & -9.48e3 & 0 & 0 \\ 0 & -9.48e3 & 20.42e3 & -10.94e3 & 0 \\ 0 & 0 & -10.94e3 & 23.34e3 & -12.4e3 \\ 0 & 0 & 0 & -12.4e3 & 26.25e3 \end{bmatrix} \begin{Bmatrix} u_1 \\ u_2 \\ u_3 \\ u_4 \\ u_5 \end{Bmatrix} = \begin{Bmatrix} 203.2791 \\ 9.9673 \\ 12.7912 \\ 14.6824 \\ 16.1750 \end{Bmatrix}.$$
(4.86)

Equation (4.86) can be written in short form as

$$\mathbf{Ku} = \mathbf{F} \qquad (4.87)$$

where **K** is the global stiffness matrix, **F** is the global force vector, and **u** is the vector of the unknown nodal displacements. Equation (4.87) can be solved by using a linear solver. The advantages of the direct substitution method are summarized as follows:

1. The method is straightforward to implement.
2. The size of the final linear system, Eq. (4.87), becomes smaller after deleting the rows corresponding to the essential boundary condition nodes.
3. The symmetry of the stiffness matrix is preserved.

The main disadvantage of the method is the requirement of additional manipulation of the stiffness matrix and the force vector.

The second method to apply essential boundary conditions is called the penalty method. In the penalty method, a large number (called the penalty number, which is a positive number several orders of magnitude larger than the largest number in the system equations given in Eq. (4.81)), is chosen and placed on the diagonal of the rows corresponding to the nodes with essential boundary conditions. In addition, the prescribed displacements are multiplied by this large number and placed in the rows of the force vector corresponding to these nodes. In this example, we set the large number to be 10^{10} and the penalty method described above leads to:

General Procedure of FEA: 1-D Problems

$$\begin{bmatrix} 802e3 & -8.02e3 & 0 & 0 & 0 & 0 \\ -8.02e3 & 17.5e3 & -9.48e3 & 0 & 0 & 0 \\ 0 & -9.48e3 & 20.42e3 & -10.94e3 & 0 & 0 \\ 0 & 0 & -10.94e3 & 23.34e3 & -12.4e3 & 0 \\ 0 & 0 & 0 & -12.4e3 & 26.25e3 & -13.85e3 \\ 0 & 0 & 0 & 0 & -13.85e3 & 1e10 \end{bmatrix} \begin{Bmatrix} u_1 \\ u_2 \\ u_3 \\ u_4 \\ u_5 \\ u_6 \end{Bmatrix}$$

$$= \begin{Bmatrix} 203.279 \\ 9.967 \\ 12.791 \\ 14.682 \\ 16.175 \\ 0 \times 1e10 \end{Bmatrix}.$$

(4.88)

It can be observed that the penalty method modifies the last row of Eq. (4.88). The last equation of the linear system becomes

$$-13.85e3 \times u_5 + 1e10 \times u_6 = 0 \times 1e10. \qquad (4.89)$$

Since the magnitude of the penalty number, 10^{10}, is six orders of magnitude larger than the coefficient -13.85×10^3, and at the same time u_5 and u_6 are expected to be of the same order of magnitude because they are the displacements of two adjacent nodes, the first term on the left hand side of Eq. (4.89) is then much smaller than the second term. Therefore, the effect of u_5 can be regarded as negligible in Eq. (4.89) and, effectively, Eq. (4.89) is just

$$10^{10} \times u_6 = 0 \times 10^{10}$$
$$\Rightarrow u_6 = 0. \qquad (4.90)$$

Thus, Eq. (4.90) is simply the essential boundary condition itself. It should be noted that the penalty number should not be too small or too large. If it is too small, the effect of u_5 becomes significant in Eq. (4.89), and u_6 will move away from its prescribed value. If the penalty number is very large, the condition number of the global stiffness matrix becomes large too and the matrix becomes ill-conditioned. In practice, the penalty number is set to be

$$\text{penalty number} = \left(10^6 \sim 10^{10}\right) \times max(abs(K(i,j))) \qquad i,j = 1, 2, \ldots, N \qquad (4.91)$$

where $abs(K(i,j))$ denotes the absolute value of the entries of the stiffness matrix, $\mathbf{K}$, N is the size of the linear system, and the factor of $10^6 \sim 10^{10}$ is called the penalty parameter. As in the direct substitution method, Eq. (4.88) can be written in short form as

$$\mathbf{Ku} = \mathbf{F} \qquad (4.92)$$

and can be solved by using a linear solver. The advantages of the penalty method include:

1. The method is straightforward to implement.
2. The symmetry of the stiffness matrix is preserved.
3. There is no manipulation of the rows or columns of the stiffness matrix.

The disadvantages of the method are

1. Large penalty numbers increase the condition number of the stiffness matrix.
2. The solution of u_6 is not "exact."

Step 9: Solve the Global System Equations

The global system of equations, $\mathbf{Ku} = \mathbf{F}$, obtained after applying the concentrated loads and the essential boundary conditions can be solved by using a variety of numerical methods as discussed in Chapter 3. In MATLAB, the command "$u = K\backslash F$" solves the linear system $\mathbf{Ku} = \mathbf{F}$ and stores the result in the u vector. For this example, the solution of the global linear system is obtained to be

$$\mathbf{u} = \begin{Bmatrix} 0.10646 \\ 0.08112 \\ 0.05862 \\ 0.03796 \\ 0.01855 \\ 0 \end{Bmatrix} \text{ in.} \tag{4.93}$$

Step 10: Perform Post-processing

The nodal displacements are obtained in Step 9. In the post-processing, other physical quantities of interest such as the strain and the stress can be obtained by using the nodal displacements. These results are also visualized via plots and/or animations in this step. The strain ϵ and stress σ in a given element ⓔ are calculated by:

$$\epsilon = \frac{du}{dx} = \sum_{i=1}^{2} N_i'^{(e)}(x) u_i^{(e)} \tag{4.94}$$

$$\sigma = E\epsilon \tag{4.95}$$

where ⓔ denotes an element, ⓔ = ①, ②, ..., ⑤. The displacement and strain plots for the 1-D elastic bar are shown in Fig. 4.12.

4.3.2 Computer Implementation

In previous sections, the finite element analysis of a 1-D elastic bar has been carried out manually. In this section, techniques involved in the computer implementation of the finite element analysis using MATLAB are introduced.

Program Structure

As discussed in Section 4.1, a finite element analysis consists of three main stages: (1) formulation and pre-processing, (2) solving, and (3) post-processing. Each stage

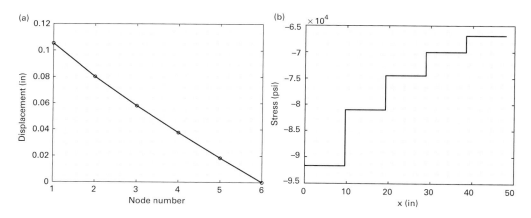

Figure 4.12 Displacement (a) and stress (b) of the elastic bar.

completes a set of tasks that are relatively independent from the tasks in the other two stages. The separation of the entire FEA of a given problem into relatively independent stages/parts is for the purpose of modularization. The relatively independent modules can communicate with each other through data files. The FEA analysis can be completed by using the modules created by different developers or even in different software packages. As long as the input and output data files have the same format, modules from different vendors can be combined to accomplish a larger task, allowing flexibility of FEA. A flow chart of a typical linear static finite element analysis is shown in Fig. 4.13.

Data Files and Data Structures

The formulation and pre-processing (Steps 1–4) of FEA are the problem statement, theoretical development, finite element formulation, geometric modeling, and domain discretization. A number of data files can be generated from the pre-processing tasks. As an example, Table 4.2 shows a set of data files and the information they store. These files are neither "standard" nor a complete set required for finite element analysis. Depending on the type and complexity of the physical problem, the number of data files needed for describing the physical system varies. However, among the files, the node and element files are the two essential files for any finite element analysis. As shown in the table, the node file stores the coordinates of the nodes and the element file contains the connectivity information of the elements. These two files provide the complete geometric information of the mesh. For this reason, the content and format of the two files have largely been accepted as de facto standard for storing mesh information and are used by most commercial FEA software packages today.

For the 1-D elastic bar problem, six files are generated, namely, "nodes.dat," "elements.dat," "materials.dat," "bcs.dat," "bfs.dat," "ndprops.dat." The files are shown in Table 4.3. In "nodes.dat," the first column contains the global node indices and the second column is the x-coordinate of the nodes. In "elements.dat," the first column lists the element indices, the second and third columns are the global node indices

4.3 1-D Elasticity

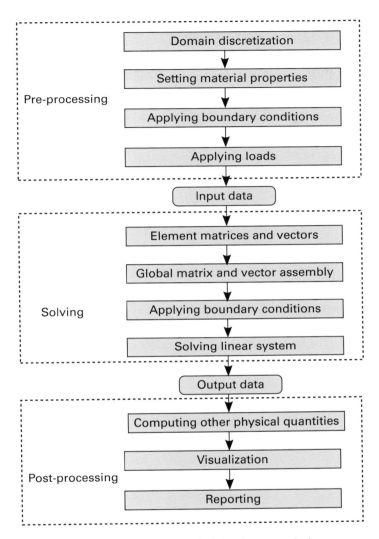

Figure 4.13 Flow chart of a linear static finite element analysis.

of the elements' first and second nodes, respectively. "bcs.dat" contains the boundary conditions at the end nodes. In "bcs.dat," the first column is the global node index, the second column is the type of the boundary conditions, where "0" represents "none," "1" denotes the displacement or essential boundary condition, and "2" denotes the surface traction or natural boundary condition. The units of the displacement and surface traction are inch and psi, respectively. The third column is the value of the corresponding boundary condition. "bfs.dat" contains the body loads at the nodes. In "bfs.dat," the first column is the global node index, the second column is the value of the body force (lb/in) acting on the nodes. "ndprops.dat" stores the location dependent nodal properties such as the varying cross sectional area in the 1-D bar

Table 4.2 Typical data files for finite element analysis

node file	list of nodes with their coordinates
element file	list of elements with their nodes and how they are connected in each element
essential BC file	list of nodes/edges/faces with essential boundary conditions
nodal loads file	list of nodes with nodal loads
surface loads file	list of element edges/faces with corresponding surface loads
body loads file	list of nodes with corresponding body loads
material properties file	list of global material properties
nodal properties file	list of nodes with location dependent geometric and/or material properties
element properties file	list of elements with location dependent geometric and/or material properties
options file	other global options and settings

Table 4.3 Data files for analysis

nodes.dat		elements.dat		
1	$0.00e+0$	1	1	2
2	$9.60e+0$	2	2	3
3	$1.92e+1$	3	3	4
4	$2.88e+1$	4	4	5
5	$3.84e+1$	5	5	6
6	$4.80e+1$			

bcs.dat			bfs.dat		ndprops.dat	
1	2	$1.0e+5$	1	$0.00e+0$	1	$2.0e+0$
2	0	0	2	$1.0627e+3$	2	$2.4e+0$
3	0	0	3	$1.3389e+3$	3	$2.8e+0$
4	0	0	4	$1.5326e+3$	4	$3.2e+0$
5	0	0	5	$1.6869e+3$	5	$3.6e+0$
6	1	0	6	$1.8171e+3$	6	$4.0e+0$

materials.dat

% Young's modulus (psi)
3.500e+7

problem. In "ndprops.dat," the first column is the node index, and the second column is the value of the cross sectional area (in^2) at the nodes. Finally, "materials.dat" stores the material properties: Young's modulus in this case. While the "nodes.dat" and "elements.dat" are considered as the standard input files, the rest of the data files are listed as an example of the result of pre-processing and the input data to be used in the finite element solver. One can generate or use input data files with different styles, data structures, and/or formats for boundary conditions, loads, geometric, and material properties, as long as all the necessary information is included and can be easily retrieved.

1-D Meshing

For 1-D problems, it is straightforward to produce a uniform mesh using MATLAB. The following is a function defined to generate a 1-D mesh.

```
1  % Generate 1-D mesh
2  % Input : start_x , end_x : x- coordinates of the starting and ending nodes
3  % Input : n_elements : number of elements
4  % Output : saved data files : nodes.dat and elements.dat
5  function CreateMesh1D(start_x, end_x, n_elements)
6  dl=(end_x-start_x)/n_elements;         % element length
7  nodes=zeros(n_elements+1,2);           % empty nodes matrix
8  nodes (:,1) =1:n_elements+1;           % 1st column: global node index
9  nodes (:,2) = start_x : dl : end_x;    % 2nd column: x- coordinates
10 % Save to nodes.dat
11 dlmwrite('nodes.dat',nodes,'delimiter','\t','precision','%.6f');
12 elements=zeros(n_elements,3) ;         % empty elements matrix
13 elements (:,1) =1:n_elements;          % 1st column: element index
14 elements (:,2) =1:n_elements;          % global node index of 1st node
15 elements (:,3) =2:n_elements+1;        % global node index of 2nd node
16 % Save to elements.dat
17 dlmwrite('elements.dat',elements,'delimiter','\t','precision','%d');
```

Shape Functions

The 1-D shape functions and their derivatives are needed in the computation. The shape functions can be implemented in a computer program by defining a function. The following shows a MATLAB function that computes the shape functions and their first derivatives at a given set of points stored in "x-vector" in a given element (start_x, end_x).

```
1  % Compute the value of the 1-D linear shape function and its first
2  % derivative at a set of input points .
3  % Input : start_x , end_x : starting and ending points of the element
4  % Input : x_vector : a list of input positions where the shape function
5  %          and its derivative should be evaluated .
6  % Output : N, Nx: shape function and its x- derivative are stored in N
7  %          and Nx, respectively , in the format shown as follows
8  %          N=[ N1(x1) N1(x2) N1(x3) ...
9  %              N2(x1) N2(x2) N2(x3) ...]
10 %          Nx=[N'1(x1) N'1(x2) N'1(x3) ...
```

```
11  %                       N'2(x1)  N'2(x2)  N'2(x3)   ...]
12  function [N,Nx]=CompElementShapeLinear1D(start_x, end_x, x_vector)
13  n=size(x_vector,1);              % obtain the size of x_vector
14  [N, Nx]= deal(zeros(2,n));       % setup empty N and Nx
15  for i=1:n                        % loop over each point x
16      N(1,i)=(x_vector(i) - end_x)/(start_x - end_x);   % N1(x)
17      N(2,i)=(x_vector(i) - start_x)/(end_x - start_x); % N2(x)
18      Nx(1,i)=1/(start_x - end_x);                       % N'1(x)
19      Nx(2,i)=1/(end_x - start_x);                       % N'2(x)
20  end
```

Assembly of Global Stiffness Matrix and Force Vector

The assembly of the global stiffness matrix and force vector was illustrated in previous sections by expanding the element matrices and vectors. However, in computer implementation, element matrices and vectors are directly assembled into the global stiffness matrix and force vector. The connectivity information of the elements and nodes is the key in the assembly process. The element indices, local element node indices, and global node indices need to be linked to each other so that the location of a given local matrix/vector entry in the global stiffness matrix and vector can be determined. As discussed previously, the connectivity information can be obtained from the "elements.dat" file, i.e.,

elements.dat

1	1	2
2	2	3
3	3	4
4	4	5
5	5	6

where the first column is the element index, the second and third columns are the global node indices. At the same time, the second and third columns are implicitly the first and second local nodes of the elements. Therefore, given an element, the global node indices of its first and second local nodes can be readily obtained.

Example 4.1 We have obtained the element matrix and vector for the element ③,

$$\mathbf{k}^{(3)} = \begin{bmatrix} k_{11}^{(3)} & k_{12}^{(3)} \\ k_{21}^{(3)} & k_{22}^{(3)} \end{bmatrix} \qquad \mathbf{f}^{(3)} = \begin{Bmatrix} f_1^{(3)} \\ f_2^{(3)} \end{Bmatrix}. \tag{4.96}$$

Where in the global matrix and global vector should we insert the entries of the element matrix and vector, respectively?

4.3 1-D Elasticity

Solution

From "elements.dat" file, we obtain that the local node index 1 (first local node) is corresponding to the global node index 3, and the local node index 2 (second local node) is corresponding to the global node index 4. Therefore, the entries of the local element matrix and vector can be "mapped" to the global matrix and vector as

$$\begin{bmatrix} k_{11}^{(3)} & k_{12}^{(3)} \\ k_{21}^{(3)} & k_{22}^{(3)} \end{bmatrix} \rightarrow \begin{bmatrix} k_{33} & k_{34} \\ k_{43} & k_{44} \end{bmatrix} \quad \begin{Bmatrix} f_1^{(3)} \\ f_2^{(3)} \end{Bmatrix} \rightarrow \begin{Bmatrix} f_3 \\ f_4 \end{Bmatrix} \quad (4.97)$$

and the entries of the local matrix and vector are then added to the global matrix and vector as shown in Fig. 4.14.

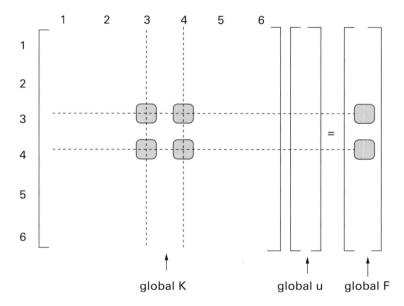

Figure 4.14 Global stiffness matrix assembly.

Note that when the local matrices and vectors are assembled in the global matrix and vector, the local entries are added on to the existing values in the global matrix and vector. A MATLAB function for assembling the element stiffness matrix **k** and force vector **f** into the global stiffness matrix **K** and force vector **F** is given as follows

```
1 % Assemble element stiffness matrix k and force vector f into the
2 % global matrix K and vector F
3 % Input : K, F: (empty) global stiffness matrix and force vector
4 % Input : eid : elemend index in the mesh
5 % Input : elements : the element matrix loaded from "elements.dat"
6 % Input : k,f: local stiffness matrix and force vector
```

```
7  % Output: K, F: Assembled global stiffness matrix and force vector
8  function [K,F]=AssembleGlobalLinear1D(K, F, eid, elements, k, f)
9  gids=[elements(eid,2) elements(eid,3) ]';  % global indices of the 1st
10                                             % and 2nd nodes of element eid
11 for m=1:2                                   % loop over rows
12     F(gids(m))=F(gids(m))+f(m);             % f(m) added onto global F
13     for n=1:2                               % loop over columns
14         K(gids(m),gids(n))=K(gids(m),gids(n))+k(m,n);  % k(m,n) added
15     end                                     % onto K
16 end
```

Isoparametric Mapping

In a computer program, the integrals in the element matrices and vectors need to be evaluated by using numerical integration. In Chapter 3, numerical integration techniques are discussed. It is shown that Gauss quadrature is an efficient method for this purpose. Using the 1-D elastic bar example, for an element ⓔ, the components of the element stiffness matrix are given by

$$k_{ij}^{\,\textcircled{e}} = \int_{x_1}^{x_2} A^{\textcircled{e}}(x) E N_i'^{\,\textcircled{e}}(x) N_j'^{\,\textcircled{e}}(x) dx \qquad i,j = 1, 2. \tag{4.98}$$

Since the nodal coordinates of the physical element are x_1 and x_2, the standard Gauss quadrature rules for integration over the interval $[-1, 1]$ can not be used. One has to scale the weights and Gauss point locations for each physical element. The purpose of isoparametric mapping is to map a standard "master" element to the physical elements so that the standard Gauss quadrature rules over $[-1, 1]$ can be directly used.

As shown in Fig. 4.15, for a linear 1-D element, the isoparametric mapping is given by

$$x = N_1(\xi)x_1 + N_2(\xi)x_2 \tag{4.99}$$

$$u = N_1(\xi)u_1 + N_2(\xi)u_2 \tag{4.100}$$

where ξ is referred to as the reference coordinate system axis. Equation (4.99) is the mapping of the element geometry (i.e., the position $\xi \to x$). Equation (4.100) is the mapping of the unknown quantity (i.e., $u(\xi) \to u(x)$). In addition, for linear elements, the shape functions are also mapped, $N(\xi) \to N(x)$, in the approximations. When the same set of $N_i(\xi)$ are used for both the element geometry (x) and the unknown quantity (u), the mapping is referred to as the isoparametric mapping.

For the linear elements shown in Fig. 4.15, it is obvious that

$$N_1(\xi) = \frac{1}{2}(1 - \xi) \tag{4.101}$$

$$N_2(\xi) = \frac{1}{2}(1 + \xi). \tag{4.102}$$

Figure 4.16 shows the isoparametric mapping for a quadratic element,

$$x = N_1(\xi)x_1 + N_2(\xi)x_2 + N_3(\xi)x_3 \tag{4.103}$$

4.3 1-D Elasticity

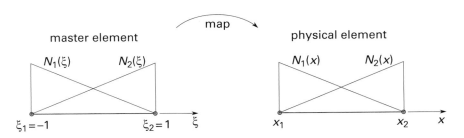

Figure 4.15 A 1-D linear isoparametric element.

$$u = N_1(\xi)u_1 + N_2(\xi)u_2 + N_3(\xi)u_3 \tag{4.104}$$

where

$$N_1(\xi) = \frac{1}{2}\xi(\xi - 1) \tag{4.105}$$

$$N_2(\xi) = (1 + \xi)(1 - \xi) \tag{4.106}$$

$$N_3(\xi) = \frac{1}{2}\xi(1 + \xi). \tag{4.107}$$

Note that, the node ξ_2 is always at the center of the master element, i.e. $\xi = 0$. However, the node x_2 is not necessarily at the center of the physical element.

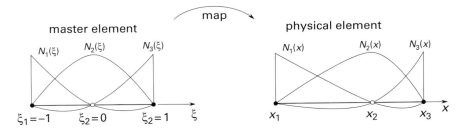

Figure 4.16 A 1-D quadratic isoparametric element.

In general, the equations of a 1-D isoparametric mapping can be written as

$$x = \sum_{i=1}^{n} N_i(\xi)x_i \tag{4.108}$$

$$u = \sum_{i=1}^{n} N_i(\xi)u_i \tag{4.109}$$

where n is the number of nodes in the element.

Example 4.2 For the mapping shown in Fig. 4.16, assuming $x_1 = 0$, $x_2 = 4$, and $x_3 = 6$, for any given ξ in the master element, the corresponding x in the physical element is calculated as

$$x = N_1(\xi)x_1 + N_2(\xi)x_2 + N_3(\xi)x_3$$
$$= \frac{1}{2}\xi(\xi - 1) \times 0 + (1 + \xi)(1 - \xi) \times 4 + \frac{1}{2}\xi(1 + \xi) \times 6$$
$$= 4 + 3\xi - \xi^2. \qquad (4.110)$$

The inverse map from x to ξ can be obtained as

$$\xi = \frac{1}{2}(3 - \sqrt{25 - 4x}). \qquad (4.111)$$

The mapped shape functions in the physical element can be obtained by substituting the expression of $\xi(x)$ into the shape functions of the master element. For example, the third shape function can be obtained as

$$N_3(x) = \frac{1}{2}\xi(x)(1 + \xi(x))$$
$$= \frac{1}{4}(3 - \sqrt{25 - 4x})\left(1 + \frac{1}{2}(3 - \sqrt{25 - 4x})\right)$$
$$= \frac{1}{2}\left(10 - 2\sqrt{25 - 4x} - x\right). \qquad (4.112)$$

It can be verified that $N_3(x_3) = 1$ and $N_3(x_1) = N_3(x_2) = 0$. Note that although it is possible to obtain the inverse mapping from the physical element to the master element, there is rarely a need to do so in practice.

A MATLAB function for calculating the 1-D linear isoparametric shape functions and their first derivatives over the master element is given below.

```
% Function to compute the value of the 1-D linear isoparametric shape
% functions and their first derivatives. The input vector xi_vector
% contains a list of points [xi] in the master element (between -1
% and 1) where the shape functions should be evaluated. The values of
% the shape functions and their x- derivative are stored in N and Nx,
% respectively, in the format shown as follows
% N=[ N1(xi1) N1(xi2) N1(xi3) ...
%     N2(xi1) N2(xi2) N2(xi3) ...]
% Nx=[N'1(xi1) N'1(xi2) N'1(xi3) ...
%     N'2(xi1) N'2(xi2) N'2(xi3) ...]
function [N,Nx]=CompShapeLinear1D(N, Nx, xi_vector)
n=size( xi_vector ,1) ;           % n xi points
for i=1:n
  N(1,i)=(1 - xi_vector(i)) /2.0;   % N1(xi)
  N(2,i)=(1 + xi_vector(i)) /2.0;   % N2(xi)
  Nx(1:2,i)= [-0.5  0.5]';          % Nx(xi)
end
```

Derivative Approximations Using Isoparametric Mapping

Since the unknown u is approximated in isoparametric elements as

$$u = \sum_{i=1}^{n} N_i(\xi) u_i \qquad (4.113)$$

the derivative of u is then obtained as

$$\begin{aligned}
\frac{du}{dx} &= \frac{d}{dx}\left(\sum_{i=1}^{n} N_i(\xi) u_i\right) \\
&= \sum_{i=1}^{n} \frac{dN_i(\xi)}{dx} u_i \\
&= \sum_{i=1}^{n} \frac{dN_i(\xi)}{d\xi} \frac{d\xi}{dx} u_i \\
&= \left(\sum_{i=1}^{n} \frac{dN_i(\xi)}{d\xi} u_i\right) \frac{d\xi}{dx}. \qquad (4.114)
\end{aligned}$$

The inverse of $\frac{d\xi}{dx}$ (i.e., $\frac{dx}{d\xi}$) is called a Jacobian, which is denoted as $J(\xi)$. From the mapping of x, the Jacobian can be calculated by

$$J(\xi) = \frac{dx}{d\xi} = \frac{d\left(\sum_{i=1}^{n} N_i(\xi) x_i\right)}{d\xi} = \sum_{i=1}^{n} \frac{dN_i(\xi)}{d\xi} x_i. \qquad (4.115)$$

Physically the Jacobian, $J(\xi)$, represents the stretch or shrink ratio of the master element at the point ξ, as given in the definition of $J(\xi)$. A MATLAB function for computing the Jacobian given in Eq. (4.115) at a given set of points in the master element ("xi_vector") is shown below.

```
1  % Compute the Jacobian of an isoparametric element mapping
2  % Input : x_vector : vector containing x- coordinates of nodes
3  % Input : Nx, shape function derivatives, from "CompShapeLinear1D"
4  % Output: J, Jacobian vector J( xi_1, xi_2, ...) ^T
5  function J=CompJacobian1D(x_vector,Nx)
6      J=(x_vector'*Nx)';   % compute vector of Jacobian
```

Uniform and Non-uniform Mapping

For 1-D elements, the isoparametric mapping is a uniform mapping when the relative positions of the corresponding nodes are the same in the master and physical elements, as shown in Fig. 4.17. Otherwise, the mapping is called non-uniform as shown in Fig. 4.18.

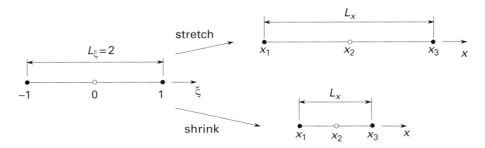

Figure 4.17 Uniform mapping.

Figure 4.18 Non-Uniform mapping.

In uniform mapping, the master element is uniformly stretched or compressed to the physical element. The Jacobian is a constant.

$$J(\xi) = \frac{L_x}{L_\xi} = \frac{L_x}{2}. \tag{4.116}$$

For non-uniform mapping, $J(\xi)$ is a function of ξ.

Example 4.3 In Example 4.2, we obtained from the isoparametric mapping that

$$x = 4 + 3\xi - \xi^2.$$

Therefore, the Jacobian can be obtained as

$$J(\xi) = \frac{dx}{d\xi} = 3 - 2\xi. \tag{4.117}$$

It is obvious that the Jacobian is a function of ξ.

For a given point ξ in the master element, if $J(\xi) = 0$, then the infinitesimal segment at ξ is mapped to a point in the physical element. If $J(\xi) < 0$, then the infinitesimal segment at ξ is flipped over on the physical element. In this case, the mapping is considered overly distorted.

Isoparametric Mapping in Element Matrices

With the isoparametric mapping, the element matrix component k_{ij} of an element e,

$$k_{ij}^{e} = \int_{x_1}^{x_2} A^{e}(x) E N_i^{'e}(x) N_j^{'e}(x) dx \qquad i,j = 1,2 \tag{4.118}$$

can be transformed to the master element by

$$x_1 \rightarrow \xi = -1 \tag{4.119}$$

$$x_2 \rightarrow \xi = 1 \tag{4.120}$$

$$dx \rightarrow J(\xi)d\xi \tag{4.121}$$

$$N_i^{'(e)}(x) = \frac{dN_i^{(e)}(x)}{dx} \rightarrow \frac{dN_i^{(e)}(\xi)}{d\xi}\frac{d\xi}{dx} = \frac{dN_i^{(e)}(\xi)}{d\xi}\frac{1}{J^{(e)}(\xi)} \tag{4.122}$$

$$A(x) \rightarrow \sum_{i=1}^{n} N_i^{(e)}(\xi) A_i^{(e)}. \tag{4.123}$$

Substituting Eqs. (4.119–4.123) into Eq. (4.118), one obtains

$$k_{ij}^{(e)} = \int_{-1}^{1} \left(\sum_{i=1}^{n} N_i^{(e)}(\xi) A_i^{(e)} \right) E N_i^{'(e)}(\xi) \frac{1}{J^{(e)}(\xi)} N_j^{'(e)}(\xi) \frac{1}{J^{(e)}(\xi)} J^{(e)}(\xi) d\xi$$

$$= \int_{-1}^{1} \left(\sum_{i=1}^{n} N_i^{(e)}(\xi) A_i^{(e)} \right) E N_i^{'(e)}(\xi) N_j^{'(e)}(\xi) \frac{1}{J^{(e)}(\xi)} d\xi \cdot \qquad i,j = 1, 2$$

$$\tag{4.124}$$

The element integral is transformed onto the master element, where a standard Gauss quadrature can be used to perform the numerical integration, i.e.,

$$k_{ij}^{(e)} \approx \sum_{g=1}^{ng} \left(\sum_{i=1}^{n} N_i^{(e)}(\xi_g) A_i^{(e)} \right) E N_i^{'(e)}(\xi_g) N_j^{'(e)}(\xi_g) \frac{1}{J^{(e)}(\xi_g)} w_g \tag{4.125}$$

where ξ_g denotes the Gauss integration points and w_g is the weight. Note that, since any element can be transformed to the master element as shown above, and the master element does not vary, ξ_g, w_g, $N_i^{(e)}(\xi_g)$, and $N_i^{'(e)}(\xi_g)$ have fixed values and can be repeatedly used for all the elements.

MATLAB Code

The MATLAB program for solving the 1-D axially loaded elastic bar is listed below. Note that the numerical results obtained from the MATLAB program will be slightly different from that given in EQ. (4.93). Why?

```
1  % Input elements, nodes, BCs and material properties
2  % (from steps 2, 3, 4 and 5)
3  load elements.dat;     % load elements
4  load nodes.dat;        % load nodes
5  load materials.dat;    % load material properties
6  load bcs.dat;          % load boundary conditions
7  load bfs.dat;          % load body forces
8  load ndprops.dat;      % load nodal properties : cross section area
```

```
9   E=materials(1);            % get Young's modulus
10  n_elements=size(elements,1);   % number of elements
11  n_nodes=size(nodes,1);         % number of nodes
12  % Create empty global matrices and vectors for later use
13  K=zeros(n_nodes,n_nodes);      % empty K matrix
14  F=zeros(n_nodes,1);            % empty RHS F vector
15  % Locations and weights of the Gauss points in the master element
16  n_gauss_points=2;              % number of Gauss points
17  [gauss_xi, gauss_w] = Get1DGauss(n_gauss_points);
18  % N, Nxi: set up empty matrices storing the value of the shape
19  % functions N and their derivatives Nxi (dN/dxi) evaluated at the
20  % Guass points in the master element
21  N=zeros(2,n_gauss_points);
22  Nxi=zeros(2,n_gauss_points);
23  % Compute the shape functions and their derivatives at each Gauss point
24  [N,Nxi]=CompShapeLinear1D(N, Nxi, gauss_xi);
25  sv=zeros(2,1);    % set up body force vector
26  area=zeros(2,1);  % set up cross section area vector
27  % The for loop: compute element matrices and vectors and aseemble them
28  % into the global matrix and RHS vector (Steps 6,7)
29  for i=1:n_elements   % loop over the elements
30      start_x = nodes(elements(i,2),2);  % x- coordinate of the start node
31      end_x= nodes(elements(i,3),2);     % x- coordinate of the end node
32      jacobian=CompJacobian1D([start_x end_x]', Nxi);
33      for g=1:n_gauss_points
34          sv(g)=bfs(i,2)*N(1,g)+ bfs(i+1,2)*N(2,g);
35          area(g)=ndprops(i,2)*N(1,g)+ ndprops(i+1,2)*N(2,g);
36      end
37      % Compute element force vector
38      f=zeros(2,1);
39      for m=1:2
40          for g=1:n_gauss_points
41              f(m,1)=f(m,1)+sv(g)*N(m,g)*gauss_w(g)*jacobian(g);
42          end
43      end
44      % Compute element stiffness matrix
45      k=zeros(2,2);
46      for m=1:2
47          for n=1:2
48              for g=1:n_gauss_points
49                  k(m,n)=k(m,n) +area(g)*E ...
50                      *Nxi(m,g)*Nxi(n,g)*gauss_w(g)/jacobian(g);
51              end
52          end
53      end
54      [K,F]=AssembleGlobalLinear1D(K, F, i, elements, k, f);
55  end
56  % for loop: apply loads, boundary conditions (Step 8)
57  for i=1:n_nodes-1:n_nodes
```

```
58      if bcs(i,2)==1
59          penalty=abs(K(i,i)+1)*1e7;   % penalty method
60          K(i,i)=penalty;
61          F(i)=bcs(i,3)*penalty;
62      end
63      if bcs(i,2)==2
64          F(i)=F(i)+bcs(i,3)*ndprops(i,2);
65      end
66  end
67  % Solve the global system equations (step 9)
68  u_nodes=K\F
69  % Post processing (step 10)
70  figure(1);
71  plot(nodes(:,1), u_nodes,'b-o','linewidth',2); % plot displacements
72  set(gca,'fontsize',16);
73  xlabel('node number','fontsize',18);
74  ylabel('Displacement (in)','fontsize',18);
75
76  k=1; % set up index of the points for computing the stress
77  for i=1:n_elements    % loop over the elements
78      node1=elements(i,2); % global node number of the local node 1
79      node2=elements(i,3); % global node number of the local node 2
80      for x=-1:0.1:1  % for points with spacing =0.1 in the master element
81          [N,Nxi]=CompShapeLinear1D(zeros(2,1),zeros(2,1),x);
82          u=u_nodes(node1)*N(1)+u_nodes(node2)*N(2); % displacement u(x)
83          J=CompJacobian1D([nodes(node1,2) nodes(node2,2)]',Nxi); % Jacobian
84          ux=u_nodes(node1)*Nxi(1)/J+u_nodes(node2)*Nxi(2)/J;  % u'(x)
85          re(k,1)=nodes(node1,2)*N(1)+nodes(node2,2)*N(2);
86          re(k,2)=u;
87          re(k,3)=E*ux;
88          k=k+1;                                            %
89      end
90  end
91  % plot stress
92  figure(2);
93  plot(re(:,1), re(:,3),'r-','linewidth',2);
94  set(gca,'fontsize',16);
95  xlabel('x (in)','fontsize',18);
96  ylabel(' Stress (psi)','fontsize',18);
```

In the MATLAB program listed above, the functions which have already been discussed previously are not repeated here. They are listed below for your reference:

function [gauss_xi, gauss_w] = Get1DGauss(n_gauss_points): Section 3.4.
function [N,Nx]=CompShapeLinear1D(N, Nx, xi_vector): Section 4.3.2.
function J=CompJacobian1D(x_vector,Nx): Section 4.3.2.
function [K,F]=AssembleGlobalLinear1D(K, F, eid, elements, k, f): Section 4.3.2.

4.4 1-D Steady State Heat Transfer

In this section, we apply the finite element procedure to 1-D steady state heat transfer problems. Similar to the 1-D elastic bar problem, we go through the FEA procedure step by step to solve an example problem of steady state heat transfer. Note that, details of the relevant numerical techniques and calculations that have been discussed for the 1-D elastic bar problem are not repeated here. Instead, they are presented here concisely for the sake of completeness.

4.4.1 Steady State Heat Transfer in a 1-D Rod

The finite element analysis procedure is applied to a 1-D steady state heat transfer problem. A thin, 1 m-long cylindrical rod composed of two different materials: a 50 cm-long section made of copper on the left and a 50 cm long section made of steel on the right (see Fig. 4.19). The cross section is circular with a radius of 0.5 cm. Heat is flowing into the left end at a steady rate of 0.1 cal/s·cm². The temperature of the right end is maintained at a constant 0 °C. The rod is in contact with air at an ambient temperature of 20 °C. There is a free convection from the lateral surface. We want to know the temperature and axial heat flux distributions along the rod.

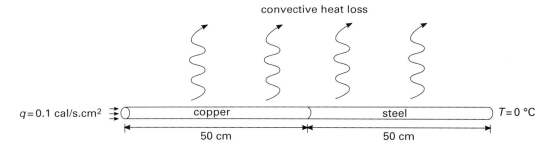

Figure 4.19 Steady state heat transfer in a 1-D rod.

Step 1: Physical Model
The problem of finding the temperature and heat flux along the length of the rod is a heat transfer problem. Since the rod is thin and long, the temperature and heat flux on any cross section of the rod is considered uniform. In addition, we are interested in the temperature and flux of the rod after steady state is reached. Therefore, this problem is regarded as a one-dimensional steady state heat transfer problem.

Step 2: Mathematical Model
Two fundamental laws in heat transfer can be used to obtain the mathematical description of the problem. The first is the energy conservation law and the second is Fourier's law. The energy conservation law in the context of heat transfer states that

$$\text{heat in} - \text{heat out} = \text{net heat causing temperature increase} \tag{4.126}$$

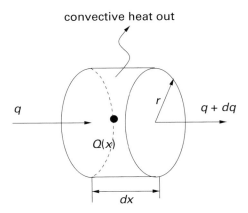

Figure 4.20 Energy conservation in an infinitesimal element.

Figure 4.20 shows a heat flow in an infinitesimal element of the rod, where q is the heat flux (J/s·m² or cal/s·m²), $Q(x)$ is the heat generated per unit volume per second by the internal heat source (J/s·m³ or cal/s·m³). Heat added in the element during a time period dt can be written as the inward heat flux multiplied by the cross sectional area of the infinitesimal element (πr^2), and multiplied by the time period dt, plus the heat generated by the heat source:

$$q\pi r^2 dt + Q\pi r^2 dx dt. \tag{4.127}$$

Heat flowing out can be written as

$$(q+dq)\pi r^2 dt + h(T - T_\infty)2\pi r dx dt. \tag{4.128}$$

Note that the second term represents the convective heat loss, where T_∞ is the environment temperature and h is called the coefficient of convective heat transfer (W/m²·K or cal/s·m²·°C). The net heat causing temperature increase on the right hand side of Eq. (4.126) is given by

$$c\rho(dT)\pi r^2 dx \tag{4.129}$$

where dT is the temperature increase, c is the specific heat (J/kg·K), and ρ is the mass density of the material (kg/m³). Thus the energy conservation law gives

$$q\pi r^2 dt + Q\pi r^2 dx dt - (q+dq)\pi r^2 dt - h(T - T_\infty)2\pi r dx dt = c\rho(dT)\pi r^2 dx. \tag{4.130}$$

Dividing all the terms by $\pi r^2 dx dt$, we obtain

$$Q - \frac{dq}{dx} - h(T - T_\infty)\frac{2}{r} = c\rho\frac{dT}{dt}. \tag{4.131}$$

Fourier's law states that the time rate of heat transfer through a material (heat flux) is proportional to the negative gradient in the temperature, i.e.,

$$q = -k(x)\frac{dT}{dx} \tag{4.132}$$

where $k(x)$ is the thermal conductivity (W/m·K). Therefore, Eq. (4.131) can be rewritten as

$$\frac{d}{dx}\left(k(x)\frac{dT}{dx}\right) + Q - \frac{2h(T-T_\infty)}{r} = c\rho\frac{dT}{dt}. \tag{4.133}$$

Note that if the temperature is a function of both x and t, Eq. (4.133) should be rewritten more appropriately with partial derivatives

$$\frac{\partial}{\partial x}\left(k(x)\frac{\partial T}{\partial x}\right) + Q - \frac{2h(T-T_\infty)}{r} = c\rho\frac{\partial T}{\partial t}. \tag{4.134}$$

In our 1-D rod problem, $Q = 0$, and since the problem is steady state, we have $\frac{dT}{dt} = 0$. Therefore, Eq. (4.133) becomes

$$\frac{d}{dx}\left(k(x)\frac{dT}{dx}\right) - \frac{2h(T-T_\infty)}{r} = 0. \tag{4.135}$$

Combining the above descriptions, we obtain the mathematical model of the 1-D heat transfer problem.

$$-\frac{d}{dx}\left(k(x)\frac{dT}{dx}\right) + \frac{2hT}{r} - \frac{2hT_\infty}{r} = 0 \tag{4.136}$$

where the material properties are given by

$$k(x) = \begin{cases} 0.12 & \text{cal/s}\cdot\text{cm}\cdot°\text{C (steel)} & 0 \leq x \leq 50 \text{ cm} \\ 0.92 & \text{cal/s}\cdot\text{cm}\cdot°\text{C (copper)} & 50 \leq x \leq 100 \text{ cm} \end{cases} \tag{4.137}$$

and $h = 3.75 \times 10^{-5}$ cal/s· cm·°C. The boundary conditions are given by

$$q_{x=0} = 0.1 \text{ cal/s}\cdot\text{cm}^2 \quad \text{(natural)} \tag{4.138}$$

$$T_{x=100} = 0\,°\text{C} \quad \text{(essential)}. \tag{4.139}$$

Equations (4.136, 4.138, 4.139) are the strong form of the 1-D steady state heat transfer problem.

Step 3: Weak Form

For heat transfer problems, the weak form can be obtained from either the variational principle or the Galerkin weighted residual method.

Variational Principle

From the variational calculus, the functional statement of the strong form can be written as (see the discussions on variational calculus in Chapter 2)

$$\Pi = \frac{1}{2}\int_0^{100cm} k\left(\frac{dT}{dx}\right)^2 dx + \frac{1}{2}\int_0^{100cm} \frac{2h}{r}T^2 dx - \int_0^{100cm} \frac{2h}{r}T_\infty T dx - T(0)q(0). \tag{4.140}$$

From the stationary condition

$$\delta\Pi = 0 \tag{4.141}$$

we obtain

$$\int_0^{100cm} k\left(\frac{d\delta T}{dx}\right)\left(\frac{dT}{dx}\right)dx + \int_0^{100cm} \frac{2h}{r}T\delta T dx - \int_0^{100cm} \frac{2hT_\infty}{r}\delta T dx$$
$$-\delta T(0)q(0) = 0. \quad (4.142)$$

Equation (4.142) is the weak form for the 1-D steady state heat transfer problem.

Galerkin Weighted Residual Method
The weak form given in Eq. (4.142) can be obtained by using the Galerkin weighted residual method as discussed in 1-D elasticity (Section 4.3.1). Multiplying an arbitrary continuous weight function $w(x)$ to the left hand side (residual) of the strong form Eq. (4.136), and integrating over the 1-D computational domain, we obtain

$$\int_0^{100cm} w(x)\left[-\frac{d}{dx}\left(k(x)\frac{dT}{dx}\right) + \frac{2hT}{r} - \frac{2hT_\infty}{r}\right]dx = 0. \quad (4.143)$$

Applying integration by parts to the second derivative term of the integral, we have

$$-w(x)k(x)\frac{dT}{dx}\bigg|_0^{100cm} + \int_0^{100cm} \frac{dw}{dx}k(x)\frac{dT}{dx} + \int_0^{100cm} w(x)\frac{2hT}{r}dx$$
$$- \int_0^{100cm} w(x)\frac{2hT_\infty}{r}dx = 0. \quad (4.144)$$

Since $w(x) = 0$ at the right end where the essential boundary condition is given, and writing $w(x)$ and $k(x)$ in short as w and k, respectively, Eq. (4.144) can be written as

$$-w(0)q(0) + \int_0^{100cm} \frac{dw}{dx}k\frac{dT}{dx} + \int_0^{100cm} w\frac{2hT}{r}dx - \int_0^{100cm} w\frac{2hT_\infty}{r}dx = 0. \quad (4.145)$$

Equation (4.145) is the weak form which is equivalent to Eq. (4.142). The two weak form expressions become identical if the weight function is chosen to be the variation of temperature.

Step 4: Discretization
For illustration purposes, the entire 1-D domain of the heat transfer is divided into four equal-length elements. Each element has a length of 25 cm. As shown in Fig. 4.21, if linear elements are used, there are five nodes in total. Figure 4.22 shows a discretization with five quadratic elements. There is an interior node in each quadratic element, making the total number of nodes nine. We continue our solution to the given heat transfer problem by using linear elements. The solution of the problem using quadratic elements is left to the reader as an exercise.

Step 5: Approximate the Temperature over the Elements
As discussed in the 1-D elastic bar example, the unknown quantity, which is the temperature in this case, is approximated as polynomial functions over each element. This approximation is carried out by using the Lagrangian interpolation through the temperatures at the nodes. The temperatures at the nodes are referred to as nodal temperatures. For a linear element, the polynomial approximation of the temperature is a

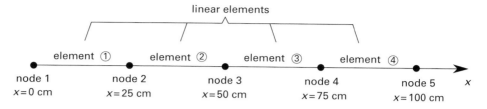

Figure 4.21 Discretization of the 1-D rod into four linear elements.

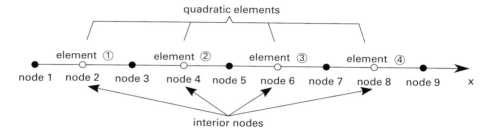

Figure 4.22 Discretization of the 1-D rod into four quadratic elements.

linear function, which can be uniquely determined by using the nodal temperatures. For a linear element shown in Fig. 4.21

$$T(x) = \sum_{i=1}^{2} N_i(x) T_i$$
$$= N_1(x) T_1 + N_2(x) T_2$$
$$= \frac{(x - x_2)}{(x_1 - x_2)} T_1 + \frac{(x - x_1)}{(x_2 - x_1)} T_2$$
$$= \frac{x - x_2}{-25 \text{ cm}} T_1 + \frac{x - x_1}{25 \text{ cm}} T_2. \quad (4.146)$$

For a quadratic element,

$$T(x) = \sum_{i=1}^{3} N_i(x) T_i$$
$$= N_1(x) T_1 + N_2(x) T_2 + N_3(x) T_3$$
$$= \frac{(x - x_2)(x - x_3)}{(x_1 - x_2)(x_1 - x_3)} T_1 + \frac{(x - x_1)(x - x_3)}{(x_2 - x_1)(x_2 - x_3)} T_2 + \frac{(x - x_1)(x - x_2)}{(x_3 - x_1)(x_3 - x_2)} T_3. \quad (4.147)$$

In the Galerkin weighted residual method, the weight function in the weak form is equivalent to the variation of the temperature which is approximated the same way the temperature is approximated. That is,

$$w(x) = \delta T(x) = N_1(x) \delta T_1 + N_2(x) \delta T_2 + \cdots = \sum_{i=1}^{n} N_i(x) \delta T_i \quad (4.148)$$

where δT_i is the variation of the temperature T at node i, n is the number of nodes in the element: $n = 2$ for a linear element and $n = 3$ for a quadratic element, $N_i(x)$, $i = 1, 2, \ldots, n$, are the shape functions the same as those in Eq. (4.146) and Eq. (4.147), respectively.

Step 6: Discretize the Weak Form over the Elements

Based on the spatial discretization, the integrals of the weak form (Eq. (4.142) or Eq. (4.145)) can be decomposed into element integrals as

$$\int_0^{100cm} f(x)dx = \int_0^{25cm} f(x)dx + \int_{25cm}^{50cm} f(x)dx + \int_{50cm}^{75cm} f(x)dx + \int_{75cm}^{100cm} f(x)dx. \quad (4.149)$$

For a given element e, denoted as "$^{(e)}$", the element matrices and vectors can be obtained by substituting the finite element approximation into the element integrals. The first integral on the right hand side of the weak form Eq. (4.142) can be rewritten as

$$\int^{(e)} \left(\sum_{i=1}^n N_i^{'(e)}(x) \delta T_i^{(e)} \right) k \left(\sum_{i=1}^n N_i^{'(e)}(x) T_i^{(e)} \right) dx \quad (4.150)$$

or written in matrix form

$$\int^{(e)} \begin{bmatrix} \delta T_1^{(e)} & \delta T_2^{(e)} & \ldots & \delta T_n^{(e)} \end{bmatrix} \begin{Bmatrix} N_1^{'(e)}(x) \\ N_2^{'(e)}(x) \\ \vdots \\ N_n^{'(e)}(x) \end{Bmatrix} k \begin{bmatrix} N_1^{'(e)}(x) & N_2^{'(e)}(x) & \ldots & N_n^{'(e)}(x) \end{bmatrix} \begin{Bmatrix} T_1^{(e)} \\ T_2^{(e)} \\ \vdots \\ T_n^{(e)} \end{Bmatrix} dx$$

$$= \begin{bmatrix} \delta T_1^{(e)} & \delta T_2^{(e)} & \ldots & \delta T_n^{(e)} \end{bmatrix} \int^{(e)} \begin{Bmatrix} N_1^{'(e)}(x) \\ N_2^{'(e)}(x) \\ \vdots \\ N_n^{'(e)}(x) \end{Bmatrix} k \begin{bmatrix} N_1^{'(e)}(x) & N_2^{'(e)}(x) & \ldots & N_n^{'(e)}(x) \end{bmatrix} dx \begin{Bmatrix} T_1^{(e)} \\ T_2^{(e)} \\ \vdots \\ T_n^{(e)} \end{Bmatrix}$$

$$= \begin{bmatrix} \delta T_1^{(e)} & \delta T_2^{(e)} & \ldots & \delta T_n^{(e)} \end{bmatrix} \mathbf{K}_c^{(e)} \begin{Bmatrix} T_1^{(e)} \\ T_2^{(e)} \\ \vdots \\ T_n^{(e)} \end{Bmatrix}. \quad (4.151)$$

Note that the vectors of δT and T are taken out of the integral since they are not functions of x. The vectors inside the integral multiply out to be a matrix, denoted as $\mathbf{K}_c^{(e)}$,

$$\mathbf{K}_c^{(e)} = \int^{(e)} k \begin{bmatrix} N_1'(x)N_1'(x) & N_1'(x)N_2'(x) & \cdots & N_1'(x)N_n'(x) \\ N_2'(x)N_1'(x) & N_2'(x)N_2'(x) & \cdots & N_2'(x)N_n'(x) \\ \vdots & \vdots & \ddots & \vdots \\ N_n'(x)N_1'(x) & N_n'(x)N_2'(x) & \cdots & N_n'(x)N_n'(x) \end{bmatrix} dx$$

$$= \begin{bmatrix} \int kN_1'(x)N_1'(x)dx & \int kN_1'(x)N_2'(x)dx & \cdots & \int kN_1'(x)N_n'(x)dx \\ \int kN_2'(x)N_1'(x)dx & \int kN_2'(x)N_2'(x)dx & \cdots & \int kN_2'(x)N_n'(x)dx \\ \vdots & \vdots & \ddots & \vdots \\ \int kN_n'(x)N_1'(x)dx & \int kN_n'(x)N_2'(x)dx & \cdots & \int kN_n'(x)N_n'(x)dx \end{bmatrix}^{(e)}.$$

(4.152)

Note that, for simplicity, the superscript (e) of the individual element shape functions, their derivatives, and the element integrals are omitted in the matrix expressions in Eq. (4.152). Instead, we just use a single superscript (e) for the entire matrix to denote that the functions are element-wise. Equation (4.152) is a general expression of the element conductivity matrix. The dimension of the matrix, n, and the expression of the derivatives of the shape functions depend on the order of the element. For linear elements, $n = 2$ and the matrix is 2×2. For quadratic elements, $n = 3$ and the matrix is 3×3. For the 1-D heat transfer problem with four linear elements, $\mathbf{K}_c^{(e)}$, e=1, 2, 3, 4, are computed as

$$\mathbf{K}_c^{①} = \begin{bmatrix} 0.0368 & -0.0368 \\ -0.0368 & 0.0368 \end{bmatrix} \quad \mathbf{K}_c^{②} = \begin{bmatrix} 0.0368 & -0.0368 \\ -0.0368 & 0.0368 \end{bmatrix}$$

$$\mathbf{K}_c^{③} = \begin{bmatrix} 0.0048 & -0.0048 \\ -0.0048 & 0.0048 \end{bmatrix} \quad \mathbf{K}_c^{④} = \begin{bmatrix} 0.0048 & -0.0048 \\ -0.0048 & 0.0048 \end{bmatrix}. \quad (4.153)$$

Next, substituting the finite element approximations of the temperature T and its variation δT into the second integral on the left hand side of the weak form Eq. (4.142), we have

$$\int^{(e)} \left(\sum_{i=1}^{n} N_i^{(e)}(x) \delta T_i^{(e)} \right) k \left(\sum_{i=1}^{n} N_i^{(e)}(x) T_i^{(e)} \right) dx. \quad (4.154)$$

By following the same procedure illustrated through Eqs. (4.150–4.152), the element temperature matrix is written as

$$\mathbf{K}_t^{(e)} = \begin{bmatrix} \int \hbar N_1(x)N_1(x)dx & \int \hbar N_1(x)N_2(x)dx & \cdots & \int \hbar N_1(x)N_n(x)dx \\ \int \hbar N_2(x)N_1(x)dx & \int \hbar N_2(x)N_2(x)dx & \cdots & \int \hbar N_2(x)N_n(x)dx \\ \vdots & \vdots & \ddots & \vdots \\ \int \hbar N_n(x)N_1(x)dx & \int \hbar N_n(x)N_2(x)dx & \cdots & \int \hbar N_n(x)N_n(x)dx \end{bmatrix}^{(e)}$$

(4.155)

where $\hbar = 2h/r$. For the 1-D heat transfer example, $\mathbf{K}_t^{(e)}$, e=1, 2, 3, 4 are computed as

$$\mathbf{K}_t^{①} = \mathbf{K}_t^{②} = \mathbf{K}_t^{③} = \mathbf{K}_t^{④} = \begin{bmatrix} 0.00125 & 0.000625 \\ 0.000625 & 0.00125 \end{bmatrix}. \quad (4.156)$$

Note that the convection coefficient and radius of the rod are all constant in this case. In addition, the length of the elements are identical. Therefore, the integrals of the element matrix for temperature have the same value for all the elements. Written in matrix form, it is easy to see that the first and second integrals on the left hand side of the weak form can be combined:

$$\begin{bmatrix} \delta T_1^{(e)} & \delta T_2^{(e)} & \ldots & \delta T_n^{(e)} \end{bmatrix} \mathbf{K}_c^{(e)} \begin{Bmatrix} T_1^{(e)} \\ T_2^{(e)} \\ \vdots \\ T_n^{(e)} \end{Bmatrix} + \begin{bmatrix} \delta T_1^{(e)} & \delta T_2^{(e)} & \ldots & \delta T_n^{(e)} \end{bmatrix} \mathbf{K}_t^{(e)} \begin{Bmatrix} T_1^{(e)} \\ T_2^{(e)} \\ \vdots \\ T_n^{(e)} \end{Bmatrix}$$

$$= \begin{bmatrix} \delta T_1^{(e)} & \delta T_2^{(e)} & \ldots & \delta T_n^{(e)} \end{bmatrix} \left(\mathbf{K}_c^{(e)} + \mathbf{K}_t^{(e)} \right) \begin{Bmatrix} T_1^{(e)} \\ T_2^{(e)} \\ \vdots \\ T_n^{(e)} \end{Bmatrix}$$

$$= \begin{bmatrix} \delta T_1^{(e)} & \delta T_2^{(e)} & \ldots & \delta T_n^{(e)} \end{bmatrix} \mathbf{K}^{(e)} \begin{Bmatrix} T_1^{(e)} \\ T_2^{(e)} \\ \vdots \\ T_n^{(e)} \end{Bmatrix}. \tag{4.157}$$

For the 1-D heat transfer example, $\mathbf{K}^{(e)}$, e=1, 2, 3, 4 are computed as

$$\mathbf{K}^{①} = \begin{bmatrix} 0.03805 & -0.036175 \\ -0.036175 & 0.03805 \end{bmatrix} \quad \mathbf{K}^{②} = \begin{bmatrix} 0.03805 & -0.036175 \\ -0.036175 & 0.03805 \end{bmatrix}$$

$$\mathbf{K}^{③} = \begin{bmatrix} 0.00605 & -0.004175 \\ -0.004175 & 0.00605 \end{bmatrix} \quad \mathbf{K}^{④} = \begin{bmatrix} 0.00605 & -0.004175 \\ -0.004175 & 0.00605 \end{bmatrix}. \tag{4.158}$$

The last integral of the weak form Eq. (4.142), $\int_0^{100 \text{cm}} \delta T \frac{2h T_\infty}{r} dx$, does not contain the unknown variable T, and only the variation of T is unknown. Therefore, substituting the approximation of the weight function into the integral produces a vector instead of a matrix. The vector can be obtained as

$$\int^{(e)} \left(\sum_{i=1}^{n} N_i^{(e)}(x) \delta T_i^{(e)} \right) \frac{h T_\infty}{r} dx$$

$$= \begin{bmatrix} \delta T_1^{(e)} & \delta T_2^{(e)} & \ldots & \delta T_n^{(e)} \end{bmatrix} \int^{(e)} \begin{Bmatrix} N_1(x) \\ N_2(x) \\ \vdots \\ N_n(x) \end{Bmatrix}^{(e)} \frac{h T_\infty}{r} dx$$

$$= \begin{bmatrix} \delta T_1^{(e)} & \delta T_2^{(e)} & \ldots & \delta T_n^{(e)} \end{bmatrix} \begin{Bmatrix} \int N_1(x)\frac{hT_\infty}{r}dx \\ \int N_2(x)\frac{hT_\infty}{r}dx \\ \vdots \\ \int N_n(x)\frac{hT_\infty}{r}dx \end{Bmatrix}^{(e)}$$

$$= \left(\delta \mathbf{T}^{(e)}\right)^T \mathbf{f}^{(e)}. \tag{4.159}$$

For the 1-D heat transfer example, $\mathbf{f}^{(e)}$, e=1, 2, 3, 4 are computed as

$$\mathbf{f}^{(1)} = \mathbf{f}^{(2)} = \mathbf{f}^{(3)} = \mathbf{f}^{(4)} = \begin{Bmatrix} 0.0375 \\ 0.0375 \end{Bmatrix}. \tag{4.160}$$

Next, the last term in the weak form Eq. (4.142), which is the term containing the natural (heat flux) boundary condition of the problem, is not an integral. However, this term can also be worked out by substituting the finite element approximation of $\delta T = \sum_{i=1}^{n} N_i(x)\delta T_i$. For a linear element, the term can then be written as

$$\begin{bmatrix} \delta T_1^{①} & \delta T_2^{①} \end{bmatrix} \begin{Bmatrix} N_1^{①}(0) \\ N_2^{①}(0) \end{Bmatrix} q(0)$$

$$= \begin{bmatrix} \delta T_1^{①} & \delta T_2^{①} \end{bmatrix} \begin{Bmatrix} 1 \\ 0 \end{Bmatrix} 0.1$$

$$= \begin{bmatrix} \delta T_1^{①} & \delta T_2^{①} \end{bmatrix} \begin{Bmatrix} 0.1 \\ 0 \end{Bmatrix}. \tag{4.161}$$

In Fig. 4.19, it is shown that the heat flux boundary condition is at the left end of the rod which belongs to the first element. In fact the location is at the first node (node 1) of element ①. This is the reason why Eq. (4.161) for this term is applied to element ① and the shape functions $N_1^{①}(0)$ and $N_2^{①}(0)$ are evaluated at $x=0$ in the element's local coordinate system. From the expression of the shape functions, it can be shown that $N_1^{①}(0) = 1$ and $N_2^{①}(0) = 0$.

Step 7: Assemble the Global System Equations

As illustrated in the elastic bar deformation case study, for each element, the element matrices and vectors can be expanded into a global matrix, and then these globalized matrices and vectors can be summed up to form a linear system. While such an approach shows the mathematical process of assembling the element matrices and vectors together, it is not efficient from an implementation point of view. Having understood the underlying logic of the so called assembly procedure, one can directly "assemble" the computed element matrices and vectors into the coefficient matrix and right hand side vector of the global linear system.

Following the assembly procedure illustrated in Example 4.1, the assembled matrix for the 1-D heat transfer example is obtained to be

$$\begin{bmatrix} 0.0381 & -0.0362 & 0 & 0 & 0 \\ -0.0362 & 0.0761 & -0.0362 & 0 & 0 \\ 0 & -0.0362 & 0.0441 & -0.0042 & 0 \\ 0 & 0 & -0.0042 & 0.0121 & -0.0042 \\ 0 & 0 & 0 & -0.0042 & 0.0060 \end{bmatrix} \begin{Bmatrix} T_1 \\ T_2 \\ T_3 \\ T_4 \\ T_5 \end{Bmatrix} = \begin{Bmatrix} 0.0375 \\ 0.0750 \\ 0.0750 \\ 0.0750 \\ 0.0375 \end{Bmatrix}. \quad (4.162)$$

Step 8: Apply the Loads and Boundary Conditions

For the 1-D heat transfer problem, there are two types of basic loads: concentrated heat flux and distributed heat flux. When a concentrated heat flux is applied at an end of the domain, it becomes a natural boundary condition, which is enforced directly in the weak form. The distributed heat flux is typically defined in the strong form as shown Eq. (4.136), therefore, the distributed heat flux is naturally included in the weak form and further goes into the element vectors in Step 6. As such, for this heat transfer in 1-D rod example, the convective heat loss load (distributed out-going heat flux) is included in the weak form. The heat flux at the right end is a natural boundary condition which is also included in the weak form in Eq. (4.142) (last term in the equation).

In the 1-D heat transfer problem, there are two boundary conditions: (1) heat flux at the left end and (2) temperature at the right end. As discussed above, the heat flux boundary condition has been included in the weak form. The temperature boundary condition is an essential boundary condition. That means one can apply any of the boundary condition application methods discussed in Section 4.3.1. For this example, the penalty method is used. Assuming the penalty number is taken as 10^7, the final linear system is obtained as

$$\begin{bmatrix} 0.0381 & -0.0362 & 0 & 0 & 0 \\ -0.0362 & 0.0761 & -0.0362 & 0 & 0 \\ 0 & -0.0362 & 0.0441 & -0.0042 & 0 \\ 0 & 0 & -0.0042 & 0.0121 & -0.0042 \\ 0 & 0 & 0 & -0.0042 & 10^7 \end{bmatrix} \begin{Bmatrix} T_1 \\ T_2 \\ T_3 \\ T_4 \\ T_5 \end{Bmatrix} = \begin{Bmatrix} 0.0375 \\ 0.0750 \\ 0.0750 \\ 0.0750 \\ 0 \times 10^7 \end{Bmatrix}. \quad (4.163)$$

Step 9: Solve the Global System

Equation (4.163) is solved using MATLAB's internal linear solver. The solution of the linear system is obtained as

$$\begin{Bmatrix} T_1 \\ T_2 \\ T_3 \\ T_4 \\ T_5 \end{Bmatrix} = \begin{Bmatrix} 28.7162 \\ 26.4036 \\ 24.7549 \\ 14.7398 \\ 0.0000 \end{Bmatrix} °C. \quad (4.164)$$

Table 4.4 Data files for analysis

nodes.dat		elements.dat			bcs.dat			elprops.dat	
1	0	1	1	2	1	2	0.1	1	0.92
2	25	2	2	3	2	0	0	2	0.92
3	50	3	3	4	3	0	0	3	0.12
4	75	4	4	5	4	0	0	4	0.12
5	100				5	1	0		

Step 10: Perform Post-processing

As discussed before, the numerical results are visualized in a meaningful way for the analysis of the behavior of the physical system. Physical quantities of interest (e.g. heat flux) are computed using the nodal temperature results. Figure 4.23 shows the plots of the temperature and heat flux profiles calculated from the finite element analysis.

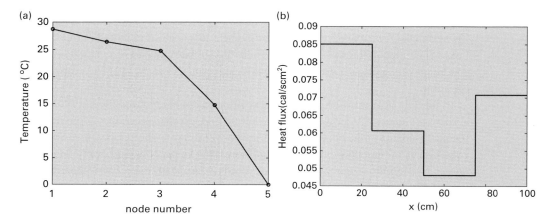

Figure 4.23 Steady state heat transfer in a 1-D bar: (a) temperature profile; (b) heat flux distribution.

4.4.2 Computer Implementation

Data Files

The input data files for the analysis are listed below. "nodes.dat" and "elements.dat" are the standard files with the format as explained in Section 4.3.2. "bcs.dat" has the boundary conditions on the end nodes. "elprops.dat" contains the element-wise properties. In this case, the property is the thermal conductivity. "globals.dat" contains the environment temperature, convection coefficient, and the radius of the rod.

MATLAB Code

The MATLAB programs for solving the 1-D steady state heat transfer in a rod are listed below. The code for the heat transfer analysis is quite similar to the one for the

4.4 1-D Steady State Heat Transfer

---
globals.dat

% global parameters
% Environment temperature (C)
20
% radius (cm)
0.5
% convection coefficient (cal/scmC)
3.75e-5

---

elastic bar analysis. Attention should be drawn to the differences and similarities of the two codes.

```
1  clear;
2  % Input elements, nodes, BCs and material properties
3  % (from steps 2, 3, 4 and 5)
4  load elements.dat;      % load elements
5  load nodes.dat;         % load nodes
6  load globals.dat;       % load global properties
7  load bcs.dat;           % load boundary conditions
8  load elprops.dat;       % load element properties: conductivity
9  hlA=2*globals(3)/globals(2);  % define 2h/r
10 hlAT_inf=hlA*globals(1);      % define 2hT_inf/r
11 n_elements=size(elements,1);  % number of elements
12 n_nodes=size(nodes,1);        % number of nodes
13 K=zeros(n_nodes,n_nodes);     % Set up empty K matrix
14 F=zeros(n_nodes,1);           % Set up empty RHS F vector
15 n_gauss_points=2;             % number of Gauss points
16 [gauss_xi, gauss_w] = Get1DGauss(n_gauss_points); % Get Gauss points
17 % N, Nxi: set up empty matrices storing the value of the shape
18 % functions N and their derivatives Nxi (dN/dxi) evaluated at the
19 % Gauss points in the master element
20 N=zeros(2,n_gauss_points);
21 Nxi=zeros(2,n_gauss_points);
22 % Compute the shape functions and their derivatives at Gauss points
23 [N,Nxi]=CompShapeLinear1D(N, Nxi, gauss_xi);
24 % The for loop: compute element matrices and vectors and assemble them
25 % into the global matrix and RHS vector (Steps 6,7)
26 for i=1:n_elements  %loop over elements
27    start_x= nodes(elements(i,2),2);  % x-coordinate of the start node
28    end_x= nodes(elements(i,3),2);    % x-coordinate of the end node
29    jacobian=CompJacobian1D([start_x end_x]', Nxi);
30    f=hlAT_inf * N*(gauss_w.*jacobian);
31    k=zeros(2,2); kt=zeros(2,2);   % Set up empty element matrices
32    % for loop: compute k and kt matrices
33    for m=1:2
34       for n=1:2
```

```
35      for g=1:n_gauss_points
36        k(m,n)=k(m,n) +elprops(i,2) ...
37            *Nxi(m,g)*Nxi(n,g)*gauss_w(g)/jacobian(g);
38        kt(m,n)=kt(m,n)+ hlA*N(m,g)*N(n,g)*gauss_w(g)*jacobian(g);
39      end
40      end
41    end
42    [K,F]=AssembleGlobalLinear1D(K, F, i, elements, k+kt, f);
43  end
44  % Apply loads and boundary conditions (step 8)
45  for i=1:n_nodes-1:n_nodes
46    if bcs(i,2)==1
47      penalty= abs(K(i,i)+1)*1e7;
48      K(i,i)=penalty;
49      F(i)=bcs(i,3)*penalty;
50    end
51    if bcs(i,2)==2
52      F(i)=F(i)+bcs(i,3);
53    end
54  end
55  T_nodes=K\F % solve the global system equations (step 9)
56  % Post-processing (step 10)
57  figure(1);
58  plot(nodes(:,1), T_nodes,'b-o','linewidth',2); % plot temperature
59  set(gca,'fontsize',16);
60  xlabel('node number','fontsize',18);
61  ylabel('Temperature (^oC)','fontsize',18);
62
63  k=1; % set up index of the points for computing heat flux
64  for i=1:n_elements        % loop over the elements
65    node1=elements(i,2); % global node number of the local node 1
66    node2=elements(i,3); % global node number of the local node 2
67    for x=-1:0.1:1
68      [N,Nxi]=CompShapeLinear1D(zeros(2,1),zeros(2,1),x);
69      J=CompJacobian1D([nodes(node1,2) nodes(node2,2)]',Nxi); % Jacobian
70      Tx=T_nodes(node1)*Nxi(1)/J+T_nodes(node2)*Nxi(2)/J; % dT(x)/dx
71      re(k,1)=nodes(node1,2)*N(1)+nodes(node2,2)*N(2);
72      re(k,2)=-elprops(i,2)*Tx;
73      k=k+1;
74    end
75  end
76  % plot heat flux
77  figure(2);
78  plot(re(:,1), re(:,2) ,'r-','linewidth',2);
79  set(gca, 'fontsize',16);
80  xlabel('x (cm)','fontsize',18);
81  ylabel('Heat flux','fontsize',18);
```

In the MATLAB programs listed above, the functions which have already been discussed previously are not repeated here. They are listed below for your reference:
function [gauss_xi, gauss_w] = Get1DGauss(n_gauss_points): Section 3.4.
function [N,Nx]=CompShapeLinear1D(N, Nx, xi_vector): Section 4.3.2.
function J=CompJacobian1D(x_vector,Nx): Section 4.3.2.
function [K,F]=AssembleGlobalLinear1D(K, F, eid, elements, k, f): Section 4.3.2.

4.5 1-D Fluid Flow

A Newtonian fluid flow can be described by the general Navier–Stokes equations, which is a set of coupled highly nonlinear partial differential equations (PDEs). However, in many cases, a fluid flow can be described by using simplified models depending on its material properties and flow conditions. Figure 4.24 shows a classification of fluid flow according to the viscosity, compressibility, rotationality (vorticity), and time dependency, and the governing equations corresponding to each type. It is shown in the figure that, depending on the fluid and flow conditions, the complex Navier–Stokes equations can be reduced to partial differential equations of simpler forms. In this section, we use an advection–diffusion problem to illustrate the solution procedure of a mass transporting fluid flow in one dimension. More fluid dynamics problems will be discussed in Chapter 7.

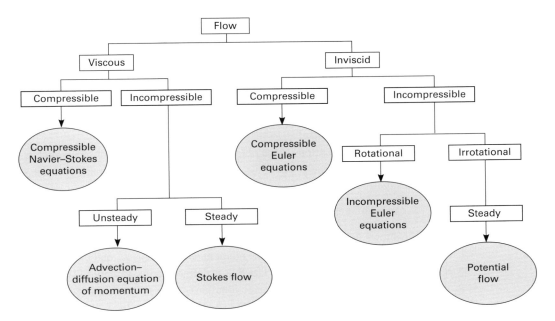

Figure 4.24 Fluid flow classification and governing equations.

4.5.1 Steady State Mass Transport by 1-D Fluid Flow

As shown in Fig. 4.25, an underground water stream of 1 km in length flows slowly in the x-direction through a porous layer of rocks at a velocity of 10^{-7} m/s. Along the length of the stream, there is a source of a contaminant continuously generating the substance at a rate of 4×10^{-10} mol/m³·s. At the right end, the underground stream flows into a large lake where the contaminant is diluted and its concentration becomes zero. At the left end, the concentration of the contaminant is measured to be 0.01 mol/m³. Determine the contaminant concentration and flux variation profiles along the length of the stream. It is known that the diffusivity of the contaminant in water is 10^{-3} m²/s.

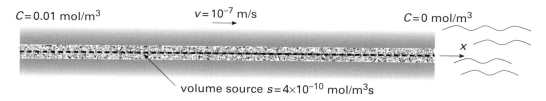

Figure 4.25 Contaminant concentration in an underground water stream.

Step 1: Physical Model

The situation of the problem is assumed to be fully developed and steady state. There are two transport mechanisms in the water stream. First, the solute (contaminant) is moved by the flow with a velocity, which is referred to as advection. Second, the solute is spread via molecular dispersion, which is referred to as diffusion. Assuming the contaminant concentration is uniform on any lateral cross section of the steam, this problem can be regarded as a one-dimensional steady state advection–diffusion problem.

Step 2: Mathematical Model

The advection–diffusion equation can be obtained by using the mass conservation law and Fick's first law of diffusion. The mass conservation law in the context of mass transport states that

$$\text{mass in} + \text{mass generated} - \text{mass out} = \text{mass increase.} \quad (4.165)$$

Figure 4.26 shows the flow passing through in an infinitesimal element of the channel. The incoming flux at the left face of the element is denoted as J_{in} and the outgoing flux at the right face is denoted as J_{out}. The flux is defined as the amount of the solute (e.g., mole) passing through a unit cross sectional area (m²) per unit time (s). Therefore, the unit of the flux is mol/m²·s. $s(x)$ is the source generating the solute measured as the amount generated per unit volume per unit time (mol/m³·s). The incoming flux J_{in} contains two components: an advection component and a diffusion component, i.e.,

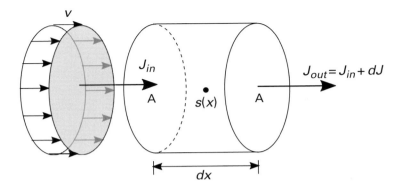

Figure 4.26 Flow in an infinitesimal element.

$$J_{in} = vC + q \quad (4.166)$$

where vC is the advection component, v is the velocity (m/s), C is the concentration of solute (mol/m^3), q is the flux due to diffusion. Therefore, the solute flowing into the element during a time period dt can be written as

$$\text{mass in} = (vC + q)Adt \quad (4.167)$$

where A is the cross sectional area of the channel. The outgoing flux can then be written as

$$J_{out} = J_{in} + d(vC) + dq \quad (4.168)$$

where $d(vC)$ and dq are the incremental change of the advective and diffusive fluxes, respectively. The solute flowing out can be written as

$$\text{mass out} = \left[vC + d(vC) + q + dq\right]Adt. \quad (4.169)$$

The solute generated inside the element is simply $sAdxdt$. The amount of increase of the solute in the element during dt is given by $dCAdx$, where dC is the increase of solute per unit volume. Substituting Eqs. (4.167, 4.169) into the mass conservation law gives

$$(vC + q)Adt + sAdxdt - (vC + d(vC) + q + dq)Adt = dCAdx. \quad (4.170)$$

Dividing all the terms by $Adxdt$, we obtain

$$s - \frac{d(vC)}{dx} - \frac{dq}{dx} = \frac{dC}{dt}. \quad (4.171)$$

Next we employ Fick's first law of diffusion given by

$$q = -D\frac{dC}{dx} \quad (4.172)$$

where D is the diffusivity (m²/s). Noting the concentration is a multi-variable function of x and t, Eq. (4.171) should be rewritten as

$$\frac{\partial}{\partial x}\left(D\frac{\partial C}{\partial x}\right) - \frac{\partial (vC)}{\partial x} + s = \frac{\partial C}{\partial t}. \tag{4.173}$$

If the flow is steady state, then $\frac{dC}{dt} = 0$ and C is function of x only. Equation (4.173) becomes

$$\frac{d}{dx}\left(D\frac{dC}{dx}\right) - \frac{d(vC)}{dx} + s = 0. \tag{4.174}$$

For incompressible fluid flow, not only is the mass of the solute conserved, the mass of fluid itself is also conserved, i.e.,

$$\frac{dv}{dx} = 0. \tag{4.175}$$

Using Eq. (4.175), Eq. (4.174) can be further written as (Fish and Belytschko 2007)

$$\frac{d}{dx}\left(D\frac{dC}{dx}\right) - v\frac{dC}{dx} + s = 0. \tag{4.176}$$

For the 1-D advection–diffusion model problem, the boundary conditions are given by

$$C_{x=0} = 0.01 \text{ mol/m}^3 \quad \text{(essential)} \tag{4.177}$$

$$C_{x=1000m} = 0 \text{ mol/m}^3 \quad \text{(essential).} \tag{4.178}$$

Equations (4.176, 4.177, 4.178) are the strong form of the 1-D steady state advection–diffusion problem.

Step 3: Weak Form

We use the Galerkin weighted residual method to obtain the weak form of Eq. (4.176). As already discussed in Section 4.3.1, we multiply the weight function $w(x)$ to the left hand side (residual) of Eq. (4.176), and integrate over the 1-D computational domain, to obtain

$$\int_0^{1km} w(x)\left[\frac{d}{dx}\left(D\frac{dC}{dx}\right)dx - v\frac{dC}{dx} + s\right]dx = 0. \tag{4.179}$$

Applying integration by parts to the second derivative term of the integral, we have

$$w(x)D(x)\frac{dC}{dx}\bigg|_0^{1km} - \int_0^{1km}\frac{dw}{dx}D(x)\frac{dC}{dx}dx - \int_0^{1km}wv\frac{dC}{dx}dx$$
$$+ \int_0^{1km}wsdx = 0. \tag{4.180}$$

Since $w(x) = 0$ at both the left and right ends where the essential boundary conditions are given, Eq. (4.180) can be written as

$$\int_0^{1km}\frac{dw}{dx}D(x)\frac{dC}{dx}dx + \int_0^{1km}wv\frac{dC}{dx}dx - \int_0^{1km}wsdx = 0. \tag{4.181}$$

4.5 1-D Fluid Flow

Equation (4.181) is the weak form of the 1-D steady state advection–diffusion equation.

Step 4: Discretization
For illustration purposes, the 1-D domain is discretized into four equal-length elements. Each element has a length of 250 m. In this example, linear elements are used. As shown in Fig. 4.27, there are five nodes in total.

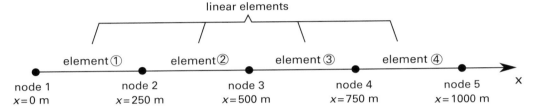

Figure 4.27 Discretization of the 1-D domain into four linear elements.

Step 5: Approximate Solute Concentration over the Elements
As linear elements are used in the discretization step, the unknown quantity, solute concentration, is approximated using linear shape functions as

$$c^{(e)}(x) = N_1^{(e)}(x)C_1^{(e)} + N_2^{(e)}(x)C_2^{(e)}$$
$$= \frac{(x-x_2)}{(x_1-x_2)}C_1^{(e)} + \frac{(x-x_1)}{(x_2-x_1)}C_2^{(e)}. \quad (4.182)$$

where x_1 and x_2 are the x-coordinates of the two nodes of the element $^{(e)}$. The weight function in the weak form is approximated the same way the solute concentration is approximated. That is,

$$w(x) = N_1^{(e)}(x)w_1^{(e)} + N_2^{(e)}(x)w_2^{(e)}. \quad (4.183)$$

The shape functions are the same as those in Eq. (4.182).

Step 6: Discretize the Weak Form over the Elements
According to the spatial discretization, the integrals of the weak form (Eq. (4.181)) can be decomposed into element integrals. That is,

$$\int_0^{1000m} f(x)dx$$
$$= \int_0^{250m} f(x)dx + \int_{250m}^{500m} f(x)dx + \int_{500m}^{750m} f(x)dx + \int_{750m}^{1000m} f(x)dx. \quad (4.184)$$

For a given element ⓔ, substituting the finite element approximation into the element integrals, the element matrices and vectors can be obtained. The first integral on the right hand side of the weak form Eq. (4.181) can be rewritten as

$$\int^{(e)} \left(\sum_{i=1}^{2} N_i^{'(e)}(x) \delta C_i^{(e)} \right) D \left(\sum_{i=1}^{2} N_i^{'(e)}(x) C_i^{(e)} \right) dx \qquad (4.185)$$

or written in matrix form

$$\int^{(e)} [w_1 \ w_2]^{(e)} \begin{Bmatrix} N_1'(x) \\ N_2'(x) \end{Bmatrix}^{(e)} D \left[N_1'(x) N_2'(x) \right]^{(e)} \begin{Bmatrix} C_1 \\ C_2 \end{Bmatrix}^{(e)} dx$$

$$= [w_1 \ w_2]^{(e)} \left\{ \int^{(e)} \begin{Bmatrix} N_1'(x) \\ N_2'(x) \end{Bmatrix}^{(e)} D \left[N_1'(x) N_2'(x) \right]^{(e)} dx \right\} \begin{Bmatrix} C_1 \\ C_2 \end{Bmatrix}^{(e)}$$

$$= [w_1 \ w_2]^{(e)} \mathbf{D}^{(e)} \begin{Bmatrix} C_1 \\ C_2 \end{Bmatrix}^{(e)}. \qquad (4.186)$$

Note that the vectors of w and C are taken out of the integral since they are not functions of x. The vectors inside the integral multiply out to be a matrix, denoted as $\mathbf{D}^{(e)}$,

$$\mathbf{D}^{(e)} = \int^{(e)} D \begin{bmatrix} N_1'(x)N_1'(x) & N_1'(x)N_2'(x) \\ N_2'(x)N_1'(x) & N_2'(x)N_2'(x) \end{bmatrix} dx$$

$$= \begin{bmatrix} \int DN_1'(x)N_1'(x)dx & \int DN_1'(x)N_2'(x)dx \\ \int DN_2'(x)N_1'(x)dx & \int DN_2'(x)N_2'(x)dx \end{bmatrix}^{(e)}. \qquad (4.187)$$

For the 1-D advection–diffusion model problem with four linear elements, $\mathbf{D}^{(e)}$, e=1, 2, 3, 4 are computed as

$$\mathbf{D}^{(1)} = \mathbf{D}^{(2)} = \mathbf{D}^{(3)} = \mathbf{D}^{(4)} = \begin{bmatrix} 4.0 & -4.0 \\ -4.0 & 4.0 \end{bmatrix} \times 10^{-6} \text{ m/s}. \qquad (4.188)$$

Note that since the elements are identical with the same material properties, the diffusivity matrices are the same for all the elements. Next, substituting the finite element approximations of the concentration C and the weight function w, the second integral on the left hand side of the weak form Eq. (4.181) can be rewritten as

$$\int^{(e)} \left(\sum_{i=1}^{n} N_i^{(e)}(x) w_i^{(e)} \right) v \left(\sum_{i=1}^{n} N_i^{'(e)}(x) C_i^{(e)} \right) dx. \qquad (4.189)$$

By following the same procedure illustrated through Eqs. (4.185–4.187), the element advection matrix is written as

$$\mathbf{A}^{(e)} = \begin{bmatrix} \int v N_1(x) N'_1(x) dx & \int v N_1(x) N'_2(x) dx \\ \int v N_2(x) N'_1(x) dx & \int v N_2(x) N'_2(x) dx \end{bmatrix}^{(e)}. \qquad (4.190)$$

where v is constant flow velocity along the 1-D domain. Note that the $\mathbf{A}$ matrix is asymmetric as the integral contains the first derivative of the concentration and there is no differentiation of the weight function. For the 1-D advection–diffusion model problem, $\mathbf{A}^{(e)}$, e=1, 2, 3, 4 are computed as

$$\mathbf{A}^{(1)} = \mathbf{A}^{(2)} = \mathbf{A}^{(3)} = \mathbf{A}^{(4)} = \begin{bmatrix} -0.05 & 0.05 \\ -0.05 & 0.05 \end{bmatrix} \times 10^{-6} \text{ m/s}. \qquad (4.191)$$

Summing the diffusivity matrix and the advection matrix, the combined coefficient matrix of an element is

$$\mathbf{K}^{(e)} = \mathbf{D}^{(e)} + \mathbf{A}^{(e)}. \qquad (4.192)$$

For our example problem,

$$\mathbf{K}^{(1)} = \mathbf{K}^{(2)} = \mathbf{K}^{(3)} = \mathbf{K}^{(4)} = \begin{bmatrix} 3.95 & -3.95 \\ -4.05 & 4.05 \end{bmatrix} \times 10^{-6} \text{ m/s}. \qquad (4.193)$$

Next, substituting the finite element approximation of the weight function into the last integral of the weak form Eq. (4.181), we obtain

$$\int^{(e)} \left(\sum_{i=1}^{2} N_i^{(e)}(x) w_i^{(e)} \right) s dx$$

$$= [w_1 \; w_2]^{(e)} \int^{(e)} \begin{Bmatrix} N_1(x) \\ N_2(x) \end{Bmatrix}^{(e)} s dx$$

$$= [w_1 \; w_2]^{(e)} \begin{Bmatrix} \int N_1(x) s dx \\ \int N_2(x) s dx \end{Bmatrix}^{(e)}$$

$$= \left(\mathbf{w}^{(e)} \right)^T \mathbf{f}^{(e)}. \qquad (4.194)$$

For our 1-D advection–diffusion example, $\mathbf{f}^{(e)}$, e=1, 2, 3, 4 are computed as

$$\mathbf{f}^{(1)} = \mathbf{f}^{(2)} = \mathbf{f}^{(3)} = \mathbf{f}^{(4)} = \begin{Bmatrix} 5 \\ 5 \end{Bmatrix} \times 10^{-8} \text{ m}^2/\text{s}. \qquad (4.195)$$

While the flux boundary condition directly goes into the weak form, in this example, the mass flux is not specified at either end of the domain. Both ends have an essential boundary condition. Therefore, there is no boundary term in Eq. (4.181). If there was a nonzero flux, as shown in the heat transfer case, the boundary flux would be included in the right hand side vector at the position corresponding to the node where the flux is specified.

Step 7: Assemble the Global System Equations

By following the assembly procedure illustrated in Section 4.3, the assembled matrix for the 1-D advection–diffusion example is computed to be

$$\begin{bmatrix} 3.95 & -3.95 & 0 & 0 & 0 \\ -4.05 & 8.00 & -3.95 & 0 & 0 \\ 0 & -4.05 & 8.00 & -3.95 & 0 \\ 0 & 0 & -4.05 & 8.00 & -3.95 \\ 0 & 0 & 0 & -4.05 & 4.05 \end{bmatrix} \times 10^{-6} \begin{Bmatrix} C_1 \\ C_2 \\ C_3 \\ C_4 \\ C_5 \end{Bmatrix} = \begin{Bmatrix} 0.5 \\ 1 \\ 1 \\ 1 \\ 0.5 \end{Bmatrix} \times 10^{-7}. \quad (4.196)$$

Note that the multiplying factors on the two sides of the equation, 10^{-6} and 10^{-7}, can be divided by a common factor of 10^{-6}.

Step 8: Apply the Loads and Essential Boundary Conditions

For the 1-D advection–diffusion problem, the essential boundary conditions are given by Eqs. (4.177, 4.178). For this example, the penalty method illustrated in Section 4.3.1 is used. After transformation, the final linear system is obtained as

$$\begin{bmatrix} 3.95 \times 10^7 & -3.95 & 0 & 0 & 0 \\ -4.05 & 8.00 & -3.95 & 0 & 0 \\ 0 & -4.05 & 8.00 & -3.95 & 0 \\ 0 & 0 & -4.05 & 8.00 & -3.95 \\ 0 & 0 & 0 & -4.05 & 4.05 \times 10^7 \end{bmatrix} \begin{Bmatrix} C_1 \\ C_2 \\ C_3 \\ C_4 \\ C_5 \end{Bmatrix} = \begin{Bmatrix} 3.95 \times 10^5 \\ 0.1 \\ 0.1 \\ 0.1 \\ 0 \end{Bmatrix}. \quad (4.197)$$

Step 9: Solve the Global System

Equation (4.197) is solved using MATLAB's internal linear solver. The solution of the linear system is obtained as

$$\begin{Bmatrix} C_1 \\ C_2 \\ C_3 \\ C_4 \\ C_5 \end{Bmatrix} = \begin{Bmatrix} 0.0100 \\ 0.0448 \\ 0.0551 \\ 0.0404 \\ 0.0000 \end{Bmatrix} \text{mol/m}^3. \quad (4.198)$$

Step 10: Perform Post-processing

After the solute concentration is computed at the nodes, physical quantities of interest (e.g., solute flux) are computed using the nodal concentration results. Figure 4.28 shows the plots of the concentration and flux profiles calculated from the finite element analysis.

4.5.2 Computer Implementation

Data Files

The input data files for the analysis are listed below. "nodes.dat" and "elements.dat" are the standard files with standard format as discussed in the 1-D elastic bar and

Table 4.5 Data files for analysis

nodes.dat		elements.dat			bcs.dat			bfs.dat	
1	0	1	1	2	1	1	0.01	1	4e − 10
2	250	2	2	3	2	0	0	2	4e − 10
3	500	3	3	4	3	0	0	3	4e − 10
4	750	4	4	5	4	0	0	4	4e − 10
5	1000				5	1	0	5	4e − 10

globals.dat

% global parameters
% velocity (m/s)
1e-7
% diffusivity (m²/s)
1e-3

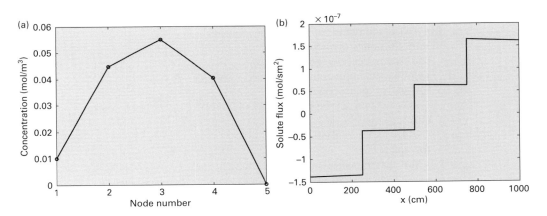

Figure 4.28 Steady state 1-D advection–diffusion of solute: (a) solute concentration profile; (b) solute flux profile.

1-D heat transfer cases. "bcs.dat" has the boundary conditions on the end nodes. "bfs.dat" contains the nodal values of the volume solute source. "globals.dat" contains the constant flow velocity and the diffusivity of the solute in water.

MATLAB Code

The MATLAB code for solving the 1-D advection–diffusion example problem is listed below. Once again, the code is similar to the one for the elastic bar analysis and the one for the heat transfer. However, while the program structure remains the same, there are differences reflecting the different physics.

```matlab
1   clear;
2   load elements.dat;    % load elements
3   load nodes.dat;       % load nodes
4   load globals.dat;     % load global properties
5   load bcs.dat;         % load boundary conditions
6   load bfs.dat;         % load volume source
7   v=globals(1);         % flow velocity
8   diffusivity =globals(2); % diffusivity
9   n_elements=size(elements,1);  % number of elements
10  n_nodes=size(nodes,1);        % number of nodes
11  K=zeros(n_nodes,n_nodes);                       % set up empty K
        matrix
12  F=zeros(n_nodes,1);           % set up empty RHS F vector
13  n_gauss_points=2;             % number of Gauss points
14  [gauss_xi, gauss_w] = Get1DGauss(n_gauss_points); % Get Gauss points
15  N=zeros(2,n_gauss_points);    % set up shape function matrix
16  Nxi=zeros(2,n_gauss_points);  % set up dN/dxi matrix
17  [N,Nxi]=CompShapeLinear1D(N, Nxi, gauss_xi); % compute N, dN/dxi
18                                               % at Gauss points
19  % compute element matrices and vectors and assemble the global system
20  for i=1:n_elements    % loop over elements
21      start_x= nodes(elements(i,2),2);
22      end_x= nodes(elements(i,3),2);
23      jacobian=CompJacobian1D([start_x end_x]', Nxi);
24      for g=1:n_gauss_points
25          sv(g)=bfs(i,2)*N(1,g)+ bfs(i+1,2)*N(2,g);  % body source
26      end
27      f=zeros(2,1); % RHS vector
28      for m=1:2
29          for g=1:n_gauss_points
30              f(m,1)=f(m,1)+sv(g)*N(m,g)*gauss_w(g)*jacobian(g);
31          end
32      end
33      k=zeros(2,2);    % diffusivity matrix
34      ka=zeros(2,2);   % advection matrix
35      for m=1:2
36          for n=1:2
37              for g=1:n_gauss_points
38                  k(m,n)=k(m,n) + diffusivity    ...
39                      *Nxi(m,g)*Nxi(n,g)*gauss_w(g)/jacobian(g);
40                  ka(m,n)=ka(m,n)+ v*N(m,g)*Nxi(n,g)*gauss_w(g);
41              end
42          end
43      end
44      [K,F]=assembleGlobalLinear1D(i, elements, k+ka, f, K, F);
45  end
46  % apply loads, boundary conditions
47  for i=1:n_nodes-1:n_nodes
48      if bcs(i,2)==1
49          penalty= abs(K(i,i))*1e7;
```

```
50      K(i,i)=penalty;
51      F(i)=bcs(i,3)*penalty;
52    end
53    if bcs(i,2)==2
54      F(i)=F(i)+bcs(i,3);
55    end
56  end
57  C_nodes=K\F      % solve the global system equations
58
59  % post-processing (step 10)
60  figure(1);
61  plot(nodes(:,1), C_nodes,'b-o','linewidth',2);
62  set(gca,'fontsize',16);
63  xlabel('node number','fontsize',18);
64  ylabel('Concentration (mole/m^3)','fontsize',18);
65  k=1; % set up index of the points for computing solute flux
66  for i=1:n_elements      % loop over the elements
67    node1=elements(i,2); % global node number of the local node 1
68    node2=elements(i,3); % global node number of the local node 2
69    for x=-1:0.1:1
70      [N,Nxi]=CompShapeLinear1D(zeros(2,1),zeros(2,1),x);
71      C=C_nodes(node1)*N(1)+C_nodes(node2)*N(2); % concentration
72      J=CompJacobian1D([nodes(node1,2) nodes(node2,2)]',Nxi); % Jacobian
73      Cx=C_nodes(node1)*Nxi(1)/J+C_nodes(node2)*Nxi(2)/J; % C'(X)
74      re(k,1)=nodes(node1,2)*N(1)+nodes(node2,2)*N(2);
75      re(k,2)=v*C-diffusivity*Cx;
76      k=k+1;
77    end
78  end
79  figure(2);
80  plot(re(:,1), re(:,2),'r-','linewidth',2);
81  set(gca,'fontsize',16);
82  xlabel('x (cm)','fontsize',18);
83  ylabel('Solute flux (m^2/s)','fontsize',18);
```

In the MATLAB programs listed above, the functions which have already been discussed previously are not repeated here. They are listed below for your reference:
function [gauss_xi, gauss_w] = Get1DGauss(n_gauss_points): Section 3.4.
function [N,Nx]=CompShapeLinear1D(N, Nx, xi_vector): Section 4.3.2.
function J=CompJacobian1D(x_vector,Nx): Section 4.3.2.
function [K,F]=AssembleGlobalLinear1D(K, F, eid, elements, k, f): Section 4.3.2.

4.6 Summary

Upon completion of this chapter, you should be able to:

- understand the general finite element procedure (steps) for linear static analysis
- understand the fundamental theories of 1-D elasticity, heat transfer, mass transport fluid flow and know how to translate these theories to finite element formulations

- derive the weak forms by using, when applicable, variational calculus, the Galerkin weighted residual method, the principle of virtual work, and minimization of potential energy
- implement a MATLAB code to discretize a 1-D domain into a mesh; generate data files of elements and nodes
- derive the 1-D shape functions and use these shape functions to approximate an unknown function and its derivatives over 1-D elements
- use Gauss quadrature to integrate 1-D functions over a given interval
- derive the expressions of element matrices and vectors from a weak form
- assemble element matrices and vectors into the global matrix and vector
- apply the essential boundary conditions by using the direct substitution and the penalty methods
- calculate the derivatives of the unknown quantities after the nodal solution is obtained
- understand the concept of continuity and its implications in the finite element analysis
- implement MATLAB functions to perform the pre-processing and post-processing steps of FEA
- know how to implement a complete MATLAB program to perform FEA of 1-D elasticity, steady state heat transfer, and mass transport problems.

4.7 Problems

4.1 The differential equation and the boundary conditions given below are defined over the 1-D domain $[0, 1]$. Derive the weak form by using both the variational calculus and the method of weighted residuals. Use δu as the weighting function.

$$-\frac{d}{dx}\left(x\frac{du}{dx}\right) + 2u = 0$$

$$u(0) = 0.5 \qquad \frac{du}{dx}(1) = -0.5.$$

4.2 Consider the Euler–Bernoulli beam problem as shown in Fig. 4.29. The differential equation of the deflection $w(x)$ is given by

$$\frac{d^2}{dx^2}\left(EI\frac{d^2w}{dx^2}\right) - q(x) = 0 \qquad \text{for } 0 < x < L$$

where $w(x)$ is the deflection of the beam, E is the Young's modulus, I is the moment of inertia, $q(x)$ is the transverse distributed load, and L is the length of the beam. Note that $dM/dx = V$, $dV/dx = -q$, and $-EI(d^2w/dx^2) = M$, where M is the moment and V is the shear force.

(a) Find the boundary conditions of the beam shown in Fig. 4.29.

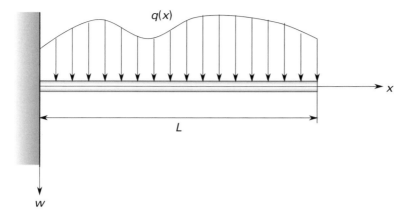

Figure 4.29 Euler–Bernoulli beam.

(b) Derive the weak form by using the method of weighted residuals so that the highest derivatives in the weak form are the second-order derivatives. Use δw as the weighting function.

4.3 Given a 4 node element of length L as shown in Fig. 4.30
(a) Derive the cubic shape functions by using the Lagrange interpolation.
(b) Assuming that $L = 3$ and $a = 0$, plot the shape functions.
(c) Assuming that $L = 3$, $a = 0$, and the nodal values are 3.0, 5.0, 1.0, and 2.0 for nodes 1, 2, 3, and 4, respectively, calculate the interpolating cubic polynomial by using both the monomial basis and the Lagrange basis. Plot the polynomials you obtain.
(d) Are the two cubic polynomials you obtained from (c) the same? Why?

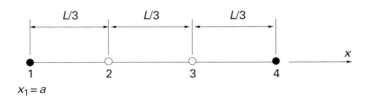

Figure 4.30 A 4 node element.

4.4 We have a function $y(x) = 1/(1 + 25x^2)$.
(a) Use MATLAB to compute the polynomial of degree 4 that interpolates the five data points of $y(x)$ at $x = [-1, -0.5, 0, 0.5, 1]$. Plot the resulting polynomial as well as the original function $y(x) = 1/(1 + 25x^2)$ over the domain $[-1, 1]$ (use a fine grid for the plots).

(b) Compute the polynomial of degree 8 that interpolates the nine data points of $y(x)$ at $x = [-1, -0.75, -0.5, -0.25, 0, 0.25, 0.5, 0.75, 1]$. Plot the resulting polynomial as well as the original function $y(x) = 1/(1 + 25x^2)$ over the domain $[-1, 1]$.

(c) Divide the domain $[-1, 1]$ into four equal length elements and construct a quadratic interpolation function on each element (note: you need to put one extra node at the center of each element). Plot the resulting quadratic interpolation functions as well as the original function $y(x) = 1/(1 + 25x^2)$ over the domain $[-1, 1]$.

(d) Which one of the above interpolation functions is the best fit to the original function?

4.5 Consider the 1-D Poisson equation with a natural boundary condition on the left and a homogeneous essential boundary condition on the right

$$-\frac{d^2u}{dx^2} = 2x \qquad 0 < x < 1$$
$$\frac{du}{dx}(0) = 1.0$$
$$u(1) = 0.$$

(a) Derive the weak form for the 1-D Poisson equation by using the Galerkin weighted residual method.

(b) Discretize the domain into two equal length linear elements, and manually calculate the element matrix and the right hand side element vector for each element.

(c) Manually assemble the global linear system.

(d) Apply the essential boundary condition by using the penalty method and solve the final linear system in MATLAB; print out your result.

(e) What is the exact solution of the problem?

(f) Plot your finite element solution and the exact solution in one figure and comment on your result.

4.6 Write a MATLAB function to compute the shape functions and their first derivatives at a given point in a given quadratic element.

(a) For the quadratic element shown in Fig. 4.31, use your MATLAB function to evaluate the shape functions and their first derivatives at $x = 0.35$.

(b) Write a MATLAB code to plot the shape functions and their first derivatives over the element by calling your MATLAB function. Note: use a fine grid of x within the element when you plot the shape functions and their first derivatives. For comparison, put all the shape function curves in one figure and all the curves of the first derivatives in another figure.

4.7 Derive the 1-point Gaussian Quadrature rule for a 1-D integration interval $[2, 3]$. Calculate $\int_2^3 2x\,dx$ by using the Gaussian Quadrature rule you have derived.

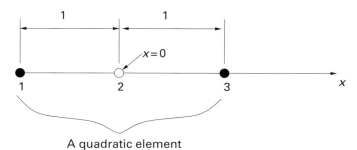

Figure 4.31 A quadratic element.

4.8 Since $\int_0^1 \frac{4}{1+x^2} dx = \pi$, one can compute an approximate value for π using numerical integration of the given function.
  (a) Use the 2-point Gaussian Quadrature rule to calculate the integral. Find the relative error in percentage.
  (b) Use the 3-point Gaussian Quadrature rule to calculate the integral. Find the relative error in percentage.

4.9 Use FEM to solve the 1-D Poisson equation with a natural boundary condition on the left and a homogeneous essential boundary condition on the right,

$$-\frac{d^2u}{dx^2} = x+1 \qquad 0 < x < 1$$

$$\frac{du}{dx}(0) = 2.0$$

$$u(1) = 0.$$

  (a) Assuming linear elements are used, for an element e (assuming it is not on the boundary), write down the analytical expressions of the element matrix (matrices) and vector(s).
  (b) Assuming linear elements are used, determine how many Gauss points you need for the Gaussian quadrature so that the integrals in the (1) element matrix and (2) element vector can be calculated exactly.
  (c) Write a MATLAB code to solve the problem by using (1) two equal length linear elements, (2) five equal length linear elements, (3) 50 equal length linear elements. Compute both u and du/dx.
  (d) Plot your finite element solutions obtained from the 2-element, 5-element, and 50-element meshes and the exact solutions together (one figure for u and one figure for du/dx) and comment on your results.

4.10 Write a MATLAB program to solve the 1-D heat transfer problem discussed in this chapter.

(a) Assuming quadratic elements are used, for an element e (assuming it is not on the boundary), write down the analytical expressions of the element matrix(matrices) and vector(s).

(b) Assuming quadratic elements are used, determine the minimum number of Gauss points you need for exact calculation of each element matrix and element vector.

(c) Use your MATLAB program to solve the problem by using (1) five equal length quadratic elements, (2) 50 equal length quadratic elements. Compute both the temperature T and dT/dx.

(d) Plot your finite element solutions on two figures. One figure should show the FEA results for T obtained from the 5-element and 50-element meshes and the other one should show dT/dx obtained from the two different meshes. Comment on your results.

References

Fish, J. & Belytschko, T. (2007), *A first course in finite elements*, Wiley & Sons.

Sokolnikoff, I. S., Specht, R. D. et al. (1956), *Mathematical theory of elasticity*, Vol. 83, McGraw-Hill.

5 FEA for Multi-Dimensional Scalar Field Problems

5.1 Overview

In this chapter, we discuss the general finite element analysis procedure for 2-D and 3-D linear scalar field problems. A scalar is a physical quantity having only magnitude, not direction. A scalar field problem is a problem whose primary unknown physical quantity is a scalar at any spatial location in the domain. For example, heat transfer problems with temperature as the primary physical quantity are scalar field problems, since temperature has only magnitude, no direction.

In this chapter, we demonstrate the finite element analysis procedure for scalar field problems by solving 2-D and 3-D steady state heat transfer problems. In Section 5.2, a steady state heat transfer problem is solved step by step in the same fashion as solving 1-D problems. Strong and weak forms of the governing equations are derived from the law of energy conservation and the method of weighted residual. 2-D finite element approximations and elements are described in detail. Numerical integration over 2-D elements is also described in detail. Convergence considerations are discussed. Numerical results obtained from each step are also provided to the reader for verification. The 2-D finite element analysis is then extended to 3-D in Section 5.3. A 3-D heat sink problem is demonstrated. In the step-by-step solution of the problem, approximation and integration for 3-D elements are introduced. At the end of each section, MATLAB codes for solving these problems are presented.

5.2 2-D Steady State Heat Transfer

In this section, 2-D steady-state heat transfer analysis is illustrated by demonstrating the step-by-step procedure of solving a heat transfer problem in a 2-D plate as shown in Fig. 5.1.

5.2.1 Steady State Heat Transfer in a Two-Dimensional Plate

Step 1: Physical Problem

We consider a thin plate as shown in Fig. 5.1. The temperature and the outward normal heat flux are specified on the four sides of the plate. In addition, there is convection heat transfer between the front and back surfaces of the plate and the ambient environment. The steady state temperature distribution in the plate is to be determined.

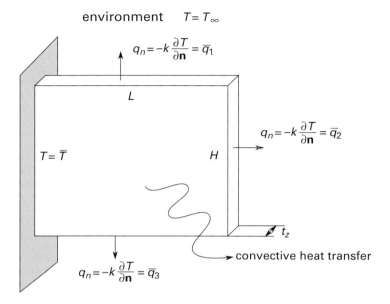

Figure 5.1 A 2-D heat transfer problem.

The physical dimensions of the plate are given by

$$L = 2 \text{ m}, \qquad H = 2 \text{ m}, \qquad t_z = 0.05 \text{ m}. \tag{5.1}$$

The relevant material properties include the thermal conductivity $k = 5$ W/m·K, and the convection coefficient of the ambient environment $h = 10$ W/m²·K. The temperature and normal heat flux at the sides are

$$\bar{T} = 80 \,°\text{C}, \tag{5.2}$$

$$\bar{q}_1 = 500 \text{ W/m}^2 \qquad \bar{q}_2 = 1000 \text{ W/m}^2 \qquad \bar{q}_3 = 1500 \text{ W/m}^2. \tag{5.3}$$

Note that, according to Fourier's law, the outward normal heat flux is the negative of the product of the thermal conductivity and the normal derivatives of the temperature. In addition, the temperature of the ambient environment is $T_\infty = 30\,°\text{C}$.

Since the plate is thin, which means the temperature can be assumed uniform in the thickness direction of the plate, and the steady state temperature distribution is to be computed, the problem can be regarded as a 2-D steady state heat transfer problem.

Step 2: Mathematical Model
As discussed in the previous chapter, the energy conservation law states that

$$\text{heat in} - \text{heat out} = \text{heat causing temperature increase.} \tag{5.4}$$

5.2 2-D Steady State Heat Transfer

In steady state, T does not change with time, therefore

$$\text{heat in} - \text{heat out} = 0. \tag{5.5}$$

To obtain the governing equation of our heat transfer problem, we cut out an infinitesimal element ($dx \times dy \times t_z$) from the plate, as shown in Fig. 5.2. Note that dx and dy are infinitesimal lengths and t_z is the thickness of the plate. The heat that goes into the infinitesimal element due to heat conduction is given by

$$q_x dy t_z + q_y dx t_z \tag{5.6}$$

where q_x and q_y are the heat fluxes in the x- and y-directions, respectively. Taking into account the heat loss due to convection, the heat out is given by

$$(q_x + dq_x)dy t_z + (q_y + dq_y)dx t_z + 2h(T - T_\infty)dxdy \tag{5.7}$$

where dq_x and dq_y are the incremental heat fluxes in the x- and y-directions, respectively, and the last term is the heat out due to surface convection. Note that h includes the convection on both the front and back surfaces.

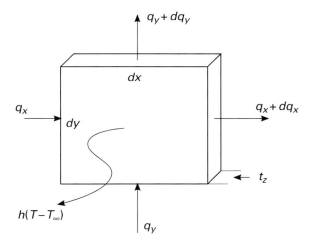

Figure 5.2 Heat flow in a differential element of the plate.

Substituting Eqs. (5.6, 5.7) into Eq. (5.5), we obtain

$$q_x dy t_z + q_y dx t_z - (q_x + dq_x)dy t_z - (q_y + dq_y)dx t_z - 2h(T - T_\infty)dxdy = 0 \tag{5.8}$$

which can be simplied to

$$dq_x dy t_z + dq_y dx t_z + 2h(T - T_\infty)dxdy = 0. \tag{5.9}$$

Dividing both sides of Eq. (5.9) by $dxdy t_z$, we obtain

$$\frac{dq_x}{dx} + \frac{dq_y}{dy} + \frac{2h}{t_z}(T - T_\infty) = 0. \tag{5.10}$$

In the differential (infinitesimal) element,

$$dq_x = \frac{\partial q_x}{\partial x}dx, \qquad dq_y = \frac{\partial q_y}{\partial y}dy \qquad (5.11)$$

and letting $2h/t_z$ be denoted as h_z, Eq. (5.12) can be rewritten as

$$\frac{\partial q_x}{\partial x} + \frac{\partial q_y}{\partial y} + h_z(T - T_\infty) = 0. \qquad (5.12)$$

For isotropic materials, the heat fluxes are given by Fourier's law as

$$q_x = -k\frac{\partial T}{\partial x}, \qquad q_y = -k\frac{\partial T}{\partial y}. \qquad (5.13)$$

Therefore, Eq. (5.12) can be further rewritten as

$$-\frac{\partial}{\partial x}\left(k\frac{\partial T}{\partial x}\right) - \frac{\partial}{\partial y}\left(k\frac{\partial T}{\partial y}\right) + h_z(T - T_\infty) = 0. \qquad (5.14)$$

Equation (5.14) is the governing partial differential equation for the 2-D steady state heat transfer problem. From Fig. 5.1, the essential boundary condition is given by

$$T = \overline{T} \qquad \text{at} \qquad x = 0 \qquad (5.15)$$

and the natural boundary condition is given by

$$-k\frac{\partial T}{\partial \mathbf{n}} = \begin{cases} \overline{q}_1 & y = 2\ \mathrm{m} \\ \overline{q}_2 & x = 2\ \mathrm{m} \\ \overline{q}_3 & y = 0. \end{cases} \qquad (5.16)$$

The equations (5.14–5.16) combined is the strong form of the 2-D heat transfer problem.

Step 3: Weak Form

The strong form given in Eqs. (5.14–5.16) can be converted to the weak form by using the variational principle (see Chapter 2). In this section, we derive the weak form by using the Galerkin weighted residual method. From the partial differential equation given in Eq. (5.14), the residual is obtained by moving all the nonzero terms to the left-hand side, i.e.,

$$R = -\frac{\partial}{\partial x}\left(k\frac{\partial T}{\partial x}\right) - \frac{\partial}{\partial y}\left(k\frac{\partial T}{\partial y}\right) + h_z(T - T_\infty) = 0. \qquad (5.17)$$

Or equivalently

$$R = \frac{\partial}{\partial x}\left(k\frac{\partial T}{\partial x}\right) + \frac{\partial}{\partial y}\left(k\frac{\partial T}{\partial y}\right) - h_z(T - T_\infty) = 0. \qquad (5.18)$$

Multiplying the residual by the variation of the temperature, δT, and integrating the product over the 2-D domain (surface), Ω, we obtain

$$\int_\Omega \delta T\left[\frac{\partial}{\partial x}\left(k\frac{\partial T}{\partial x}\right) + \frac{\partial}{\partial y}\left(k\frac{\partial T}{\partial y}\right) - h_z(T - T_\infty)\right]d\Omega = 0 \qquad (5.19)$$

5.2 2-D Steady State Heat Transfer

or in short form,

$$\int_\Omega \delta T \left[\nabla \cdot (k\nabla T) - h_z(T - T_\infty)\right] d\Omega = 0. \tag{5.20}$$

Equation (5.20) can be rewritten as

$$\int_\Omega \delta T \left(\nabla \cdot (k\nabla T)\right) d\Omega - \int_\Omega h_z(T - T_\infty)\delta T d\Omega = 0. \tag{5.21}$$

From vector calculus, it is easy to verify that

$$\nabla \cdot (f\mathbf{v}) = f \nabla \cdot \mathbf{v} + \mathbf{v} \cdot \nabla f. \tag{5.22}$$

In Eq. (5.22), let $\mathbf{v} = k\nabla T$ and $f = \delta T$, we obtain

$$\nabla \cdot (\delta T(k\nabla T)) = \delta T \left(\nabla \cdot (k\nabla T)\right) + k\nabla T \cdot \nabla \delta T. \tag{5.23}$$

That is

$$\delta T \left(\nabla \cdot (k\nabla T)\right) = \nabla \cdot (\delta T(k\nabla T)) - k\nabla T \cdot \nabla \delta T. \tag{5.24}$$

Substituting Eq. (5.24) into Eq. (5.21) gives

$$\int_\Omega [\nabla \cdot (\delta T(k\nabla T)) - k\nabla T \cdot \nabla \delta T] d\Omega - \int_\Omega h_z(T - T_\infty)\delta T d\Omega = 0 \tag{5.25}$$

i.e.

$$\int_\Omega \nabla \cdot (\delta T(k\nabla T)) d\Omega - \int_\Omega k\nabla T \cdot \nabla \delta T d\Omega - \int_\Omega h_z(T - T_\infty)\delta T d\Omega = 0. \tag{5.26}$$

The next step is to apply the divergence theorem to the first term in Eq. (5.26). The divergence theorem states that the volume integral of the divergence of a vector function is equal to the integral over the surface of the function component normal to the surface, i.e.,

$$\int_\Omega \nabla \cdot \mathbf{v} d\Omega = \int_\Gamma \mathbf{v} \cdot \mathbf{n} d\Gamma. \tag{5.27}$$

Applying Eq. (5.27) to the first term in Eq. (5.26), Eq. (5.26) can be rewritten as

$$\int_\Gamma (\delta T(k\nabla T)) \cdot \mathbf{n} d\Gamma - \int_\Omega k\nabla T \cdot \nabla \delta T d\Omega - \int_\Omega h_z(T - T_\infty)\delta T d\Omega = 0 \tag{5.28}$$

or

$$\int_\Gamma \delta T k(\nabla T \cdot \mathbf{n}) d\Gamma - \int_\Omega k\nabla T \cdot \nabla \delta T d\Omega - \int_\Omega h_z(T - T_\infty)\delta T d\Omega = 0. \tag{5.29}$$

From vector calculus, it can be shown that

$$\nabla T \cdot \mathbf{n} = \frac{\partial T}{\partial \mathbf{n}}. \tag{5.30}$$

Therefore, Eq. (5.29) can be rewritten as

$$\int_\Gamma \delta T k \frac{\partial T}{\partial \mathbf{n}} d\Gamma - \int_\Omega k\nabla T \cdot \nabla \delta T d\Omega - \int_\Omega h_z(T - T_\infty)\delta T d\Omega = 0. \tag{5.31}$$

According to the boundary conditions specified, the boundary of the domain Γ is composed of two parts: (1) the portion of the boundary where the essential boundary condition is specified (denoted as Γ_T) and (2) the portion where the natural boundary condition is specified (denoted as Γ_q), as shown in Fig. 5.3.

Figure 5.3 Notations for the domain and its boundary.

Since $\delta T = 0$ on Γ_T, the first integral in Eq. (5.31) vanishes on Γ_T. Therefore, Eq. (5.31) can be rewritten as

$$\int_{\Gamma_q} \delta T k \frac{\partial T}{\partial \mathbf{n}} d\Gamma - \int_{\Omega} k \nabla T \cdot \nabla \delta T d\Omega - \int_{\Omega} h_z(T - T_\infty) \delta T d\Omega = 0. \quad (5.32)$$

Equation (5.32) is the weak form of the 2-D steady state heat transfer problem. It is necessary to emphasize that, the process of deriving the weak form transfers a differentiation operation from T to δT so that the derivative orders of T and δT are balanced and the continuity requirement for T is reduced.

In vector calculus, the derivation from Eq. (5.21) to Eq. (5.29) gives a formula called Green's first identity, i.e.,

$$\int_{\Omega} w(\nabla \cdot \mathbf{v}) d\Omega = \int_{\Gamma} w(\mathbf{v} \cdot \mathbf{n}) d\Gamma - \int_{\Omega} \nabla w \cdot \mathbf{v} d\Omega. \quad (5.33)$$

Green's first identity can be regarded as the integration by parts in 2-D or 3-D. By using Green's first identity, one can obtain Eq. (5.29) directly from Eq. (5.21).

Step 4: Discretization

The 2-D physical domain can be discretized into a set of elements as shown in Fig. 5.4. In two dimensions, quadrilateral and triangular shaped elements with straight or curved edges are typically used. The process of generating the elements for a given physical domain is called the meshing process. While mesh generation is a field in itself, we will introduce a few basic mesh generation methods and their implementation in Chapter 6. It should be noted that, for domains with regular geometries such as the square plate example problem we are working on, it is straightforward to discretize the domain into square or rectangular elements, which are special kinds of quadrilateral elements. For complex geometries, however, the meshing process can be involved and computationally expensive. In addition, sometimes it can be difficult to ensure high quality for all the elements. More details will be discussed in Chapter 6. For the simplicity of illustration, the plate in the current example problem is discretized into four equal-size square elements as shown in Fig. 5.5.

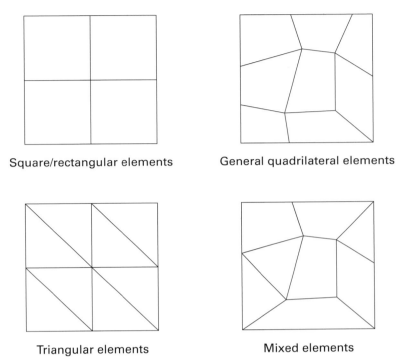

Figure 5.4 Mesh examples for the plate.

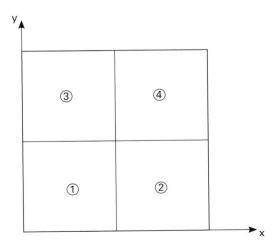

Figure 5.5 Elements of the plate.

Step 5: Finite Element Approximation

The approximation of the temperature in a 2-D element has the same form as that in the 1-D heat transfer problem described in Section 4.4.1, i.e.,

$$T = \sum_{i=1}^{n} N_i T_i \qquad (5.34)$$

where n is the number of nodes in the element, T_i are the nodal temperatures, and N_i are shape functions of the element. The difference, however, is the dimensionality of the temperature and shape functions. The unknown temperature and the shape functions are now all 2-D functions, i.e.

$$T(x, y) = \sum_{i=1}^{n} N_i(x, y) T_i. \qquad (5.35)$$

The partial derivatives of the temperature are then expressed as

$$\frac{\partial T}{\partial x} = \sum_{i=1}^{n} \frac{\partial N_i(x, y)}{\partial x} T_i. \qquad (5.36)$$

$$\frac{\partial T}{\partial y} = \sum_{i=1}^{n} \frac{\partial N_i(x, y)}{\partial y} T_i. \qquad (5.37)$$

Depending on the type of the elements, the shape function can have different forms. Typically, 2-D elements are triangular, rectangular, or general quadrilateral shaped with straight or curved edges as shown in Fig. 5.6. Nodes are placed at specific locations of the elements depending on the order of the finite element approximation. In the following, we describe in detail various types of elements that are used in 2-D finite element analysis.

Linear Rectangular Elements

The shape functions of the Lagrange family rectangular elements are constructed by using the product of 1-D Lagrange interpolation functions (i.e. the 1-D shape functions) in x- and y-directions. For example, the construction of the shape function of node 2 in the rectangular element is shown in Fig. 5.7.

The shape function $N_2(x, y)$ associated with node 2 is constructed as the product of two 1-D shape functions associated with the same node in two 1-D elements: one is the 1-D element of horizontal edge from node 2 to node 1 (referred to as edge 2-1), and the other is the element of vertical edge from node 2 to node 3 (edge 2-3). In the 1-D element of edge 2-1, the 1-D shape function associated with node 2 is clearly $(a - x)/2a$. Likewise, the 1-D shape function associated with node 2 in the element of edge 2-3 is $(b + y)/2b$. The product of the 1-D shape functions (shown in the figure) gives

$$N_2(x, y) = \frac{a - x}{2a} \times \frac{b + y}{2b} = \frac{(a - x)(b + y)}{4ab}. \qquad (5.38)$$

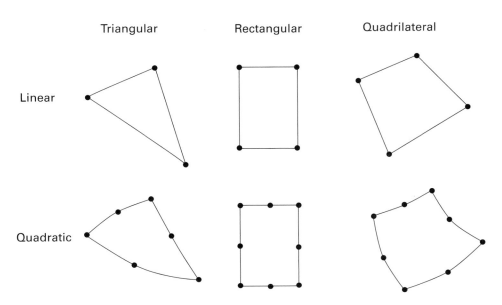

Figure 5.6 Types of 2-D elements.

By using the same procedure, the shape functions of the linear rectangular element can be obtained

$$N_1(x, y) = \frac{1}{4ab}(a + x)(b + y) \tag{5.39}$$

$$N_2(x, y) = \frac{1}{4ab}(a - x)(b + y) \tag{5.40}$$

$$N_3(x, y) = \frac{1}{4ab}(a - x)(b - y) \tag{5.41}$$

$$N_4(x, y) = \frac{1}{4ab}(a + x)(b - y). \tag{5.42}$$

The shape functions above are called bilinear (linear in x and linear in y) shape functions.

Higher Order Rectangular Elements: Lagrange Family

The shape functions of higher order rectangular elements of Lagrange type or Lagrange family can be constructed in the same way as that for the linear elements. That is, the shape function associated with a node is obtained by the product of two 1-D quadratic shape functions associated with the same node in two 1-D quadratic elements in the x- and y-directions. For example, Fig. 5.8 illustrates the shape functions for corner (node 2), mid-side (node 7), and center (node 9) nodes.

It should be noted that, quadratic or higher order rectangular elements of Lagrange type have a uniform grid of nodes. For the quadratic element shown in Fig. 5.8, there is an interior node at the center of the element. Each of the corner nodes, such as

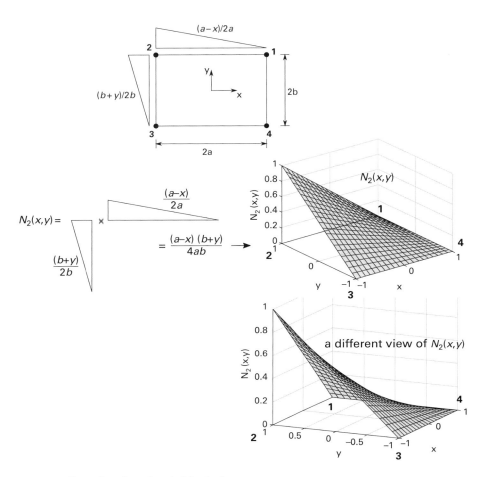

Figure 5.7 Shape function of node 2 in the linear rectangular element.

node 2, is shared by a horizontal and a vertical edge each considered as a 3-node 1-D quadratic element. The 2-D shape functions associated with these corner nodes are the product of the 1-D quadratic shape functions of the same node obtained from the 1-D quadratic edge elements sharing the node, as shown in Fig. 5.8. Each of the mid-side nodes lies on an edge of the element, either horizontal or vertical. Taking node 7 as an example, it is in the middle of edge 3-7-4 which is regarded as a 1-D quadratic element. In the vertical direction, although there is no edge containing node 7, an implicit vertical element with nodes 7, 9, and 5 as its nodes is assumed. The 2-D shape function of node 7 is then the product of the shape function of node 7 in the 1-D element 3-7-4 and the shape function of node 7 in the 1-D element 7-9-5, as shown in Fig. 5.8. Finally, for the center node 9, implicit 1-D quadratic elements 6-9-8 and 5-9-7 are assumed, and the 2-D shape function of node 9 is the product of the two 1-D shape functions of node 9 in these two 1-D elements. This procedure of constructing the shape functions for Lagrange type rectangular elements can be applied to higher order elements with $m \times n$ nodes.

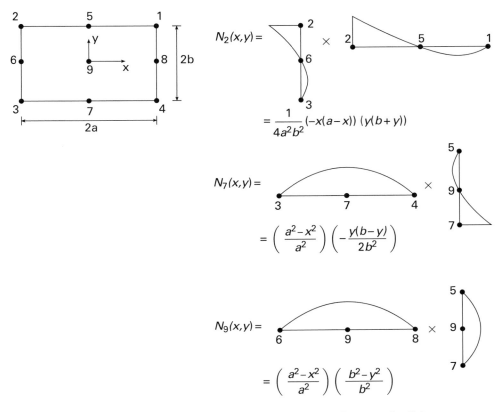

Figure 5.8 Shape functions of a quadratic rectangular element (Lagrange family).

Higher Order Rectangular Elements: Serendipity Family

It can be shown that the interior node of the quadratic Lagrange type rectangular element can be omitted without affecting the element's approximation capability. This will be demonstrated later in this chapter. Higher order elements with reduced internal nodes are called serendipity (type) elements. For example, Fig. 5.9 shows the quadratic, cubic, and quartic rectangular elements of the serendipity family. In comparison to the Lagrange type elements, the serendipity elements contain fewer nodes and fewer shape functions for the finite element approximation. Thus, the serendipity elements are computationally more efficient. In practice, 8-node quadratic serendipity elements are employed in most commercial FEA packages.

The procedure of constructing the shape functions for the 8-node rectangular serendipity elements is as follows:

1 For the mid-side nodes, the shape functions are obtained by taking the product of the second and first-order 1-D shape functions as shown in Fig. 5.10.
2 For a given corner node, we take the associated bilinear shape function shown in Eqs. (5.39–5.42) and subtract half of the mid-side node shape functions (shown in

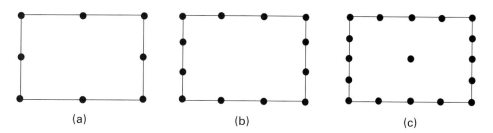

Figure 5.9 Rectangular elements of serendipity family: (a) quadratic, (b) cubic, (c) quartic.

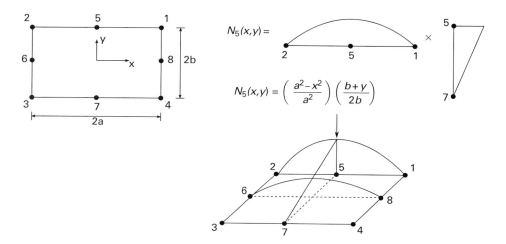

Figure 5.10 Shape functions of a quadratic rectangular element of serendipity type: mid-side nodes.

Fig. 5.10) neighboring the corner node. For example, the shape function of node 2 is constructed as shown in Fig. 5.11. The subtraction is due to the requirement that a shape function $N_i(x, y)$ should be equal to 1 at node i and zero at all the other nodes in the element. The bilinear shape functions shown in Eqs. (5.39–5.42) are equal to 0.5 at the adjacent mid-side nodes. Since the 2-D mid-side node shape functions are equal to 1 at the associated mid-side nodes, subtracting half of the corresponding mid-side node shape functions makes the resultant shape function equal to zero at the mid-side nodes while keeping the value of the shape function equal to 1 at node i.

This procedure can be easily applied to construct shape functions for elements with different numbers of nodes on the edges. This kind of element can be used to achieve a transition between elements of different orders, as shown in Fig. 5.12.

For example, the same procedure illustrated in the 8-node quadratic serendipity element is applied to generate the shape function N_1 for the cubic-linear transition elements as shown in Fig. 5.13.

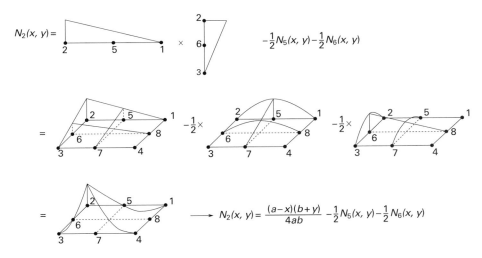

Figure 5.11 Shape functions of a quadratic rectangular element of serendipity type.

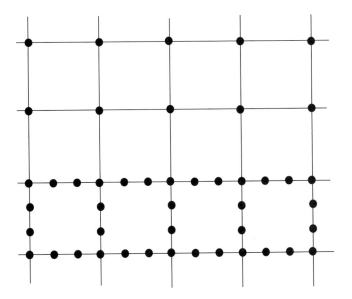

Figure 5.12 A mesh with a transition from cubic to linear elements.

Triangular Elements

Similar to rectangular elements, the unknown function (e.g. temperature in our example) is approximated as polynomials in triangular elements. The polynomials that can be represented by different triangular elements are listed in Table 5.1. The shape functions of the triangular elements can be constructed by using different methods. In the following, we discuss the methods of monomial basis and area coordinates.

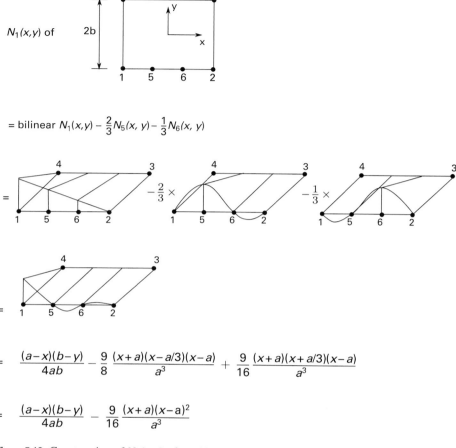

Figure 5.13 Construction of $N_1(x, y)$ of a cubic-linear transition element.

Method 1: Monomial basis. The shape functions for triangular elements can be obtained by directly solving for the unknown coefficients in the polynomials listed in Table 5.1. This approach has been illustrated for 1-D polynomial interpolation in Chapter 3. Here we take the heat transfer problem as an example to illustrate the 2-D interpolation of the temperature field within a triangular element. Assuming the nodal temperatures of the linear triangular element are T_1, T_2, and T_3 at nodes 1, 2, and 3, respectively, and since the temperature field in the element is $T = a_1 + a_2 x + a_3 y$, we obtain a set of linear equations

$$\begin{bmatrix} 1 & x_1 & y_1 \\ 1 & x_2 & y_2 \\ 1 & x_3 & y_3 \end{bmatrix} \begin{Bmatrix} a_1 \\ a_2 \\ a_3 \end{Bmatrix} = \begin{Bmatrix} T_1 \\ T_2 \\ T_3 \end{Bmatrix}. \qquad (5.43)$$

5.2 2-D Steady State Heat Transfer

Table 5.1 Triangular elements

Element type	Element with nodes	Polynomials represented
Linear	(triangle with nodes 1, 2, 3)	$T = a_1 + a_2 x + a_3 y$
Quadratic	(triangle with nodes 1–6)	$T = a_1 + a_2 x + a_3 y$ $+ a_4 x^2 + a_5 xy + a_6 y^2$
Cubic	(triangle with nodes 1–9)	$T = a_1 + a_2 x + a_3 y$ $+ a_4 x^2 + a_5 xy + a_6 y^2$ $+ a_7 x^3 + a_8 x^2 y + a_9 xy^2 + a_{10} y^3$

The coefficients a_1, a_2 and a_3 can be obtained by solving Eq. (5.43) as

$$\left\{ \begin{array}{c} a_1 \\ a_2 \\ a_3 \end{array} \right\} = \frac{1}{2A} \left[\begin{array}{ccc} x_2 y_3 - x_3 y_2 & x_3 y_1 - x_1 y_3 & x_1 y_2 - x_2 y_1 \\ y_2 - y_3 & y_3 - y_1 & y_1 - y_2 \\ x_3 - x_2 & x_1 - x_3 & x_2 - x_1 \end{array} \right] \left\{ \begin{array}{c} T_1 \\ T_2 \\ T_3 \end{array} \right\} \quad (5.44)$$

where A is the area of the triangular element which is half of the determinant of the coefficient matrix in Eq. (5.43), i.e.,

$$A = \frac{1}{2} \left| \begin{array}{ccc} 1 & x_1 & y_1 \\ 1 & x_2 & y_2 \\ 1 & x_3 & y_3 \end{array} \right|. \quad (5.45)$$

Thus, the temperature T can be rewritten as

$$T = a_1 + a_2 x + a_3 y$$

$$= [1 \ x \ y] \begin{Bmatrix} a_1 \\ a_2 \\ a_3 \end{Bmatrix}$$

$$= \frac{[1 \ x \ y]}{2A} \begin{bmatrix} x_2 y_3 - x_3 y_2 & x_3 y_1 - x_1 y_3 & x_1 y_2 - x_2 y_1 \\ y_2 - y_3 & y_3 - y_1 & y_1 - y_2 \\ x_3 - x_2 & x_1 - x_3 & x_2 - x_1 \end{bmatrix} \begin{Bmatrix} T_1 \\ T_2 \\ T_3 \end{Bmatrix}. \quad (5.46)$$

Written in the standard finite element approximation form, the temperature field is

$$T(x, y) = \sum_{i=1}^{3} N_i(x, y) T_i \quad (5.47)$$

where the shape functions are then

$$[N_1 \ N_2 \ N_3] = \frac{[1 \ x \ y]}{2A} \begin{bmatrix} x_2 y_3 - x_3 y_2 & x_3 y_1 - x_1 y_3 & x_1 y_2 - x_2 y_1 \\ y_2 - y_3 & y_3 - y_1 & y_1 - y_2 \\ x_3 - x_2 & x_1 - x_3 & x_2 - x_1 \end{bmatrix}. \quad (5.48)$$

Since the monomial basis approach requires solving a linear system, it becomes inconvenient for higher order elements for which the coefficient matrix in Eq. (5.43) is 6×6 or larger.

Method 2: Area coordinates. Given a point (x, y) in the element, we define its area coordinates, (l_1, l_2, l_3), as shown in Fig. 5.14, to be

$$l_1 = \frac{A_1}{A}, \quad l_2 = \frac{A_2}{A}, \quad l_3 = \frac{A_3}{A} \quad (5.49)$$

where A is the total area of the element and A_i, $i = 1, 2, 3$, are the areas of the triangles formed by the point (x, y) and the nodes of an edge as shown in the figure. Note that A_i is labeled by using the node i that is located opposite to the triangle. In addition, it is clear that the area coordinates are functions of x and y.

By using the formula for the area of a triangle, Eq. (5.45), it can be verified that

$$l_1(x, y) = \frac{A_1}{A} = \frac{(x_2 y_3 - x_3 y_2) + (y_2 - y_3)x + (x_3 - x_2)y}{2A} \quad (5.50)$$

$$l_2(x, y) = \frac{A_2}{A} = \frac{(x_3 y_1 - x_1 y_3) + (y_3 - y_1)x + (x_1 - x_3)y}{2A} \quad (5.51)$$

$$l_3(x, y) = \frac{A_3}{A} = \frac{(x_1 y_2 - x_2 y_1) + (y_1 - y_2)x + (x_2 - x_1)y}{2A}. \quad (5.52)$$

Comparing Eqs. (5.50–5.52) and Eq. (5.48) reveals that, for linear triangular elements, the shape functions are the same as the area coordinates, i.e.,

$$N_1(x, y) = l_1(x, y), \quad N_2(x, y) = l_2(x, y), \quad N_3(x, y) = l_3(x, y). \quad (5.53)$$

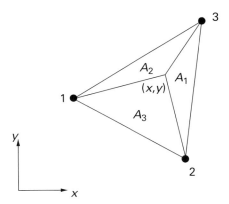

Figure 5.14 Area coordinates.

There are several properties of the area coordinates. It can be shown that

$$l_1(x, y) + l_2(x, y) + l_3(x, y) = 1 \tag{5.54}$$

$$l_1(x, y)x_1 + l_2(x, y)x_2 + l_3(x, y)x_3 = x \tag{5.55}$$

$$l_1(x, y)y_1 + l_2(x, y)y_2 + l_3(x, y)y_3 = y. \tag{5.56}$$

Equations (5.54–5.56) are known as the completeness conditions for linear triangular elements. Completeness conditions are discussed in detail later in this chapter. Another property of the area coordinates is shown in Fig. 5.15. On lines that are parallel to the edge/side opposite node i, the area coordinate l_i is constant, and $l_i = 1$ on node i and $l_i = 0$ on the edge/side opposite node i. Figure 5.15 shows an example for l_1.

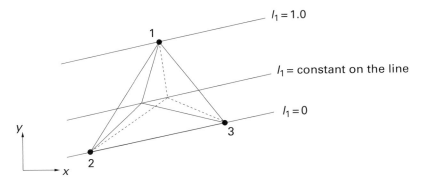

Figure 5.15 Area coordinates perpendicular to an edge.

The three-dimensional view of the area coordinates (i.e., shape functions of the linear triangular elements) are shown in Fig. 5.16.

By using the area coordinates, it is convenient to construct shape functions for higher order triangular elements. For the quadratic triangular element shown in

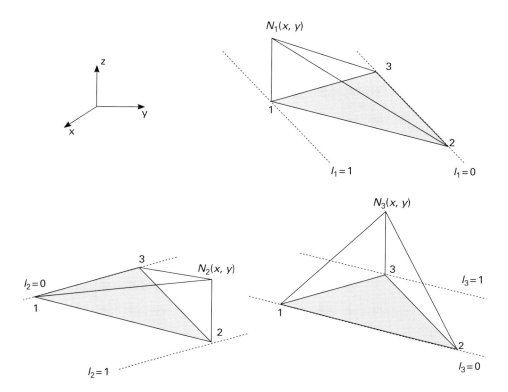

Figure 5.16 Shape functions of linear triangular elements.

Fig. 5.17, the shape functions of the three corner nodes are constructed by using the 1-D Lagrange interpolation of the area coordinates. To construct the shape functions for the side nodes, one needs to use a required property of shape functions:

$$N_i(x_j, y_j) = \begin{cases} 1 & i = j \\ 0 & i \neq j. \end{cases} \tag{5.57}$$

Equation (5.57) is called the Kronecker delta property. The procedure of constructing the side-node shape functions is illustrated below via an example shown in Fig. 5.17. Shown in Fig. 5.17 is a 6-node triangular element. The values of the area coordinates for points along the dashed lines are shown in the figure. For $N_1(x, y)$, the shape function is the quadratic Lagrange interpolation in the direction perpendicular to the parallel lines of constant l_1. The shape function is equal to 1 at node 1, zero at nodes 4 and 6 ($l_1 = 1/2$), and zero at nodes 2 3 and 5 ($l_1 = 0$). The Lagrange interpolation in l_1 generates a surface as shown in Fig. 5.18. The part of this surface located in the triangular elements is the shape function $N_1(x, y)$, i.e.,

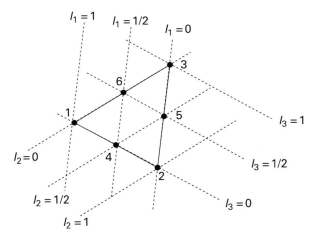

Figure 5.17 Shape functions of quadratic triangular elements.

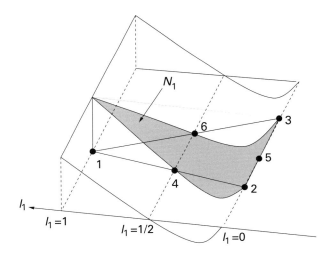

Figure 5.18 Shape functions of quadratic triangular elements.

$$N_1 = \frac{(l_1 - \frac{1}{2})(l_1 - 0)}{\frac{1}{2} \cdot 1} = 2(l_1 - \frac{1}{2})l_1. \qquad (5.58)$$

Note that we omit "(x, y)" here for brevity, in fact N_i and l_i, $i = 1, 2, \ldots$, are functions of (x, y). The shape functions $N_2(x, y)$ and $N_3(x, y)$ can be obtained in the same way:

$$N_2 = 2(l_2 - \frac{1}{2})l_2 \qquad (5.59)$$

$$N_3 = 2(l_3 - \frac{1}{2})l_3. \qquad (5.60)$$

Next we construct the shape functions for the side nodes. Taking node 4 as an example, from the Kronecker delta property, it is required that $N_4 = 0$ at nodes 1, 2, 3, 5, and 6. It is known that $l_2 = 0$ at nodes 1, 6, and 3. Therefore, if we include l_2 in N_4, then $N_4 = 0$ is satisfied at nodes 1, 6, and 3. Similarly, if we include l_1 in N_4, then $N_4 = 0$ at the nodes 2, 5, and 3. Thus we can write

$$N_4 = f(l_1, l_2, l_3) l_1 l_2. \tag{5.61}$$

As shown in Table 5.1, the 6-node triangular element is quadratic. $l_1 l_2$ in Eq. (5.61) is already quadratic (note: l_1 and l_2 are both linear). Therefore, $f(l_1, l_2, l_3)$ in Eq. (5.61) must be a constant number, i.e.,

$$N_4 = C l_1 l_2. \tag{5.62}$$

The Kronecker delta property of shape functions also requires that

$$N_4 = 1 \qquad \text{at node 4} \tag{5.63}$$

Therefore,

$$\begin{aligned} N_4(\text{node 4}) &= C \cdot l_1(\text{node 4}) \cdot l_2(\text{node 4}) \\ &= C \cdot \frac{1}{2} \cdot \frac{1}{2} \\ &= 1. \end{aligned} \tag{5.64}$$

This gives $C = 4$ and

$$N_4 = 4 l_1 l_2. \tag{5.65}$$

By following the same procedure, we obtain

$$N_5 = 4 l_2 l_3, \qquad N_6 = 4 l_1 l_3. \tag{5.66}$$

The shape functions of other higher order triangular elements (e.g. 10-node triangular elements) can be constructed in the same manner.

Isoparametric Elements

While rectangular elements are straightforward to understand and implement, to form a mesh over a complex geometry the elements must be allowed to take more general shapes, as shown in Fig. 5.19.

However, approximation of the unknown physical quantities over elements with irregular geometries becomes more difficult. Therefore, it is desirable for us to mesh the physical domain using elements with general shapes and do the approximations over elements with simple regular geometries. This is done by using the so called master (or parent) elements for approximation and transforming the geometry and approximations between the master elements and the physical elements through some kind of mapping. The basic idea underlying this centers on the mapping of the simple geometric shape in a reference coordinate system into more general shapes in the

 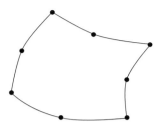

Linear quadrilateral element   Element with curved edges

Figure 5.19 General quadrilateral elements for complex geometries.

physical/global Cartesian coordinate system. The mapping from reference to physical coordinates takes the form

$$x = \sum_{i=1}^{n} N_i(\xi, \eta) x_i \qquad y = \sum_{i=1}^{n} N_i(\xi, \eta) y_i \qquad (5.67)$$

and

$$T(\xi, \eta) = \sum_{i=1}^{n} N_i(\xi, \eta) T_i \qquad (5.68)$$

where n is number of nodes, (x_i, y_i) is the position of node i of the physical element in the Cartesian coordinate system, T_i is the nodal temperature at node i, (ξ, η) is the pair of coordinates in the reference coordinate system whose axes are ξ and η, and $N_i(\xi, \eta)$ is the shape function of the master/parent element associated with node i. As shown in Fig. 5.20, the mapping process can be described as follows. First, the nodes, nodal temperatures, and nodal (x, y) coordinates are transferred to the nodes of the master element. For example, the temperature of node 1 is transferred to node 1 of the master element so that the temperature of node 1 of the master element is T_1. Therefore, in the master element the temperature can be approximated by using the standard finite element approximation. The temperature at an arbitrary point (ξ, η) in the master element is then given by Eq. (5.68). Next, given the point (ξ, η) in the master element, the location of the corresponding point (x, y) in the real element can be obtained from Eq. (5.67). Thus, $T(\xi, \eta)$ is the temperature of the corresponding point (x, y) in the physical element, i.e., $T(x, y)$.

When the same shape functions are used to specify the relation between the physical (x, y) and reference (ξ, η) coordinates as well as to approximate the unknown physical quantities (such as temperature), the element mapping is called isoparametric mapping. 2-D master elements for several popular quadrilateral and triangular elements are shown in Fig. 5.21.

The shape functions $N_i(\xi, \eta)$ in Eqs. (5.67, 5.68) are constructed on the master elements in the reference coordinate system by using the procedure discussed

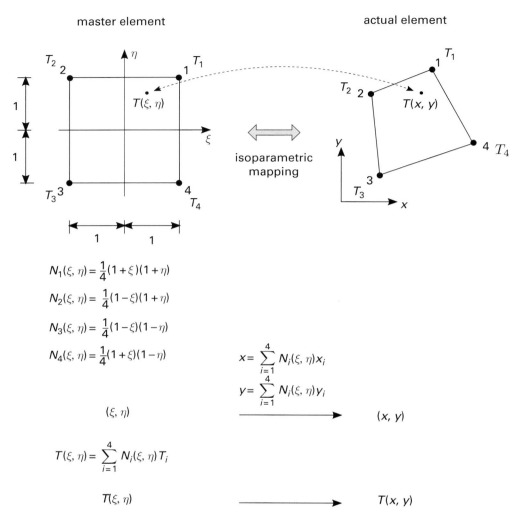

Figure 5.20 Isoparametric mapping.

previously for the rectangular and triangular elements. For example, for 4-node quadrilateral elements, the shape functions of the master element can be obtained from Eqs. (5.39–5.42) by applying $a = b = 1$ and replacing x, y by ξ, η, i.e.,

$$N_1 = \frac{1}{4}(1+\xi)(1+\eta) \tag{5.69}$$

$$N_2 = \frac{1}{4}(1-\xi)(1+\eta) \tag{5.70}$$

$$N_3 = \frac{1}{4}(1-\xi)(1-\eta) \tag{5.71}$$

5.2 2-D Steady State Heat Transfer

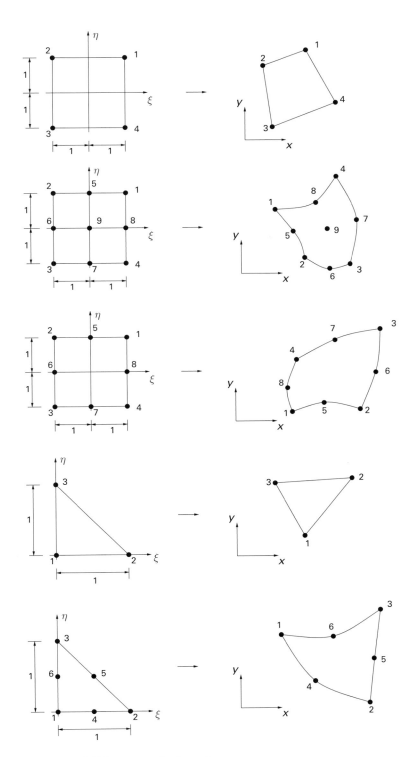

Figure 5.21 2-D isoparametric elements.

$$N_4 = \frac{1}{4}(1+\xi)(1-\eta). \tag{5.72}$$

Similarly, the shape functions for quadratic quadrilateral and triangular elements shown in Fig. 5.21 can be obtained. These shape functions are listed in Tables 5.2 and 5.3. A MATLAB function computing the shape functions and their derivatives for the linear square master element shown in Table 5.2 is listed below.

```
1  % Compute the shape functions and their first derivatives at a set
2  % of points defined by ( xi_vector , eta_vector ) in a 2-D 4-node square
3  % master element. Input is a single point when the length of
4  % xi _vector and eta _vector is one
5  % Input : xi_vector , eta_vector : coordinates of the input points
6  % Output: N: matrix storing shape function values with the format
7  %          [N1(xi_1 , eta_1 )   N1(xi_2 , eta_2 )   N1(xi_3 , eta_3 ) ...
8  %           N2(xi_1 , eta_1 )   N2(xi_2 , eta_2 )   N2(xi_3 , eta_3 ) ...
9  %           N3(xi_1 , eta_1 )   N3(xi_2 , eta_2 )   N3(xi_3 , eta_3 ) ...
10 %            ...                 ...                 ...              ...]
11 % Output: Nx, Ny: matrice of dNi/dxi ( xi , eta ) and dNi/deta ( xi , eta )
12 %          respectively , format is the same as N
13 function [N,Nx,Ny]=CompNDNatPointsQuad4(xi_vector, eta_vector)
14 np=size( xi_vector ,1);
15 N=zeros(4,np); Nx=zeros(4,np); Ny=zeros(4,np);  % set up empty matrices
16 master_nodes=[1 1; -1 1; -1 -1; 1 -1];  % coordinates of the nodes
17                                         % of the master element
18 % for loop : compute N, Nx, Ny
19 for j=1:np                              % columns for point 1,2 ...
20     xi=xi_vector(j);                    % xi-coordinate of point j
21     eta=eta_vector(j);                  % eta-coordinate of point j
22     for i=1:4                           % rows for N1, N2, ...
23         nx=master_nodes(i,1);
24         ny=master_nodes(i,2);
25         N(i,j)=(1.0 + nx*xi)*(1.0 + ny*eta) /4.0;   % Ni( xi , eta )
26         Nx(i,j)= nx*(1.0 + ny*eta) /4.0;            % dNi/dxi ( xi , eta )
27         Ny(i,j)= ny*(1.0 + nx*xi ) /4.0;            % dNi/deta ( xi , eta )
28     end
29 end
```

The following code lists the function computing the shape functions and their derivatives for the 6-node triangular master element shown in Table 5.3.

```
1  % Compute the shape functions and their derivatives at points defined
2  % by ( xi_vec , eta_vec ) in a 2-D 6-node triangular master element with
3  % master_nodes =[0 0; 1 0; 0 1; 0.5 0; 0.5 0.5; 0 0.5];
4  % Input : xi_vec , eta_vec : coordinates of the input points
5  % Output : N: matrix storing shape function values with the format
6  %          [N1(xi_1 , eta_1 )   N1(xi_2 , eta_2 )   N1(xi_3 , eta_3 ) ...
```

```
7  %             N2(xi_1, eta_1)   N2(xi_2, eta_2)   N2(xi_3, eta_3)   ...
8  %             N3(xi_1, eta_1)   N3(xi_2, eta_2)   N3(xi_3, eta_3)   ...
9  %                 ...               ...               ...          ...]
10 % Output: Nx, Ny: matrix of dNi/dxi(xi, eta) and dNi/deta(xi, eta)
11 %                 respectively, format is the same as N
12 function [N,Nx,Ny]=CompNDNatPointsTri6(xi_vec, eta_vec)
13 np=size(xi_vec,1);     % number of points
14 N=zeros(6,np);  Nx=zeros(6,np);  Ny=zeros(6,np);  % empty matrices
15 % for loop: compute N, Nx, Ny
16 for j=1:np                          % columns for point 1,2 ...
17    xi=xi_vec(j);                    % xi-coordinate of point j
18    eta=eta_vec(j);                  % eta-coordinate of point j
19    % next 3 lines: N1(xi, eta), dN1/dxi(xi, eta), dN1/deta(xi, eta)
20    N(1,j)= 2*(1-xi-eta)*(1-xi-eta -0.5);
21    Nx(1,j)= -2*(1-xi-eta - 0.5) -2*(1-xi-eta);
22    Ny(1,j)=-2*(1-xi-eta - 0.5) -2*(1-xi-eta);
23    % next 3 lines: N2 and associated derivatives
24    N(2,j)= 2*xi*(xi - 0.5);
25    Nx(2,j)= 2*(xi - 0.5) + 2*xi;
26    Ny(2,j)= 0;
27    % next 3 lines: N3 and associated derivatives
28    N(3,j)= 2*eta*(eta - 0.5);
29    Nx(3,j)= 0;
30    Ny(3,j)= 2*(eta - 0.5) + 2*eta;
31    % next 3 lines: N4 and associated derivatives
32    N(4,j)= 4*(1-xi-eta)*xi;
33    Nx(4,j)= -4*xi + 4*(1-xi-eta);
34    Ny(4,j)= -4*xi;
35    % next 3 lines: N5 and associated derivatives
36    N(5,j)= 4*xi*eta;
37    Nx(5,j)= 4*eta;
38    Ny(5,j)= 4*xi;
39    % next 3 lines: N6 and associated derivatives
40    N(6,j)= 4*eta*(1-xi-eta);
41    Nx(6,j)= -4*eta;
42    Ny(6,j)= 4*(1-xi-eta) -4*eta;
43 end
```

Given the map from the reference coordinates to the physical coordinates, Eq. (5.67), it is possible to obtain the inverse map from (x, y) to (ξ, η), i.e.,

$$\xi = \xi(x, y) \qquad \eta = \eta(x, y). \tag{5.73}$$

However, depending on the shape of the physical elements, the expression of (ξ, η) in terms of (x, y) can be complex. More importantly, the inverse map is usually not used in a finite element analysis. Therefore, it is unnecessary to explicitly obtain the inverse map.

Table 5.2 2-D isoparametric quadrilateral elements

Element	Shape functions	Order
(4-node square, nodes 1,2,3,4 at corners)	$N_1 = \frac{1}{4}(1+\xi)(1+\eta)$ $N_2 = \frac{1}{4}(1-\xi)(1+\eta)$ $N_3 = \frac{1}{4}(1-\xi)(1-\eta)$ $N_4 = \frac{1}{4}(1+\xi)(1-\eta)$	2
(9-node square, nodes 1–4 corners, 5–8 midsides, 9 center)	$N_1 = \frac{1}{4}\xi(1+\xi)\eta(1+\eta)$ $N_2 = -\frac{1}{4}\xi(1-\xi)\eta(1+\eta)$ $N_3 = \frac{1}{4}\xi(1-\xi)\eta(1-\eta)$ $N_4 = -\frac{1}{4}\xi(1+\xi)\eta(1-\eta)$ $N_5 = \frac{1}{2}(1-\xi^2)\eta(1+\eta)$ $N_6 = -\frac{1}{2}\xi(1-\xi)(1-\eta^2)$ $N_7 = -\frac{1}{2}(1-\xi^2)\eta(1-\eta)$ $N_8 = \frac{1}{2}\xi(1+\xi)(1-\eta^2)$ $N_9 = (1-\xi^2)(1-\eta^2)$	3
(8-node serendipity square)	$N_1 = \frac{1}{4}(1+\xi)(1+\eta) - \frac{1}{2}N_8 - \frac{1}{2}N_5$ $N_2 = \frac{1}{4}(1-\xi)(1+\eta) - \frac{1}{2}N_5 - \frac{1}{2}N_6$ $N_3 = \frac{1}{4}(1-\xi)(1-\eta) - \frac{1}{2}N_6 - \frac{1}{2}N_7$ $N_4 = \frac{1}{4}(1+\xi)(1-\eta) - \frac{1}{2}N_7 - \frac{1}{2}N_8$ $N_5 = \frac{1}{2}(1-\xi^2)(1+\eta)$ $N_6 = \frac{1}{2}(1-\xi)(1-\eta^2)$ $N_7 = \frac{1}{2}(1-\xi^2)(1-\eta)$ $N_8 = \frac{1}{2}(1+\xi)(1-\eta^2)$	3

5.2 2-D Steady State Heat Transfer

Table 5.3 2-D isoparametric triangular elements

Element	Area coordinates	Shape functions	Order
(triangle with nodes 1, 2, 3 at $(0,0)$, $(1,0)$, $(0,1)$)	$l_1 = 1 - \xi - \eta$ $l_2 = \xi$ $l_3 = \eta$	$N_1 = l_1$ $N_2 = l_2$ $N_3 = l_3$	2
(triangle with nodes 1, 2, 3, 4, 5, 6)	$l_1 = 1 - \xi - \eta$ $l_2 = \xi$ $l_3 = \eta$	$N_1 = 2l_1(l_1 - 1/2)$ $N_2 = 2l_2(l_2 - 1/2)$ $N_3 = 2l_3(l_3 - 1/2)$ $N_4 = 4l_1 l_2$ $N_5 = 4l_2 l_3$ $N_6 = 4l_3 l_1$	3

Example 5.1 For 4-node quadrilateral isoparametric elements, what are the (x, y) coordinates of the point mapped from $(\xi = 0, \eta = 0)$? What is the temperature at that point?

Solution

$$x = \sum_{i=1}^{4} N_i(\xi, \eta)|_{(\xi=0, \eta=0)} x_i$$

$$= \frac{1}{4}x_1 + \frac{1}{4}x_2 + \frac{1}{4}x_3 + \frac{1}{4}x_4 = \frac{1}{4}(x_1 + x_2 + x_3 + x_4)$$

$$y = \sum_{i=1}^{n} N_i(\xi, \eta)|_{(\xi=0, \eta=0)} y_i = \frac{1}{4}(y_1 + y_2 + y_3 + y_4)$$

$$T = \sum_{i=1}^{n} N_i(\xi, \eta)|_{(\xi=0, \eta=0)} T_i = \frac{1}{4}(T_1 + T_2 + T_3 + T_4).$$

Therefore, $(\xi = 0, \eta = 0)$ is mapped to a point in the physical element with its x- and y-coordinates being the average of the x- and y-coordinates of the four nodes. The temperature at this point is the average of the nodal temperatures.

FEA for Multi-Dimensional Scalar Field Problems

When evaluating the derivatives of the unknown quantities, by using the chain rule, we have

$$\left\{\begin{array}{c} \dfrac{\partial}{\partial \xi} \\ \dfrac{\partial}{\partial \eta} \end{array}\right\} = \left[\begin{array}{cc} \dfrac{\partial x}{\partial \xi} & \dfrac{\partial y}{\partial \xi} \\ \dfrac{\partial x}{\partial \eta} & \dfrac{\partial y}{\partial \eta} \end{array}\right] \left\{\begin{array}{c} \dfrac{\partial}{\partial x} \\ \dfrac{\partial}{\partial y} \end{array}\right\} = \mathbf{J} \left\{\begin{array}{c} \dfrac{\partial}{\partial x} \\ \dfrac{\partial}{\partial y} \end{array}\right\} \quad (5.74)$$

where the matrix $\mathbf{J}$ is the Jacobian matrix relating the master element coordinate derivatives to the physical coordinate derivatives, and

$$\frac{\partial x}{\partial \xi} = \sum_i \frac{\partial N_i(\xi, \eta)}{\partial \xi} x_i, \quad \frac{\partial y}{\partial \xi} = \sum_i \frac{\partial N_i(\xi, \eta)}{\partial \xi} y_i \quad (5.75)$$

$$\frac{\partial x}{\partial \eta} = \sum_i \frac{\partial N_i(\xi, \eta)}{\partial \eta} x_i, \quad \frac{\partial y}{\partial \eta} = \sum_i \frac{\partial N_i(\xi, \eta)}{\partial \eta} y_i, \quad (5.76)$$

Note that, mathematically, the Jacobian matrix is defined as the matrix of all first-order partial derivatives of a vector-valued function. By using Eq. 5.74, when we compute $\dfrac{\partial}{\partial x}$ and $\dfrac{\partial}{\partial y}$, we can use

$$\left\{\begin{array}{c} \dfrac{\partial}{\partial x} \\ \dfrac{\partial}{\partial y} \end{array}\right\} = \mathbf{J}^{-1} \left\{\begin{array}{c} \dfrac{\partial}{\partial \xi} \\ \dfrac{\partial}{\partial \eta} \end{array}\right\}. \quad (5.77)$$

Example 5.2 A 4-node quadrilateral isoparametric element is shown in Fig. 5.22. Obtain the Jacobian matrix and its determinant.

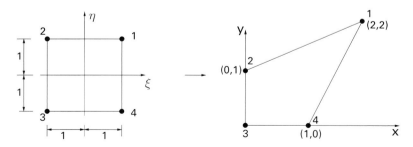

Figure 5.22 A mapped 4-node quadrilateral element.

5.2 2-D Steady State Heat Transfer

Solution

The Jacobian matrix is given by

$$\mathbf{J} = \begin{bmatrix} \dfrac{\partial x}{\partial \xi} & \dfrac{\partial y}{\partial \xi} \\ \dfrac{\partial x}{\partial \eta} & \dfrac{\partial y}{\partial \eta} \end{bmatrix} \tag{5.78}$$

and

$$\frac{\partial x}{\partial \xi} = \sum_{i=1}^{4} \frac{\partial N_i(\xi, \eta)}{\partial \xi} x_i$$

$$= \frac{\partial}{\partial \xi} \left[\frac{1}{4}(1+\xi)(1+\eta) \quad \frac{1}{4}(1-\xi)(1+\eta) \quad \frac{1}{4}(1-\xi)(1-\eta) \quad \frac{1}{4}(1+\xi)(1-\eta) \right] \begin{Bmatrix} 2 \\ 0 \\ 0 \\ 1 \end{Bmatrix}$$

$$= \left[\frac{1}{4}(1+\eta) \quad -\frac{1}{4}(1+\eta) \quad -\frac{1}{4}(1-\eta) \quad \frac{1}{4}(1-\eta) \right] \begin{Bmatrix} 2 \\ 0 \\ 0 \\ 1 \end{Bmatrix}$$

$$= \frac{3}{4} + \frac{1}{4}\eta,$$

$$\frac{\partial y}{\partial \xi} = \sum_{i=1}^{4} \frac{\partial N_i(\xi, \eta)}{\partial \xi} y_i$$

$$= \frac{\partial}{\partial \xi} \left[\frac{1}{4}(1+\xi)(1+\eta) \quad \frac{1}{4}(1-\xi)(1+\eta) \quad \frac{1}{4}(1-\xi)(1-\eta) \quad \frac{1}{4}(1+\xi)(1-\eta) \right] \begin{Bmatrix} 2 \\ 1 \\ 0 \\ 0 \end{Bmatrix}$$

$$= \left[\frac{1}{4}(1+\eta) \quad -\frac{1}{4}(1+\eta) \quad -\frac{1}{4}(1-\eta) \quad \frac{1}{4}(1-\eta)\right] \begin{Bmatrix} 2 \\ 1 \\ 0 \\ 0 \end{Bmatrix}$$

$$= \frac{1}{4}(1+\eta),$$

$$\frac{\partial x}{\partial \eta} = \sum_{i=1}^{4} \frac{\partial N_i(\xi,\eta)}{\partial \eta} x_i$$

$$= \frac{\partial}{\partial \eta}\left[\frac{1}{4}(1+\xi)(1+\eta) \quad \frac{1}{4}(1-\xi)(1+\eta) \quad \frac{1}{4}(1-\xi)(1-\eta) \quad \frac{1}{4}(1+\xi)(1-\eta)\right] \begin{Bmatrix} 2 \\ 0 \\ 0 \\ 1 \end{Bmatrix}$$

$$= \left[\frac{1}{4}(1+\xi) \quad \frac{1}{4}(1-\xi) \quad -\frac{1}{4}(1-\xi) \quad -\frac{1}{4}(1+\xi)\right] \begin{Bmatrix} 2 \\ 0 \\ 0 \\ 1 \end{Bmatrix}$$

$$= \frac{1}{4}(1+\xi),$$

$$\frac{\partial y}{\partial \eta} = \sum_{i=1}^{4} \frac{\partial N_i(\xi,\eta)}{\partial \eta} y_i$$

$$= \left[\frac{1}{4}(1+\xi) \quad \frac{1}{4}(1-\xi) \quad -\frac{1}{4}(1-\xi) \quad -\frac{1}{4}(1+\xi)\right] \begin{Bmatrix} 2 \\ 1 \\ 0 \\ 0 \end{Bmatrix}$$

$$= \frac{3}{4} + \frac{1}{4}\xi.$$

Thus,

$$\mathbf{J} = \begin{bmatrix} \frac{3}{4} + \frac{1}{4}\eta & \frac{1}{4}(1+\eta) \\ \frac{1}{4}(1+\xi) & \frac{3}{4} + \frac{1}{4}\xi \end{bmatrix},$$

and

$$\det(\mathbf{J}) = \left(\frac{3}{4} + \frac{1}{4}\eta\right)\left(\frac{3}{4} + \frac{1}{4}\xi\right) - \frac{1}{16}(1+\eta)(1+\xi).$$

In the following, we present a MATLAB function for the calculation of a Jacobian matrix at a point (xi, eta) in a 2-D master element.

```
% Compute Jacobian matrix at a point (xi, eta) in a 2-D master element
% Input: element_nodes: the physical coordinates of a 2-D element
%        in the format of [x1, y1; x2 y2; x3, y3; ...]
% Input: Nx, Ny: dN/dxi, dN/deta vector at the point
% Output: Jacobian matrix at the point
function J= CompJacobian2DatPoint(element_nodes, Nxi, Neta)
J=zeros(2,2);
for j=1:2
    J(1,j) = Nxi' * element_nodes(:,j);
    J(2,j) = Neta' * element_nodes(:,j);
end
```

In two dimensions, we also need to map an infinitesimal area in the master element to the physical element. As shown in Fig. 5.23, an infinitesimal area, $d\xi\,d\eta$, in the master element is mapped to the shaded area in the physical element. The infinitesimal area in the physical element, $d\Omega$, can be obtained as

$$d\Omega = \left\| \begin{Bmatrix} dx_1 \\ dy_1 \\ 0 \end{Bmatrix} \times \begin{Bmatrix} dx_2 \\ dy_2 \\ 0 \end{Bmatrix} \right\|$$

$$= dx_1 dy_2 - dy_1 dx_2. \tag{5.79}$$

By using the chain rule, the infinitesimal lengths dx and dy can be written as

$$dx = \frac{\partial x}{\partial \xi}d\xi + \frac{\partial x}{\partial \eta}d\eta \tag{5.80}$$

$$dy = \frac{\partial y}{\partial \xi}d\xi + \frac{\partial y}{\partial \eta}d\eta. \tag{5.81}$$

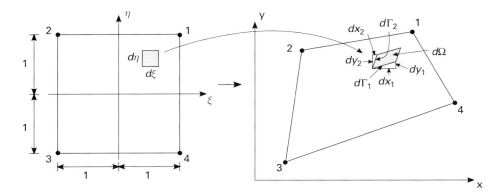

Figure 5.23 Mapping an infinitesimal area.

In fact, the above equations can be written in matrix form as

$$\left\{ \begin{array}{c} dx \\ dy \end{array} \right\} = \left[\begin{array}{cc} \frac{\partial x}{\partial \xi} & \frac{\partial x}{\partial \eta} \\ \frac{\partial y}{\partial \xi} & \frac{\partial y}{\partial \eta} \end{array} \right] \left\{ \begin{array}{c} d\xi \\ d\eta \end{array} \right\} = \mathbf{J}^T \left\{ \begin{array}{c} d\xi \\ d\eta \end{array} \right\}. \quad (5.82)$$

Since the infinitesimal line segment vector $\{dx_1 \ dy_1\}^T$ in the physical element is corresponding to the horizontal line segment $d\xi$ in the master element which has no variation in η (i.e., $d\eta = 0$), we obtain

$$dx_1 = \frac{\partial x}{\partial \xi} d\xi \quad (5.83)$$

$$dy_1 = \frac{\partial y}{\partial \xi} d\xi. \quad (5.84)$$

For the line segment vector $\{dx_2 \ dy_2\}^T$, its counterpart in the master element is $d\eta$ which has no variation in ξ (i.e., $d\xi = 0$), we have

$$dx_2 = \frac{\partial x}{\partial \eta} d\eta \quad (5.85)$$

$$dy_2 = \frac{\partial y}{\partial \eta} d\eta. \quad (5.86)$$

Substituting Eqs. (5.83–5.86), we obtain

$$d\Omega = dx_1 dy_2 - dy_1 dx_2$$

$$= \frac{\partial x}{\partial \xi} d\xi \frac{\partial y}{\partial \eta} d\eta - \frac{\partial y}{\partial \xi} d\xi \frac{\partial x}{\partial \eta} d\eta$$

$$= \left(\frac{\partial x}{\partial \xi} \frac{\partial y}{\partial \eta} - \frac{\partial y}{\partial \xi} \frac{\partial x}{\partial \eta} \right) d\xi d\eta$$

$$= det(\mathbf{J}) d\xi d\eta. \quad (5.87)$$

As the infinitesimal line segments $\{d\xi \ 0\}^T$ and $\{0 \ d\eta\}^T$ are mapped to $\{dx_1 \ dy_1\}^T$ and $\{dx_2 \ dy_2\}^T$ in the physical element, respectively, the segment length of $d\xi$ and $d\eta$ become $d\Gamma_1$ and $d\Gamma_2$ as shown in Fig. 5.23, and we have

$$d\Gamma_1 = \sqrt{dx_1^2 + dy_1^2} = \left[\sqrt{\left(\frac{\partial x}{\partial \xi}\right)^2 + \left(\frac{\partial y}{\partial \xi}\right)^2}\right] d\xi \tag{5.88}$$

and

$$d\Gamma_2 = \sqrt{dx_2^2 + dy_2^2} = \left[\sqrt{\left(\frac{\partial x}{\partial \eta}\right)^2 + \left(\frac{\partial y}{\partial \eta}\right)^2}\right] d\eta. \tag{5.89}$$

In general, we can write the mapped infinitesimal line segment length as

$$d\Gamma = \begin{cases} \left[\sqrt{\left(\frac{\partial x}{\partial \xi}\right)^2 + \left(\frac{\partial y}{\partial \xi}\right)^2}\right] d\xi & \text{if } d\Gamma \text{ is corresponding to } d\xi \\ \left[\sqrt{\left(\frac{\partial x}{\partial \eta}\right)^2 + \left(\frac{\partial y}{\partial \eta}\right)^2}\right] d\eta & \text{if } d\Gamma \text{ is corresponding to } d\eta. \end{cases} \tag{5.90}$$

Step 6: Element Matrices and Vectors

In the 1-D problems, we have illustrated the derivation of the element matrices and vectors and their assembly into the global system equations by expanding the discretized global weak form into a set of summable matrices and vectors. The global system is then obtained by summing the expanded element matrices and vectors. This procedure explains the mathematical base of the assembly process of FEA. For efficiency of implementation and computation, however, in practice the global system is often obtained by directly assembling the unexpanded element matrices and vectors of the "element weak form."

For our heat transfer problem, the 2-D plate is discretized into four square elements as shown in Fig. 5.5. We further decide the order of approximation to be linear. That is, we use 4-node linear quadrilateral elements for the finite element analysis. Note that, while the shape of the quadrilateral elements happens to be square in our example, isoparametric mapping is employed as if they were general quadrilateral elements. Therefore, this choice of mesh only makes illustration clearer without loss of generality. It is shown in Fig. 5.24 that there are a total of nine nodes in the mesh.

Note that the governing differential equation of the problem is valid everywhere in the entire domain. For our problem, the governing differential equation is given by

$$\frac{\partial}{\partial x}\left(k\frac{\partial T}{\partial x}\right) + \frac{\partial}{\partial y}\left(k\frac{\partial T}{\partial y}\right) - h_z(T - T_\infty) = 0. \tag{5.91}$$

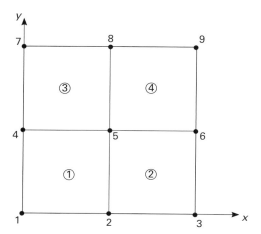

Figure 5.24 Elements and nodes of the plate.

Taking element 3 as an example, by using the method of weighted residual, we obtain the weak form for element 3 as

$$\int_{\Gamma_q}^{(3)} \delta T k \frac{\partial T}{\partial \mathbf{n}} d\Gamma - \int_{\Omega}^{(3)} k\nabla T \cdot \nabla \delta T d\Omega - \int_{\Omega}^{(3)} h_z(T - T_\infty)\delta T d\Omega = 0. \quad (5.92)$$

Note that Eq. (5.92) has the same form as the global weak form because the PDE and the derivation are the same except the domain and boundary here are element 3 and its boundary (denoted as ③ in Eq. (5.92)). In general, for a given element e, the weak form can be written as

$$\int_{\Gamma_q}^{(e)} \delta T k \frac{\partial T}{\partial \mathbf{n}} d\Gamma - \int_{\Omega}^{(e)} k\nabla T \cdot \nabla \delta T d\Omega - \int_{\Omega}^{(e)} h_z(T - T_\infty)\delta T d\Omega = 0. \quad (5.93)$$

Equation (5.93) is called an element weak form. When we examine Eq. (5.93) more closely, the first term appears to be problematic. In the global weak form the normal derivative of the temperature, $\frac{\partial T}{\partial \mathbf{n}}$ is known on Γ_q as the natural boundary condition. However, as shown in Fig. 5.25, $\frac{\partial T}{\partial \mathbf{n}}$ is not known on the interior boundary edges. In this case, additional unknowns are introduced in the element weak form. Fortunately, $\frac{\partial T}{\partial \mathbf{n}}$ is typically continuous across the elements, and it can be shown that the unknown normal derivatives $\frac{\partial T}{\partial \mathbf{n}}$ on the interior boundary of an element will be canceled by the interior boundary $\frac{\partial T}{\partial \mathbf{n}}$ of the neighboring elements when the global system is assembled. Therefore, $\frac{\partial T}{\partial \mathbf{n}}$ can be omitted for the interior boundary edges of the elements and Γ_q in the element weak form only includes the element boundary that coincides with the portion of the boundary of the global physical domain with natural boundary conditions.

5.2 2-D Steady State Heat Transfer

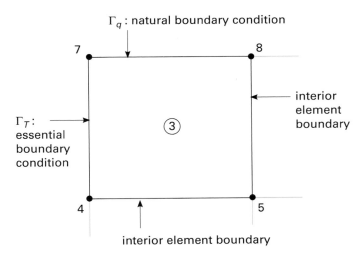

Figure 5.25 Boundary of element ③.

Having obtained the element weak form, the finite element approximations can be written as

$$T = \begin{bmatrix} N_1 & N_2 & N_3 & N_4 \end{bmatrix} \begin{Bmatrix} T_1 \\ T_2 \\ T_3 \\ T_4 \end{Bmatrix} \quad (5.94)$$

$$\nabla T = \begin{Bmatrix} \dfrac{\partial T}{\partial x} \\ \dfrac{\partial T}{\partial y} \end{Bmatrix} = \begin{bmatrix} \dfrac{\partial N_1}{\partial x} & \dfrac{\partial N_2}{\partial x} & \dfrac{\partial N_3}{\partial x} & \dfrac{\partial N_4}{\partial x} \\ \dfrac{\partial N_1}{\partial y} & \dfrac{\partial N_2}{\partial y} & \dfrac{\partial N_3}{\partial y} & \dfrac{\partial N_4}{\partial y} \end{bmatrix} \begin{Bmatrix} T_1 \\ T_2 \\ T_3 \\ T_4 \end{Bmatrix} \quad (5.95)$$

$$\delta T = \begin{bmatrix} \delta T_1 & \delta T_2 & \delta T_3 & \delta T_4 \end{bmatrix} \begin{Bmatrix} N_1 \\ N_2 \\ N_3 \\ N_4 \end{Bmatrix} \quad (5.96)$$

$$\nabla \delta T = \begin{Bmatrix} \dfrac{\partial \delta T}{\partial x} \\ \dfrac{\partial \delta T}{\partial y} \end{Bmatrix} = \begin{bmatrix} \dfrac{\partial N_1}{\partial x} & \dfrac{\partial N_2}{\partial x} & \dfrac{\partial N_3}{\partial x} & \dfrac{\partial N_4}{\partial x} \\ \dfrac{\partial N_1}{\partial y} & \dfrac{\partial N_2}{\partial y} & \dfrac{\partial N_3}{\partial y} & \dfrac{\partial N_4}{\partial y} \end{bmatrix} \begin{Bmatrix} \delta T_1 \\ \delta T_2 \\ \delta T_3 \\ \delta T_4 \end{Bmatrix}. \tag{5.97}$$

Note that all the approximations are over individual elements. The shape functions and nodal variables are all "local" (i.e., defined element-wise). For the sake of brevity, from now on, we omit the superscript "ⓔ" of the shape functions, nodal temperatures, and temperature variations. However, we keep "ⓔ" for integrals to denote element integrations. In addition, we omit "ⓔ" when the situation is self-explanatory.

Substituting the approximations into the element weak form, we obtain for the first term

$$\int_{\Gamma_q}^{(e)} \delta T k \frac{\partial T}{\partial \mathbf{n}} d\Gamma = \begin{bmatrix} \delta T_1 & \delta T_2 & \delta T_3 & \delta T_4 \end{bmatrix} \left(\int_{\Gamma_q}^{(e)} \begin{Bmatrix} N_1 \\ N_2 \\ N_3 \\ N_4 \end{Bmatrix} k \frac{\partial T}{\partial \mathbf{n}} d\Gamma \right) \tag{5.98}$$

the second term

$$\int_{\Omega}^{(e)} k \nabla T \cdot \nabla \delta T d\Omega$$

$$= \int_{\Omega}^{(e)} (\nabla \delta T)^T (k \nabla T) d\Omega$$

$$= \begin{bmatrix} \delta T_1 & \delta T_2 & \delta T_3 & \delta T_4 \end{bmatrix} \times$$

$$\left(\int_{\Omega}^{(e)} \begin{bmatrix} \dfrac{\partial N_1}{\partial x} & \dfrac{\partial N_1}{\partial y} \\ \dfrac{\partial N_2}{\partial x} & \dfrac{\partial N_2}{\partial y} \\ \dfrac{\partial N_3}{\partial x} & \dfrac{\partial N_3}{\partial y} \\ \dfrac{\partial N_4}{\partial x} & \dfrac{\partial N_4}{\partial y} \end{bmatrix} k \begin{bmatrix} \dfrac{\partial N_1}{\partial x} & \dfrac{\partial N_2}{\partial x} & \dfrac{\partial N_3}{\partial x} & \dfrac{\partial N_4}{\partial x} \\ \dfrac{\partial N_1}{\partial y} & \dfrac{\partial N_2}{\partial y} & \dfrac{\partial N_3}{\partial y} & \dfrac{\partial N_4}{\partial y} \end{bmatrix} d\Omega \right) \begin{Bmatrix} T_1 \\ T_2 \\ T_3 \\ T_4 \end{Bmatrix}$$

(5.99)

5.2 2-D Steady State Heat Transfer

the third term

$$\int_\Omega^{(e)} h_z(T - T_\infty)\delta T d\Omega$$

$$= \int_\Omega^{(e)} h_z T \delta T d\Omega - \int_\Omega^{(e)} h_z T_\infty \delta T d\Omega$$

$$= \begin{bmatrix} \delta T_1 & \delta T_2 & \delta T_3 & \delta T_4 \end{bmatrix} \left(\int_\Omega^{(e)} h_z \begin{bmatrix} N_1 \\ N_2 \\ N_3 \\ N_4 \end{bmatrix} \begin{bmatrix} N_1 & N_2 & N_3 & N_4 \end{bmatrix} d\Omega \right) \begin{Bmatrix} T_1 \\ T_2 \\ T_3 \\ T_4 \end{Bmatrix}$$

$$- \begin{bmatrix} \delta T_1 & \delta T_2 & \delta T_3 & \delta T_4 \end{bmatrix} \left(\int_\Omega^{(e)} h_z T_\infty \begin{bmatrix} N_1 \\ N_2 \\ N_3 \\ N_4 \end{bmatrix} d\Omega \right). \quad (5.100)$$

Combining all the terms, the discretized element weak form can be written as

$$\begin{bmatrix} \delta T_1 & \delta T_2 & \delta T_3 & \delta T_4 \end{bmatrix} \left(\int_{\Gamma_q}^{(e)} \begin{Bmatrix} N_1 \\ N_2 \\ N_3 \\ N_4 \end{Bmatrix} k\frac{\partial T}{\partial \mathbf{n}} d\Gamma \right) + \begin{bmatrix} \delta T_1 & \delta T_2 & \delta T_3 & \delta T_4 \end{bmatrix} \left(\int_\Omega^{(e)} h_z T_\infty \begin{bmatrix} N_1 \\ N_2 \\ N_3 \\ N_4 \end{bmatrix} d\Omega \right)$$

$$- \begin{bmatrix} \delta T_1 & \delta T_2 & \delta T_3 & \delta T_4 \end{bmatrix} \left(\int_\Omega^{(e)} \begin{bmatrix} \frac{\partial N_1}{\partial x} & \frac{\partial N_1}{\partial y} \\ \frac{\partial N_2}{\partial x} & \frac{\partial N_2}{\partial y} \\ \frac{\partial N_3}{\partial x} & \frac{\partial N_3}{\partial y} \\ \frac{\partial N_4}{\partial x} & \frac{\partial N_4}{\partial y} \end{bmatrix} k \begin{bmatrix} \frac{\partial N_1}{\partial x} & \frac{\partial N_2}{\partial x} & \frac{\partial N_3}{\partial x} & \frac{\partial N_4}{\partial x} \\ \frac{\partial N_1}{\partial y} & \frac{\partial N_2}{\partial y} & \frac{\partial N_3}{\partial y} & \frac{\partial N_4}{\partial y} \end{bmatrix} d\Omega \right) \begin{Bmatrix} T_1 \\ T_2 \\ T_3 \\ T_4 \end{Bmatrix}$$

$$- \begin{bmatrix} \delta T_1 & \delta T_2 & \delta T_3 & \delta T_4 \end{bmatrix} \left(\int_\Omega^{(e)} h_z \begin{bmatrix} N_1 \\ N_2 \\ N_3 \\ N_4 \end{bmatrix} \begin{bmatrix} N_1 & N_2 & N_3 & N_4 \end{bmatrix} d\Omega \right) \begin{Bmatrix} T_1 \\ T_2 \\ T_3 \\ T_4 \end{Bmatrix} = 0. \quad (5.101)$$

Canceling out $[\delta T_1 \ \delta T_2 \ \delta T_3 \ \delta T_4]$, Eq. (5.101) can be rewritten as

$$\left(\int_{\Gamma_q}^{(e)} \begin{Bmatrix} N_1 \\ N_2 \\ N_3 \\ N_4 \end{Bmatrix} k\frac{\partial T}{\partial \mathbf{n}} d\Gamma\right) - \left(\int_{\Omega}^{(e)} \begin{bmatrix} \frac{\partial N_1}{\partial x} & \frac{\partial N_1}{\partial y} \\ \frac{\partial N_2}{\partial x} & \frac{\partial N_2}{\partial y} \\ \frac{\partial N_3}{\partial x} & \frac{\partial N_3}{\partial y} \\ \frac{\partial N_4}{\partial x} & \frac{\partial N_4}{\partial y} \end{bmatrix} k \begin{bmatrix} \frac{\partial N_1}{\partial x} & \frac{\partial N_2}{\partial x} & \frac{\partial N_3}{\partial x} & \frac{\partial N_4}{\partial x} \\ \frac{\partial N_1}{\partial y} & \frac{\partial N_2}{\partial y} & \frac{\partial N_3}{\partial y} & \frac{\partial N_4}{\partial y} \end{bmatrix} d\Omega\right) \begin{Bmatrix} T_1 \\ T_2 \\ T_3 \\ T_4 \end{Bmatrix}$$

$$- \left(\int_{\Omega}^{(e)} h_z \begin{bmatrix} N_1 \\ N_2 \\ N_3 \\ N_4 \end{bmatrix} \begin{bmatrix} N_1 & N_2 & N_3 & N_4 \end{bmatrix} d\Omega\right) \begin{Bmatrix} T_1 \\ T_2 \\ T_3 \\ T_4 \end{Bmatrix} + \left(\int_{\Omega}^{(e)} h_z T_\infty \begin{bmatrix} N_1 \\ N_2 \\ N_3 \\ N_4 \end{bmatrix} d\Omega\right) = 0. \quad (5.102)$$

We keep all the terms containing the vector of nodal temperature on the left hand side and move the remaining terms to the right hand side. The resultant equation is given by

$$\left(\int_{\Omega}^{(e)} \begin{bmatrix} \frac{\partial N_1}{\partial x} & \frac{\partial N_1}{\partial y} \\ \frac{\partial N_2}{\partial x} & \frac{\partial N_2}{\partial y} \\ \frac{\partial N_3}{\partial x} & \frac{\partial N_3}{\partial y} \\ \frac{\partial N_4}{\partial x} & \frac{\partial N_4}{\partial y} \end{bmatrix} k \begin{bmatrix} \frac{\partial N_1}{\partial x} & \frac{\partial N_2}{\partial x} & \frac{\partial N_3}{\partial x} & \frac{\partial N_4}{\partial x} \\ \frac{\partial N_1}{\partial y} & \frac{\partial N_2}{\partial y} & \frac{\partial N_3}{\partial y} & \frac{\partial N_4}{\partial y} \end{bmatrix} d\Omega\right) \begin{Bmatrix} T_1 \\ T_2 \\ T_3 \\ T_4 \end{Bmatrix}$$

$$+ \left(\int_{\Omega}^{(e)} h_z \begin{bmatrix} N_1 \\ N_2 \\ N_3 \\ N_4 \end{bmatrix} \begin{bmatrix} N_1 & N_2 & N_3 & N_4 \end{bmatrix} d\Omega\right) \begin{Bmatrix} T_1 \\ T_2 \\ T_3 \\ T_4 \end{Bmatrix}$$

$$= \left(\int_{\Gamma_q}^{(e)} \begin{Bmatrix} N_1 \\ N_2 \\ N_3 \\ N_4 \end{Bmatrix} k\frac{\partial T}{\partial \mathbf{n}} d\Gamma\right) + \left(\int_{\Omega}^{(e)} h_z T_\infty \begin{bmatrix} N_1 \\ N_2 \\ N_3 \\ N_4 \end{bmatrix} d\Omega\right). \quad (5.103)$$

In short form,

$$\mathbf{K}^{(e)}\begin{Bmatrix} T_1 \\ T_2 \\ T_3 \\ T_4 \end{Bmatrix} + \mathbf{K}_t^{(e)}\begin{Bmatrix} T_1 \\ T_2 \\ T_3 \\ T_4 \end{Bmatrix} = \mathbf{f}^{(e)} \qquad (5.104)$$

where $\mathbf{K}^{(e)}$ is the element matrix of conductivity (or element conductivity matrix), $\mathbf{K}_t^{(e)}$ is the element matrix of convection (or element convection matrix), and $\mathbf{f}^{(e)}$ is the element load vector.

Although for simple elements such as the rectangular elements, the integrals in the element matrices and vectors can be evaluated directly on the physical element in the xy-coordinate system, in general, the integrations are performed on the master element through the isoparametric mapping. For the element matrix $\mathbf{K}^{(e)}$

$$\mathbf{K}^{(e)} = \int_\Omega^{(e)} \begin{bmatrix} \frac{\partial N_1}{\partial x} & \frac{\partial N_1}{\partial y} \\ \frac{\partial N_2}{\partial x} & \frac{\partial N_2}{\partial y} \\ \frac{\partial N_3}{\partial x} & \frac{\partial N_3}{\partial y} \\ \frac{\partial N_4}{\partial x} & \frac{\partial N_4}{\partial y} \end{bmatrix} k \begin{bmatrix} \frac{\partial N_1}{\partial x} & \frac{\partial N_2}{\partial x} & \frac{\partial N_3}{\partial x} & \frac{\partial N_4}{\partial x} \\ \frac{\partial N_1}{\partial y} & \frac{\partial N_2}{\partial y} & \frac{\partial N_3}{\partial y} & \frac{\partial N_4}{\partial y} \end{bmatrix} d\Omega \qquad (5.105)$$

the mapping between the physical element and the master element is given by

$$\int_\Omega^{(e)} \quad \rightarrow \quad \int_{-1}^{1}\int_{-1}^{1} \qquad (5.106)$$

and, from Eq. (5.87),

$$d\Omega = det(\mathbf{J})d\xi\, d\eta. \qquad (5.107)$$

As shown in Eq. (5.77), the x- and y-derivatives can be transformed to derivatives with respect to ξ and η by

$$\begin{Bmatrix} \frac{\partial}{\partial x} \\ \frac{\partial}{\partial y} \end{Bmatrix} = \mathbf{J}^{-1}\begin{Bmatrix} \frac{\partial}{\partial \xi} \\ \frac{\partial}{\partial \eta} \end{Bmatrix}. \qquad (5.108)$$

Therefore,

$$\begin{bmatrix} \frac{\partial}{\partial x} & \frac{\partial}{\partial y} \end{bmatrix} = \begin{bmatrix} \frac{\partial}{\partial \xi} & \frac{\partial}{\partial \eta} \end{bmatrix}\mathbf{J}^{-T}. \qquad (5.109)$$

Applying these relations to the shape functions, we obtain

$$\begin{bmatrix} \frac{\partial N_1}{\partial x} & \frac{\partial N_1}{\partial y} \\ \frac{\partial N_2}{\partial x} & \frac{\partial N_2}{\partial y} \\ \frac{\partial N_3}{\partial x} & \frac{\partial N_3}{\partial y} \\ \frac{\partial N_4}{\partial x} & \frac{\partial N_4}{\partial y} \end{bmatrix} = \begin{bmatrix} \frac{\partial N_1}{\partial \xi} & \frac{\partial N_1}{\partial \eta} \\ \frac{\partial N_2}{\partial \xi} & \frac{\partial N_2}{\partial \eta} \\ \frac{\partial N_3}{\partial \xi} & \frac{\partial N_3}{\partial \eta} \\ \frac{\partial N_4}{\partial \xi} & \frac{\partial N_4}{\partial \eta} \end{bmatrix} \mathbf{J}^{-T} \quad (5.110)$$

and

$$\begin{bmatrix} \frac{\partial N_1}{\partial x} & \frac{\partial N_2}{\partial x} & \frac{\partial N_3}{\partial x} & \frac{\partial N_4}{\partial x} \\ \frac{\partial N_1}{\partial y} & \frac{\partial N_2}{\partial y} & \frac{\partial N_3}{\partial y} & \frac{\partial N_4}{\partial y} \end{bmatrix} = \mathbf{J}^{-1} \begin{bmatrix} \frac{\partial N_1}{\partial \xi} & \frac{\partial N_2}{\partial \xi} & \frac{\partial N_3}{\partial \xi} & \frac{\partial N_4}{\partial \xi} \\ \frac{\partial N_1}{\partial \eta} & \frac{\partial N_2}{\partial \eta} & \frac{\partial N_3}{\partial \eta} & \frac{\partial N_4}{\partial \eta} \end{bmatrix}. \quad (5.111)$$

Substituting Eqs. (5.106, 5.107, 5.110, 5.111) into Eq. (5.105), we obtain the mapped element matrix

$$\mathbf{K}^{(e)} = \int_{-1}^{1} \int_{-1}^{1} \begin{bmatrix} \frac{\partial N_1}{\partial \xi} & \frac{\partial N_1}{\partial \eta} \\ \frac{\partial N_2}{\partial \xi} & \frac{\partial N_2}{\partial \eta} \\ \frac{\partial N_3}{\partial \xi} & \frac{\partial N_3}{\partial \eta} \\ \frac{\partial N_4}{\partial \xi} & \frac{\partial N_4}{\partial \eta} \end{bmatrix} \mathbf{J}^{-T} k \mathbf{J}^{-1} \begin{bmatrix} \frac{\partial N_1}{\partial \xi} & \frac{\partial N_2}{\partial \xi} & \frac{\partial N_3}{\partial \xi} & \frac{\partial N_4}{\partial \xi} \\ \frac{\partial N_1}{\partial \eta} & \frac{\partial N_2}{\partial \eta} & \frac{\partial N_3}{\partial \eta} & \frac{\partial N_4}{\partial \eta} \end{bmatrix} det(\mathbf{J}) d\xi d\eta.$$

(5.112)

Note that the constant thermal conductivity k remains the same in the mapped form. We take element 2 shown in Fig. 5.24 to illustrate the calculation of $\mathbf{K}^{(e)}$. The nodal coordinates of element 2 is shown in Fig. 5.26.

From the shape functions of the master element, we have

$$\frac{\partial N_1}{\partial \xi} = \frac{1}{4}(1+\eta), \quad \frac{\partial N_2}{\partial \xi} = -\frac{1}{4}(1+\eta), \quad \frac{\partial N_3}{\partial \xi} = -\frac{1}{4}(1-\eta), \quad \frac{\partial N_4}{\partial \xi} = \frac{1}{4}(1-\eta)$$

$$\frac{\partial N_1}{\partial \eta} = \frac{1}{4}(1+\xi), \quad \frac{\partial N_2}{\partial \eta} = \frac{1}{4}(1-\xi), \quad \frac{\partial N_3}{\partial \eta} = -\frac{1}{4}(1-\xi), \quad \frac{\partial N_4}{\partial \eta} = -\frac{1}{4}(1+\xi).$$

(5.113)

Therefore,

$$\frac{\partial x}{\partial \xi} = \sum_{i=1}^{4} \frac{\partial N_i}{\partial \xi} x_i$$

$$= \frac{1}{4}(1+\eta)x_1 - \frac{1}{4}(1+\eta)x_2 - \frac{1}{4}(1-\eta)x_3 + \frac{1}{4}(1-\eta)x_4.$$

5.2 2-D Steady State Heat Transfer

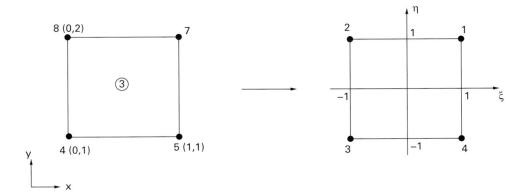

Figure 5.26 Mapping of element ③.

$$= \frac{1}{4}(1+\eta) \cdot 1 - \frac{1}{4}(1+\eta) \cdot 0 - \frac{1}{4}(1-\eta) \cdot 0 + \frac{1}{4}(1-\eta) \cdot 1$$
$$= \frac{1}{2} \tag{5.114}$$

$$\frac{\partial x}{\partial \eta} = \sum_{i=1}^{4} \frac{\partial N_i}{\partial \eta} x_i$$
$$= \frac{1}{4}(1+\xi) \cdot 1 + \frac{1}{4}(1-\xi) \cdot 0 - \frac{1}{4}(1-\xi) \cdot 0 - \frac{1}{4}(1+\xi) \cdot 1 = 0 \tag{5.115}$$

$$\frac{\partial y}{\partial \xi} = \sum_{i=1}^{4} \frac{\partial N_i}{\partial \xi} y_i$$
$$= \frac{1}{4}(1+\eta) \cdot 2 - \frac{1}{4}(1+\eta) \cdot 2 - \frac{1}{4}(1-\eta) \cdot 1 + \frac{1}{4}(1-\eta) \cdot 1 = 0 \tag{5.116}$$

$$\frac{\partial y}{\partial \eta} = \sum_{i=1}^{4} \frac{\partial N_i}{\partial \eta} y_i$$
$$= \frac{1}{4}(1+\xi) \cdot 2 + \frac{1}{4}(1-\xi) \cdot 2 - \frac{1}{4}(1-\xi) \cdot 1 - \frac{1}{4}(1+\xi) \cdot 1 = \frac{1}{2}. \tag{5.117}$$

The Jacobian matrix can be obtained as

$$\mathbf{J} = \begin{bmatrix} \dfrac{\partial x}{\partial \xi} & \dfrac{\partial y}{\partial \xi} \\ \dfrac{\partial x}{\partial \eta} & \dfrac{\partial y}{\partial \eta} \end{bmatrix} = \begin{bmatrix} \dfrac{1}{2} & 0 \\ 0 & \dfrac{1}{2} \end{bmatrix}. \tag{5.118}$$

Therefore det($\mathbf{J}$) = 1/4. It is easy to check that the area of element 2 is 1/4 of the area of the master element. Therefore,

$$\mathbf{K}^{\circledS} = \int_{-1}^{1}\int_{-1}^{1} \begin{bmatrix} \frac{1}{4}(1+\eta) & \frac{1}{4}(1+\xi) \\ -\frac{1}{4}(1+\eta) & \frac{1}{4}(1-\xi) \\ -\frac{1}{4}(1-\eta) & -\frac{1}{4}(1-\xi) \\ \frac{1}{4}(1-\eta) & -\frac{1}{4}(1+\xi) \end{bmatrix} \begin{bmatrix} 2 & 0 \\ 0 & 2 \end{bmatrix}$$

$$k \begin{bmatrix} 2 & 0 \\ 0 & 2 \end{bmatrix} \begin{bmatrix} \frac{1}{4}(1+\eta) & -\frac{1}{4}(1+\eta) & -\frac{1}{4}(1-\eta) & \frac{1}{4}(1-\eta) \\ \frac{1}{4}(1+\xi) & \frac{1}{4}(1-\xi) & -\frac{1}{4}(1-\xi) & -\frac{1}{4}(1+\xi) \end{bmatrix} \frac{1}{4} d\xi d\eta$$

$$= \frac{k}{16} \int_{-1}^{1}\int_{-1}^{1} \begin{bmatrix} (1+\eta)^2+(1+\xi)^2 & -(1+\eta)^2+(1-\xi^2) & -(1-\eta^2)-(1-\xi^2) & (1-\eta^2)-(1+\xi)^2 \\ -(1+\eta)^2+(1-\xi^2) & (1+\eta)^2+(1-\xi)^2 & (1-\eta^2)-(1-\xi)^2 & -(1-\eta^2)-(1-\xi^2) \\ -(1-\eta^2)-(1-\xi^2) & (1-\eta^2)-(1-\xi)^2 & (1-\eta)^2+(1-\xi)^2 & -(1-\eta)^2+(1-\xi^2) \\ (1-\eta^2)-(1+\xi)^2 & -(1-\eta^2)-(1-\xi^2) & -(1-\eta)^2+(1-\xi^2) & (1+\eta)^2+(1+\xi)^2 \end{bmatrix} d\xi d\eta. \quad (5.119)$$

Equation (5.119) can be evaluated by using the Gauss quadrature. In 2-D, the Gauss quadrature can be constructed by using two successive 1-D Gauss quadrature processes, as shown below (also see Chapter 3)

$$I = \int_{-1}^{1}\int_{-1}^{1} F(\xi,\eta)d\xi d\eta = \int_{-1}^{1}\left(\int_{-1}^{1} F(\xi,\eta)d\xi\right) d\eta$$

$$\approx \int_{-1}^{1}\left(\sum_{k=1}^{nk} F(\xi_k,\eta)w_k\right) d\eta \approx \sum_{h=1}^{nh}\left(\sum_{k=1}^{nk} F(\xi_k,\eta_h)w_k\right) w_h$$

$$= \sum_{h=1}^{nh}\sum_{k=1}^{nk} F(\xi_k,\eta_h)w_k w_h. \quad (5.120)$$

Equation (5.120) can be rewritten in short form as

$$I \approx \sum_{g=1}^{ng} F(\xi_g,\eta_g)w_g \quad (5.121)$$

where the total number of Gauss points is the product of the 1-D Gauss points, i.e., $ng = nh \times nk$, the Gauss points (ξ_g, η_g) has the ξ-coordinate from ξ_k, $k = 1, 2, \ldots, nk$ and the the η-coordinate from η_h, $h = 1, 2, \ldots, nh$, respectively, and $w_g = w_k \times w_h$. Tables 5.4–5.6 provide the 2-D Gauss quadrature rules and recommended usage for quadrilateral and triangular elements. Note that the numerical quadrature is performed on the master elements. A MATLAB function obtaining the Gauss points and weights for the 4-node square master element is listed below. Note that the function only returns the results for 1×1, 2×2 or 3×3 Gauss quadrature rules. However, it is straightforward to add more Gauss quadrature rules in the function, which is left as an exercise to the reader.

```
1  % Get Gauss points and weights in the 2x2 square master element
2  % Input: rows, cols: the rows x cols point Gauss quadrature
3  % Output: gauss_points : vectors stores the locations of Gauss points
4  % Output: gauss_weights : the corresponding weights of the Gauss points
5  function [gauss_points, gauss_weights]=GetQuadGauss(rows, cols)
6  % set up empty return variables
7  gauss_points=zeros(rows*cols,2);
8  gauss_weights=zeros(rows*cols,1);
9  % if-else block: assign Gauss points and weights for
10 % the given input parameters
11 % currently only allows 1x1, 2x2 and 3x3 point Gauss quadrature
12 % can be extended if needed
13 if rows==1 && cols==1                  % 1x1 point Gauss quadrature
14    x=0;                                % 1-D Gauss points
15    w=4;                                % 1-D weights
16 elseif rows==2 && cols==2              % 2x2 point Gauss quadrature
17    x=[-1.0/sqrt(3) 1.0/sqrt(3) ];      % 1-D Gauss points
18    w=[1 1]'*[1 1];                     % 1-D weights
19 elseif rows==3 && cols==3              % 3x3 point Gauss quadrature
20    x=[-sqrt(3/5) 0 sqrt(3/5) ];        % 1-D Gauss points
21    w=[5/98/9 5/9]'*[5/9 8/9 5/9];      % 1-D weights
22 else
23    fprintf('Error calling GetQuadGauss\n'); % error message
24    return;
25 end
26 % for-loop block: set up the 2-D Gauss points and weights
27 k=1;
28 for i=1:rows
29    for j=1:cols
30       gauss_points(k,1:2)=[x(i) x(j)];
31       gauss_weights(k)=w(i,j);
32       k=k+1;
33    end
34 end
```

By following the same procedure of calculating $\mathbf{K}^{(e)}$, $\mathbf{K}_t^{(e)}$ in Eq. (5.104) can be obtained.

$$\mathbf{K}_t^{(e)} = \int_\Omega^{(e)} h_z \begin{Bmatrix} N_1 \\ N_2 \\ N_3 \\ N_4 \end{Bmatrix} \begin{bmatrix} N_1 & N_2 & N_3 & N_4 \end{bmatrix} d\Omega. \quad (5.122)$$

Table 5.4 Gauss quadrature rules for quadrilateral elements

Gauss points	Degree of precision	Gauss points	ξ-coordinates	η-coordinates	Weights
1 (1×1)	1		$\xi_1 = 0$	$\eta_1 = 0$	$w_1 = 4$
4 (2×2)	3		$\xi_1 = \sqrt{1/3}$ $\xi_2 = -\sqrt{1/3}$ $\xi_3 = -\sqrt{1/3}$ $\xi_4 = \sqrt{1/3}$	$\eta_1 = \sqrt{1/3}$ $\eta_2 = \sqrt{1/3}$ $\eta_3 = -\sqrt{1/3}$ $\eta_4 = -\sqrt{1/3}$	$w_{1,2,3,4} = 1$
9 (3×3)	5		$\xi_1 = \sqrt{3/5}$ $\xi_2 = 0$ $\xi_3 = -\sqrt{3/5}$ $\xi_4 = -\sqrt{3/5}$ $\xi_5 = -\sqrt{3/5}$ $\xi_6 = 0$ $\xi_7 = \sqrt{3/5}$ $\xi_8 = \sqrt{3/5}$ $\xi_9 = 0$	$\eta_1 = \sqrt{3/5}$ $\eta_2 = \sqrt{3/5}$ $\eta_3 = \sqrt{3/5}$ $\eta_4 = 0$ $\eta_5 = -\sqrt{3/5}$ $\eta_6 = -\sqrt{3/5}$ $\eta_7 = -\sqrt{3/5}$ $\eta_8 = 0$ $\eta_9 = 0$	$w_{1,3,5,7} = \dfrac{5}{9} \times \dfrac{5}{9}$ $w_{2,4,6,8} = \dfrac{5}{9} \times \dfrac{8}{9}$ $w_9 = \dfrac{8}{9} \times \dfrac{8}{9}$

Table 5.5 Gauss quadrature rules on triangular elements: $\int\int F(\xi,\eta)d\xi\,d\eta = \frac{1}{2}\sum w_i F(\xi_i,\eta_i)$

Gauss points	Degree of precision	Gauss points	ξ-coordinates	η-coordinates	Weights
1	1		$\xi_1 = 1/3$	$\eta_1 = 1/3$	$w_1 = 1$
3	2		$\xi_1 = \dfrac{1}{6}$ $\xi_2 = \dfrac{2}{3}$ $\xi_3 = \dfrac{1}{6}$	$\eta_1 = \dfrac{1}{6}$ $\eta_2 = \dfrac{1}{6}$ $\eta_3 = \dfrac{2}{3}$	$w_{1,2,3} = \dfrac{1}{3}$
7	5		$\xi_1 = 0.10128\ 65073\ 235$ $\xi_2 = 0.79742\ 69853\ 531$ $\xi_3 = \xi_1$ $\xi_4 = 0.47014\ 20641\ 051$ $\xi_5 = \xi_4$ $\xi_6 = 0.05971\ 58717\ 898$ $\xi_7 = \dfrac{1}{3}$	$\eta_1 = \xi_1$ $\eta_2 = \xi_1$ $\eta_3 = \xi_2$ $\eta_4 = \xi_6$ $\eta_5 = \xi_4$ $\eta_6 = \xi_4$ $\eta_7 = \xi_7$	$w_{1,2,3} = 0.12593\ 91805\ 448$ $w_{4,5,6} = 0.13239\ 41527\ 885$ $w_7 = 0.225$

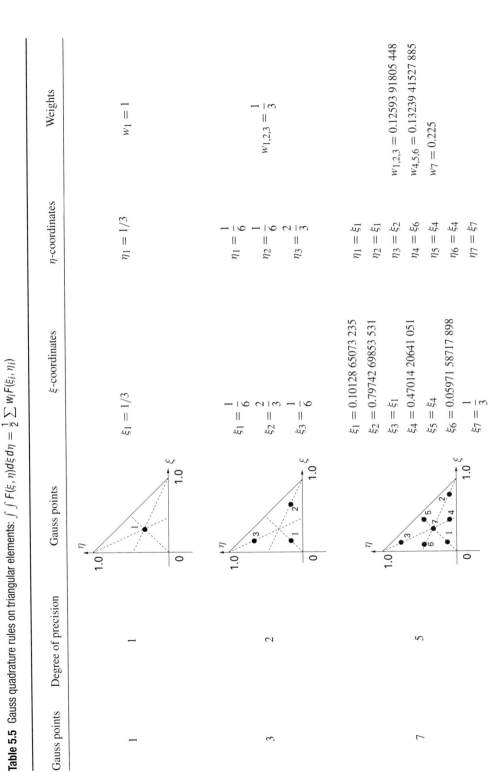

Table 5.6 Recommended Gauss quadrature order for the evaluation of isoparametric element matrices

2-D elements	Gauss points
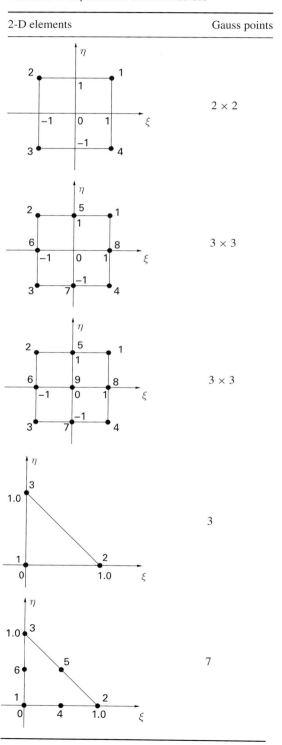	2 × 2
	3 × 3
	3 × 3
	3
	7

5.2 2-D Steady State Heat Transfer

Similar to $\mathbf{K}^{(e)}$, $\mathbf{K}_t^{(e)}$ is mapped to the master element as

$$\mathbf{K}_t^{(e)} = \int_{-1}^{1}\int_{-1}^{1} h_z \begin{Bmatrix} N_1 \\ N_2 \\ N_3 \\ N_4 \end{Bmatrix} \begin{bmatrix} N_1 & N_2 & N_3 & N_4 \end{bmatrix} det(\mathbf{J}) d\xi d\eta$$

$$= \int_{-1}^{1}\int_{-1}^{1} h_z \begin{bmatrix} N_1^2 & N_1N_2 & N_1N_3 & N_1N_4 \\ N_2N_1 & N_2^2 & N_2N_3 & N_2N_4 \\ N_3N_1 & N_3N_2 & N_3^2 & N_3N_4 \\ N_4N_1 & N_4N_2 & N_4N_3 & N_4^2 \end{bmatrix} det(\mathbf{J}) d\xi d\eta$$

$$= \int_{-1}^{1}\int_{-1}^{1} \frac{h_z}{16} \begin{bmatrix} (1+\xi)^2(1+\eta)^2 & (1-\xi^2)(1+\eta)^2 & (1-\xi^2)(1-\eta^2) & (1+\xi)^2(1-\eta^2) \\ (1-\xi^2)(1+\eta)^2 & (1-\xi)^2(1+\eta)^2 & (1-\xi)^2(1-\eta^2) & (1-\xi^2)(1-\eta^2) \\ (1-\xi^2)(1-\eta^2) & (1-\xi)^2(1-\eta^2) & (1-\xi)^2(1-\eta)^2 & (1-\xi^2)(1-\eta)^2 \\ (1+\xi)^2(1-\eta^2) & (1-\xi^2)(1-\eta^2) & (1-\xi^2)(1-\eta)^2 & (1+\xi)^2(1-\eta)^2 \end{bmatrix} det(\mathbf{J}) d\xi d\eta. \quad (5.123)$$

Taking element 3 as an example, the 2×2 Gauss quadrature of Eq. (5.119) and Eq. (5.123) gives, respectively,

$$\mathbf{K}^{(3)} = \begin{bmatrix} 3.3333 & -0.8333 & -1.6667 & -0.8333 \\ -0.8333 & 3.3333 & -0.8333 & -1.6667 \\ -1.6667 & -0.8333 & 3.3333 & -0.8333 \\ -0.8333 & -1.6667 & -0.8333 & 3.3333 \end{bmatrix} \quad (5.124)$$

and

$$\mathbf{K}_t^{(3)} = \begin{bmatrix} 44.4444 & 22.2222 & 11.1111 & 22.2222 \\ 22.2222 & 44.4444 & 22.2222 & 11.1111 \\ 11.1111 & 22.2222 & 44.4444 & 22.2222 \\ 22.2222 & 11.1111 & 22.2222 & 44.4444 \end{bmatrix}. \quad (5.125)$$

Note that, as the elements in our example are all of the same shape and equal size, the element $\mathbf{K}^{(e)}$ and $\mathbf{K}_t^{(e)}$ matrices are the same for all the elements. While this is convenient for this special mesh, it is not the case in general.

On the right hand side of Eq. (5.104) is the element vector, which is the sum of the last two terms of Eq. (5.103).

$$\mathbf{f}^{(e)} = \int_{\Gamma_q}^{(e)} \begin{Bmatrix} N_1 \\ N_2 \\ N_3 \\ N_4 \end{Bmatrix} k \frac{\partial T}{\partial \mathbf{n}} d\Gamma + \int_{\Omega}^{(e)} h_z T_\infty \begin{Bmatrix} N_1 \\ N_2 \\ N_3 \\ N_4 \end{Bmatrix} d\Omega. \qquad (5.126)$$

The second term on the right hand side contains the element integrals which can be calculated by using 2-D Gauss quadrature, i.e.,

$$\int_{\Omega}^{(e)} h_z T_\infty \begin{Bmatrix} N_1 \\ N_2 \\ N_3 \\ N_4 \end{Bmatrix} d\Omega = \int_{-1}^{1} \int_{-1}^{1} h_z T_\infty \begin{Bmatrix} N_1 \\ N_2 \\ N_3 \\ N_4 \end{Bmatrix} det(\mathbf{J}) d\xi d\eta$$

$$= \int_{-1}^{1} \int_{-1}^{1} \frac{h_z T_\infty}{4} \begin{Bmatrix} (1+\xi)(1+\eta) \\ (1-\xi)(1+\eta) \\ (1-\xi)(1-\eta) \\ (1+\xi)(1-\eta) \end{Bmatrix} det(\mathbf{J}) d\xi d\eta.$$

(5.127)

The first term, however, contains boundary integrals over Γ_q, i.e.,

$$\int_{\Gamma_q}^{(e)} \begin{Bmatrix} N_1 \\ N_2 \\ N_3 \\ N_4 \end{Bmatrix} k \frac{\partial T}{\partial \mathbf{n}} d\Gamma. \qquad (5.128)$$

In the isoparametric mapping, the boundary integration (line integral) is required to be transformed to the boundary of the master element, i.e., the boundary Γ_q is required to be mapped to the boundary of the master element. As shown in Fig. 5.23 and Eq. (5.90), the mapping of an infinitesimal line segment between the physical and the master elements can be written as

$$d\Gamma = \begin{cases} \sqrt{\left(\dfrac{\partial x}{\partial \xi}\right)^2 + \left(\dfrac{\partial y}{\partial \xi}\right)^2}\, d\xi & \text{if } d\Gamma \text{ is corresponding to } d\xi \\[2ex] \sqrt{\left(\dfrac{\partial x}{\partial \eta}\right)^2 + \left(\dfrac{\partial y}{\partial \eta}\right)^2}\, d\eta & \text{if } d\Gamma \text{ is corresponding to } d\eta. \end{cases}$$

(5.129)

As shown in Fig. 5.27, the boundary edge of the physical element can be corresponding to either a horizontal or a vertical edge of the master element. The boundary integrals can then be transformed to the master element edge as

$$\int_{\Gamma_q}^{(e)} \begin{Bmatrix} N_1(x,y) \\ N_2(x,y) \\ N_3(x,y) \\ N_4(x,y) \end{Bmatrix} k\frac{\partial T}{\partial n}(x,y) d\Gamma$$

$$= \int_{-1}^{1} \begin{Bmatrix} N_1(\xi,\eta) \\ N_2(\xi,\eta) \\ N_3(\xi,\eta) \\ N_4(\xi,\eta) \end{Bmatrix} k\frac{\partial T}{\partial n}(\xi,\eta) \sqrt{\left(\frac{\partial x}{\partial \xi}\right)^2 + \left(\frac{\partial y}{\partial \xi}\right)^2}\, d\xi \qquad (5.130)$$

when the Γ_q edge of the physical element is mapped to a horizontal edge of the master element, and

$$\int_{\Gamma_q}^{(e)} \begin{Bmatrix} N_1(x,y) \\ N_2(x,y) \\ N_3(x,y) \\ N_4(x,y) \end{Bmatrix} k\frac{\partial T}{\partial n}(x,y) d\Gamma$$

$$= \int_{-1}^{1} \begin{Bmatrix} N_1(\xi,\eta) \\ N_2(\xi,\eta) \\ N_3(\xi,\eta) \\ N_4(\xi,\eta) \end{Bmatrix} k\frac{\partial T}{\partial n}(\xi,\eta) \sqrt{\left(\frac{\partial x}{\partial \eta}\right)^2 + \left(\frac{\partial y}{\partial \eta}\right)^2}\, d\eta \qquad (5.131)$$

when the heat flux edge of the physical element is mapped to a vertical edge of the master element. The transformed integrals in Eqs. (5.130, 5.131) can be calculated in the master element by using 1-D Gauss quadrature as shown in Fig. 5.27. For example, 2-point Gauss quadrature of Eq. (5.130) on the horizontal edge connecting node 1 and node 2 of the master element shown in Fig. 5.27 gives

$$I \approx \sum_{g=1}^{2} \begin{Bmatrix} N_1(\xi_g,1) \\ N_2(\xi_g,1) \\ N_3(\xi_g,1) \\ N_4(\xi_g,1) \end{Bmatrix} k\frac{\partial T}{\partial n}(\xi_g,1) \sqrt{\left(\frac{\partial x}{\partial \xi}\bigg|_{\xi=\xi_g,\eta=1}\right)^2 + \left(\frac{\partial y}{\partial \xi}\bigg|_{\xi=\xi_g,\eta=1}\right)^2}\, w_g.$$

(5.132)

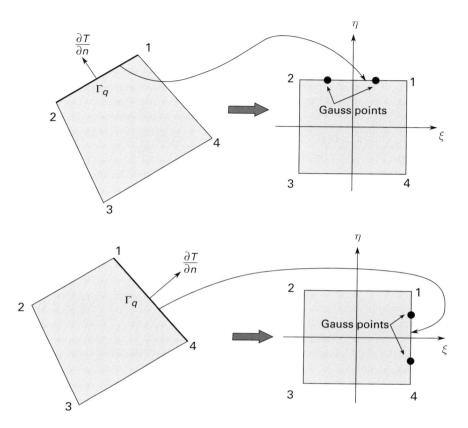

Figure 5.27 Surface heat flux calculation.

Note that the shape functions in Eq. (5.132) are the 2-D shape functions of the master element evaluated at the two Gauss points on the top edge connecting node 1 and node 2. Typically, boundary condition $\frac{\partial T}{\partial \mathbf{n}}$ is given at the nodes. The term $\frac{\partial T}{\partial \mathbf{n}}(\xi_g, 1)$ in Eq. (5.132) can be calculated by using 1-D approximation on the the top edge connecting node 1 and node 2. This approach, however, requires extra 1-D shape functions. A convenient approach is to use the same set of 2-D shape functions to approximate $\frac{\partial T}{\partial \mathbf{n}}$ on the edge. In this example, we can obtain

$$\frac{\partial T}{\partial \mathbf{n}}(\xi_g, 1) = N_1(\xi_g, 1) \left.\frac{\partial T}{\partial \mathbf{n}}\right|_{node1} + N_2(\xi_g, 1) \left.\frac{\partial T}{\partial \mathbf{n}}\right|_{node2}. \tag{5.133}$$

As the locations and weights of the Gauss points on the edges of the master element are needed in the calculation of the surface heat flux, a MATLAB function obtaining the Gauss points and weights on the edges of the master elements is listed below.

```
1  % Get Gauss points and weights on edges of the square master element
2  % Input: neg: number of Gauss points on each edge
3  % Output: gauss_points: vectors stores the locations of Gauss points
4  % Output: gauss_weights: the corresponding weights of the Gauss points
```

```
5   function [gauss_points, gauss_weights]=GetQuadEdgeGauss(neg)
6   % set up empty return variables
7   gauss_points=zeros(neg*4,2);
8   gauss_weights=zeros(neg*4,1);
9   % switch block, assign 1-D Gauss points and weights for
10  % the given input parameter
11  switch neg
12    case 1                           % 1 Gauss point per edge
13      x=0;
14      w=2.0;
15    case 2                           % 2 Gauss points per edge
16      x=[-1.0/sqrt(3) 1.0/sqrt(3)]';
17      w=[1 1]';
18    case 3                           % 3 Gauss points per edge
19      x=[-sqrt(3/5) 0 sqrt(3/5)]';
20      w=[5/9 8/9 5/9]';
21    case 4                           % 4 Gauss points per edge
22      x=[-0.861136 -0.339981 0.339981 0.861136]';
23      w=[0.34785 0.652145 0.652145 0.34785]';
24    otherwise                        % max 4 Gauss points here
25      fprintf('Error calling GetQuadEdgeGauss\n') % add more if needed
26  end
27  % set up the Gauss points and weights
28  plus_ones=ones(neg,1);              % neg x 1 vector of 1.0
29  minus_ones=-1*ones(neg,1);          % neg x 1 vector of -1.0
30  gauss_points (:,1) =[-x' minus_ones' x' plus_ones ']';
31  gauss_points (:,2) =[plus_ones' -x' minus_ones' x ']';
32  gauss_weights (:,1) =[w' w' w' w']';
```

Following the Gauss quadrature procedure described above, the right hand side vector Eq. (5.126) can be evaluated for all the elements. For the plate heat transfer example, the vectors for the four elements are obtained to be

$$\mathbf{f}^{①} = \begin{Bmatrix} 0 \\ 0 \\ -750 \\ -750 \end{Bmatrix} + \begin{Bmatrix} 3000 \\ 3000 \\ 3000 \\ 3000 \end{Bmatrix} \quad \mathbf{f}^{②} = \begin{Bmatrix} -500 \\ 0 \\ -750 \\ -1250 \end{Bmatrix} + \begin{Bmatrix} 3000 \\ 3000 \\ 3000 \\ 3000 \end{Bmatrix} \quad (5.134)$$

$$\mathbf{f}^{③} = \begin{Bmatrix} -250 \\ -250 \\ 0 \\ 0 \end{Bmatrix} + \begin{Bmatrix} 3000 \\ 3000 \\ 3000 \\ 3000 \end{Bmatrix} \quad \mathbf{f}^{④} = \begin{Bmatrix} -750 \\ -250 \\ 0 \\ -500 \end{Bmatrix} + \begin{Bmatrix} 3000 \\ 3000 \\ 3000 \\ 3000 \end{Bmatrix} \quad (5.135)$$

where the first vector on the right hand side is the outgoing heat at the sides of the plate, and the second vector is the ambient temperature part of the convection heat loss.

Step 7: Global Matrix and Vector Assembly

The mapping between the local node index and the global node index is the key in the assembly of the global matrices. Assuming the global node indices have been assigned in the meshing process as shown in Fig. 5.28, the local node indices, 1, 2, 3, and 4 of element 3 correspond to the global node indices 8, 7, 4, and 5, respectively. The element matrix entries should be added to the proper positions corresponding to the global node indices in the global matrices. The global vector is assembled in the same fashion, as shown in Fig. 5.28. The mapping of the local and global node indices is implicitly stored in the "elements.dat" input file and can be readily retrieved. A MATLAB function that prepares the vector of global node indices of the local nodes, as well as the physical coordinates of a given element is shown below.

```
1  % Set up (1) the physical coordinates of an element and (2) global
2  % node IDs of the nodes of the element
3  % Input: ele: element ID
4  % Input: nodes, elements: input nodes and elements matrices
5  %             (i.e. nodes.dat and elements.dat)
6  % Output: element_nodes: the physical coordinates of a 2-D element
7  %             in the format of [x1, y1; x2 y2; x3, y3; ...]
8  % Output: node_id_map: a column vector storing the global node IDs of
9  %             local nodes 1, 2, 3, ...
10 function [element_nodes, node_id_map]= SetElementNodes(ele, nodes, elements)
11 n_nodes_per_element=size(elements,2)-1;  % number of nodes in element
12 dimension=size(nodes,2)-1;                % dimension of the element
13 element_nodes=zeros(n_nodes_per_element,dimension);  % result matrix
14 node_id_map=zeros(n_nodes_per_element,1);            % result vector
15 % for-loop block: set up element_nodes and node_id_map
16 for i=1:n_nodes_per_element
17     global_node_id=elements(ele,i+1);
18     for j=1:dimension
19         element_nodes(i,j) = nodes(global_node_id,j+1);
20     end
21     node_id_map(i,1) = global_node_id;
22 end
```

In general, the dimension of the global coefficient matrix is $(n \times nodal\_DOF) \times (n \times nodal\_DOF)$ where n is the number of nodes in the mesh and $nodal\_DOF$ is the nodal degrees of freedom. For scalar field problems, $nodal\_DOF = 1$. The size of the global right hand side vector is $(n \times nodal\_DOF) \times 1$. Therefore, the global matrix and vector dimension of the heat transfer problem are simply $n \times n$ and $n \times 1$, respectively. For the 9-node mesh, $n = 9$. The following MATLAB code completes the assembly process of the global coefficient matrix.

```matlab
% Assemble element matrix ke into the global matrix K
% Input : K, ke: global and local matrices, respectively
%         node_id_map: column vector of global node IDs corresponding
%         to the local nodes
% Input: ndDOF: nodal degrees of freedom
% Output: assembled global matrix K
function [K]= AssembleGlobalMatrix(K,ke,node_id_map,ndDOF)
n_nodes_per_element= size(node_id_map,1); % number of nodes in element
% for-loop block: assembly process
for i = 1:n_nodes_per_element            % loop over the nodes
  row_node = node_id_map(i,1);           % global ID for row node
  row=ndDOF*row_node − (ndDOF−1);        % row number in the global K
  for j = 1:n_nodes_per_element          % loop over the nodes
    col_node = node_id_map(j,1);         % global ID for col node
    col=ndDOF*col_node − (ndDOF−1);      % column number in the global K
    K(row:row+ndDOF−1, col:col+ndDOF−1)= ... % add entry or
      K(row:row+ndDOF−1, col:col+ndDOF−1) + ... % block of ke onto
      ke((i−1)*ndDOF+1:i*ndDOF,(j−1)*ndDOF+1:j*ndDOF); % K
  end
end
```

The following MATLAB code completes the assembly process of the global right hand side vector. Note that both assembly functions are sufficiently general for assembling other global matrices and vectors, as long as appropriate input and output are specified.

```matlab
% Assemble element vector fe into the global vector F
% Input: F, fe : global and local vectors, respectively
%         node_id_map: column vector of global node IDs corresponding
%         to the local nodes
% Input: ndDOF: nodal degrees of freedom
% Output: assembled global vector F
function [F]= AssembleGlobalVector(F,fe,node_id_map,ndDOF)
n_nodes_per_element= size(node_id_map,1); % number of nodes in element
% for-loop block: assembly process
for i = 1:n_nodes_per_element            % loop over the nodes
  row_node = node_id_map(i,1);           % global ID for row node
  row=ndDOF*row_node − (ndDOF−1);        % row number in the global F
  F(row:row+ndDOF−1,1) = F(row:row+ndDOF−1,1) + ... % add fe to F
                        fe((i−1)*ndDOF+1:i*ndDOF,1);
end
```

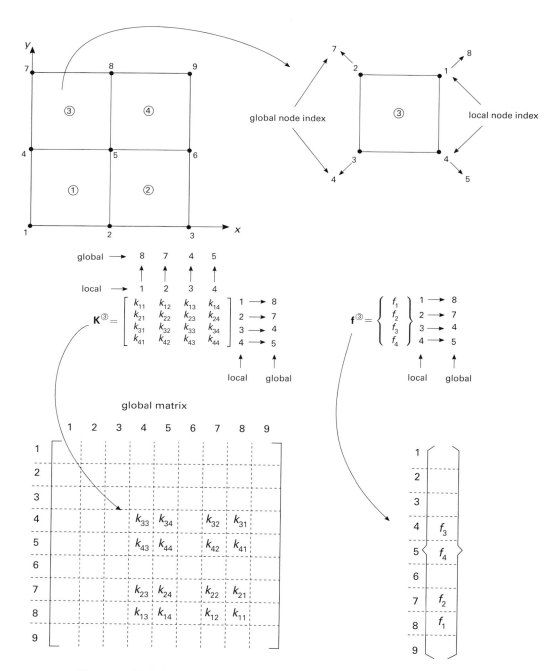

Figure 5.28 Global matrix and vector assembly.

Step 8: Apply the Boundary Conditions

The natural (heat flux) boundary condition is already included in the weak form boundary integral. The application of the essential (temperature) boundary condition can be implemented by using several approaches. We have discussed the direct substitution method and the penalty method in 1-D FEA. These methods are equally applicable in 2-D analysis.

Equation (5.136) shows the assembled global linear system after the temperature boundary condition is applied using the penalty method.

$$\begin{bmatrix} 10^8 & 21.3889 & 0 & 21.3889 & 9.4444 & 0 & 0 & 0 & 0 \\ 21.3889 & 95.5556 & 21.3889 & 9.4444 & 42.7778 & 9.4444 & 0 & 0 & 0 \\ 0 & 21.3889 & 10^8 & 0 & 9.4444 & 21.3889 & 0 & 0 & 0 \\ 21.3889 & 9.4444 & 0 & 95.5556 & 42.7778 & 0 & 21.3889 & 9.4444 & 0 \\ 9.4444 & 42.7778 & 9.4444 & 42.7778 & 191.1111 & 42.7778 & 9.4444 & 42.7778 & 9.4444 \\ 0 & 9.4444 & 21.3889 & 0 & 42.7778 & 95.5556 & 0 & 9.4444 & 21.3889 \\ 0 & 0 & 0 & 21.3889 & 9.4444 & 0 & 10^8 & 21.3889 & 0 \\ 0 & 0 & 0 & 9.4444 & 42.7778 & 9.4444 & 21.3889 & 95.5556 & 21.3889 \\ 0 & 0 & 0 & 0 & 9.4444 & 21.3889 & 0 & 21.3889 & 47.7778 \end{bmatrix} \begin{Bmatrix} T_1 \\ T_2 \\ T_3 \\ T_4 \\ T_5 \\ T_6 \\ T_7 \\ T_8 \\ T_9 \end{Bmatrix} = \begin{Bmatrix} 80 \times 10^8 \\ 4500 \\ 1750 \\ 80 \times 10^8 \\ 12000 \\ 5000 \\ 80 \times 10^8 \\ 5500 \\ 2250 \end{Bmatrix}. \quad (5.136)$$

Step 9: Solve the Global System

Once the essential boundary conditions are applied, the final linear system can be solved easily in MATLAB. The unknown nodal temperature can be computed. The solution of the linear system, Eq. (5.136), is obtained as

$$\begin{Bmatrix} T_1 \\ T_2 \\ T_3 \\ T_4 \\ T_5 \\ T_6 \\ T_7 \\ T_8 \\ T_9 \end{Bmatrix} = \begin{Bmatrix} 80.0000 \\ 3.6165 \\ 16.2762 \\ 80.0000 \\ 24.4624 \\ 31.0430 \\ 80.0000 \\ 12.6437 \\ 22.7000 \end{Bmatrix} \ °C. \quad (5.137)$$

Step 10: Post-processing

Similar to the calculation of strain and stress in the 1-D elastic bar problem, in the post-processing stage, the temperature and heat flux distribution in the domain can be calculated by using the nodal temperature at the nodes. To visualize the temperature and heat flux distribution over the computational domain, we can generate surface and/or contour plots. For rectangular domains with a uniform mesh, the sur-

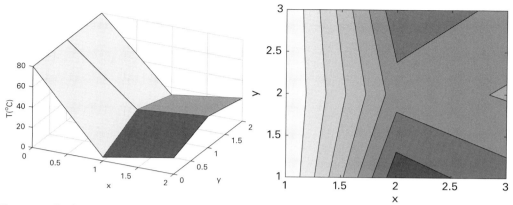

Figure 5.29 Surface plot of the temperature result. **Figure 5.30** Contour plot of the temperature result.

face and contour plots can be straightforwardly plotted in MATLAB. Figures 5.29 and 5.30 show the surface and contour plots of the temperature. The calculation and visualization of the heat flux is left to the reader as an exercise.

5.2.2 Computer Implementation

The implementation of 2-D geometric models, meshing, and visualization is more involved than 1-D cases. By using the plate heat transfer problem as an example, we illustrate the implementation of the 2-D FEA procedure.

Data Files and Data Structures

The input data files are similar to those in the 1-D analysis. The nodes, elements, material properties, loads, boundary conditions, and other model options are stored in forms of arrays and matrices. Note that the third column of "edgeFlux.dat" stores $k\frac{\partial t}{\partial n}$.

Data Visualization

For 2-D finite element analysis, the results can typically be viewed by using 3-D surface plots. The xy-plane holds the 2-D computational domain, and the magnitude of the result is represented by using the z-coordinates. The variation of the computed physical quantity (temperature in our problem) is visualized as a wavy surface over the 2-D computational domain. In MATLAB, several basic functions are used to generate surface plots: "meshgrid," "mesh," "surf," and "contourf." The following code generates a surface plot of the temperature distribution over the 2-D computational domain.

Table 5.7 Data files for analysis

nodes.dat			elements.dat				
1	0.0	0.0	1	5	4	1	2
2	1.0	0.0	2	6	5	2	3
3	2.0	0.0	3	8	7	4	5
4	0.0	1.0	4	9	8	5	6
5	1.0	1.0					
6	2.0	1.0					
7	0.0	2.0					
8	1.0	2.0					
9	2.0	2.0					

edgeFlux.dat			nodalTemp.dat		materials.dat
1	3	−1500.0	1	80.0	% Thermal conductivity (W/mK)
2	3	−1500.0	4	80.0	5
2	4	−1000.0	7	80.0	% Convection coefficient (W/(m² K)
3	1	−500.0			10
4	1	−500.0			
4	4	−1000.0			

options.dat

% Dimensions
2
% Thickness of the structure (used only in 2-D analysis)
0.05
% Element type. Format: No. of edges or faces, No. of nodes
4 4
% Environmental temperature for surface convection
30

```
1  clear all ;
2  [x,y]=meshgrid (0:1:2,0:1:2) ;   % 2–D grid coordinates of the mesh
3  load feT.dat;                    % load the results from previous analysis
4  sx=x(1,1);     sy=y(1,1);        % lower–left corner of the domain
```

```
5   dx=x(1,2)-sx; dy=y(2,1)-sy;   % size of the elements in x and y
6   T=x*0;                         % empty T matrix
7   n_nodes=size(feT,1);           % total number of nodes
8   % for-loop: set the temperature for each of the mesh grid points
9   for i=1:n_nodes
10      xp=feT(i,2);               % x-coordinate of the node
11      yp=feT(i,3);               % y-coordinate of the node
12      col=(xp-sx)/dx+1;
13      row=(yp-sy)/dy+1;
14      T(row,col)=feT(i,4);       % temperature at (x, y)
15  end
16  % surface plot of temperature
17  figure(1);
18  surf(x,y,T);
19  set(gca,'fontsize',16);
20  xlabel('x');
21  ylabel('y');
22  zlabel('T(^oC)');
23  % contour plot of temperature
24  figure(2)
25  colormap( flipud(gray(256)) );  % set up gray scale colormap
26  contourf(T);
27  set(gca,'fontsize',16);
28  xlabel('x');
29  ylabel('y');
```

Memory Management

In the implementation of FEA codes, it is important to optimize the memory usage. An obvious reason is that memory usage is directly proportional to the size of the problem, which is typically the degrees of freedom of the system. A good implementation minimizes the memory usage for a given problem, which also means the ability to simulate larger problems for a given computer memory capacity.

As a high level programming language, MATLAB has an effective memory management system which allocates and initializes new memory blocks as needed and collects memory garbage behind the scenes. However, in the FEA implementation, two major causes of large memory usage remain: (1) problems with large degrees of freedom lead to large matrices, and (2) functions and loops produce many copies of large matrices.

The solution to the first problem is using sparse matrices. Fortunately, due to the localized influence of the nodal variables in a finite element mesh, the coefficient matrices generated from the weak form are typically sparse. MATLAB function

$$S = \text{sparse}(m,n);$$

generates an m-by-n all zero sparse matrix. The value assignment and element accessing have the same syntax as the regular (dense) matrix operations, e.g.,

S(i,j)=val;

v=S(i,j);

Internally, MATLAB sparse matrix only keeps the nonzero elements, effectively reducing the memory usage of storing the matrix.

The second cause of large memory usage is more implicit. In a FEA code, functions involving modifying or manipulating large matrices (e.g. assembling the global matrices) are often necessary. When matrices are passed into a function as input variables and then modified inside the function, a copy is typically made. Thus large chunks of memory are needed if the input matrices to be duplicated are large. To make the situation worse, such modifications or manipulations can be performed repeatedly in various loops. MATLAB's solution to this problem relies on its built-in memory optimization mechanisms: copy-on-write (COW) and in-place data manipulation or in-place modification. The copy-on-write mechanism transparently allocates a temporary copy of the data only when it sees that the input data is modified. That is, the actual memory block allocation is delayed until absolutely necessary. In the case that a function does not modify an input matrix, MATLAB does not make a copy of the input matrix, thus avoiding unnecessary memory usage. Another mechanism is the in-place modification: you can assign the output of an operation on an input variable to the same input variable when you do not need to preserve the original input values. The in-place modification has two variants: during regular variable copy operations such as "x=x+1," and when passing data as input parameters into a function: "function x= DoThings(x)." In order to achieve in-place data manipulation operation of the function, we must both use the same variable in the caller workspace ("x = DoThings(x)") and also ensure that the called function is of the form "function x = DoThings(x)." No duplication of "x" is performed when these two requirements are met. The following is an example.

```
1  x=0:1e-7:1;     % Define a long array
2  x=DoThings(x);  % In-place function call, no memory copy
3
4  % function definition
5  function x=DoThings(x)
6      x=sin(x);   % In-place data manipulation, no memory copy
7  end
```

MATLAB Code

The MATLAB code for solving the 2-D heat transfer example problem is listed below.

```
1  % next 6 lines : read the input files
2  load options.dat;
3  load nodes.dat;
4  load elements.dat;
```

```
5   load materials.dat;
6   load edgeFlux.dat;
7   load nodalTemp.dat;
8   n_nodes=size(nodes,1);                  % number of nodes
9   n_nodalTemp=size(nodalTemp,1);          % number of nodal temperature
10  % global material properties and constants
11  kappa=materials(3,1);
12  thickness=options(2,1);
13  convection_coeffcient=materials(4,1);
14  hz=convection_coeffcient*2/thickness;
15  extTemp=options(4,1);
16  % compute K and F
17  K=CompK(nodes, elements, kappa, hz);     % compute global K matrix
18  F=CompF(nodes, elements, kappa, hz, extTemp, edgeFlux);  % global F
19  % apply displacement boundary condition
20  coeff=abs(max(K(nodalTemp(1,1),:)))*1e7; % penalty number
21  for i=1:n_nodalTemp
22      node_id=nodalTemp(i,1);
23      K(node_id, node_id)=coeff;
24      F(node_id, 1) = nodalTemp(i,2)*coeff;
25  end
26  % solve the global linear system
27  T=K\F;
28  % save the temperature results in file
29  TOut=zeros(n_nodes,4);
30  for n=1:n_nodes
31      TOut(n,1:3)=nodes(n,1:3);
32      TOut(n,4)=T(n,1);
33  end
34  TOut
35  save -ascii -double feT.dat TOut;
```

```
1   % Compute K matrix
2   % Input: nodes, elements, kappa (thermal conducticity),
3   %         hz (surface convection related parameter),
4   % Output: K matrix
5   function K=CompK(nodes, elements, kappa, hz)
6   n_nodes = size(nodes,1);
7   n_elements = size(elements,1);
8   n_nodes_per_element = size(elements,2)-1;
9   K=zeros(n_nodes,n_nodes);
10  DN=zeros(2,4);
11  Nv=zeros(1,4);
12  [gauss_points, gauss_weights]=GetQuadGauss(2,2);
13  n_gauss_points=size(gauss_points,1);
14  [N,Nx,Ny]=CompNDNatPointsQuad4(gauss_points(:,1), gauss_points(:,2));
15
```

5.2 2-D Steady State Heat Transfer

```
16  % for-loop block : compute K matrix, loop over all the elements
17  for e=1:n_elements
18      ke=zeros(n_nodes_per_element, n_nodes_per_element);
19      kte=ke;
20      [element_nodes, node_id_map]= SetElementNodes(e, nodes, elements);
21      % next 10 lines : compute element stiffness matrix ke
22      for g=1:n_gauss_points
23          J=CompJacobian2DatPoint(element_nodes, Nx(:,g), Ny(:,g));
24          detJ=det(J);
25          Jinv=inv(J);
26          DN(1,:)=Nx(:,g);
27          DN(2,:)=Ny(:,g);
28          Nv(1,:)=N(:,g);
29          ke=ke+DN'*Jinv'*kappa*Jinv*DN*detJ*gauss_weights(g);
30          kte=kte+ Nv'*hz*Nv*detJ*gauss_weights(g);
31      end
32      % assemble ke into global K
33      K= AssembleGlobalMatrix(K,ke+kte,node_id_map,1);
34  end
```

```
1   % Compute F vector
2   % Input: nodes, elements, kappa (thermal conducticity),
3   %         hz ( surface convection related parameter),
4   %         extTemp (environmental temperatre ), edgeFlux ( flux on edges )
5   % Output: F vector
6   function F=CompF(nodes, elements, kappa, hz, extTemp, edgeFlux)
7   n_nodes = size(nodes,1);
8   n_elements = size(elements,1);
9   n_flux_edges = size(edgeFlux,1);
10  n_nodes_per_element = size(elements,2)-1;
11  F=zeros(n_nodes,1);
12
13  % next 23 lines : apply edge flux
14  n_edge_gauss_points=2;
15  [gauss_points, gauss_weights]=GetQuadEdgeGauss(n_edge_gauss_points);
16  [N,Nx,Ny]=CompNDNatPointsQuad4(gauss_points(:,1), gauss_points(:,2));
17  % for-loop block : loop over the edges affected by flux and apply flux
18  for t=1:n_flux_edges
19      fe=zeros(n_nodes_per_element, 1);
20      eid=edgeFlux(t,1);
21      [element_nodes,node_id_map]= SetElementNodes(eid,nodes,elements);
22      edge=edgeFlux(t,2);
23      % loop over the edge Gauss points to compute boundary integrals
24      for g=1:n_edge_gauss_points
25          gid=2*edge-2+g;  % Gauss point ID in edge Gauss points
26          J=CompJacobian2DatPoint(element_nodes,Nx(:,gid),Ny(:,gid));
27          if (edge==1) | (edge==3)
```

```matlab
28          lengthJ =sqrt(J(1,1)^2+J(1,2)^2);
29        else
30          lengthJ =sqrt(J(2,1)^2+J(2,2)^2);
31        end
32        Nv=N(:,gid);
33        fe =fe+ Nv*edgeFlux(t,3)*lengthJ *gauss_weights(gid);
34      end
35      F= AssembleGlobalVector(F,fe,node_id_map,1); % assemble  F
36  end
37
38  % next 16 lines: apply surface convection
39  [ gauss_points , gauss_weights]=GetQuadGauss(2,2);
40  n_gauss_points=size( gauss_points ,1) ;
41  [N,Nx,Ny]=CompNDNatPointsQuad4(gauss_points(:,1), gauss_points(:,2));
42  % for−loop block
43  for e=1:n_elements
44    fe=zeros(n_nodes_per_element, 1);
45    [element_nodes,node_id_map]= SetElementNodes(e,nodes,elements);
46    % next 6 lines : compute element  convection   vector
47    for g=1:n_gauss_points
48      J=CompJacobian2DatPoint(element_nodes,Nx(:,g), Ny(:,g));
49      detJ=det(J);
50      Nv=N(:,g);
51      fe=fe+Nv*hz*extTemp*detJ*gauss_weights(g);
52    end
53    F= AssembleGlobalVector(F,fe,node_id_map,1); % assemble  F
54  end
```

In the MATLAB programs listed above, the functions which have already been discussed previously are not repeated here. They are listed below for your reference:
function [gauss_points, gauss_weights]=GetQuadGauss(rows, cols): Section 5.2.1.
function [gauss_points, gauss_weights]=GetQuadEdgeGauss(neg): Section 5.2.1.
function [N,Nx,Ny]=CompNDNatPointsQuad4(xi_vector, eta_vector): Section 5.2.1.
function [element_nodes, node_id_map]= SetElementNodes(ele, nodes, elements): Section 5.2.1.
function J= CompJacobian2DatPoint(element_nodes, Nxi, Neta): Section 5.2.1.
function K= AssembleGlobalMatrix(K,ke,node_id_map,ndDOF): Section 5.2.1.
function F= AssembleGlobalVector(F,fe,node_id_map,ndDOF): Section 5.2.1.

5.3   3-D Steady State Heat Transfer

In this section, 3-D steady state heat transfer analysis is illustrated by demonstrating the step-by-step procedure of solving a 3-D heat sink problem. We begin with a derivation of the basic differential equation of steady state heat transfer in three dimensions. We then use the same procedure as in the 2-D heat transfer problem to obtain the weak form. In the steps of discretization and approximation, we introduce various 3-D elements and their shape functions. Numerical integration and element equations are then described for the 3-D elements. The assembly process illustrated in the 2-D heat transfer problem is extended to the 3-D case and the example problem is solved by using the MATLAB linear system solver. Finally, computer implementation and associated programs are presented for the analysis.

5.3.1 3-D Heat Transfer in a Heat Sink

Step 1: Physical Problem

We consider a 3-D heat sink as shown in Fig. 5.31. The temperature of the bottom surface (hot side) is 100 °C and the top surface has a temperature gradient due to convection. For simplicity of illustration, all the other faces are assumed adiabatic (i.e. the heat flux is zero) in this example. Other heat flux boundary conditions such as convective heat transfer between the heat sink surface and the ambient environment can be applied as well. The dimensions of the heat sink are shown in the figure. The steady state temperature distribution in the plate is to be determined by FEA. The relevant material properties include the thermal conductivity $k = 5$ W/m · K and the convection coefficient of ambient environment $h = 10$ W/(m^2 · K).

Step 2: Mathematical Model

The mathematical model can be derived by following the same procedure discussed for the 2-D heat transfer problem. The governing partial differential equations, written in vector form, is the same for two and three dimensions.

$$\nabla \cdot (k\nabla T) = 0. \tag{5.138}$$

Note that, different from the 2-D heat transfer example, here there is no internal convection term. For a constant thermal conductivity of the material, Eq. (5.138) can be rewritten as

$$\nabla^2 T = 0 \tag{5.139}$$

which is the Laplace equation. While the form of the Laplace equation shown in Eq. (5.139) is the same in two and three dimensions, expanding the equation explicitly in terms of *xyz*-coordinates gives

$$\frac{\partial^2 T}{\partial x^2} + \frac{\partial^2 T}{\partial y^2} + \frac{\partial^2 T}{\partial z^2} = 0 \tag{5.140}$$

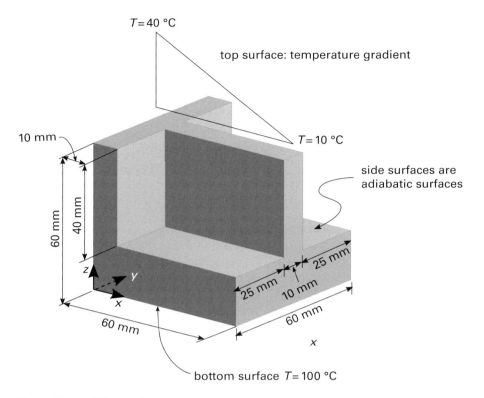

Figure 5.31 A 3-D heat sink.

where the third term vanishes for 2-D. From Fig. 5.31, the essential boundary conditions are given by

$$T = 100\,°\text{C} \quad \text{at} \quad z = 0, \tag{5.141}$$

$$T = 40 - \frac{x}{2}\,°\text{C} \quad \text{at} \quad z = 60\text{ mm} \tag{5.142}$$

and the natural boundary condition is given by

$$q_n = -k\frac{\partial T}{\partial \mathbf{n}} = 0 \quad \text{on the rest of the boundary.} \tag{5.143}$$

The equations (5.140–5.143) combined is the strong form of the 3-D steady state heat transfer problem.

Step 3: Weak Form
Since the vector form of the governing partial differential equation, Eq.(5.138), is the same in both two and three dimensions, the derivation procedure of the weak form for 3-D heat transfer is also the same as that in 2-D case. Therefore, the derivation is not repeated here and the weak form of Eqs. (5.140–5.143) is directly given below as

5.3 3-D Steady State Heat Transfer

$$\int_{\Gamma_q} \delta T k \frac{\partial T}{\partial \mathbf{n}} d\Gamma - \int_{\Omega} k \nabla T \cdot \nabla \delta T d\Omega = 0. \tag{5.144}$$

As there is no internal convection term in the 3-D heat transfer example problem, the third term in the 2-D weak form given in Eq. (5.32) does not appear here in the 3-D weak form, Eq. (5.144). In addition, while the mathematical representation of the gradient or normal derivative of the temperature is the same for 2-D and 3-D, their explicit expressions in 3-D are

$$\nabla T = \begin{Bmatrix} \frac{\partial T}{\partial x} \\ \frac{\partial T}{\partial y} \\ \frac{\partial T}{\partial z} \end{Bmatrix} \qquad \frac{\partial T}{\partial \mathbf{n}} = \begin{Bmatrix} \frac{\partial T}{\partial x} \\ \frac{\partial T}{\partial y} \\ \frac{\partial T}{\partial z} \end{Bmatrix} \cdot \begin{Bmatrix} n_x \\ n_y \\ n_z \end{Bmatrix}. \tag{5.145}$$

Step 4: Discretization

In three dimensions, a computational domain (volume) can be discretized into 3-D elements by using methods similar to the ones used in 2-D meshing. The widely used methods include the Delaunay Refinement method and the Advancing Front method. While the details of these methods for 3-D meshing are not discussed here, the main procedures of these methods remain the same as in 2-D meshing which is discussed in detail in Chapter 6. However, it is obvious that the difference of the dimensionality makes the element topology of 3-D elements very different from that of 2-D elements. Several types of 3-D elements are popular in commercial FEA packages. They include tetrahedral, prismatic, pyramidal, and hexahedral elements as shown in Fig. 5.32. As an example, Fig. 5.33 shows a mesh of tetrahedral elements for the heat sink structure. This particular mesh contains 38,380 elements.

Step 5: Finite Element Approximation

Once again, the approximation of the temperature in a 3-D element has the same form as the 1-D and 2-D approximation:

$$T = \sum_{i=1}^{n} N_i T_i \tag{5.146}$$

where n is the number of nodes in the element, T_i are the nodal temperatures, and N_i are shape functions of the element. The difference, however, is the dimensionality of the temperature and shape functions. The unknown temperature and the shape functions are now all 3-D functions, i.e.,

$$T(x, y) = \sum_{i=1}^{n} N_i(x, y, z) T_i. \tag{5.147}$$

For the approximation, the shape functions need to be constructed for the various types of elements shown in Fig. 5.32. Similar to the 2-D scenarios, the technique

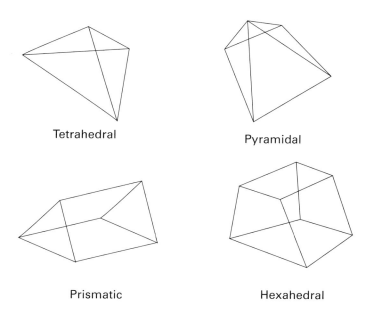

Figure 5.32 3-D element shapes.

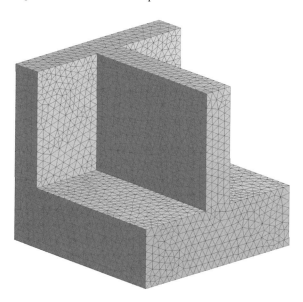

Figure 5.33 A mesh of tetrahedral elements for the heat sink.

for consistent approximation in elements with irregular geometries is isoparametric mapping. Master elements are defined in a $\xi\eta\zeta$-coordinate system for each type of the 3-D elements, and nodes are placed at specific locations of the elements depending on the order of the finite element approximation. Figure 5.34 shows the master elements in the $\xi\eta\zeta$-reference coordinate system corresponding to the physical elements in the

xyz-coordinate system shown in Fig. 5.32 for isoparametric mapping. The left column shows the linear elements and in the right column are the quadratic elements.

Tetrahedral Elements

For the linear tetrahedral master element, the shape functions are given by

$$N_1 = 1 - \xi - \eta - \zeta$$
$$N_2 = \xi$$
$$N_3 = \eta$$
$$N_4 = \zeta. \tag{5.148}$$

A MATLAB function to calculate the shape functions and their derivatives at a given point (ξ, η, ζ) is shown below. Note that the derivatives of the linear shape functions are constants within the element.

```
1  % Compute the shape functions and their first derivatives at a set
2  % of points defined by (xi_vec, eta_vec, zeta_vec) in a 3-D
3  % 4-node tetrahedral master element.
4  % Input is a single point when the length of the input vectors is 1
5  % Input: xi_vec, eta_vec, zeta_vec: coordinates of the input points
6  % Output: N: matrix storing shape function values with the format
7  %           [N1(xi_1, eta_1, zeta_1)  N1(xi_2, eta_2, zeta_2) ...
8  %            N2(xi_1, eta_1, zeta_1)  N2(xi_2, eta_2, zeta_2) ...
9  %            N3(xi_1, eta_1, zeta_1)  N3(xi_2, eta_2, zeta_2) ...
10 %            ...                      ...                     ...]
11 % Output: Nx, Ny, Nz: matrice of dNi/dxi(xi, eta, zeta),
12 %           dNi/deta(xi, eta, zeta), dNi/dzeta(xi, eta, zeta),
13 %           respectively, format is the same as N
14 function [N,Nx,Ny,Nz]=CompNDNatPointsTetra4(xi_vec,eta_vec,zeta_vec)
15 np=size(xi_vec,1);
16 N=zeros(4,np);   Nx=zeros(4,np);
17 Ny=zeros(4,np);  Nz=zeros(4,np);   % set up empty matrices
18
19 % for loop: compute N, Nx, Ny, Nz
20 for j=1:np              % columns for point 1,2 ...
21     xi=xi_vec(j);       % xi- coordinate of point j
22     eta=eta_vec(j);     % eta - coordinate of point j
23     zeta=zeta_vec(j);   % zeta - coordinate of point j
24     N(1,j)=1.0-xi-eta-zeta;  % N1(xi, eta, zeta)
25     N(2,j)=xi;          % N2(xi, eta, zeta)
26     N(3,j)=eta;         % N3(xi, eta, zeta)
27     N(4,j)=zeta;        % N4(xi, eta, zeta)
28     Nx(1,j)=-1.0;       % dN1/dxi(xi, eta, zeta)
29     Nx(2,j)=1.0;        % dN2/dxi(xi, eta, zeta)
30     Ny(1,j)=-1.0;       % dN1/deta(xi, eta, zeta)
31     Ny(3,j)=1.0;        % dN3/deta(xi, eta, zeta)
32     Nz(1,j)=-1.0;       % dN1/dzeta(xi, eta, zeta)
```

222 FEA for Multi-Dimensional Scalar Field Problems

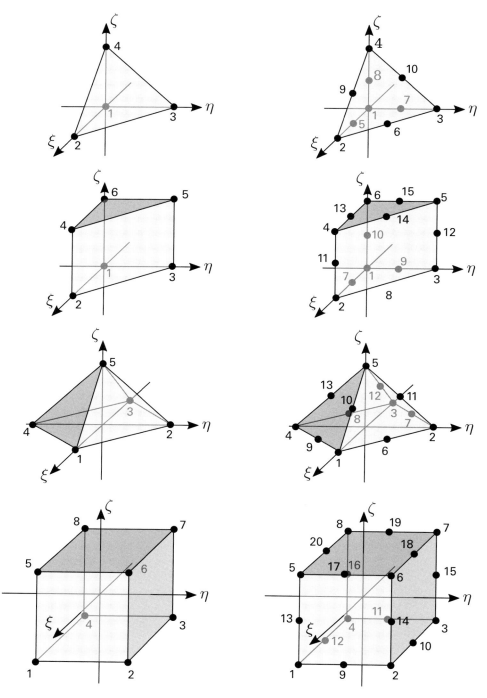

Figure 5.34 3-D master elements.

```
33      Nz(4,j)=1.0;                % dN4/dzeta ( xi , eta , zeta )
34   end
```

While a Gauss quadrature rule for precise integration over the volumetric element would depend on the order of integrand, 1-point Gauss quadrature is typically employed for the linear tetrahedral element. The position of the Gauss point and its weight are given by

$$(\xi_1, \eta_1, \zeta_1) = \left(\frac{1}{4}, \frac{1}{4}, \frac{1}{4}\right) \qquad w_1 = \frac{1}{6}. \tag{5.149}$$

For the quadratic tetrahedral master element, there are 10 nodes and therefore 10 shape functions. The shape functions are given as

$$\lambda = 1 - \xi - \eta - \zeta$$
$$N_1 = \lambda(2\lambda - 1)$$
$$N_2 = \xi(2\xi - 1)$$
$$N_3 = \eta(2\eta - 1)$$
$$N_4 = \zeta(2\zeta - 1)$$
$$N_5 = 4\xi\lambda$$
$$N_6 = 4\xi\eta$$
$$N_7 = 4\eta\lambda$$
$$N_8 = 4\zeta\lambda$$
$$N_9 = 4\xi\zeta$$
$$N_{10} = 4\eta\zeta. \tag{5.150}$$

For integration, 5-point Gauss quadrature is typically adopted. The locations of the Gauss points and their weights are given in Table 5.8.

While it is desirable to store all the Gauss quadrature points and weights in a set of global data containers (e.g. a set of arrays or matrices), one can also define these rules

Table 5.8 5-point Gauss quadrature for quadratic tetrahedral elements

Point	Location (ξ, η, ζ)	Weight
1	$\left(\frac{1}{4}, \frac{1}{4}, \frac{1}{4}\right)$	$-\frac{2}{15}$
2	$\left(\frac{1}{6}, \frac{1}{6}, \frac{1}{6}\right)$	$\frac{3}{40}$
3	$\left(\frac{1}{6}, \frac{1}{6}, \frac{1}{2}\right)$	$\frac{3}{40}$
4	$\left(\frac{1}{6}, \frac{1}{2}, \frac{1}{6}\right)$	$\frac{3}{40}$
5	$\left(\frac{1}{2}, \frac{1}{6}, \frac{1}{6}\right)$	$\frac{3}{40}$

using functions. It leads to more functions, but more convenient to call and manage. For example, a MATLAB function returning the Gauss point coordinates and weights for the 1-point quadrature rule of tetrahedral elements (linear or quadratic) is given below. It is easy to see that expanding the function to return the Gauss points and weights of the 5-point quadrature rule is straightforward.

```
% Get Gauss points and weights in a tetrahedral master element
% Input: ng: number of Gauss points
% Output: gauss_points : vectors store the locations of Gauss points
% Output: gauss_weights : the corresponding weights of the Gauss points
function [ gauss_points , gauss_weights]=GetTetraGauss(ng)
gauss_points =zeros(ng,3); % set up empty return variables
gauss_weights=zeros(ng,1);
% if-else block: assign Gauss points and weights for
% the given input parameters
% currently only allows 1 point Gauss quadrature .
% can be extended if needed
if ng==1
    gauss_points (:,:) =1.0/4.0;
    gauss_weights (1,1) =1.0/6.0;
else
    fprintf('Error calling GetTetraGauss\n'); % error message
    return;
end
```

Brick (Hexahedral) Elements

As shown in Fig. 5.34, two types of hexahedral elements (also called brick elements) are typically used in practice: 8-node linear and 20-node quadratic elements. The shape functions of the linear brick element in the reference coordinate system are given by

$$N_i(\xi, \eta, \zeta) = \frac{1}{8}(1 + \xi_i\xi)(1 + \eta_i\eta)(1 + \zeta_i\zeta) \qquad i = 1,\ldots,8 \qquad (5.151)$$

where ξ_i, η_i, and ζ_i are the coordinates of node i of the master element, which are given in Table 5.9.

From Eq. (5.151), it is easy to obtain the derivatives of the shape functions as

$$\frac{\partial N_i}{\partial \xi} = \frac{1}{8}\xi_i(1 + \eta_i\eta)(1 + \zeta_i\zeta)$$

$$\frac{\partial N_i}{\partial \eta} = \frac{1}{8}\eta_i(1 + \xi_i\xi)(1 + \zeta_i\zeta)$$

$$\frac{\partial N_i}{\partial \zeta} = \frac{1}{8}\zeta_i(1 + \xi_i\xi)(1 + \eta_i\eta). \qquad (5.152)$$

Table 5.9 Coordinates of the master element nodes

i	ξ_i	η_i	ζ_i
1	-1	-1	-1
2	1	-1	-1
3	1	1	-1
4	-1	1	-1
5	-1	-1	1
6	1	-1	1
7	1	1	1
8	-1	1	1

A MATLAB function for calculating Eqs. (5.151 and 5.152) is shown below.

```
1  % Compute the shape functions and their first derivatives at a set
2  % of points defined by ( xi_v, eta_v, zeta_v ) in a 3-D
3  % 8-node hexahedral master element.
4  % Input: xi_vec, eta_vec, zeta_vec : coordinates of the input points
5  % Output: N: matrix storing shape function values with the format
6  %              [N1(xi_1, eta_1, zeta_1)  N1(xi_2, eta_2, zeta_2) ...
7  %               N2(xi_1, eta_1, zeta_1)  N2(xi_2, eta_2, zeta_2) ...
8  %               N3(xi_1, eta_1, zeta_1)  N3(xi_2, eta_2, zeta_2) ...
9  %               ...                       ...                    ...]
10 % Output: Nx, Ny, Nz: matrice of dNi/dxi ( xi, eta, zeta ),
11 %              dNi/ deta ( xi, eta, zeta ), dNi/ dzeta ( xi, eta, zeta ),
12 %              respectively, format is the same as N
13 function [N,Nx,Ny,Nz]=CompNDNatPointsHexa8(xi_v, eta_v, zeta_v)
14 np=size(xi_v,1);
15 N=zeros(8,np);  Nx=zeros(8,np);
16 Ny=zeros(8,np); Nz=zeros(8,np);  % set up empty matrices
17
18 master_nodes=[-1 -1 -1; 1 -1 -1; 1 1 -1; -1 1 -1;...
19               -1 -1 1; 1 -1 1; 1 1 1; -1 1 1];
20 % for loop: compute N, Nx, Ny
21 for j=1:np                        % columns for point 1,2 ...
22     xi=xi_v(j);                   % xi- coordinate of point j
23     eta=eta_v(j);                 % eta- coordinate of point j
24     zeta=zeta_v(j);               % zeta- coordinate of point j
25     for i=1:8
26         nx=master_nodes(i,1);
27         ny=master_nodes(i,2);
28         nz=master_nodes(i,3);
29         N(i,j)= (1.0 + nx*xi)*(1.0 + ny*eta)*(1.0 + nz*zeta)/8.0;
```

```
30      Nx(i,j)= nx*(1.0 + ny*eta)*(1.0 + nz*zeta) /8.0;
31      Ny(i,j)= ny *(1.0 + nx*xi)*(1.0 + nz*zeta) /8.0;
32      Nz(i,j)= nz *(1.0 + nx*xi)*(1.0 + ny*eta) /8.0;
33    end
34  end
```

Similar to the 2-D Gauss quadrature rule for quadrilateral elements, the Gauss quadrature rule for brick elements can be constructed using the cross product of the 1-D Gauss quadrature rules. For the eight-node linear elements, the 3-D Gauss quadrature rule is typically $2 \times 2 \times 2$. The locations of the Gauss points and their weights are given in Table 5.10, also as shown in Fig. 5.35.

Table 5.10 $2 \times 2 \times 2$ Gauss quadrature for brick elements

Point	Location (ξ, η, ζ)	Weight
1	$\left(-\frac{1}{\sqrt{3}}, -\frac{1}{\sqrt{3}}, -\frac{1}{\sqrt{3}}\right)$	1
2	$\left(\frac{1}{\sqrt{3}}, -\frac{1}{\sqrt{3}}, -\frac{1}{\sqrt{3}}\right)$	1
3	$\left(-\frac{1}{\sqrt{3}}, \frac{1}{\sqrt{3}}, -\frac{1}{\sqrt{3}}\right)$	1
4	$\left(-\frac{1}{\sqrt{3}}, -\frac{1}{\sqrt{3}}, \frac{1}{\sqrt{3}}\right)$	1
5	$\left(\frac{1}{\sqrt{3}}, \frac{1}{\sqrt{3}}, -\frac{1}{\sqrt{3}}\right)$	1
6	$\left(-\frac{1}{\sqrt{3}}, \frac{1}{\sqrt{3}}, \frac{1}{\sqrt{3}}\right)$	1
7	$\left(\frac{1}{\sqrt{3}}, -\frac{1}{\sqrt{3}}, \frac{1}{\sqrt{3}}\right)$	1
8	$\left(\frac{1}{\sqrt{3}}, \frac{1}{\sqrt{3}}, \frac{1}{\sqrt{3}}\right)$	1

A MATLAB function returning the Gauss points and weights for the $2 \times 2 \times 2$ Gauss quadrature for brick elements (Table 5.10 and Fig. 5.35) is given by

```
1  % Get Gauss points and weights in a brick master element
2  % Input : n_xi, n_eta, n_zeta : number of Gauss points in three directions
3  % Output: gauss_points : vectors store the locations of Gauss points
4  % Output: gauss_weights : the corresponding weights of the Gauss points
5  function [ gauss_points , gauss_weights]=GetHexaGauss(n_xi,n_eta,n_zeta)
6  ng=n_xi*n_eta*n_zeta;       % total number of Gauss points
7  gauss_points=zeros(ng,3);   % set up empty return variables
8  gauss_weights=zeros(ng,1);
```

5.3 3-D Steady State Heat Transfer

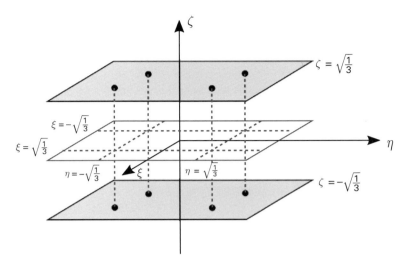

Figure 5.35 2 × 2 × 2 Gauss quadrature.

```
 9  % if-else block : assign Gauss points and weights
10  % currently only allows 8 point Gauss quadrature,
11  % can be extended if needed
12  if ng==8
13      pos=sqrt(3.0) /3.0;
14      gauss_points (:,:) =pos;
15      gauss_points (1,1:3) = -pos;
16      gauss_points (2,2:3) = -pos;
17      gauss_points (3,3) = -pos;
18      gauss_points (4,1) = -pos;
19      gauss_points (4,3) = -pos;
20      gauss_points (5,1:2) = -pos;
21      gauss_points (6,2) = -pos;
22      gauss_points (8,1) = -pos;
23      gauss_weights (:,:) =1.0;
24  else
25      fprintf ('Error calling GetHexaGaussn'); % error message
26      return;
27  end
```

For the 20-node quadratic brick element, the nodes are numbered such that the first eight nodes are the corner nodes which are the same as the nodes of the linear element. Node 9 to node 20 are the mid-edge nodes as shown in the Fig. 5.34. The shape functions for the corner nodes are

$$N_i(\xi, \eta, \zeta) = \frac{1}{8}(1 + \xi_i\xi)(1 + \eta_i\eta)(1 + \zeta_i\zeta)(\xi_i\xi + \eta_i\eta + \zeta_i\zeta - 2)$$

$$i = 1, \ldots, 8. \quad (5.153)$$

For the mid-side nodes with $\xi_i = 0$, the shape functions are

$$N_i(\xi, \eta, \zeta) = \frac{1}{4}(1 - \xi^2)(1 + \eta_i\eta)(1 + \zeta_i\zeta) \qquad i = 10, 18, 20, 12. \qquad (5.154)$$

Similarly, for the mid-side nodes with $\eta_i = 0$,

$$N_i(\xi, \eta, \zeta) = \frac{1}{4}(1 - \eta^2)(1 + \xi_i\xi)(1 + \zeta_i\zeta) \qquad i = 9, 11, 19, 17. \qquad (5.155)$$

For the mid-side nodes with $\zeta_i = 0$,

$$N_i(\xi, \eta, \zeta) = \frac{1}{4}(1 - \zeta^2)(1 + \xi_i\xi)(1 + \eta_i\eta) \qquad i = 13, 14, 15, 16. \qquad (5.156)$$

Integrations over the 20-node quadratic brick element can be performed by using a $3 \times 3 \times 3$ Gauss quadrature rule. Same as the $2 \times 2 \times 2$ rule, the $3 \times 3 \times 3$ quadrature rule can be obtained by using the product of the 3-point 1-D Gauss quadrature rule along ξ-, η-, and ζ- axes, as shown in Fig. 5.36. There are a total of 27 Gauss points in this case. The positions of the points are shown in the figure. The weight of a point is the product of the 1-D 3-point Gauss quadrature weights of its ξ-, η-, and ζ- coordinates. For example, for the Gauss point $(-\sqrt{\frac{3}{5}}, \sqrt{\frac{3}{5}}, 0)$, the weights for the points $-\sqrt{\frac{3}{5}}, \sqrt{\frac{3}{5}}$, and 0 in the 1-D 3-point Gauss quadrature rule are $\frac{5}{9}, \frac{5}{9}$, and $\frac{8}{9}$, respectively. The weight for the 3-D Gauss point is then $w = \frac{5}{9} \times \frac{5}{9} \times \frac{8}{9} = \frac{200}{729}$. Note that there are other integration rules for numerical integration in the $2 \times 2 \times 2$ cubic master element such as the 4-point, 6-point, and 14-point rules. They are not discussed in detail here.

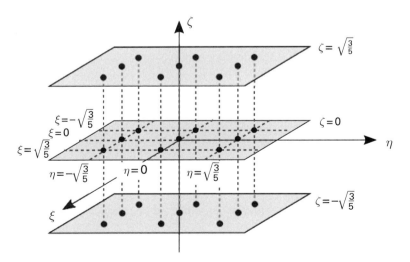

Figure 5.36 $3 \times 3 \times 3$ Gauss quadrature for a quadratic brick element.

5.3 3-D Steady State Heat Transfer

Prismatic Elements
As shown in Fig. 5.34, two types of prismatic elements are typically used: 6-node linear and 15-node quadratic elements. The shape functions of the 15-node prismatic element in the reference coordinate system are given by

$$\phi = 1 - \xi - \eta$$

$$N_1 = \xi(1-\zeta)(\xi - 1 - \frac{1}{2}\zeta)$$

$$N_2 = \eta(1-\zeta)(\eta - 1 - \frac{1}{2}\zeta)$$

$$N_3 = \phi(1-\zeta)(\phi - 1 - \frac{1}{2}\zeta)$$

$$N_4 = 2\eta\phi(1-\zeta)$$

$$N_5 = 2\xi\phi(1-\zeta)$$

$$N_6 = 2\xi\eta(1-\zeta)$$

$$N_7 = 2\xi(1-\zeta^2)$$

$$N_8 = 2\eta(1-\zeta^2) \qquad (5.157)$$

$$N_9 = 2\phi(1-\zeta^2)$$

$$N_{10} = \xi(1+\zeta)(\xi - 1 + \frac{1}{2}\zeta)$$

$$N_{11} = \eta(1+\zeta)(\eta - 1 + \frac{1}{2}\zeta)$$

$$N_{12} = \phi(1+\zeta)(\phi - 1 + \frac{1}{2}\zeta)$$

$$N_{13} = 2\eta\phi(1+\zeta)$$

$$N_{14} = 2\xi\phi(1+\zeta)$$

$$N_{15} = 2\xi\eta(1+\zeta).$$

From Eq. (5.157), it is straightforward to obtain the derivatives of the shape functions as:

$$\frac{\partial N_1}{\partial \xi} = (1-\zeta)(2\xi - 1 - \frac{1}{2}\zeta) \qquad \frac{\partial N_1}{\partial \eta} = 0 \qquad \frac{\partial N_1}{\partial \zeta} = \xi(\zeta - \xi + \frac{1}{2})$$

$$\frac{\partial N_2}{\partial \xi} = 0 \qquad \frac{\partial N_2}{\partial \eta} = (1-\zeta)(2\eta - 1 - \frac{1}{2}\zeta) \qquad \frac{\partial N_2}{\partial \zeta} = \eta(\zeta - \eta + \frac{1}{2})$$

$$\frac{\partial N_3}{\partial \xi} = (1-\zeta)(1 + \frac{1}{2}\zeta - 2\phi) \qquad \frac{\partial N_3}{\partial \eta} = (1-\zeta)(1 + \frac{1}{2}\zeta - 2\phi) \qquad \frac{\partial N_3}{\partial \zeta} = \phi(\zeta - \phi + \frac{1}{2})$$

$$\frac{\partial N_4}{\partial \xi} = -2\eta(1-\zeta) \qquad \frac{\partial N_4}{\partial \eta} = 2(1-\zeta)(\phi - \eta) \qquad \frac{\partial N_4}{\partial \zeta} = -2\eta\phi$$

$$\frac{\partial N_5}{\partial \xi} = 2(1-\zeta)(\phi - \xi) \qquad \frac{\partial N_5}{\partial \eta} = -2\xi(1-\zeta) \qquad \frac{\partial N_5}{\partial \zeta} = -2\xi\phi$$

$$\frac{\partial N_6}{\partial \xi} = 2\eta(1-\zeta) \qquad \frac{\partial N_6}{\partial \eta} = 2\xi(1-\zeta) \qquad \frac{\partial N_6}{\partial \zeta} = -2\xi\eta$$

$$\frac{\partial N_7}{\partial \xi} = 1 - \zeta^2 \qquad \frac{\partial N_7}{\partial \eta} = 0 \qquad \frac{\partial N_7}{\partial \zeta} = -2\xi\eta$$

$$\frac{\partial N_8}{\partial \xi} = 0 \qquad \frac{\partial N_8}{\partial \eta} = 1 - \zeta^2 \qquad \frac{\partial N_8}{\partial \zeta} = -2\eta\zeta$$

$$\frac{\partial N_9}{\partial \xi} = \zeta^2 - 1 \qquad \frac{\partial N_9}{\partial \eta} = \zeta^2 - 1 \qquad \frac{\partial N_9}{\partial \zeta} = -2\phi\zeta$$

$$\frac{\partial N_{10}}{\partial \xi} = (1+\zeta)(2\xi - 1 + \frac{1}{2}\zeta) \qquad \frac{\partial N_{10}}{\partial \eta} = 0 \qquad \frac{\partial N_{10}}{\partial \zeta} = \xi(\zeta + \xi - \frac{1}{2})$$

$$\frac{\partial N_{11}}{\partial \xi} = 0 \qquad \frac{\partial N_{11}}{\partial \eta} = (1+\zeta)(2\eta - 1 + \frac{1}{2}\zeta) \qquad \frac{\partial N_{11}}{\partial \zeta} = \eta(\zeta + \eta - \frac{1}{2})$$

$$\frac{\partial N_{12}}{\partial \xi} = (1+\zeta)(1 - \frac{1}{2}\zeta - 2\phi) \qquad \frac{\partial N_{12}}{\partial \eta} = (1+\zeta)(1 - \frac{1}{2}\zeta - 2\phi) \qquad \frac{\partial N_{12}}{\partial \zeta} = \phi(\zeta + \phi - \frac{1}{2})$$

$$\frac{\partial N_{13}}{\partial \xi} = -2\eta(1+\zeta) \qquad \frac{\partial N_{13}}{\partial \eta} = 2(1+\zeta)(\phi - \eta) \qquad \frac{\partial N_{13}}{\partial \zeta} = 2\eta\phi$$

$$\frac{\partial N_{14}}{\partial \xi} = 2(1+\zeta)(\phi - \xi) \qquad \frac{\partial N_{14}}{\partial \eta} = -2\xi(1+\zeta) \qquad \frac{\partial N_{14}}{\partial \zeta} = 2\xi\phi$$

$$\frac{\partial N_{15}}{\partial \xi} = 2\eta(1+\zeta) \qquad \frac{\partial N_{15}}{\partial \eta} = 2\xi(1+\zeta) \qquad \frac{\partial N_{15}}{\partial \zeta} = 2\xi\eta.$$

(5.158)

For the 15-node prismatic elements, the numerical integration is typically performed using an 8-point 3-D Gauss quadrature rule. The locations of the Gauss points and their weights are given in Table 5.11, also as shown in Fig. 5.35. The polynomial order of the quadrature rule is 3.

Compared to the first three types of elements, pyramidal elements are less used due to its unusual properties (non-polynomial shape functions). For this reason, this type of element is not discussed here. The interested reader is referred to (Zgainski et al. 1996) for more details.

Step 6: Element Matrices and Vectors
The element version of the weak form Eq. (5.144) can be written as

$$\int_{\Gamma_q}^{(e)} \delta T k \frac{\partial T}{\partial \mathbf{n}} d\Gamma - \int_{\Omega}^{(e)} k \nabla T \cdot \nabla \delta T d\Omega = 0 \qquad (5.159)$$

where "e" represents an element. Since the mesh contains four-node tetrahedral elements, we substitute tetrahedral element shape functions, Eq. (5.148), into the finite element approximation Eq. (5.147), for the gradient of the temperature and temperature variation. We have

5.3 3-D Steady State Heat Transfer

Table 5.11 8-point Gauss quadrature for 15-point prismatic elements

Point	Location (ξ, η, ζ)	Weight
1	$\left(\dfrac{1}{3}, \dfrac{1}{3}, -\dfrac{1}{\sqrt{3}}\right)$	$-\dfrac{27}{96}$
2	$\left(\dfrac{1}{5}, \dfrac{1}{5}, -\dfrac{1}{\sqrt{3}}\right)$	$\dfrac{25}{96}$
3	$\left(\dfrac{3}{5}, \dfrac{1}{5}, -\dfrac{1}{\sqrt{3}}\right)$	$\dfrac{25}{96}$
4	$\left(\dfrac{1}{5}, \dfrac{3}{5}, -\dfrac{1}{\sqrt{3}}\right)$	$\dfrac{25}{96}$
5	$\left(\dfrac{1}{3}, \dfrac{1}{3}, \dfrac{1}{\sqrt{3}}\right)$	$-\dfrac{27}{96}$
6	$\left(\dfrac{1}{5}, \dfrac{1}{5}, \dfrac{1}{\sqrt{3}}\right)$	$\dfrac{25}{96}$
7	$\left(\dfrac{3}{5}, \dfrac{1}{5}, \dfrac{1}{\sqrt{3}}\right)$	$\dfrac{25}{96}$
8	$\left(\dfrac{1}{5}, \dfrac{3}{5}, \dfrac{1}{\sqrt{3}}\right)$	$\dfrac{25}{96}$

$$\nabla T = \begin{Bmatrix} \dfrac{\partial T}{\partial x} \\ \dfrac{\partial T}{\partial y} \\ \dfrac{\partial T}{\partial z} \end{Bmatrix} = \begin{bmatrix} \dfrac{\partial N_1}{\partial x} & \dfrac{\partial N_2}{\partial x} & \dfrac{\partial N_3}{\partial x} & \dfrac{\partial N_4}{\partial x} \\ \dfrac{\partial N_1}{\partial y} & \dfrac{\partial N_2}{\partial y} & \dfrac{\partial N_3}{\partial y} & \dfrac{\partial N_4}{\partial y} \\ \dfrac{\partial N_1}{\partial z} & \dfrac{\partial N_2}{\partial z} & \dfrac{\partial N_3}{\partial z} & \dfrac{\partial N_4}{\partial z} \end{bmatrix} \begin{Bmatrix} T_1 \\ T_2 \\ T_3 \\ T_4 \end{Bmatrix} \quad (5.160)$$

and

$$\nabla \delta T = \begin{Bmatrix} \dfrac{\partial \delta T}{\partial x} \\ \dfrac{\partial \delta T}{\partial y} \\ \dfrac{\partial \delta T}{\partial z} \end{Bmatrix} = \begin{bmatrix} \dfrac{\partial N_1}{\partial x} & \dfrac{\partial N_2}{\partial x} & \dfrac{\partial N_3}{\partial x} & \dfrac{\partial N_4}{\partial x} \\ \dfrac{\partial N_1}{\partial y} & \dfrac{\partial N_2}{\partial y} & \dfrac{\partial N_3}{\partial y} & \dfrac{\partial N_4}{\partial y} \\ \dfrac{\partial N_1}{\partial z} & \dfrac{\partial N_2}{\partial z} & \dfrac{\partial N_3}{\partial z} & \dfrac{\partial N_4}{\partial z} \end{bmatrix} \begin{Bmatrix} \delta T_1 \\ \delta T_2 \\ \delta T_3 \\ \delta T_4 \end{Bmatrix}. \quad (5.161)$$

Substituting the approximations into the element weak form, we obtain for the first term

$$\int_{\Gamma_q}^{(e)} \delta T k \frac{\partial T}{\partial \mathbf{n}} d\Gamma = \begin{bmatrix} \delta T_1 & \delta T_2 & \delta T_3 & \delta T_4 \end{bmatrix} \left(\int_{\Gamma_q}^{(e)} \begin{Bmatrix} N_1 \\ N_2 \\ N_3 \\ N_4 \end{Bmatrix} k \frac{\partial T}{\partial \mathbf{n}} d\Gamma \right), \quad (5.162)$$

the second term

$$\int_{\Omega}^{(e)} k \nabla T \cdot \nabla \delta T d\Omega$$

$$= \int_{\Omega}^{(e)} (\nabla \delta T)^T (k \nabla T) d\Omega$$

$$= \begin{bmatrix} \delta T_1 & \delta T_2 & \delta T_3 & \delta T_4 \end{bmatrix} \times$$

$$\left(\int_{\Omega}^{(e)} \begin{bmatrix} \frac{\partial N_1}{\partial x} & \frac{\partial N_1}{\partial y} & \frac{\partial N_1}{\partial z} \\ \frac{\partial N_2}{\partial x} & \frac{\partial N_2}{\partial y} & \frac{\partial N_2}{\partial z} \\ \frac{\partial N_3}{\partial x} & \frac{\partial N_3}{\partial y} & \frac{\partial N_3}{\partial z} \\ \frac{\partial N_4}{\partial x} & \frac{\partial N_4}{\partial y} & \frac{\partial N_4}{\partial z} \end{bmatrix} k \begin{bmatrix} \frac{\partial N_1}{\partial x} & \frac{\partial N_2}{\partial x} & \frac{\partial N_3}{\partial x} & \frac{\partial N_4}{\partial x} \\ \frac{\partial N_1}{\partial y} & \frac{\partial N_2}{\partial y} & \frac{\partial N_3}{\partial y} & \frac{\partial N_4}{\partial y} \\ \frac{\partial N_1}{\partial z} & \frac{\partial N_2}{\partial z} & \frac{\partial N_3}{\partial z} & \frac{\partial N_4}{\partial z} \end{bmatrix} d\Omega \right) \begin{Bmatrix} T_1 \\ T_2 \\ T_3 \\ T_4 \end{Bmatrix}.$$

(5.163)

Canceling out the vector of nodal temperature variation, the first term becomes a vector and the second term is the product of a matrix and the nodal temperature vector, i.e.,

$$\left(\int_{\Omega}^{(e)} \begin{bmatrix} \frac{\partial N_1}{\partial x} & \frac{\partial N_1}{\partial y} & \frac{\partial N_1}{\partial z} \\ \frac{\partial N_2}{\partial x} & \frac{\partial N_2}{\partial y} & \frac{\partial N_2}{\partial z} \\ \frac{\partial N_3}{\partial x} & \frac{\partial N_3}{\partial y} & \frac{\partial N_3}{\partial z} \\ \frac{\partial N_4}{\partial x} & \frac{\partial N_4}{\partial y} & \frac{\partial N_4}{\partial z} \end{bmatrix} k \begin{bmatrix} \frac{\partial N_1}{\partial x} & \frac{\partial N_2}{\partial x} & \frac{\partial N_3}{\partial x} & \frac{\partial N_4}{\partial x} \\ \frac{\partial N_1}{\partial y} & \frac{\partial N_2}{\partial y} & \frac{\partial N_3}{\partial y} & \frac{\partial N_4}{\partial y} \\ \frac{\partial N_1}{\partial z} & \frac{\partial N_2}{\partial z} & \frac{\partial N_3}{\partial z} & \frac{\partial N_4}{\partial z} \end{bmatrix} d\Omega \right) \begin{Bmatrix} T_1 \\ T_2 \\ T_3 \\ T_4 \end{Bmatrix}$$

$$= \left(\int_{\Gamma_q}^{(e)} \begin{Bmatrix} N_1 \\ N_2 \\ N_3 \\ N_4 \end{Bmatrix} k \frac{\partial T}{\partial \mathbf{n}} d\Gamma \right). \tag{5.164}$$

In short form,

$$\mathbf{K}^{(e)} \begin{Bmatrix} T_1 \\ T_2 \\ T_3 \\ T_4 \end{Bmatrix} = \mathbf{f}^{(e)} \tag{5.165}$$

where $\mathbf{K}^{(e)}$ is the element conductivity matrix and $\mathbf{f}^{(e)}$ is the element load vector.

For a given tetrahedral element, evaluating $\mathbf{K}^{(e)}$ and $\mathbf{f}^{(e)}$ requires isoparametric mapping of the variables. In 3-D, the transformation between the derivatives in the physical coordinate system and those in the master element reference coordinate system is given by

$$\begin{Bmatrix} \frac{\partial}{\partial \xi} \\ \frac{\partial}{\partial \eta} \\ \frac{\partial}{\partial \zeta} \end{Bmatrix} = \begin{bmatrix} \frac{\partial x}{\partial \xi} & \frac{\partial y}{\partial \xi} & \frac{\partial z}{\partial \xi} \\ \frac{\partial x}{\partial \eta} & \frac{\partial y}{\partial \eta} & \frac{\partial z}{\partial \eta} \\ \frac{\partial x}{\partial \zeta} & \frac{\partial y}{\partial \zeta} & \frac{\partial z}{\partial \zeta} \end{bmatrix} \begin{Bmatrix} \frac{\partial}{\partial x} \\ \frac{\partial}{\partial y} \\ \frac{\partial}{\partial z} \end{Bmatrix} = \mathbf{J} \begin{Bmatrix} \frac{\partial}{\partial x} \\ \frac{\partial}{\partial y} \\ \frac{\partial}{\partial z} \end{Bmatrix} \tag{5.166}$$

where $\mathbf{J}$ is the Jacobian matrix. With this relation, we have

$$\begin{Bmatrix} \frac{\partial}{\partial x} \\ \frac{\partial}{\partial y} \\ \frac{\partial}{\partial z} \end{Bmatrix} = \mathbf{J}^{-1} \begin{Bmatrix} \frac{\partial}{\partial \xi} \\ \frac{\partial}{\partial \eta} \\ \frac{\partial}{\partial \zeta} \end{Bmatrix} \tag{5.167}$$

and therefore,

$$\begin{bmatrix} \frac{\partial N_1}{\partial x} & \frac{\partial N_2}{\partial x} & \frac{\partial N_3}{\partial x} & \frac{\partial N_4}{\partial x} \\ \frac{\partial N_1}{\partial y} & \frac{\partial N_2}{\partial y} & \frac{\partial N_3}{\partial y} & \frac{\partial N_4}{\partial y} \\ \frac{\partial N_1}{\partial z} & \frac{\partial N_2}{\partial z} & \frac{\partial N_3}{\partial z} & \frac{\partial N_4}{\partial z} \end{bmatrix} = \mathbf{J}^{-1} \begin{bmatrix} \frac{\partial N_1}{\partial \xi} & \frac{\partial N_2}{\partial \xi} & \frac{\partial N_3}{\partial \xi} & \frac{\partial N_4}{\partial \xi} \\ \frac{\partial N_1}{\partial \eta} & \frac{\partial N_2}{\partial \eta} & \frac{\partial N_3}{\partial \eta} & \frac{\partial N_4}{\partial \eta} \\ \frac{\partial N_1}{\partial \zeta} & \frac{\partial N_2}{\partial \zeta} & \frac{\partial N_3}{\partial \zeta} & \frac{\partial N_4}{\partial \zeta} \end{bmatrix}. \quad (5.168)$$

Next, by mapping the volume of the physical tetrahedral element to the $2 \times 2 \times 2$ master element, the $\mathbf{K}^{(e)}$ matrix can be rewritten as

$$\mathbf{K}^{(e)} = \int_{master}^{(e)} \begin{bmatrix} \frac{\partial N_1}{\partial \xi} & \frac{\partial N_1}{\partial \eta} & \frac{\partial N_1}{\partial \zeta} \\ \frac{\partial N_2}{\partial \xi} & \frac{\partial N_2}{\partial \eta} & \frac{\partial N_2}{\partial \zeta} \\ \frac{\partial N_3}{\partial \xi} & \frac{\partial N_3}{\partial \eta} & \frac{\partial N_3}{\partial \zeta} \\ \frac{\partial N_4}{\partial \xi} & \frac{\partial N_4}{\partial \eta} & \frac{\partial N_4}{\partial \zeta} \end{bmatrix} \mathbf{J}^{-T} k \mathbf{J}^{-1} \begin{bmatrix} \frac{\partial N_1}{\partial \xi} & \frac{\partial N_2}{\partial \xi} & \frac{\partial N_3}{\partial \xi} & \frac{\partial N_4}{\partial \xi} \\ \frac{\partial N_1}{\partial \eta} & \frac{\partial N_2}{\partial \eta} & \frac{\partial N_3}{\partial \eta} & \frac{\partial N_4}{\partial \eta} \\ \frac{\partial N_1}{\partial \zeta} & \frac{\partial N_2}{\partial \zeta} & \frac{\partial N_3}{\partial \zeta} & \frac{\partial N_4}{\partial \zeta} \end{bmatrix} det(\mathbf{J}) d\xi d\eta d\zeta$$

$$= \int_{master}^{(e)} \mathbf{L}^T \mathbf{J}^{-T} k \mathbf{J}^{-1} \mathbf{L} det(\mathbf{J}) d\xi d\eta d\zeta. \quad (5.169)$$

The numerical integration of the $\mathbf{K}^{(e)}$ matrix is carried out by using a quadrature rule of choice. The general expression of the integration can be written as

$$\mathbf{K}^{(e)} = \sum_{g=1}^{ng} \mathbf{L}^T(\xi_g, \eta_g, \zeta_g) \mathbf{J}^{-T}(\xi_g, \eta_g, \zeta_g) k \mathbf{J}^{-1}(\xi_g, \eta_g, \zeta_g) \mathbf{L}(\xi_g, \eta_g, \zeta_g) det(\mathbf{J}(\xi_g, \eta_g, \zeta_g)) w_g$$

(5.170)

where ng is the number of integration points, and w_g is the weight for the g-th integration point.

The right hand side vector $\mathbf{f}^{(e)}$ is evaluated on the boundary (surface) where heat flux is not zero. The boundary integral is performed over the corresponding face of the master element via isoparametric mapping. The procedure is similar to the boundary integration in the 2-D case, except the boundary is now a surface area. For the heat sink example, the side surfaces are all adiabatic. The heat flux is zero on Γ_q. Therefore, $\mathbf{f}^{(e)} = \mathbf{0}$ n this case.

Step 7: Global Matrix and Vector Assembly

The global matrix and vector assembly process is the same as that in 2-D case. For each element, the one-to-one correspondence between the local node index and the global node index is needed for the assembly process. That information is obtained from the input file "elements.dat." The sequence of the nodes listed for each element gives not only how the nodes are connected but also how the local node index (1, 2, 3, ...) is mapped to the global node index (the first, second, third, ..., node numbers listed in each row of the file). With this node index mapping, the process of assembly is essentially the same as shown in Fig. 5.28. The relevant MATLAB functions that have been discussed and used in the 2-D heat transfer analysis can be readily used for 3-D cases. The functions are "SetElementNodes" (Section 5.2.1), "AssembleGlobalMatrix" (Section 5.2.1), and "AssembleGlobalVector" (Section 5.2.1).

Step 8: Apply the Essential Boundary Condition
The natural (heat flux) boundary condition is already included in the weak form boundary integral. The essential (temperature) boundary condition is applied the same way as in 1-D and 2-D heat transfer analysis. The methods are described in detail in Section 5.2.1. The final linear system is then obtained as

$$\mathbf{KT} = \mathbf{F}. \tag{5.171}$$

Step 9: Solve the Global System
After the essential boundary conditions are applied, the final linear system can be solved directly in MATLAB. The unknown nodal temperature vector **T** can be computed.

Step 10: Post-processing
In this step, by using the temperature results, we compute secondary physical quantities such as heat flux. As in the 2-D analysis, heat flux is simply -1 times the material thermal conductivity times the temperature gradient. Having obtained the nodal temperatures, within each element, the temperature gradient can be easily calculated by using Eqs. (5.160 and 5.168). Note that, if the temperature gradient is calculated at the nodes for each element, for a node shared by multiple elements, its temperature gradient obtained from these elements may not be the same. Typically, the average of these values is taken as the temperature gradient result at that node. The calculation of the heat fluxes of the 3-D heat sink is left to the reader as an exercise.

Another task of post-processing is visualization. 3-D plots of the results are more involved to generate. For 2-D problems, a result as a function of *xy*-location can be shown as the third coordinate in a 3-D surface plot. For 3-D field problems, however, the computational domain has already occupied the 3-D space. The results (e.g. temperature) would be in the fourth dimension to plot. In this case, color can be used as a measure of the magnitude of the results. To show the 3-D results in color, we have two relatively simple approaches: (1) coloring element faces on the domain boundary, (2) painting the nodes as colored scattered points. Figure 5.37 shows the heat sink temperature result by using 3-D scatter plots. The MATLAB function for plotting 3-D scattered points is "scatter3."

5.3.2 Computer Implementation

In this section, by using the heat sink problem as an example, we present the data files and the MATLAB code for the 3-D steady state heat transfer finite element analysis.

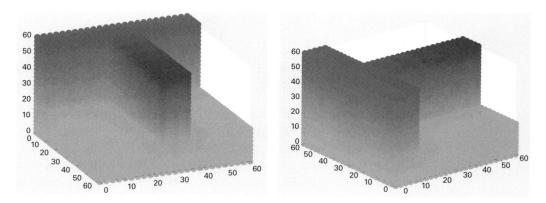

Figure 5.37 Temperature result shown using 3-D scatter plots.

Data Files

The input data files are shown in Table 5.12. The nodes, elements, material properties, loads, boundary conditions, and other model options are stored in forms of arrays and matrices. Note that there are 38,380 elements and 8279 nodes in the mesh shown in Fig. 5.33. Only the first five lines of the "nodes.dat" and "elements.dat" are shown in the tables. For the same reason, we only show the first five lines of "nodalTemp.dat" in the table.

MATLAB Code

The first code is the main program of the heat sink analysis.

```
1  % next 5 lines : read the input files
2  load modelOptions.dat;
3  load nodes.dat;
4  load elements.dat;
5  load materials.dat;
6  load nodalTemp.dat;
7  n_nodes=size(nodes,1);              % number of nodes
8  n_nodalTemp=size(nodalTemp,1);
9  %-- Global material properties and constants
10 kappa=materials (1,1);
11  convection_coeffcient =materials (2,1);
12 K=CompK(nodes, elements, kappa);    % compute global K matrix
13 heatSource =[];                      % there is no heat source
14 F=CompF(nodes, elements, heatSource); % compute global F vector
15
16 %-- apply temperature boundary condition
17 coeff=abs(max(K(nodalTemp(1,1),:)))*1e7;  % penalty number
18 for i=1:n_nodalTemp
19   node_id=nodalTemp(i,1);
20   K(node_id, node_id)=coeff;
21   F(node_id, 1) = nodalTemp(i,2)*coeff;
```

5.3 3-D Steady State Heat Transfer

Table 5.12 Data files for analysis

nodes.dat

1	60.0000	25.0000	20.0000
2	10.0000	25.0000	20.0000
3	57.4950	25.0000	20.0000
4	54.9942	25.0000	20.0000
5	52.4933	25.0000	20.0000
⋮	⋮	⋮	⋮

elements.dat

1	2679	3155	6672	5652
2	2560	3598	5820	5327
3	594	651	5331	730
4	596	707	5340	757
5	740	764	774	6417
⋮	⋮	⋮	⋮	⋮

nodalTemp.dat

373	100.0000
374	100.0000
375	35.0000
391	35.0000
392	35.0000
⋮	⋮

materials.dat

% Thermal conductivity (W/mK)
5
% Convection coefficient (W/(m^2 K)
10

options.dat

% Dimensions
3
% Element type. Format: No. of edges or faces, No. of nodes
44
% Environmental temperature for surface convection
30

```
22  end
23  % solve the global linear system
24  T=K\F;
25  % next 6 lines: save the displacement results in file
26  TOut=zeros(n_nodes,5);
27  for n=1:n_nodes
28      TOut(n,1:4)=nodes(n,1:4);
29      TOut(n,5)=T(n,1);
30  end
```

```
31  save −ascii −double feT.dat TOut
32
33  figure (1);
34  load mymap.dat;
35  colormap(mymap);
36  scatter3 (nodes (:,2) , nodes (:,3) ,nodes (:,4) ,600,T (:,1) ,' filled ');
```

The function to compute the element conductivity matrices and assembly them into the global matrix.

```
1   % Compute global K matrix
2   function K=CompK(nodes, elements, kappa)
3     n_nodes = size(nodes,1);
4     n_elements = size(elements,1);
5     n_nodes_per_element = size(elements,2) −1;
6     K=zeros(n_nodes,n_nodes);
7     DN=zeros(3,n_nodes_per_element);
8     [gauss_points , gauss_weights]=GetTetraGauss(1);
9     n_gauss_points=size( gauss_points ,1)
10    [N,Nx,Ny,Nz]=CompNDNatPointsTetra4(gauss_points(:,1),...
11                    gauss_points (:,2) , gauss_points (:,3) );
12
13    % compute K matrix: loop over all the elements
14    for e=1:n_elements
15      ke=zeros(n_nodes_per_element, n_nodes_per_element);
16      [element_nodes, node_id_map]= SetElementNodes(e, nodes, elements);
17      % next 9 lines: compute element conductivity matrix ke
18      for g=1:n_gauss_points
19        J=CompJacobian3DatPoint(element_nodes,Nx(:,g),Ny(:,g),Nz(:,g));
20        detJ=det(J);
21        Jinv=inv(J);
22        DN(1,:)=Nx(:,g);
23        DN(2,:)=Ny(:,g);
24        DN(3,:)=Nz(:,g);
25        ke=ke+DN'*Jinv'*kappa*Jinv*DN*detJ*gauss_weights(g);
26      end
27      % assemble ke into global K
28      K= AssembleGlobalMatrix(K,ke,node_id_map,1);
29    end
```

The function to compute the element vectors and assembly them into the global right hand side vector.

```
1  % Compute global RHS load vector
2  function F=CompF(nodes, elements, heatSource)
```

```
3  n_nodes = size(nodes,1);
4  n_heat_nodes = size(heatSource,1);
5  F=zeros(n_nodes,1);
6  % for-loop: apply heat source
7  for i=1:n_heat_nodes
8      row=heatSource(i,1);
9      F(row,1) = F(row,1)+ heatSource(i,2);
10 end
```

The function to compute the 3×3 Jacobian matrix at a given integration point.

```
1  % Compute Jacobian at a point (xi, eta, zeta) in a 3-D master element
2  % Input: element_nodes: the physical coordinates of a 3-D element
3  %              in the format of [x1, y1; x2 y2; x3, y3; ...]
4  % Input: dN/dxi, dN/deta, dN/dzeta vectors at the point
5  % Output: Jacobian matrix at the point
6  function J= CompJacobian3DatPoint(element_nodes, Nxi, Neta, Nzeta)
7  J=zeros(3,3);
8  for j=1:3
9      J(1,j) = Nxi' * element_nodes(:,j);
10     J(2,j) = Neta' * element_nodes(:,j);
11     J(3,j) = Nzeta' * element_nodes(:,j);
12 end
```

In the MATLAB programs listed above, the functions which have already been discussed previously are not repeated here. They are listed below for your reference:
function [N,Nx,Ny]=CompNDNatPointsTetra4(xi_vector, eta_vector, zeta_vector): Section 5.3.1.
function [gauss_points, gauss_weights]=GetTetraGauss(ng): Section 5.3.1.
function [element_nodes, node_id_map]= SetElementNodes(ele, nodes, elements): Section 5.2.1.
function K= AssembleGlobalMatrix(K,ke,node_id_map,ndDOF): Section 5.2.1.

5.4 Numerical Issues and Performance Considerations

Since finite element analysis of engineering problems is a numerical solution of the governing equations, numerical errors are introduced through the steps of the solutions. Broadly defined, the errors are mainly introduced into the solution through discretization and integration. As the numerical integration error has been discussed in the numerical analysis chapter, in this section we discuss the discretization error and its effect on the convergence of the numerical solution. In addition, mesh/element quality is important and directly affect the accuracy of the solutions. We introduce several rules of thumb as the constraints for the mesh generation process.

5.4.1 Convergence Considerations

A numerical solution exhibits convergence when the discretization error approaches zero as the mesh is made infinitely fine (i.e., as the elements approach zero size). Such kind of convergence is referred to as the discretization convergence or simply mesh convergence. How fast the discretization error disappears as the element size is decreased is measured by a quantity called rate of convergence. The discretization error is denoted by $O(h^p)$. The $O()$ notation represents the order of the error. $O(h^p)$ means the discretization error is proportional to h^p if h is small enough, where

- p: rate of convergence
- h: element size.

For example, when $p = 3$, halving h reduces the discretization error to 1/8 of the original value. Note that, for a uniform mesh, h is typically set as the (edge) length of any of the elements in the mesh. For non-uniform mesh, h can be taken as the element size in a critical region which contains the physical quantity of interest. An alternative measure of h for non-uniform 2-D meshes is $h = (A/N_m)^{1/2}$, where A is the area of the region being meshed, and N_m is the number of elements in the mesh. In this sense, h is an averaged size of the elements.

Sufficient (Not Necessary) Conditions for Monotonic Convergence

A monotonic mesh convergence is guaranteed when a finite element approximation satisfies two conditions:

1. **Compatibility**: Taking the heat transfer problem as an example, if the weak form contains derivatives of temperature T up to order l, then the approximated function T must be C^{l-1} continuous between the elements. Recall that a function is C^k continuous if all of its derivatives up to order k are continuous. For second-order partial differential equations (as in our steady state heat transfer example), the highest derivative of T in the weak form is of order 1. Therefore, we require C^0 continuity of T between the elements. A physical interpretation of this requirement is shown in Fig. 5.38. The compatibility requirement is satisfied when the approximate function T and δT along an element boundary depend on the nodal quantities along that boundary in the same way for adjacent elements. Simply speaking, there should not be a "gap" between the neighboring elements' T and δT along the shared edges.
2. **Completeness**: The elements must be capable of exactly representing a complete polynomial of order l. For example, in 2-D
   - $l = 1$: approximation must be able to represent any $a_1 + a_2 x + a_3 y$
   - $l = 2$: approximation must be able to represent any $a_1 + a_2 x + a_3 y + a_3 x^2 + a_4 xy + a_5 y^2$.

5.4 Numerical Issues and Performance Considerations

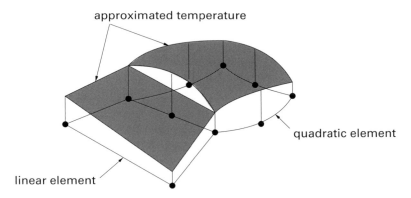

Figure 5.38 Compatibility ensures that no gaps exist between the elements.

If we want to be able to exactly represent $T(x, y) = \alpha_0 + \alpha_1 x + \alpha_2 y$, we must check whether the approximation function matches the exact solution. Using the shape functions

$$T(x, y) = \sum_{i=1}^{n} N_i(x, y) T_i$$

$$= \sum_{i=1}^{n} N_i(x, y)(\alpha_0 + \alpha_1 x_i + \alpha_2 y_i)$$

$$= \alpha_0 \left(\sum_{i=1}^{n} N_i(x, y) \right) + \alpha_1 \left(\sum_{i=1}^{n} N_i(x, y) x_i \right) + \alpha_2 \left(\sum_{i=1}^{n} N_i(x, y) y_i \right).$$

(5.172)

For the element to be complete to the order of 1, we must have $T(x, y) = T_{exact}(x, y)$, i.e.

$$\sum_{i=1}^{n} N_i(x, y) = 1 \quad (5.173)$$

$$\sum_{i=1}^{n} N_i(x, y) x_i = x \quad (5.174)$$

$$\sum_{i=1}^{n} N_i(x, y) y_i = y. \quad (5.175)$$

The above conditions are for convergence purposes. It is easy to verify that the shape functions we have constructed for the 1-D, 2-D, and 3-D elements so far satisfy the above completeness conditions. It should be noted that the completeness and compatibility conditions are sufficient but not necessary.

Prediction of Convergence Rate (p)

Taking a 2-D element for example and expanding the temperature function at a point i (i.e. (x_i, y_i)) by using Taylor's series, we have

$$T(x, y) = T(x_i, y_i) + \left(\frac{\partial T}{\partial x}\right)_i (x - x_i) + \left(\frac{\partial T}{\partial y}\right)_i (y - y_i) + \left(\frac{\partial^2 T}{\partial x^2}\right)_i \frac{(x - x_i)^2}{2}$$

$$+ \left(\frac{\partial^2 T}{\partial x \partial y}\right)_i (x - x_i)(y - y_i) + \left(\frac{\partial^2 T}{\partial y^2}\right)_i \frac{(y - y_i)^2}{2} + O(h^3) \quad (5.176)$$

where $\left(\frac{\partial T}{\partial x}\right)_i$ denotes the partial derivative of T with respect to x evaluated at point (x_i, y_i). Note that, one can imagine the point (x_i, y_i) to be the center of an element, and the area of the element to be the region of Taylor's series expansion in the neighborhood of (x_i, y_i). Then the element size h is proportional to $(x - x_i)$ and $(y - y_i)$. Therefore, the residual term in Eq. (5.176) can be written in terms of h. Since the value of $T(x_i, y_i)$ and the partial derivatives are constants, Eq. (5.176) is essentially

$$T(x, y) = a_1 + a_2(x - x_i) + a_3(y - y_i)$$

$$+ a_4(x - x_i)^2 + a_5(x - x_i)(y - y_i) + a_6(y - y_i)^2 + O(h^3). \quad (5.177)$$

If, for an element of size h, the finite approximation is complete to the second-order polynomial, the first six terms of the right hand side of Eq. (5.177) can be precisely represented by the shape functions. Therefore, the error in this case is the higher order terms represented by $O(h^3)$. That is, the discretization error is of order h^3. The reason the order is h^3 is that when h goes to infinitesimal, h^3 dominates all the higher order terms. In general, if the finite element approximation is complete to order m, the unknown quantity in the element can be locally fit up to order m. The discretization error is then $O(h^{m+1})$ and $p = m + 1$.

The polynomial terms that are represented by the various types of 2-D elements can be depicted by using Pascal's triangle. Figures 5.39–5.41 list the monomials that can be represented, the complete polynomial order, and the convergence order of the rectangular and triangular elements we have discussed so far.

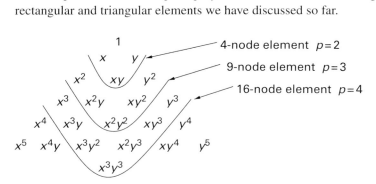

Figure 5.39 Polynomials represented by the Lagrange family rectangular elements.

5.4 Numerical Issues and Performance Considerations

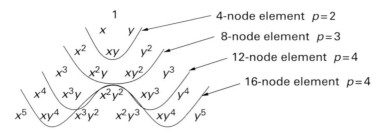

Figure 5.40 Polynomials represented by the serendipity family rectangular elements.

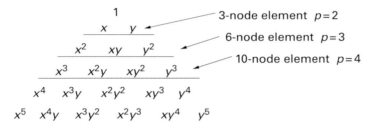

Figure 5.41 Polynomials represented by the triangular elements.

5.4.2 The Patch Test

The patch test was original proposed by Irons and Razzaque (1972) and further elaborated by Taylor et al. (1986). The patch test serves as a simple method to prove convergence of an element. In essence, the patch test checks whether a constant distribution of the primary physical quantity and its first derivatives within an arbitrary element patch can be represented exactly (completeness). It also checks whether the constant distribution holds across the elements (compatibility). The patch test was originally described for linear elasticity problems as shown in Fig. 5.42(a), in which the elements pass the patch test if a constant strain field is reproduced. For a scalar field problem such as the heat transfer problem, the patch test can be set up as shown in Fig. 5.42(b) where the exact solution of the temperature field is a linear function of x and the heat flux in the x-direction is a constant. The patch test shown in the figure is passed if the constant x-direction heat flux is reproduced by the finite element solution. After that, the patch test needs to be done to check if a constant y-direction heat flux can be reproduced. That is, temperature fields in the form of

$$T(x, y) = ax + b \qquad \frac{\partial T}{\partial x} = a \qquad (5.178)$$

and

$$T(x, y) = cy + d \qquad \frac{\partial T}{\partial y} = c \qquad (5.179)$$

should be reproduced by the finite element solution in the patch test to guarantee the convergence of the model.

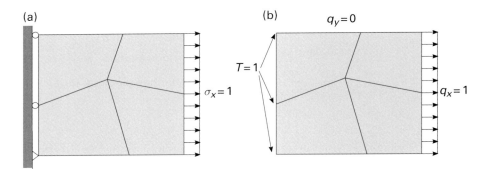

Figure 5.42 Patch test.

The patch test can be viewed as a numerical convergence condition test. In addition, it also reflects the sufficiency of the Gauss quadrature. Insufficient quadrature points can cause the model to fail the patch test.

5.4.3 Element Quality

It is understood that the convergence properties of the elements ensure that the FEA solution approaches the exact solution when the mesh is refined (smaller elements and more degrees of freedom). Nevertheless, for a given number of elements (i.e., the degrees of freedom are fixed for the model), mesh quality can have a large influence upon the accuracy of the FEA solution. A "bad" mesh can cause large errors and sometimes unrealistic results. The question is then what is a good mesh. Unfortunately, while there are various element quality measures proposed in the literature (Liu and Joe 1994, Geuzaine & Remacle 2009, Pébay and Baker 2003) and used in commercial FEA software packages, there is no "standard" definition or quantitative metrics for measuring the quality of a mesh. In general, the quality of a mesh depends on the quality of the individual elements as well as the distribution and transition of the elements in the computational domain. Here we limit our discussion to the quality of elements.

The quality of elements affects the performance of FEA in multiple ways. The most significant effects include (Knupp 2007):

- Bad shape of an element may cause ill-conditioning of the stiffness matrix. This will lead to large error in the solution of the linear systems.
- A degenerated or close-to-degenerated element gives zero value or very small value of the determinant of the Jacobian, causing problems in calculating the element matrices.
- A twisted element will lead to a negative determinant of the Jacobian, causing non-physical results.

- When element size is excessively large compared to the variation of the solution, the approximation over the element becomes insufficient. However, this issue is problem dependent. It is typically considered as a part of the global mesh quality.

For element quality measures, the commonly used metrics measure some local geometric properties that are independent of the solution to the physical problem. Two simplest measures are the angles and aspect ratio of the elements. An element with one or more vertex angle that is very small or very close to 180°, or having a large aspect ratio is considered of poor quality. Figure 5.43 shows two examples of such elements.

In practice, for 2-D elements it has been accepted that any inner vertex angle θ of the element should satisfy

$$15° < \theta < 165°. \tag{5.180}$$

For 3-D elements, the rule given in Eq. (5.180) is applied to both vertex angles and the dihedral angle between the face planes of the element. In addition, it is recommended that the largest aspect ratio of an element should not exceed 5:

$$max(\text{aspect ratio}) < 5. \tag{5.181}$$

For the element shown in Fig. 5.43(b), the aspect ratio is simply b/h. For an element (2-D or 3-D) with irregular shape, the aspect ratio can be obtained from its smallest bounding box.

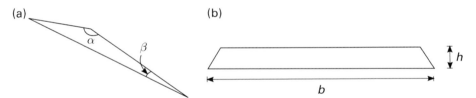

Figure 5.43 Element quality measure: (a) angles and (b) aspect ratio.

A more mathematical approach for determining the element quality is based on element radii ratio (Geuzaine & Remacle 2009). As shown in Fig. 5.44, the element radii ratio is defined as the ratio between the radii of the inscribed circle and circumcircles, $s = r/R$. For triangular elements, the radii ratio can be obtained by

$$s = 4\frac{\sin\alpha \sin\beta \sin\gamma}{\sin\alpha + \sin\beta + \sin\gamma} \tag{5.182}$$

where α, β, and γ are the three inner vertex angles of the element. From the expression, it can be shown that $s = 1$ for equilateral triangles and $s = 0$ for degenerated triangles. For tetrahedral elements, we have

$$s = \frac{6\sqrt{6}V}{\left(\sum_{i=1}^{4} A_i\right) \max_{j=1,\ldots,6}(L_j)} \tag{5.183}$$

where V is the volume of the element, A_i is the area of the i-th face, and L_j is the length of the j-th edge. Similar to triangular elements, the radii ratio of tetrahedral elements $s \in [0, 1]$ and $s = 0$ if the element is degenerated.

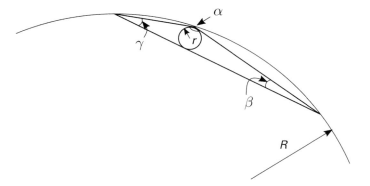

Figure 5.44 Element quality measure: element radii ratio.

Another element quality measure that is similar to the radii ratio metric is being used by commercial FEA package ANSYS. The element quality measure is straightforward and can be applied to various 2-D and 3-D elements. The metric is summarized below (ANSYS Release 17):

$$s = C \frac{A}{\sum_{i=1}^{} (L_i)^2} \quad \text{for 2-D elements} \quad (5.184)$$

and

$$s = C \frac{V}{\sqrt{\left[\sum_{i=1}^{} (L_i)^2\right]^3}} \quad \text{for 3-D elements} \quad (5.185)$$

where A is the area of 2-D elements, V is the volume of 3-D elements, and L_i is the length of i-th edge. The constant factor C is different for different elements. Table 5.13 lists the value of C for different types of elements. Once again $s \in [0, 1]$. $s = 1$ for best quality and $s = 0$ for degenerate elements.

5.5 Further Considerations

In Chapter 4, we performed finite element analysis for 1-D elasticity, steady state heat transfer, and fluid flow problems. If we look closely at the governing equations of the 1-D elasticity and heat transfer problems, we notice that they have the same form. The type of the differential equation is called Poisson's equation. In this chapter, while the heat transfer problem becomes 2-D and 3-D, its governing partial differential equation is still a Poisson's equation. Poisson's equation is an elliptic partial differential

Table 5.13 Element quality factor C for different elements

Element	C
Triangle	6.92820323
Quadrilateral	4.0
Tetrahedral	124.70765802
Hexagonal	41.56921938
Prismatic	62.35382905
Pyramidal	96

equation and it describes the physics of many different physical systems. The general form of Poisson's equation is given by

$$-\nabla \cdot (\kappa \nabla u) = f \tag{5.186}$$

with the natural boundary condition

$$\kappa \frac{\partial u}{\partial n} = \bar{q} \tag{5.187}$$

and essential boundary condition

$$u = \bar{u}. \tag{5.188}$$

Note that when the right hand side f is zero, Poisson's equation becomes Laplace's equation. Table 5.14 shows a variety of physical problems that are described by Poisson's equation and the physical meaning of the coefficients for the individual physical problems.

It is shown in the table that although these physical problems are vastly different, they are all governed by Poisson's equation. The finite element method is a general numerical method of solving differential equations. Therefore, a FEA code solving Poisson's equation for a given physical problem can be simply used to solve other physical problems governed by the same equation. In other words, there is essentially no difference in solving a heat transfer problem and an electrostatics problem by using the finite element method, except the physical meaning of the variables.

5.6 Summary

Upon completion of this chapter, you should be able to:

- understand the fundamental theory of 2-D and 3-D steady state heat transfer and derive the differential governing equations

Table 5.14 Physical problems governed by Poisson's equation (Reddy 1984)

Problem	Primary variable u	Material constant κ	Source f	Secondary variable $\frac{\partial u}{\partial x}, \frac{\partial u}{\partial y}$
Heat transfer	Temperature	Conductivity	Heat source	Heat flux $q_x = -\kappa \frac{\partial u}{\partial x}$, $q_y = -\kappa \frac{\partial u}{\partial y}$
Irrotational flow	Steam function	Density	Mass production	Velocities $v_x = -\frac{\partial u}{\partial x}$, $v_y = \frac{\partial u}{\partial y}$
Groundwater flow	Piezomatric head	Permeability	Recharge	Velocities $v_x = -\kappa \frac{\partial u}{\partial x}$, $v_y = -\kappa \frac{\partial u}{\partial y}$
Torsion	Stress function	Inverse of shear modulus	2× angle of twist rate	Shear stresses $\tau_{zy} = -\frac{\partial u}{\partial x}$, $\tau_{zx} = \frac{\partial u}{\partial y}$
Electrostatics	Electric potential	Dielectric constant	Charge density	Electric field $\mathbf{E} = -\nabla u$
Magnetostatics	Magnetic potential	Permeability	Charge density	Magnetic field $\mathbf{H} = -\nabla u$

- derive the weak forms by using the Galerkin weighted residual method
- understand the concept and methods of isoparametric mapping
- derive the 2-D shape functions for triangular and quadrilateral elements and use these shape functions to approximate an unknown function and its derivatives over 2-D elements
- use Gauss quadrature to integrate 2-D functions over 2-D elements
- use Gauss quadrature to integrate 3-D functions over 3-D elements
- derive the expressions of element matrices and vectors from a weak form for 2-D and 3-D heat transfer problems
- assemble element matrices and vectors into the global matrix and vector for 2-D and 3-D heat transfer problems
- apply the essential boundary conditions by using the direct substitution and the penalty methods for 2-D and 3-D heat transfer problems
- calculate the heat flux after the nodal temperatures are obtained
- understand the sufficient conditions (compatibility and completeness) of convergence
- predict the convergence rate based on the order of approximation
- understand the concept of patch test

- understand the element quality measures
- implement a MATLAB function to generate uniform meshes for 2-D rectangular domains
- implement MATLAB functions to perform the post-processing steps for 2-D and 3-D heat transfer problems
- implement a complete MATLAB program to perform FEA of 2-D and 3-D steady state heat transfer problems.

5.7 Problems

5.1 In our 2-D heat transfer problem, assuming there is a heat source distribution $Q(x, y)$ in the plate, the governing PDE can be written as follows:

$$-\frac{\partial}{\partial x}\left(k\frac{\partial T}{\partial x}\right) - \frac{\partial}{\partial y}\left(k\frac{\partial T}{\partial y}\right) + h_z(T - T_\infty) - Q(x, y) = 0. \qquad (5.189)$$

All the boundary conditions remain the same as shown in the text. Use the Galerkin weighted residual method to derive the weak form.

5.2 Write MATLAB code to plot all the nine shape functions for the nine-node rectangular element shown in Fig. 5.45. A sample plot of $N_1(x, y)$ is shown in Fig. 5.46.

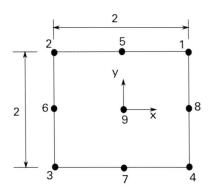

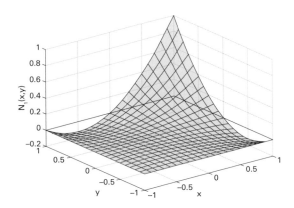

Figure 5.45 Element nodes for Problem 5.2.

Figure 5.46 Problem 5.3: shape function $N_1(x, y)$.

5.3 Establish the shape functions of the 2-D element shown in Fig. 5.47. Use MATLAB to plot all the shape functions (3-D view, similar to Fig. 5.46).

5.4 Given a nine-node rectangular element as shown in Fig. 5.48.
(a) Construct the shape functions for the element.
(b) If the temperature field at nodes A and B is 1 °C and zero at all other nodes, what is the temperature at $x = y = 1$?

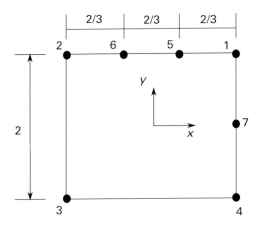

Figure 5.47 Element nodes for Problem 5.3.

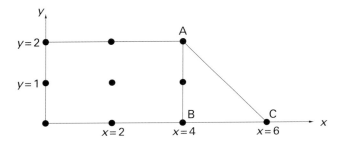

Figure 5.48 Element nodes for Problem 5.4.

(c) Consider the 3-node triangular element ABC located to the right of the nine-node rectangular element. Will the temperature function be continuous across the edge AB? Explain.

5.5 Consider two triangular elements as shown in Fig. 5.49. If the exact temperature field is x^2, can the two elements represent the exact solution? Explain.

5.6 Prove the properties of the area coordinates of a triangular element shown in Equations (5.54–5.56).

5.7 Derive expressions for the Jacobian matrix $\mathbf{J}$ in terms of ξ and η for the 2-D elements shown in Fig. 5.50.

5.8 Derive expressions for the Jacobian matrix $\mathbf{J}$, $det(\mathbf{J})$, and $\dfrac{\partial N_2}{\partial x}$ in terms of ξ and η for the mapped element shown in Fig. 5.51.

5.9 Assume the 4-node square master element is mapped to a given 4-node quadrilateral element. Write MATLAB code to compute, for a given point (ξ, η) in the master element, (1) the values of the shape functions $N_i(\xi, \eta)$, $i = 1, 2, 3, 4$, (2)

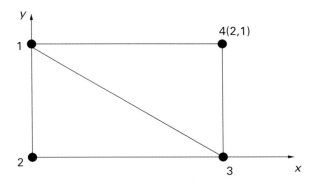

Figure 5.49 Element nodes for Problem 5.5.

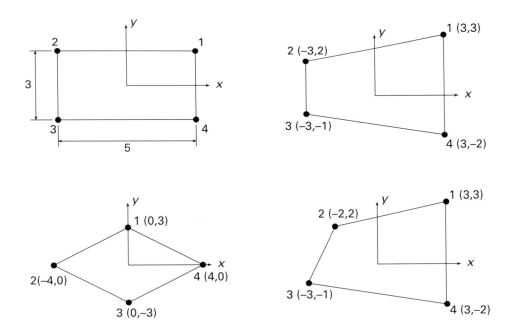

Figure 5.50 Element nodes for Problem 5.7.

the x- and y-derivatives of the shape functions, and (3) the Jacobian **J** at (ξ, η). Your input parameters would be the x- and y-coordinates of the four nodes of the quadrilateral element, and the (ξ, η) coordinate of the point in the master element where the Jacobian, the shape functions, and their derivatives are to be computed. Given the isoparametric element shown in Fig. 5.51, use your code to compute N_i, $\dfrac{\partial N_i}{\partial x}$, $\dfrac{\partial N_i}{\partial y}$, $i = 1, 2, 3, 4$, and **J** at $(\xi = 1, \eta = -1)$ and $(\xi = 0.5, \eta = 0.5)$, respectively.

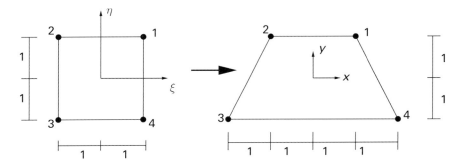

Figure 5.51 Element nodes for Problem 5.8.

5.10 Write MATLAB code to integrate the following expressions over the master element (Fig. 5.52) using 1×1, 2×2, and 3×3 Gaussian quadrature rules. Compare your results with the exact results. Explain your observations.

(a) ξ^4
(b) $\xi^2 \eta^4$
(c) $\cos \frac{\pi \xi}{2} \cos \frac{\pi \eta}{2}$

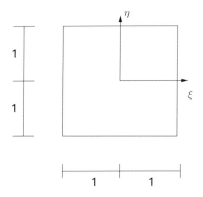

Figure 5.52 Master element for Problem 5.10.

5.11 We use the Jacobian matrix to compute the x- and y- derivatives in terms of ξ and η as follows

$$\left\{ \begin{array}{c} \frac{\partial}{\partial x} \\ \frac{\partial}{\partial y} \end{array} \right\} = \mathbf{J}^{-1} \left\{ \begin{array}{c} \frac{\partial}{\partial \xi} \\ \frac{\partial}{\partial \eta} \end{array} \right\}. \qquad (5.190)$$

Since the inverse of the Jacobian is sought, the Jacobian matrix must be non-singular, i.e., $det(\mathbf{J}) \neq 0$. In most situations, the Jacobian is naturally non-singular such as element 1 in Fig. 5.53. However, when the physical element is much distorted such as element 2 in the figure, the one-to-one correspondence between the x, y and ξ, η coordinate systems is lost. In this case, the isoparametric mapping makes the physical element effectively fold back upon itself. Then the determinant of the Jacobian can become zero and negative in the region where the element folds back upon itself. To see this, consider the isoparametric elements shown in Fig. 5.53. Use MATLAB to plot the following for each element.

(a) The lines, η as the variable and constant $\xi = -\frac{2}{3}, -\frac{1}{3}, 0, \frac{1}{3}, \frac{2}{3}$.
(b) The lines, ξ as the variable and constant $\eta = -\frac{2}{3}, -\frac{1}{3}, 0, \frac{1}{3}, \frac{2}{3}$.
(c) The determinant of the Jacobian over element 1 (in an oblique aerial view).
(d) The determinant of the Jacobian over element 2 (in an oblique aerial view).
(e) In the master element, mark the regions where $det(\mathbf{J})$ is positive, zero, and negative for element 2.

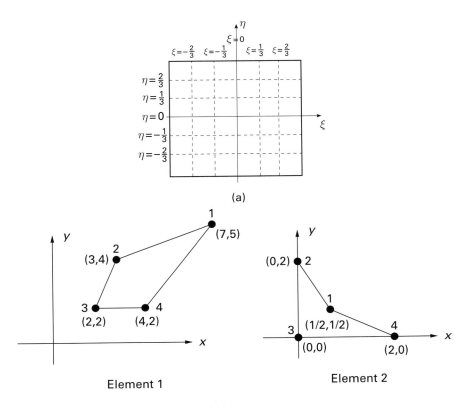

Figure 5.53 Element nodes for Problem 5.11.

5.12 Prove that for any parallelogram-shaped isoparametric element (assuming 4-node) the Jacobian determinant is constant.

5.13 Sketch a 4-node quadrilateral element for which $det(\mathbf{J})$ is a function of ξ but not of η.

5.14 The triangular element shown in Fig. 5.54 is mapped using the standard isoparametric mapping for two-dimensional elements.

Given that the shape functions are

$$N_1(\xi, \eta) = \xi$$
$$N_2(\xi, \eta) = \eta$$
$$N_3(\xi, \eta) = 1 - \xi - \eta$$

compute

(a) the Jacobian, **J**;

(b) $\dfrac{\partial N_3}{\partial x}$.

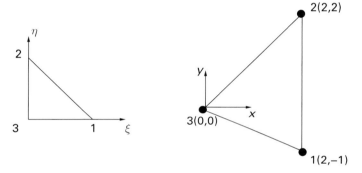

Figure 5.54 Element nodes for Problem 5.14.

5.15 Perform a patch test for the model shown in Fig. 5.55. The elements are four-node quadrilateral elements. Assuming a unit thermal conductivity, calculate the heat flux at the nodes. Note that the patch test should be performed for both x- and y-directional heat fluxes.

5.16 Consider a heat transfer problem defined over the square domain as shown in Fig. 5.56. Assume that there is no convective heat transfer on the surfaces. The thermal conductivity is $k = 1$. The exact solution of this problem is given by

$$T(x, y) = -x^3 - y^3 + 3xy^2 + 3x^2y$$

Solve the problem by using:

(a) 4 × 4 equal size linear rectangular elements
(b) 4 × 4 equal size quadratic serendipity elements
(c) 4 × 4 equal size cubic serendipity elements.

Print out the calculated nodal temperatures on $y = 0.25$ for each mesh. Plot the three finite element solutions of temperature on $y = 0.25$ and compare with the exact solution. Note that the Gauss quadrature scheme should be chosen such that the integrals can be calculated exactly.

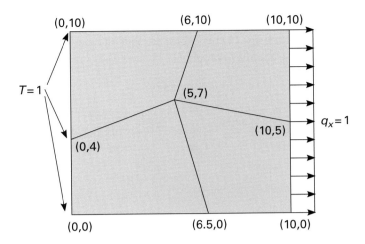

Figure 5.55 Model for Problem 5.15.

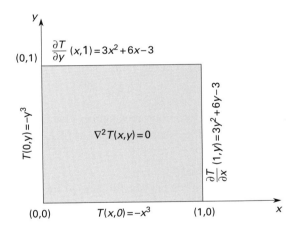

Figure 5.56 Square domain for Problem 5.16.

5.17 A series of heating cables have been placed in a conducting medium, as shown in Fig. 5.57. The medium has a conductivity of $k = 10$ W/(cm·°C), a mass density of 5 g/cm^3, and a specific heat of 1 J/g·°C . Each cable maintains a temperature of 80 °C during the operation. The lower surface is bounded by an insulating medium. The upper surface of the conducting medium is exposed to the environment at a temperature of -5 °C . Take the convection coefficient between the medium and the upper surface to be 5 W/cm^2 · K. Write a MATLAB code to compute the steady state temperature distribution in the conducting medium and obtain the temperature and heat flux at point A. Use any symmetry available in the problem. Note: (1) the insulation boundary condition implies the heat flux= 0; (2) the temperature field is symmetric about the vertical lines in the middle of any two heating cables, which implies heat flux =0 on these vertical lines; (3) the cables are thin and can be treated as points in the 2-D domain; (4) use a uniform mesh of 0.25 cm × 0.25 cm elements.

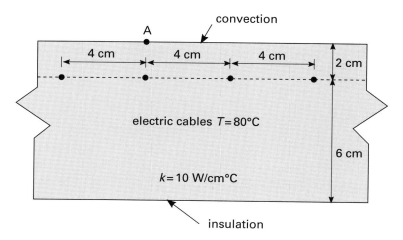

Figure 5.57 Heating cables in conducting medium for Problem 5.17.

5.18 Consider a PDE defined over a rectangular domain with boundary conditions as shown in Fig 5.58(a). The exact solution of this problem is

$$T(x, y) = x^2 - y^2 \qquad x, y \in \Omega.$$

Assuming the domain is discretized into four rectangular elements as shown in Fig. 5.58(b), we would like to obtain the exact solution from FEA with a minimum computational cost.

(a) What type of element should be used?
(b) What Gauss quadrature integration scheme should be used for the element area integrals?
(c) What Gauss quadrature integration scheme should be used for the element boundary integrals?
(d) Compute the element conductivity matrix for element 1. Assume the thermal conductivity is 1.0.

5.19 Consider a chimney constructed of two isotropic materials: dense concrete ($k = 2.0$ W/m · K) and bricks ($k = 0.9$ W/m · K). The temperature of the hot gases on the inside surface of the chimney is 140 °C, whereas the outside is exposed to the surrounding air, which is at $T = 10$ °C. The dimensions of the chimney are shown below. The convection coefficient of air is 60 W/m² · K. Write FEA code to compute the temperature and flux in the structure, refine your mesh, and compare the results obtained from different meshes, and discuss the physical meaning of your FEA results. Note that element boundaries have to coincide with the interface between the concrete and bricks.

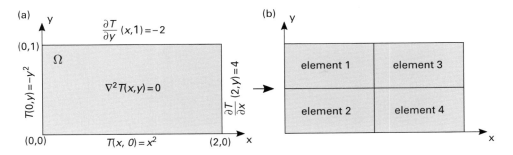

Figure 5.58 Boundary conditions for PDE for Problem 5.18.

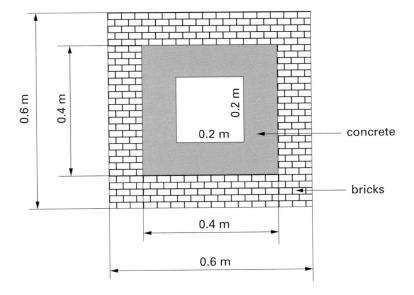

Figure 5.59 Problem 5.19: chimney cross section.

5.20 Figure 5.60 shows a 4-node tetrahedral element. The coordinates of the element vertices are given. The temperatures at the vertices are shown in the figure.

(a) Determine the temperature at $(12, 6, 16)$.
(b) Determine the normal heat flux at the centroid of the bottom triangular face.

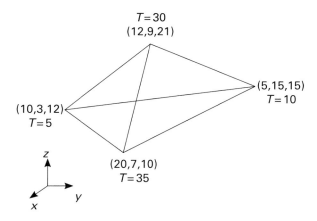

Figure 5.60 Element nodes for Problem 5.20.

References

ANSYS (Release 17), *ANSYS user manual*, ANSYS, Inc.

Geuzaine, C. & Remacle, J.-F. (2009), "Gmsh: A 3-d finite element mesh generator with built-in pre-and post-processing facilities," *International Journal for Numerical Methods in Engineering* **79**(11), 1309–1331.

Irons, B. M. & Razzaque, A. (1972), "Experience with the patch test for convergence of finite elements," in *The mathematical foundations of the finite element method with applications to partial differential equations*, Elsevier, pp. 557–587.

Knupp, P. (2007), Remarks on mesh quality, Technical report, Sandia National Lab.(SNL-NM).

Liu, A. & Joe, B. (1994), "Relationship between tetrahedron shape measures," *BIT Numerical Mathematics* **34**(2), 268–287.

Pébay, P. & Baker, T. (2003), "Analysis of triangle quality measures," *Mathematics of Computation* **72**(244), 1817–1839.

Reddy, J. N. (1984), *Energy and variational methods in applied mechanics: with an introduction to the finite element method*, Wiley.

Taylor, R., Simo, J., Zienkiewicz, O. & Chan, A. (1986), "The patch test a condition for assessing FEM convergence," *International Journal for Numerical Methods in Engineering* **22**(1), 39–62.

Zgainski, F.-X., Coulomb, J.-L., Maréchal, Y., Claeyssen, F. & Brunotte, X. (1996), "A new family of finite elements: the pyramidal elements," *IEEE Transactions on Magnetics* **32**(3), 1393–1396.

6 Mesh Generation

6.1 Overview

Discretizing a computational domain into a set of discrete elements with simple geometries is a critical step of FEA. The non-overlapping elements combined represent the geometry of a computational domain. In addition, the elements are the small volumes where the physical quantities are approximated using simple mathematical functions, and the mesh of the elements ensures that the functions are stitched together piece by piece. Depending on the characteristics of the geometry and the type of the physical problem, the computational domains and elements can be categorized into 1-D, 2-D, and 3-D types. From the dimensionality point of view, slender geometries with their size in the longitudinal direction much larger than those in the transverse direction are typically represented using 1-D elements. 2-D elements are employed for geometries large in two dimensions and small in the third. Obviously, dimensions of 3-D elements are comparable in all three directions.

The quantity of elements and quality of mesh are important for the accuracy and efficiency of a finite element analysis. The total degrees of freedom of the finite element mesh is directly proportional to the number of elements and degrees of freedom of each element. More precisely, the total degrees of freedom is equal to the product of the number of nodes and the degrees of freedom of each node. As discussed in the previous chapters, the size of the final discretized system of equations to be solved in a linear problem is proportional to the total degrees of freedom of the mesh. The computational cost of solving a linear system $\mathbf{Kd} = \mathbf{F}$ by using a state-of-the-art solver is on the order of $Nlog^2(N)$ where N is the total degrees of freedom. The cost of solving a nonlinear system is much worse than that. Therefore, the number of elements (nodes) in the mesh determines the computational cost of the analysis. On the other hand, error of a finite element solution mainly comes from (1) discretization error, (2) integration error, and (3) numerical round-off error. Among these, the first two are more closely related to the mesh. The discretization error depends on how well the domain geometry (e.g. curved boundaries) are represented by the elements and how accurately the actual variation of the physical quantity of interest is represented by the simple approximation function defined within the elements. The larger the discrepancies the larger the discretization error. We know that these discrepancies depend on size of the element and how close the approximating function is to the actual variation. Smaller discretization error requires smaller element size and higher order (or more complex) approximation; both imply more degrees of freedom of the mesh. In addition, the integration error is inversely proportional to the size of the elements and number of integration points. Therefore, smaller elements lead to smaller errors but more degrees of freedom and thus larger computational cost. There is a trade-off between the solution accuracy and computational cost.

Having understood the direct impact of a mesh on the efficiency and accuracy of a FEA solution, the importance of constructing a good mesh becomes obvious. In many cases, a carefully chosen meshing strategy enables achieving high solution accuracy with much less computational cost. Generally speaking, in constructing a mesh one should consider the following aspects: (1) element type (1-D, 2-D, 3-D, linear, quadratic, etc.) should be simple and yet reasonably accurate in representing the geometry and physical behavior of the system; (2) the shape of any element in the mesh should support a well-behaved approximation function within its boundary; (3) element size should be large as long as it is within the limit of error tolerance; (4) element size variation and transition in a mesh should follow the variation of the physical quantities in the computational domain.

In this chapter, we first introduce some of the basic modeling and meshing concepts and techniques for different types of computational domains. Next, we focus on 2-D domains and describe in detail the modeling method of planar straight line graph and the meshing approach of Delaunay refinement for generating 2-D meshes of triangular elements.

6.1.1 Geometric Modeling of Objects

A physical entity such as an object or volume can be completely and uniquely defined by its geometrical and topological data. The geometry data of an object include its shape and dimensions. The topological data describe the connectivity and associativity of its constituents (vertices, edges, surfaces, etc.). Broadly speaking, an object or volume can be described by using wire-frame, surface, and solid modeling approaches. Wire-frame modeling, also referred to as edge representation, represents the physical object by using points, lines and curves. The surface modeling approach defines the object by a set of surfaces. The surfaces do not necessarily define a closed volume or connect to their neighbors. While both the wire-frame and surface modeling approaches delineate the shape and dimension (i.e. the geometrical data) of the object, neither of them provide the topological relations of the constituents (points, lines, curves and surfaces). In comparison, solid models contains both geometrical and topological information of the object or volume. There is no ambiguity in the definition of a physical entity. Most solid modeling techniques are based on the "half-space" concept, which states that the boundary surface of a physical entity always divides the infinite space into two sides: the material (or interior) side and the empty (or exterior) side. To differentiate the two sides, the boundary surfaces are defined as directed surfaces. The boundary of the physical entity is the union of the directed boundary surfaces, and the interior (material) volume space of the physical entity is contained within the directed boundary surfaces. Therefore, the topological information (i.e., interior and exterior spaces, boundary surfaces, and their connections to the edges and vertices) is all recorded. In finite element analysis, the definition of a computational domain requires the complete representation of the object or volume. Therefore, the solid models are necessary.

There are quite a few solid modeling methods available. Two most widely used methods are the constructive solid geometry (CSG) method and the boundary representation (B-rep) method. In the CSG method, the physical entity is considered as formed via a set of consecutive Boolean operations (union, subtraction, and intersection) on a collection of pre-defined geometric primitives (box, sphere, cylinder, cone, etc.). For example, a plate with a hole can be created by subtracting a cylinder from a rectangular box. The Boolean operation steps can be conveniently defined by using a tree structure called CSG tree. Figure 6.1(b) shows an example of modeling a solid structure using a CSG tree. The boundary representation method represents a solid as a collection of directed boundary surfaces. Both the surface geometry and the topological relations among these boundary surfaces are included in the model, as shown in Fig. 6.1(c).

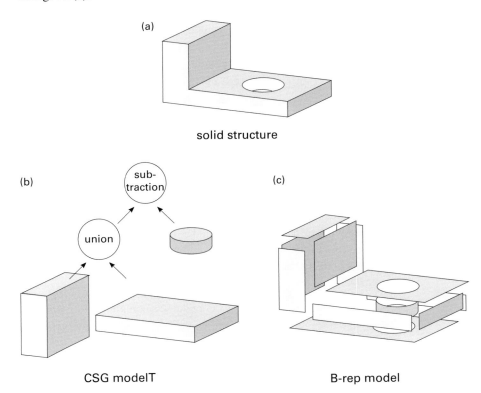

Figure 6.1 Solid modeling methods.

6.1.2 2-D Meshing Methods

The meshing process is very different in different dimensions. Creating 1-D mesh is straightforward since it typically involves dividing a line or curve into smaller segments. Meshing a 2-D/3-D domain with complex topological and geometric features requires sophisticated mathematical strategies and efficient numerical algorithms. Broadly speaking, meshing methods can be categorized into (1) uniform, (2) structured, and (3) un-structured methods. In this chapter we focus on 2-D meshing methods.

Uniform Mesh of a Regular Shaped Domain

If the computational domain is of a regular shape, such as a rectangle, a circle, a box, or a sphere, and a uniform mesh or a mesh with simple size gradient is sought, a mesh generator can be easily implemented. In the following example, we list a MATLAB function, "UniformMeshQuad4(sx, sy, ex, ey, nx, ny)," for creating a uniform mesh of 4-node quadrilateral elements for a 2-D rectangular domain. The function creates "nodes" and "elements" matrices and saves the matrices as the "nodes.dat" and "elements.dat" files which are the familiar input files for a finite element analysis.

The input variables of the function are "sx, sy, ex, ey, nx, ny." This requires the user to give the x- and y-coordinates of the lower left corner (sx and sy) and the upper right corner (ex, ey) of the rectangular domain, and the number of divisions in x- (nx) and y- (ny) directions, respectively. For example, the meshes shown in Figs. 6.2 and 6.3 are generated by using the function with

[nodes, elements]=UniformMeshQuad4(0,0,2,2,2,2);

and

[nodes, elements]=UniformMeshQuad4(0,0,3,5,4,5);

respectively. In the figures, the numbers appearing at the corners are the node numbers or IDs and those inside the elements and having parentheses are the element numbers or IDs.

For generating uniform meshes, whether it is 2-D or 3-D, the MATLAB function "meshgrid()" is particularly useful. The function directly creates the grid points (i.e. nodes) and returns the x- and y- (and z- in 3-D) coordinate matrices. These matrices are also very useful for post-processing of the FEA results. For example, MATLAB functions producing surface or contour plots of FEA results, such as "surf()" and "contourf()," etc., use the x- and y-coordinate matrices of the nodes as their input variables.

```
1  % Create a uniform mesh of a rectangular domain
2  % Input: sx,sy: x- and y- coordinates of the domain's lower left corner
3  %        ex,ey: x- and y- coordinates of the upper right corner
4  % Output: nodes, elements: nodes and elements matrices
5  function [nodes,elements]=UniformMeshQuad4(sx,sy,ex,ey,nx,ny)
6  [x,y]=meshgrid(sx:(ex-sx)/nx:ex, sy:(ey-sy)/ny:ey);    % x,y matrices of
7                                                         % the grid points
```

```
 8  nodes=zeros((nx+1)*(ny+1),3);    % empty nodes matrix
 9  elements=zeros(nx*ny,5);          % empty elements matrix
10  nids=zeros(ny+1,nx+1);            % empty node ID matrix
11
12  % next 11 lines : create nodes
13  fid=fopen('nodes.dat','w+');      % open file "nodes.dat"
14  k=1;
15  for i=1:ny+1      % rows
16    for j=1:nx+1    % columns
17      nids(i,j)=k;  % grid matrix of node IDs
18      nodes(k,1:3)=[k x(i,j) y(i,j)];
19      fprintf(fid,'%d %.10f %.10f\n',k,x(i,j),y(i,j));  % write file
20      k=k+1;
21    end
22  end
23  fclose(fid);      % close "nodes.dat"
24
25  % next 13 lines : create elements
26  fid=fopen('elements.dat','w+');   % open file "elements.dat"
27  k=1;
28  for i=1:ny        % rows
29    for j=1:nx      % columns
30      elements(k,1)=k;
31      elements(k,2:3)=[nids(i,j) nids(i,j+1)];
32      elements(k,4:5)=[nids(i+1,j+1) nids(i+1,j)];
33      fprintf(fid,'%d %d %d %d %d \n', k, nids(i,j), ...
34              nids(i,j+1),nids(i+1,j+1),nids(i+1,j));
35      k=k+1;
36    end
37  end
38  fclose(fid);      % close "elements.dat"
```

While a uniform mesh generator can be easily implemented in a simple MATLAB code as shown in the example, for 2-D and 3-D domains with arbitrary geometries, mesh generation is non-trivial and sometimes more challenging than the analysis steps. Since the very early days of the finite element method, a large number of mesh generation algorithms and methods have been developed and devised to mesh arbitrary geometries. In the following, we briefly introduce the main ideas of the so called structured and unstructured meshing methods.

Structured Mesh Template
The main idea of mesh template approaches is to map or overlap a predefined mesh of a regular geometry onto a region of irregular geometry. A representative method of this category is the mapped element generation method that most commercial FEA software packages are using. In this method, a 2-D domain with complex geometry is decomposed, often manually, into sub-domains with simple geometries such as

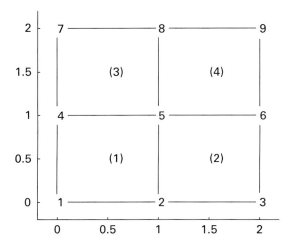

Figure 6.2 A 2-by-2 domain is discretized into 2 × 2 elements.

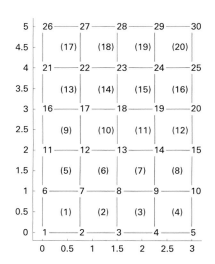

Figure 6.3 A 3-by-5 domain is discretized into 4 × 5 elements.

triangular or quadrilateral shapes. Then the predefined mesh on the regular geometries (e.g. unit square) is mapped to the triangular or quadrilateral regions by using various mapping methods. Examples of such methods include the transfinite mapping methods and isoparametric mapping methods. Figure 6.4 shows such a structured mesh (Blazek 2015).

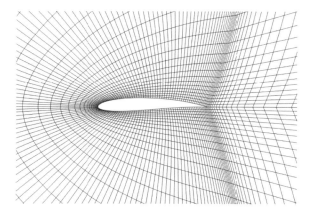

Figure 6.4 A structured mesh (obtained from Blazek 2015, p. 363).

Unstructured Meshing Methods

The two groups of well-known effective and popular meshing methods for generating unstructured meshes are (1) advancing front methods and (2) Delaunay triangulation methods. In both methods, the goal is to connect all the nodes into non-overlapping triangular elements and simultaneously satisfy the size and angle requirements for the generated elements.

Advancing Front Methods

The advancing front methods create a layer of elements at the boundary first and propagate the boundary elements into the domain. Figure 6.5 illustrates the steps of the advancing front method for meshing a 2-D domain. The individual steps of this grid generation scheme can be summarized as follows (Blazek 2015):

1. Discretize the boundaries of the physical domain and generate a list of edges which represent the front. The discretized boundary representation (B-rep) is also called a Planar Straight Line Graph (PSLG). Sort the list in the order of increasing edge size.
2. Select the first edge of the list and place a new node in the normal direction above the center of the front edge. The situation is displayed in Fig. 6.5. The distance into the domain is governed by the local mesh size and neighborhood constraints.
3. Define a circle centered at the new node. The radius of the circle depends on the local mesh size.
4. If there is no existing node or boundary vertices located inside the circle, generate a new element using the current front edge and the new node.
5. Otherwise, form elements using each of the nodes inside the circle with the current front edge, and accept the first one which does not include any other node and satisfies given quality measures.
6. Delete the current front edge and add the newly formed edges to the list. Sort the list again.
7. Continue with step 2 until the list is empty.
8. Check the quality of the mesh. Smooth the mesh and/or swap edges if needed.

Delaunay Triangulation and Refinement Methods

A Delaunay triangulation and refinement method starts with the PSLG description of the domain boundary as the input. It forms initial triangulation using boundary points. Next it splits any encroached boundary segment into two subsegments by inserting a node at its midpoint. This boundary division process continues until no boundary segment is encroached. Then it replaces any undesired element (bad or large) by inserting its circumcenter. The triangulation is then repeated with the new point. Once again, any new found undesired element is replaced by inserting its circumcenter. This process repeats until the mesh is good.

In this book, we illustrate the details of a particular Delaunay triangulation and refinement method developed and improved by Lee and Schachter (1980), Guibas and Stolfi (1985), Dwyer (1987), Shewchuk (1996), Shewchuk (2002), Chew (1993), and Ruppert (1995). We take the 2-D wrench shown in Fig. 6.6 as an example, and describe its geometrical model and meshing steps.

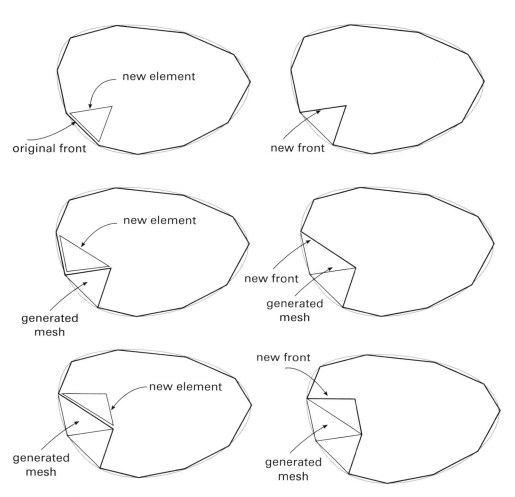

Figure 6.5 Advancing front method.

6.2 Modeling of 2-D Geometries Using Planar Straight Line Graph

As discussed in the previous section, the 2-D domain in Fig. 6.6 can be represented in multiple ways. In practice, most 2-D mesh generation algorithms operate on a boundary representation (B-rep) of 2-D shapes, which is referred to as planar straight line graph (PSLG) representation. A PSLG is a collection of boundary vertices and segments of the 2-D shape, as illustrated in Fig. 6.6. A segment is an edge of the 2-D shape's inner or outer boundary. By definition, a PSLG is required to contain both endpoints of every segment it contains, and a segment may intersect vertices and other segments only at its endpoints. Figure 6.6 shows a wrench-like domain which will be meshed later as an example. The domain has two boundaries: one outer and one inner. Each boundary is formed by a loop of sequentially connected straight segments. The

6.2 Modeling of 2-D Geometries using PSLG

segments are joined by the vertices. For the example, the vertices are numbered and their coordinates are shown in the figure.

The PSLG is the input for mesh generation algorithms. A data file is typically prepared to store the information of the PSLG. On the right side of Fig. 6.6 is such a data file representing the domain, named "shape.dat." The first number of the first row is the number of boundary loops in the domain. In this case it is two: one outer loop and one inner loop. The second number of the first row is set to be zero since MATLAB requires each row to have the same number of elements for the data file to be loaded as a matrix. This zero can also be regarded as a reserved number for future use. The first number of the following two rows stores the number of vertices of each loop. For this example, there are a total of 18 vertices in the outer loop (first number of the second row) and 8 in the inner loop (first number of the third row). After that, the rows contain the x- and y-coordinates of the vertices. For the outer loop, they are listed in the counterclockwise direction. The inner loop vertices are listed in the clockwise direction.

The domain data file "shape.dat" is the input file for mesh generation. A MATLAB code reading in the data file and plotting the domain is listed below. Note that "shape.dat" is loaded into MATLAB as a matrix. The number of loops and vertices, and the coordinates are readily accessible by other functions.

```
1  % plot the 2-D domain defined in PSLG format
2  clear all;
3  load shape.dat;
4  n_shapes=shape(1,1)              % number of shapes
5  n_outer_pts=shape(2,1)           % number of vertices in the outer loop
6  n_inner_pts (1:n_shapes-1)=shape(3:1+n_shapes,1) % inner vertices
7  start_pt=n_shapes+2;             % starting point of the outer loop
8  end_pt=n_shapes+1+n_outer_pts;   % ending point of the outer loop
9
10 plot(shape([start_pt:end_pt start_pt],1), ...   % plot the
11     shape([start_pt:end_pt start_pt],2),'-');  % outer loop
12 hold on;
13
14 % for-loop block: plot the inner loop(s)
15 for i=1:n_shapes-1               % loop over each inner loop
16     start_pt=end_pt+1;           % starting point of the current inner loop
17     end_pt= start_pt + n_inner_pts(i)-1;  % ending point of the inner loop
18     plot(shape([start_pt:end_pt start_pt],1) ,...   % plot the current
19         shape([start_pt:end_pt start_pt],2),'-');  % inner loop
20 end
21 hold off;
```

Mesh Generation

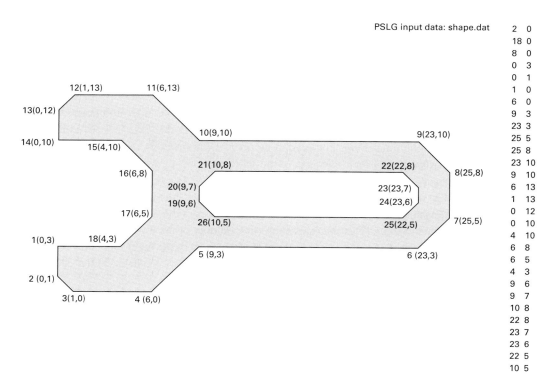

Figure 6.6 A 2-D domain defined by planar straight line graph (PSLG).

6.3 Delaunay Triangulation and Refinement Meshing

Once the PSLG is prepared for the 2-D shape, the next step is to create the mesh. In this section, we illustrate a Delaunay triangulation and refinement method following the work of Lee and Schachter (1980), Guibas and Stofi (1985), Dwyer (1987), Ruppert (1995), and Shewchuk (1996, 2002). Delaunay triangulation has a set of geometric properties that are advantageous for creating finite element meshes, namely (1) empty circumcircle property: a circle circumscribing any triangle (referred to as a circumcircle) in the Delaunay triangulation does not contain any other input node in its interior and (2) max minimum angle property: the Delaunay triangulation maximizes the minimum angle of the triangles. The main steps of the meshing method is summarized below. Note that the PSLG has already been generated in the last section. The first step is listed here for the sake of completeness. In the next sections, the rest of the steps are described in detail.

1. Discretize the boundaries of the physical domain and generate PSLG.
2. Generate an initial Delaunay triangulation which covers all boundary nodes. This is done by a triangulation of the boundary nodes themselves.
3. Delete elements outside of the domain and recover the boundaries.

4 Build a list of all bad boundary segments which are encroached by an interior node.
5 Start from the first boundary segment in the list, split the encroached boundary segment into two subsegments by inserting a node at its midpoint. The resultant subsegments may or may not be encroached themselves; splitting continues until no subsegment is encroached.
6 Build a list of all bad triangles in violation of size or quality requirement.
7 Start from the first triangle of the list, place a new point at the circumcenter and locally re-triangulate the grid. However, if the new node would encroach upon any boundary subsegment, then it is not inserted; instead, all the subsegments it would encroach upon are split.
8 Check each new element and add it to the list if not in accordance with the size/quality measure.
9 If there are still elements in the list, go to step 7.

6.3.1 Delaunay Triangulation

For a given set of discrete nodes P in a 2-D plane, a Delaunay triangulation is a triangulation $DT(P)$ such that no node in P is inside the circumcircle of any triangle in $DT(P)$. On the other hand, one can say that, given a node set P, a triangle of a triangulation $DT(P)$ is a Delaunay triangle if and only if its circumcircle does not contain any other point of P in its interior. A Voronoi diagram of the point set P, $VD(P)$, is a partition of the plane into a set of convex regions (called Voronoi cells) such that any point in each region lies nearer to the node in its interior than to any other node in the plane. For the same set of nodes, the Delaunay triangulation is the dual graph of the Voronoi diagram. Figure 6.8 shows an example of the Voronoi diagram (dashed lines) and Delaunay triangulation (solid lines) of a set of nodes. The circumcenters of Delaunay triangles are the vertices of the Voronoi diagram. Note that the circumcenter of a triangle is the center of its circumcircle. On the other hand, if two Delaunay triangles share an edge in the Delaunay triangulation, their circumcenters are to be connected in the Voronoi diagram. The Delaunay triangulation maximizes the minimum angle of the triangles.

An intuitive method to construct the Voronoi diagram is to determine the Voronoi cells individually for the nodes. For a given center node, draw the lines that perpendicularly bisect the lines between the center node and its surrounding nodes. Then the smallest convex polygon enclosing the center node is the Voronoi cell of that node. When we use this rule for every node, the 2-D plane is completely covered by Voronoi cells, constituting the Voronoi diagram. Since the circumcenters of Delaunay triangles are the vertices of the Voronoi diagram, the Voronoi vertices give the complete set of Delaunay triangles.

As shown in Fig. 6.7, the Delaunay triangulation itself looks like a good mesh of triangular elements. In fact, the good performance of Delaunay triangulation and refinement meshing method can be attributed to several favorable characteristics of Delaunay triangulation (Shewchuk 2002, Lawson 1977, Lee and Lin 1986):

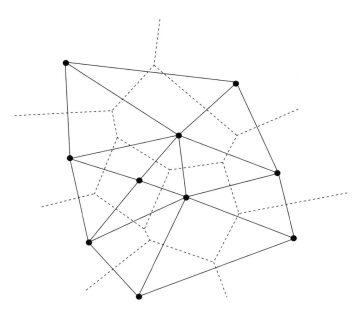

Figure 6.7 The Voronoi diagram (dashed lines) and Delaunay triangulation (solid lines) of a set of nodes.

- A Delaunay triangulation maximizes the minimum angle among all possible triangulations of a point set.
- Inserting a vertex is a local operation, resulting in efficient mesh refinement operations.
- The circumcenter of a Delaunay triangle is inherently the farthest point from its three vertices.

These advantages lead to an optimal triangulation among all the possible triangulations of a given PSLG. For this reason, Delaunay triangulation and refinement have been studied extensively and many algorithms are available in the literature. In this text, we discuss a popular algorithm for the initial Delaunay triangulation of a PSLG: divide-and-conquer algorithm (Lee & Schachter 1980, Guibas & Stolfi 1985).

In the divide-and-conquer-algorithm, we recursively partition the nodes into two halves, the left half (L) and the right half (R), compute the Delaunay triangulations for each half, and then merge the two halves into a single Delaunay triangulation. The partition of the nodes into left and right halves continues until the nodes are divided down to primitive partitions containing only two or three nodes. The two and three nodes in the primitive partitions are connected, and then the merging process begins from the primitive partitions to form larger left and right Delaunay triangulations until the Delaunay triangulation of the whole set is obtained. A pseudo code of the recursive algorithm is shown below.

6.3 Delaunay Triangulation and Refinement Meshing

```
1  function [ triangulation ]=BuildDelaunay(group of nodes)
2  if number of nodes==2
3    connect the two nodes to form an edge
4  elseif number of nodes==3
5    connect the three nodes into a triangle
6  else
7    [left half, right half]=Divide(group of nodes);
8    [left triangulation]=BuildDelaunay(left half);
9    [right triangulation]=BuildDelaunay(right half);
10   [triangulation]=Merge(left triangulatio, right triangulation);
11 end
```

As shown in Fig. 6.8, the entire set of nodes is first partitioned into the left and right halves encircled in the two dashed-line loops. The division continues in a recursive way dividing the group of nodes on the left into the smaller groups of (A,B,C) and (D, E), and the group on the right into (F,G,H) and (I,J). Then these groups are further divided into smaller groups until the number of nodes in each group is either two or three, as shown in Fig. 6.9. Once a group of two or three nodes is reached, the nodes in the group are connected to form either an edge (two nodes) or a triangle (three nodes), as shown in Fig. 6.8. Then, these primitive triangulations of the smallest left and right groups are merged with each other to form larger triangulations by tracing back along the left–right division paths from the bottom up.

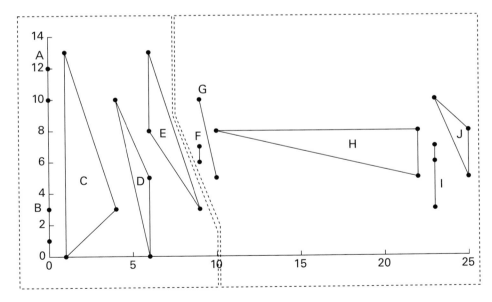

Figure 6.8 The division process on a set of nodes.

Next we illustrate the merge of two Delaunay triangulations by using an example shown in Fig. 6.10. The figure shows a certain stage in the merge process at which

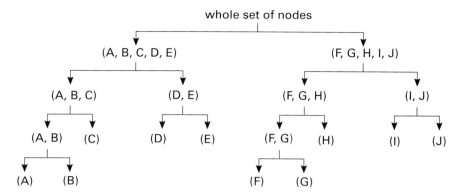

Figure 6.9 The partition of the whole set of nodes into left and right groups until the smallest groups of two or three nodes are reached.

two smaller Delaunay triangulations (encircled by the dashed line) have already been constructed. The next step is merging the two triangulations into one. In the merge step, the "lowest common tangent" of the left and right triangulations is first obtained. It can be shown that this "lowest common tangent" is a Delaunay edge in the merged triangulation. It is thus set to be the first inserted edge connecting the left and right triangulations, denoted as "(LR)" in the figure. The left and right ends of "(LR)" are denoted as "Lp" and "Rp," respectively. In the MATLAB implementation that will be shown later, the function "CompLCT" obtains this line segment.

After the lowest common tangent "(LR)" is obtained, the next step is to determine a left edge and a right edge for "(LR)." In this step, as shown in Fig. 6.11(a), at the left end of "(LR)," that is, "Lp," starting from the positive x-direction and going in the counterclockwise direction, the first two edges that have "Lp" as an end are selected. The other ends of the two edges are denoted as "lp1" and "lp2" respectively, as shown in Fig. 6.11(a). It is then tested to see if the point "Rp" falls inside the circumcircle of the triangle formed by the nodes "Lp," "lp1," and "lp2." If the answer is Yes, the edge connecting "lp1" and "Lp" is deleted and "lp1" is assigned to "lp2." Then, the next edge from "Lp" in the counterclockwise direction is selected, and "lp2" is assigned to the other end of the selected edge. The in-circle test is then carried out again. This time, "Rp" is not inside the circumcircle of "Lp," "lp1," and "lp2." Thus the edge "lp1"–"Lp" is set to be the left edge of "(LR)," as shown in Fig. 6.11(b). The same process is carried out for the right side point "Rp," as shown in Fig. 6.11(c). The only difference is that the "candidate" edges for the right edge from point "Rp" are taken in order in the clockwise direction starting from the negative x-direction. As the result of this procedure, the edge "rp1"–"Rp" is determined to be the right edge of "(LR)" as shown in Fig. 6.11(d). With both the left and right edge determined, one more in-circle test is performed as shown in Fig. 6.11(e): test whether the point "rp1" lies inside the circumcircle of "lp1," "Lp," and "Rp." If the result is No, then "lp1" and "Rp" is connected to form a new edge and the new edge is a Delaunay edge and is inserted into the triangulation. If Yes, then "lp1" must be outside of the circumcircle

6.3 Delaunay Triangulation and Refinement Meshing 273

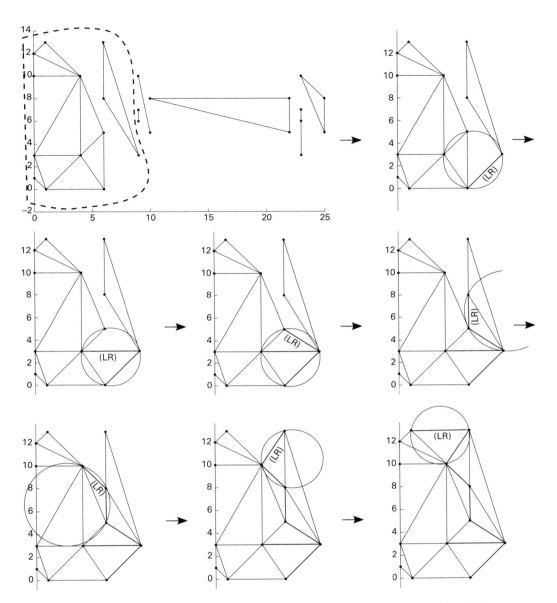

Figure 6.10 Merging two Delaunay triangulations into a single Delaunay triangulation.

of "Lp," "Rp," and "rp1." Then "rp1" and "Lp" is connected to form a new Delaunay edge and is inserted into the triangulation. As shown in Fig. 6.11(f), the newly inserted crossing edge becomes the new "(LR)" edge. The entire process of determining the next crossing edge "(LR)" repeats itself until the top most "(LR)" is obtained. The merge of the two triangulations is then accomplished, as shown in Fig. 6.10.

After all the triangulations are merged together from bottom up, the Delaunay triangulation of the entire set of nodes is shown in Fig. 6.12. It can be shown that in

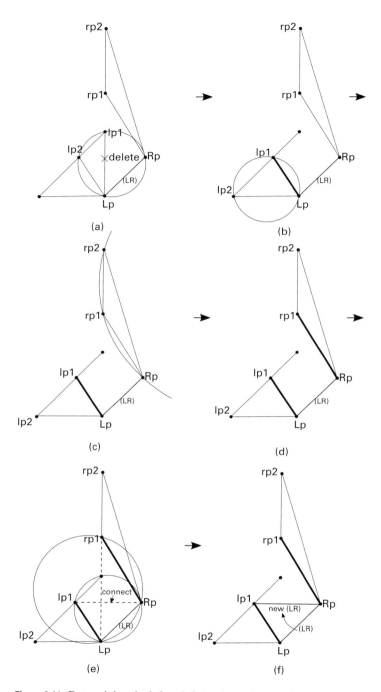

Figure 6.11 Determining the left and right edges of a crossing Delaunay edge (LR) and obtaining the next (LR).

the Delaunay triangulation no node is inside the circumcircle of any triangle (note: a fourth node on the circumcircle of a triangle is allowed here). Figure 6.12 shows that there are edges outside of the boundary of the prescribed domain. These edges are not a part of the finite element mesh we seek to create, thus are deleted from the edge list. The resultant triangulation is shown in Fig. 6.13. While in this step, the Delaunay triangulation is completed with the guaranteed geometric properties for the triangles, the triangles may not satisfy the required or desired properties of a finite element mesh, such as the element size and vertex angle requirements. In this case, the step of Delaunay refinement is necessary. For this reason, the Delaunay triangulation obtained by the divide-and-conquer algorithm is referred to as the initial triangulation.

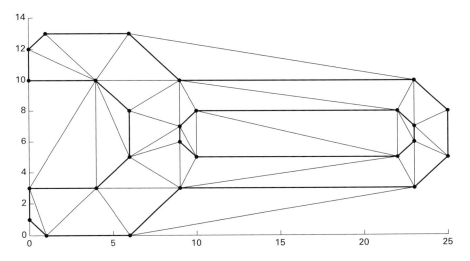

Figure 6.12 Mesh after initial Delaunay triangulation.

6.3.2 Delaunay Refinement

In the previous section, the exterior edges of the initial Delaunay triangulation are deleted. The boundary edges and nodes must be "respected" (i.e. should not be deleted or modified) since they represent the physical boundary. Such a Delaunay triangulation is called a constrained Delaunay triangulation. Delaunay refinement algorithms operate by maintaining a Delaunay or constrained Delaunay triangulation, which is refined by inserting carefully placed vertices until the mesh meets constraints on triangle quality and size (Shewchuk 1996).

Before describing the Delaunay refinement algorithm, we illustrate several geometrical properties and quantities in Fig. 6.14. As shown in Fig. 6.14(a), the diametral circle of an edge (a,b) is the smallest circle that encloses the edge, and the edge is a diameter of the circle. An edge is said to be encroached if a node (node c in the figure) other than its endpoints lies on or inside its diametral circle. A triangle is considered "skinny" when its smallest vertex angle is small. Quantitatively it can be measured by

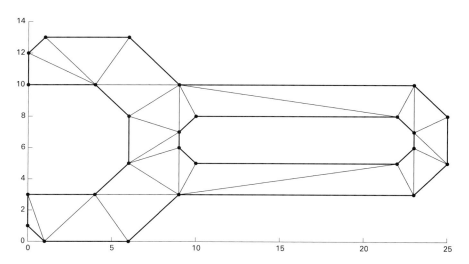

Figure 6.13 Edges outside of the prescribed domain are deleted.

using the circumradius-to-shortest edge ratio of the triangle. As shown in Fig. 6.14(b), the circumradius is the radius of the circumcircle of the triangle, and the length of the shortest edge is denoted as "l." Thus the circumradius-to-shortest edge ratio is simply r/l. It can be shown that the ratio r/l is related to the smallest angle of the triangle θ_{min} by $r/l = 1/(2 \sin \theta_{min})$. The Delaunay refinement algorithm we discuss in this section employs a bound of $r/l = \sqrt{2}$ (Ruppert 1995), which implies that $\theta_{min} = 20.7°$. This bound is guaranteed if the smallest input vertex angle is at least $60°$ (Shewchuk 1996). The main idea of the Delaunay refinement algorithm is described as follows:

- Any encroached edge is immediately split into two edges by inserting a node at its midpoint. These newly created edges have smaller diametral circles, and may or may not be encroached themselves; splitting continues until no edge is encroached.
- Each skinny triangle ($\theta_{min} > 20.7°$) is eliminated by inserting a new node at its circumcenter and the local triangulation is subsequently modified due to the insertion. However, if the new node would encroach upon any existing edge, then it is not inserted; instead, all the existing edges it would encroach upon are split.

A pseudo code of the Delaunay refinement algorithm is shown below. The pseudo code shows that the encroached edges are to be taken care of before the skinny and large triangles. This is because an encroached boundary edge may cause the situation of an associated circumcenter falling outside the domain boundary, and the edge should not be inserted into the triangulation. Once the encroached edges are eliminated by using the split operation, it can be shown that no circumcenter will lie outside the domain boundary. That is, when the triangulation has no encroached edge, all triangles and edges of the triangulation are Delaunay. Figure 6.15 shows the encroached

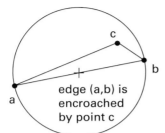
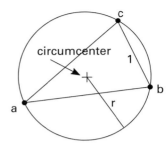

Figure 6.14 Diametral circle of an edge and circumcircle of a triangle.

edges (thick lines) in the initial Delaunay triangulation. These encroached edges are to be split into two halves. A new node is inserted at the splitting point. The local triangulation is modified in the process. The edge split process can be regarded as a special case of another triangulation operation: insertion of a new node on an edge. It is more convenient that we first describe the general node insertion operation and then illustrate the edge split operation as a special case.

```
1  function DelaunayRefinement()
2  Add the encroched edges to list A;
3  while list A is not empty
4      Get the first edge from the list;
5      Split the edge;
6  end
7
8  Add skinny and large triangles to list B;
9  while list B is not empty
10     Get the first triangle from list B;
11     Compute circumcircle of the triangle;
12     if the circumcenter is encroaching an edge
13         Split the edge;
14     else
15         Insert the circumcenter into the triangulation;
16     end
17 end
```

Figure 6.16 (a) indicates that a new node (solid circle) is to be inserted in the given triangulation at the location shown. The node is located inside a triangle. Three new edges connecting its vertices to the new node are formed. The three new edges are the "spokes" incident onto the new node and the edges of the bounding triangle are considered as the boundary edges of the new node. It can be proved that the new

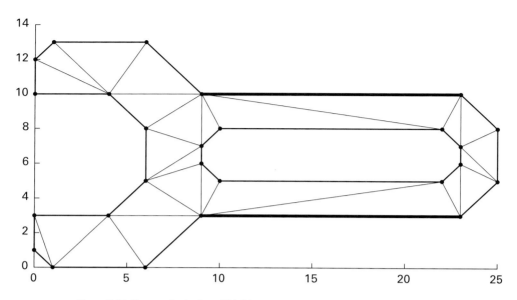

Figure 6.15 Encroached edges (thick).

"spokes" are guaranteed to be Delaunay edges. However, some of the boundary edges may now become invalid. In general, after the new "spokes" are added, the new node is surrounded by a set of triangles. The union of these triangles forms a star-shaped polygon around the new node. For the example shown in Fig. 6.16, the surrounding polygon is just the old triangle that encloses the new node. The "spokes" are Delaunay edges to be kept. The boundary edges, however, may or may not be Delaunay edges, which depends on whether they pass the empty circumcircle test or not. Figure 6.16 (b) shows the application of empty circumcircle test to the triangles formed by the three boundary edges and the new node. It is shown that one of the triangles fails the test and the associated edge is not a Delaunay edge. The other two edges are Delaunay edges and kept in the new triangulation. The failed boundary edge, as shown in Fig. 6.16 (b), is enclosed by the quadrilateral formed by the two triangles that share it. The failed boundary edge is then replaced by the other diagonal of the quadrilateral, as shown in Fig. 6.16 (c). The operation is called a "swap" operation. It can be shown that the replacing diagonal is guaranteed to be a Delaunay edge. As shown in the figure the added edge is also a "spoke" of the new node at the center. At the same time, due to the swap operation, two new boundary edges become part of the new center node's bounding polygon, as shown in Fig. 6.16 (d). Like all the other boundary edges, these two new boundary edges are to be tested for the empty circumcircle condition. Failing the test will result in more swap operations and boundary edges. The algorithm terminates when all the boundary edges pass the empty circumcircle test. Figure 6.16 (e) shows the new triangulation after the insertion of the node.

A special case arises when the new node happens to fall on an existing edge "e" in the triangulation. In that case, we delete the edge "e" and connect the new node to the

6.3 Delaunay Triangulation and Refinement Meshing

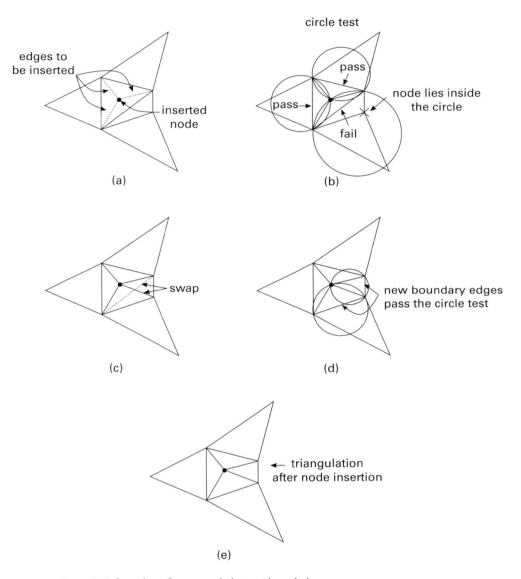

Figure 6.16 Insertion of a new node into a triangulation.

nodes surrounding the old edge "e." The split operation in the Delaunay refinement algorithm is such a special case in which the new node is to be inserted at the center of an existing edge. There are two situations to be considered. First, if the edge "e" is an edge on the domain boundary, it must belong to a single triangle. After the edge "e" is deleted, we connect the new node with the three nodes of the triangle to form two new triangles. In this case, the new node has three "spokes" and two boundary edges (two remaining edges of the bounding triangle). The second situation is that the edge "e" is an interior edge. In this case, deleting edge e creates a quadrangular hole.

Connecting the new node to the four vertices of this quadrangular hole forms four new "spokes" on the new node. The four edges of the "hole" are then the boundary edges of the new node. In either situation, the empty circumcircle test and swap operation can be performed on the boundary edges as described in the previous paragraph.

Figure 6.17 illustrates how the encroached edge shown in Fig. 6.15 is eliminated by using the split operation. The encroached edge (thick line) shown in Fig. 6.17(a) is first deleted. A new node is inserted at the mid-point of the deleted encroached edge. The node is then connected to the three vertices of the triangle which contained the deleted encroached edge, as shown in Fig. 6.17 (b). An equivalent way to describe the process is that the encroached edge is split into two halves, a node is inserted at the splitting point, and the node is then connected to the vertex of the triangle that is opposite to the encroached edge. After the node is added and new edges are formed. The added node is surrounded by two triangles which can also be viewed as three "spokes" and two boundary edges. As discussed before, the three "spokes" are Delaunay edges and the two boundary edges should be tested for empty circumcircle property. In our example, it turns out that the boundary edge on the left fails the circle test and needs to be swapped, as shown in Fig. 6.17(b). Furthermore, due to the swap, the bottom edge of the triangulation becomes encroached. This newly formed encroached edge is split in Fig. 6.17(c). The boundary edges of the new node at the bottom split point pass the circle test and no further action is needed. It should be noted that, although the top encroached edge is split into two edges, each of the edges is still encroached as shown in Fig. 6.17(c). These two edges are split again and new edges are formed and tested as shown in Fig. 6.17(d). As the result of the split operations shown in Fig. 6.17(d), the two bottom edges become encroached again and are split in Fig. 6.17(e). Finally, all the encroached edges are eliminated in the triangulation as shown in Fig. 6.17(f).

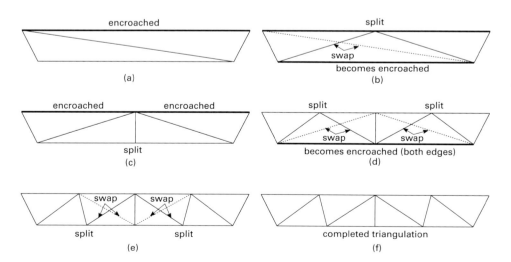

Figure 6.17 Splitting and eliminating encroached edges.

6.3 Delaunay Triangulation and Refinement Meshing

Once the encroached edges are eliminated, the skinny elements are identified in the triangulation. For a circumradius-to-shortest edge ratio $r/l = \sqrt{2}$, the allowable smallest vertex angle is $20.7°$ in any triangle. Figure 6.18 shows that four triangles (shaded) in the triangulation do not satisfy this requirement. Each skinny triangle is eliminated by inserting a new node at its circumcenter. The local triangulation is modified by following the process depicted in Fig. 6.16. Note that, in the node insertion operation, if the new node would encroach upon any existing edge, then it is not inserted; instead, all the existing edges it would encroach upon are split. Figure 6.19 shows the mesh of the domain after the skinny triangles in the initial Delaunay triangulation are eliminated.

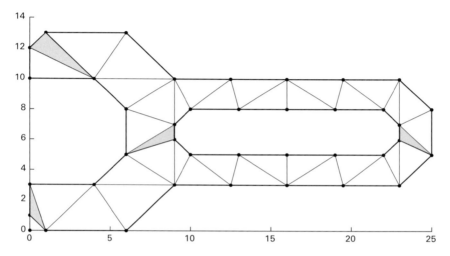

Figure 6.18 Skinny elements in the triangulation.

Other than avoiding skinny elements, another typical requirement of a finite element mesh is the element size. The mesh of Delaunay triangulation shown in Fig. 6.19 does not have any size criterion. It is simply a constrained Delaunay triangulation with "good shaped" elements. It is therefore the coarsest mesh satisfying the quality requirements (i.e., no encroached edges and skinny elements). To reduce the size of the elements and make the mesh finer, we split the "large" elements by inserting new nodes at their circumcenter, the same process depicted in Fig. 6.16. This mesh refinement process continues until no element is larger than the specified size. There are many ways to define the size of an element. Here we simply use the edge length as the measure. That is, the refinement process stops only when no edge in the mesh is larger than the given value. Figures 6.20–6.22 show the meshes obtained by setting the maximum edge length to be 2, 1, and 0.5, respectively. Note that the simple size criterion leads to a uniform refinement of the mesh as shown in the figures. Methods for local and adaptive mesh refinement are available. The reader is referred to additional references (Shewchuk 2002, Blazek 2015) for detailed description of the methods.

282   **Mesh Generation**

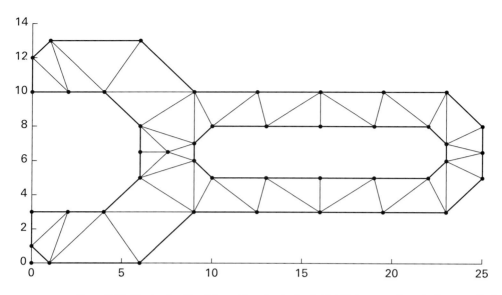

Figure 6.19 Mesh after the skinny triangles in the initial Delaunay triangulation are eliminated.

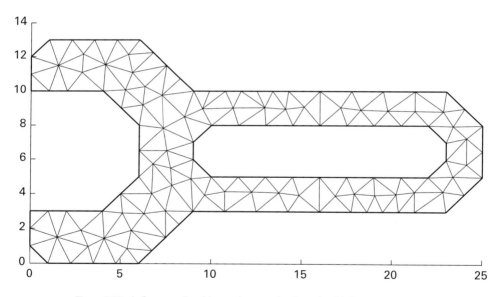

Figure 6.20 A finer mesh with maximum edge length of 2.0.

6.4   Computer Implementation

In this section we first describe the data structures that are related to the implementation of the mesh generator. We also introduce a set of auxiliary computation geometry functions that are not only used by the mesh generator but also useful for solid modeling in general.

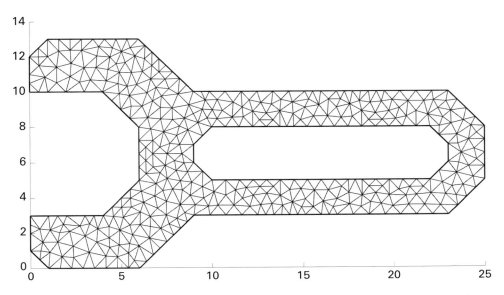

Figure 6.21 An even finer mesh of the 2-D shape with maximum edge length of 1.0.

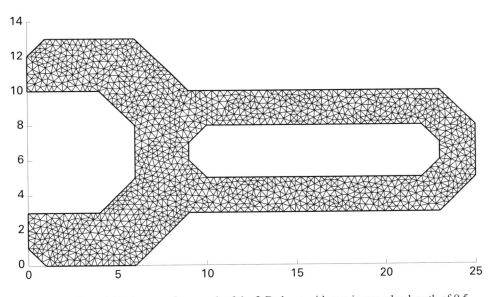

Figure 6.22 An even finer mesh of the 2-D shape with maximum edge length of 0.5.

6.4.1 Data Structures

The complexity of meshing algorithms requires utilization of data structures other than the native data structures of MATLAB: array, matrix and cell. The mesh generator implementation we present in this section utilizes two types of data structures for the calculation: (1) general container data structures including MATLAB's arrays and matrices, and an implementation of queues; and (2) a set of implicitly associated MATLAB matrices storing the information of nodes, edges, and triangles of the triangulation/mesh.

Container Data Structures

We first consider the implementation of queues. Depending on the purpose of the data processing, there are many types of queues, including simple queue, priority queue, circular queue, and other less common queues. In the 2-D meshing algorithm implementation, the simple queue is used. A simple queue is a first in/first out (FIFO) collection of elements as shown in Fig. 6.23. The elements can be any kind of object. In MATLAB, it can be a number, an array, a cell, or an instance of a class. There are two basic operations on a queue: adding an element to the rear end of the queue (also called enqueue), and removing an element from the front end of the queue (also called dequeue). While MATLAB provides tools for object-oriented (OO) implementation of queues, the array implementation is sufficient for our purpose.

In our implementation, we assumed each element in the queue is an array of length "n_col." Thus, the queue itself is a two-dimensional n×n_col matrix where n is the size of the queue (i.e. the maximum number of elements the queue can hold). Thus the queue is a container data structure that stores and moves data in a specific way. When each element array contains only one number, then the queue itself becomes an array. As shown in the code listing, the "QueueCreate" and "QueueRemove" functions are straightforward. The "QueueAdd" function, however, requires a little explanation. As shown in Fig. 6.23, the "Head" and "Tail" may move due to the "Add" (enqueue) and "Remove" (dequeue) operations. When more and more elements are added to a queue with certain size, the "Tail" may reach the end of the queue. In this case, the container queue needs to be enlarged to accommodate more data. However, due to the "Remove" operations, sometimes the actual number of elements in the queue may not be large even if the "Tail" is at the end of the queue. In that case, enlarging the queue is unnecessary and wastes memory storage. In the "QueueAdd" function, this situation is detected and the queue elements are moved to the first half of the queue.

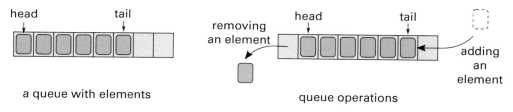

Figure 6.23 A simple queue.

```
1  function [queue, head, tail] = QueueCreate(n_col)
2      queue = zeros(10, n_col);           % initial queue size = 10
3      head = 1; tail = 0;
4  end
```

```
1  function [queue, head, tail ] = QueueAdd(queue, head, tail, element)
2  q_size=size(queue,1);
3  len= tail −head;
4  if len==−1                              % if queue is empty
5      queue (1,:) =element; head=1; tail =0;
6  end
7
8  % next 11 lines : check the size and location of the queue and adjust
9  if tail == q_size                       % if tail is close to the end
10     if tail −head < 0.5*q_size          % if not even half − filled
11        queue(1: tail −head+1,:)=queue(head: tail,:); % move queue forward
12        head=1;
13        tail =head+len;
14     else                                % if more than half  filled
15        t=zeros(2*q_size, size(queue,2));  % double the size
16        t (1: tail ,:) =queue;
17        queue=t;
18     end
19  end
20  tail = tail +1;                        % tail  location + 1
21  queue( tail ,:) =element;              % add the element
```

```
1  function [queue, head, tail , element] = QueueRemove(queue, head, tail )
2  if head> tail
3      element = nan;
4  else
5      element = queue(head ,:) ; % obtain the element
6      head = head+1;              % head location + 1
7  end
```

Data Structures for Delaunay Triangulation and Refinement

The Delaunay triangulation and refinement algorithm requires extensive calculation and manipulation of the primitive geometrical entities: nodes, edges, and triangles. Therefore, it is critical to store the information of these geometrical entities in a well-organized way in order to allow easy and efficient operations. Several well-designed data structures have been developed for Delaunay triangulation and refinement including the winged-edge representation (Baumgart 1975), quad-edge data structure (Guibas & Stolfi 1985), and triangle based data structure (Shewchuk 1996). These data structures are optimized for versatility and/or efficiency. However, such performance often comes at the cost of complexity, in both theoretical representation and implementation. In this text, we use a data structure similar to the quad-edge type but with the help of an accessory data structure for the triangles in the mesh. Figures 6.24–6.26 illustrate the data structures.

The data structure of edges is shown in Fig. 6.24. The two end nodes of the edge are called origin and destination, respectively. The edge is considered directed from the origin node to the destination node. For a given "current" edge, taking the origin node, and all the edges that are connected to it, we go around the origin node in the counterclockwise direction; the edge that is immediately following the "current" edge is called the "onext" edge, the edge that is immediately previous to the "current" edge is called "oprev" edge. Likewise, the immediate previous and next edges of the destination node are called the "dprev" and "dnext" edges, as shown in Fig. 6.24. In addition, an edge is either an edge of a triangle or shared by two triangles. The information described above is recorded in ten fields for each edge. As shown in Fig. 6.24, the first two fields are the IDs of the origin and destination nodes, respectively. The third is a flag field which is an integer of value 0, 1 or 2. The 0 value implies that the edge has been deleted. The edge is an interior edge if the flag equals to 1, or a boundary edge if it is 2. The fourth to seventh fields store the edge IDs of "oprev," "onext," "dprev," and "dnext," respectively. The eighth field is the length of the edge. The ninth and tenth fields store the IDs of the triangles associated with the edge. The field contains 0 if the triangle is "null" (i.e. it does not exist). As shown in Fig. 6.24, all the edge information is stored in the "edges" matrix. Each row of the matrix stores an edge. The matrix has 10 columns. It should be pointed out that, a linked-list data structure could be used for the storage and manipulation of edges and would be more efficient in terms of memory usages. However, linked-list implementation involves MATLAB's object-oriented programming classes and functions, which would complicate the meshing code. In addition, MATLAB's native matrix data structure is sufficiently flexible and efficient, and works well with the meshing algorithms.

Figure 6.25 shows the data structure of nodes. A node is defined by its coordinates in the xy-coordinate system, the edges connected to it, and the angles of the connected edges. We use two matrices to store the information. The "nodes" matrix stores the x- and y-coordinates of the node (first and second columns), number of connected edges (third column), and the IDs of the connected edges sorted in counterclockwise direction. Note that, the edges connected to the node are sorted in the counterclockwise direction starting from the positive side of the x-axis (angle $=0°$). The angles of the edges measured from the x-axis are stored in the "ray_angles" matrix, as shown in Fig. 6.25.

The last data structure which completes the geometric description of the Delaunay triangulation is the "tri" matrix in which each row stores the IDs of the three nodes (columns 1–3) and three edges (columns 4–6) of a triangle, as shown in Fig. 6.26.

Having defined the data structures for the meshing process, a set of functions are defined for the primitive operations that are repeatedly used in the main code. Note that, for the sake of convenience, we set a set of variables and matrices to be "global" so that they are accessible and modifiable by all functions. It is obvious that the "nodes," "edges," "ray_angles," and "tris" are all global. In the following, we list the functions that perform basic node, edge, and triangle operations.

1 **function** AddNode(x,y): add a node with x- and y-coordinates.
2 **function** AddTriangle(nids, eids): add a triangle, "nids" stores the IDs of the three nodes and "eids" stores the IDs of the three edges.

6.4 Computer Implementation

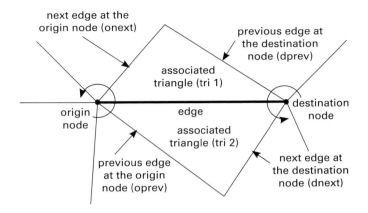

edge	1	2	3	4	5	6	7	8	9	10
	origin node	destination node	flag	oprev	onext	dprev	dnext	length	tri 1	tri 2
1										
2										
⋮										

edges matrix

Figure 6.24 Data structure of edges.

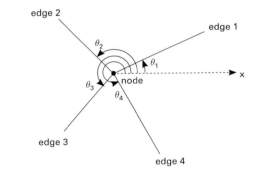

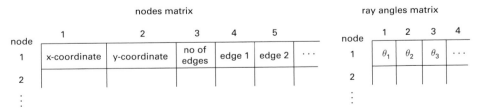

nodes matrix

node	1	2	3	4	5	
	x-coordinate	y-coordinate	no of edges	edge 1	edge 2	...
1						
2						
⋮						

ray angles matrix

node	1	2	3	4
1	θ_1	θ_2	θ_3	...
2				
⋮				

Figure 6.25 Data structure of nodes.

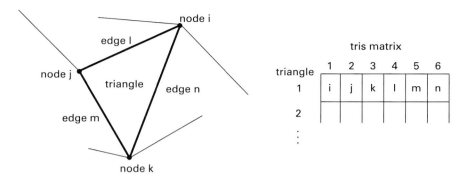

Figure 6.26 Data structure of triangles.

3 **function** re=IsExistingEdge(a,b): determine if the edge with end nodes "a" and "b" exists. If Yes, re=1, otherwise re=0. Note that "a" and "b" are node IDs.
4 **function** [oprev,onext,dprev,dnext]=NeighborEdges(a): return the four neighbor edges of edge "a." Note that "a" is an edge ID.
5 **function** [eids,nids]=NeighborEdgesNodes(eid): return the nodes ("nids") and edges ("eids") enclosing the edge "eid." It is a triangle with three nodes if edge "eid" is a boundary edge or a quadrilateral with four nodes if "eid" is an interior edge.
6 **function** re=GetRayEndNode(nd,edge_id): return the ID of the other node of an edge "edge_id" connected to node "nd."
7 **function** [eid, nid]=GetNbrRayEndNode(center_nd, ray_nd_in, flag): get the edge that is next (flag=1) or previous (flag=-1) to the input edge defined by the center node ("center_nd") and another node ("ray_nd_in"). Return the IDs of the edge and its end node that is not "center_nd."
8 **function** InsertRay(nd,type,edge_row,edge_angle): insert an edge with its edge ID "edge_row" with its angle "edge_angle" at a node "nd." "type" is either 1 (in-coming) or 2 (out-going).
9 **function** DeleteEdge(eid): delete edge "eid."
10 **function** DeleteTriByEdge(eid): delete the triangle(s) associated with edge "eid."
11 **function** Connect(a,b): connect two nodes "a" and "b" to form a new edge and then insert the edge into the triangulation.

```
1  function AddNode(x,y)
2  global nodes ray_angles last_node_row;
3  row=last_node_row+1;   % current row pointer = the last non-empty
4                         % row of the "nodes" matrix + 1
5  % if block : if the ''nodes'' matrix is full, double the size
6  if row>size(nodes,1)
7    tmp=nodes;
8    nodes=zeros(last_node_row*2,23);
9    nodes(1:last_node_row ,:) =tmp;
```

6.4 Computer Implementation

```
10    tmp=ray_angles;
11    ray_angles=zeros(last_node_row*2,20);
12    ray_angles (1:last_node_row ,:) =tmp;
13  end;
14  nodes(row,1:2)=[x y];     % add the node at the row pointed by "row"
15  last_node_row=last_node_row+1;    % the last non−empty row + 1
```

```
1   function AddTriangle(nids, eids)
2   global tris  last_tri_row  edges;
3   % next 6 lines : test if "tris" matrix is full, if Yes, double its size
4   row=last_tri_row+1;
5   if row>size(tris,1)
6     tmp=tris;
7     tris =zeros(last_tri_row *2,6);
8     tris (1: last_tri_row ,:) =tmp;
9   end;
10
11  % add the triangle information to the last non−empty row of "tris"
12  tris (row,1:3)=nids';    tris (row,4:6)=eids';
13
14  % next 9 lines : update the "edges" matrix to record the added triangle
15  for i=4:6
16    eid= tris (row,i);
17    if edges(eid,9)==0; edges(eid,9)=row;
18    elseif edges(eid,10)==0; edges(eid,10)=row;
19    else fprintf ('Add tris error\n');
20    end
21  end
22
23  last_tri_row = last_tri_row +1; % the last non−empty row + 1
```

```
1   function re=IsExistingEdge(a,b)
2   global edges nodes;
3   % for−block: test the edges connected to node "a"
4   for i=1:nodes(a,3)
5     if (edges(nodes(a,3+i),1)==b) ||( edges(nodes(a,3+i),2)==b)
6       re=nodes(a,3+i); return;    % returning the edge number
7     end
8   end
9   re=0;   % return 0 ( false )
```

```
1   function [oprev,onext,dprev,dnext]=NeighborEdges(a)
2   global edges
3   oprev=edges(a,4);  onext=edges(a,5);
4   dprev=edges(a,6);  dnext=edges(a,7);
```

```
1  function [eids,nids]=NeighborEdgesNodes(eid)
2  global edges;
3  [oprev,onext,dprev,dnext]=NeighborEdges(eid);  % get neighbor edges
4  o=edges(eid,1); d=edges(eid,2);   % o=origin node, d=destination node
5  nop=GetRayEndNode(o,oprev);       % get the "other" end node of the
6  non=GetRayEndNode(o,onext);       % connected edges
7  ndp=GetRayEndNode(d,dprev);
8  ndn=GetRayEndNode(d,dnext);
9  % next 11 lines : get neighbor edge
10 if edges(eid,3)==2                 % if it is a boundary edge
11    nids=zeros(3,1);
12    if nop==ndn
13       eids=[oprev dnext]';          % two neighbor edges
14    else
15       eids=[dprev onext]';          % two neighbor edges
16    end
17 elseif edges(eid,3)==1              % if it is an interior edge
18    nids=zeros(4,1);
19    eids=[oprev dnext dprev onext]'; % four neighbor edges
20 end
21 % next 11 lines : get the associated nodes
22 nids(1,1)=o;
23 k=2;
24 if nop==ndn
25    nids(k,1)=nop;
26    k=k+1;
27 end
28 nids(k,1)=d;
29 k=k+1;
30 if non==ndp
31    nids(k,1)=non;
32 end
```

```
1  function re=GetRayEndNode(nd,edge_id)
2  global edges;
3  % if-block: get the end node of the edge that is away from node "nd"
4  if edges(edge_id,1)==nd; re=edges(edge_id,2);
5  else re=edges(edge_id,1);
6  end
```

```
1  function [eid, nid]=GetNbrRayEndNode(center_nd, ray_nd_in, flag)
2  global edges nodes ray_angles;
3  n_rays=nodes(center_nd,3);
4  for i=1:n_rays
5     edge_id=nodes(center_nd,i+3);
6     if (edges(edge_id,1)==ray_nd_in) || (edges(edge_id,2)==ray_nd_in)
7        if flag==1                          % next ray
```

```
8        nbr_edge_id=nodes(center_nd, 4+mod(i,n_rays));
9      else                              % prev ray
10       nbr_edge_id=nodes(center_nd, 4+mod(i-2+n_rays,n_rays));
11     end
12     nid=GetRayEndNode(center_nd, nbr_edge_id);
13     eid=nbr_edge_id;
14     return;
15   end
16 end
17 nid=center_nd;
18 fprintf(' failed ray operation \n');
```

```
1  function InsertRay(nd,type,edge_row,edge_angle)
2  global edges nodes ray_angles;
3  global n_nodes n_edges n_max_edges last_edge_row;
4  % if-block: insert the ray in the node record
5  n_rays=nodes(nd,3);
6  if n_rays==0
7    nodes(nd,3)=1;
8    nodes(nd,4)=edge_row;
9    ray_angles(nd,1)=edge_angle;
10 else
11   tmp(:,1)=[nodes(nd,4:3+n_rays) edge_row];
12   tmp(:,2)=[ray_angles(nd,1:n_rays) edge_angle];
13   tmp=sortrows(tmp,2);
14   nodes(nd,4:3+n_rays+1)=tmp(:,1);
15   ray_angles(nd,1:n_rays+1)=tmp(:,2);
16   nodes(nd,3)=nodes(nd,3)+1;
17 end
18
19 % for-loop: set up the opre, onext, dpre, dnext entries
20 n_rays=nodes(nd,3);
21 for i=1:n_rays
22   if edge_row==nodes(nd,3+i)
23     if type==1
24       edges(edge_row,4)=nodes(nd, 4+mod(i-2+n_rays,n_rays));
25       edges(edge_row,5)=nodes(nd, 4+mod(i,n_rays));
26       orig=edges(edge_row,1);
27       opre_edge=edges(edge_row,4);
28       if edges(opre_edge,1)==orig
29         edges(opre_edge,5)=edge_row;
30       else
31         edges(opre_edge,7)=edge_row;
32       end
33       onext_edge=edges(edge_row,5);
34       if edges(onext_edge,1)==orig
35         edges(onext_edge,4)=edge_row;
36       else
```

```
37          edges(onext_edge,6)=edge_row;
38       end
39     else
40       edges(edge_row,6)=nodes(nd, 4+mod(i-2+n_rays,n_rays));
41       edges(edge_row,7)=nodes(nd, 4+mod(i,n_rays));
42       dest=edges(edge_row,2);
43       dprev_edge=edges(edge_row,6);
44       if edges(dprev_edge,1)==dest
45          edges(dprev_edge,5)=edge_row;
46       else
47          edges(dprev_edge,7)=edge_row;
48       end
49       dnext_edge=edges(edge_row,7);
50       if edges(dnext_edge,1)==dest
51         edges(dnext_edge,4)=edge_row;
52       else
53          edges(dnext_edge,6)=edge_row;
54       end
55     end
56   end
57 end
```

```
1  function DeleteEdge(eid)
2  global edges nodes ray_angles n_edges;
3  if edges(eid,3)==0, return, end;   % return if it is already deleted
4  edges(eid,3)=0;                    % flag to show the edge is deleted
5  orig=edges(eid,1) ; dest=edges(eid,2);
6
7  % next 17 lines : update the records of neighbor edges
8  [opre_edge, onext_edge, dprev_edge, dnext_edge]=NeighborEdges(eid);
9  if edges(opre_edge,1)==orig
10    edges(opre_edge,5)=edges(eid,5);
11 else edges(opre_edge,7)=edges(eid,5);
12 end
13 if edges(onext_edge,1)==orig
14    edges(onext_edge,4)=edges(eid,4) ;
15 else edges(onext_edge,6)=edges(eid,4) ;
16 end
17 if edges(dprev_edge,1)==dest
18    edges(dprev_edge,5)=edges(eid,7);
19 else edges(dprev_edge,7)=edges(eid,7);
20 end
21 if edges(dnext_edge,1)==dest
22    edges(dnext_edge,4)=edges(eid,6);
23 else edges(dnext_edge,6)=edges(eid,6);
24 end
```

```
25
26  % for-loop: delete the edge ray from node records
27  for i=1:2
28    node=edges(eid,i);
29    nodes(node,3)=nodes(node,3)-1;
30    for j=4:size(nodes,2)
31      if nodes(node,j)==eid
32        nodes(node,j:end)=[nodes(node,j+1:end) 0];
33        ray_angles(node,j-3:end)=[ray_angles(node,j-2:end) 0];
34        break;
35      end
36    end
37  end
38  n_edges=n_edges-1;
```

```
1   function DeleteTriByEdge(eid)
2   global edges tris ;
3   [egs,nds]=NeighborEdgesNodes(eid);  % get neighbor edges of eid
4   % for-loop: clean up neighbor edges' record of the triangles
5   for i=1:2
6     tri =edges(eid,8+i);
7     if tri ~=0
8       for j=1:size(egs,1)
9         for k=9:10
10          if edges(egs(j),k)==tri; edges(egs(j),k)=0; end;
11        end
12      end
13    end
14  end
15  % for-loop: clean up edge "eid" and delete the triangle
16  for i=1:2
17    tri =edges(eid,8+i);
18    edges(eid,8+i)=0;
19    if tri ~=0; tris(tri,6)=0; end;
20  end
```

```
1   function Connect(a,b)
2   global edges nodes ray_angles;
3   global n_nodes n_edges n_max_edges last_edge_row;
4   % next 7 lines : check if "edge" matrix is full, double the size if Yes
5   row=last_edge_row+1;
6   if row>n_max_edges
7     n_max_edges=n_max_edges*2;
8     tmp=edges;
9     edges=zeros(n_max_edges,10);
```

```
10    edges(1:n_max_edges/2,:)=tmp;
11  end;
12
13  % next 6 lines : add a new edge row
14  edges(row,1)=a;
15  edges(row,2)=b;
16  edges(row,3)=1;    % flag : 0: deleted edge, 1: interior edge
17  edges(row,8)=sqrt((nodes(b,1)-nodes(a,1))^2+(nodes(b,2)-nodes(a,2))^2);
18  n_edges=n_edges+1;
19  last_edge_row=last_edge_row+1;
20
21  % next 3 lines : insert a ray to each of a, b
22  [edge_angle_a, edge_angle_b]=TwoNodeRayAngles(a, b);
23  InsertRay(a,1,row,edge_angle_a);
24  InsertRay(b,2,row,edge_angle_b);
```

6.4.2 Auxiliary Geometry Operations

In addition to the data structures of the nodes, edges, and triangles, and the related functions, a set of auxiliary geometrical operation functions are needed in the triangulation. For example, the following function "CCW" determines whether or not three points form a counterclockwise oriented triangle. The mathematical method of the function is simply the cross product of the two vectors directed from the first to the second points, and from the second to the third, respectively. It should be noted that such functions are not only for the meshing code. They are general purpose functions reusable for other codes.

```
1   % Return True iff p1, p2, p3 form a counterclockwise oriented triangle
2   function re=CCW(p1,p2,p3)
3   global nodes;
4   x1=nodes(p1,1); y1=nodes(p1,2);
5   x2=nodes(p2,1); y2=nodes(p2,2);
6   x3=nodes(p3,1); y3=nodes(p3,2);
7   a=(x1-x3) * (y2-y3) - (x2-x3) * (y1-y3);
8   if   a > 1e-5
9       re=1;          % true
10  elseif  a<-1e-5
11      re=-1;         % false
12  else
13      re=0;          % collinear case
14  end
```

Another important geometrical calculation is to determine if a point is interior to the region of the plane that is bounded by a circle defined by three counterclockwise directed points. The function "InCircle(a, b, c, d)" is defined to be true if and only if d is inside the circle abc if the points a, b, and c define a circle and they are in counterclockwise order on the circle. If a, b, c, d are co-circular then the result is false. Mathematically the in-circle test is equivalent to Pedoe (1995):

$$\begin{vmatrix} x_a & y_a & x_a^2 + y_a^2 & 1 \\ x_b & y_b & x_b^2 + y_b^2 & 1 \\ x_c & y_c & x_c^2 + y_c^2 & 1 \\ x_d & y_d & x_d^2 + y_d^2 & 1 \end{vmatrix} > 0. \tag{6.1}$$

```
% Return True if d lies inside the circle defined by a, b, and c
function re=InCircle(a,b,c,d)
global nodes;
xa=nodes(a,1); ya=nodes(a,2);
xb=nodes(b,1); yb=nodes(b,2);
xc=nodes(c,1); yc=nodes(c,2);
xd=nodes(d,1); yd=nodes(d,2);
A=[xa ya xa^2+ya^2 1
   xb yb xb^2+yb^2 1
   xc yc xc^2+yc^2 1
   xd yd xd^2+yd^2 1];
if det(A)>0; re=1;
else ; re=0;
end;
```

In equation Eq. (6.1), if we assume all the four points are on the same circle, and take the x- and y-coordinates of point d as variables, the determinant =0 can be written as a function of x and y in the form of a circle:

$$(x - x_0)^2 + (y - y_0)^2 = r^2 \tag{6.2}$$

where

$$x_0 = -\frac{b_x}{2a} \qquad y_0 = -\frac{b_y}{2a}, \tag{6.3}$$

$$r = \frac{\sqrt{b_x^2 + b_y^2 - 4ac}}{2|a|} \tag{6.4}$$

$$a = \begin{vmatrix} x_a & y_a & 1 \\ x_b & y_b & 1 \\ x_c & y_c & 1 \end{vmatrix} \qquad c = \begin{vmatrix} x_a^2 + y_a^2 & x_a & y_a \\ x_b^2 + y_b^2 & x_b & y_b \\ x_c^2 + y_c^2 & x_c & y_c \end{vmatrix} \tag{6.5}$$

$$b_x = -\begin{vmatrix} x_a^2 + y_a^2 & y_a & 1 \\ x_b^2 + y_b^2 & y_b & 1 \\ x_c^2 + y_c^2 & y_c & 1 \end{vmatrix} \qquad b_y = \begin{vmatrix} x_a^2 + y_a^2 & x_a & 1 \\ x_b^2 + y_b^2 & x_b & 1 \\ x_c^2 + y_c^2 & x_c & 1 \end{vmatrix}. \tag{6.6}$$

Therefore, the center point (x_0, y_0) and the radius r of the circumcircle of points a, b, and c are obtained, as shown in function "CompCircumcircle."

```
1  function [x0,y0,r]=CompCircumcircle(a,b,c)
2  global nodes;
3  xa=nodes(a,1); ya=nodes(a,2);
4  xb=nodes(b,1); yb=nodes(b,2);
5  xc=nodes(c,1); yc=nodes(c,2);
6  a=det([xa ya 1
7         xb yb 1
8         xc yc 1]);
9  bx=-det([xa^2 + ya^2 ya 1
10          xb^2 + yb^2 yb 1
11          xc^2 + yc^2 yc 1]);
12 by=det([xa^2 + ya^2 xa 1
13         xb^2 + yb^2 xb 1
14         xc^2 + yc^2 xc 1]);
15 c= -det([xa^2 + ya^2 xa ya
16          xb^2 + yb^2 xb yb
17          xc^2 + yc^2 xc yc]);
18 x0=-bx/2/a;
19 y0=-by/2/a;
20 r=sqrt(bx^2 + by^2 - 4*a*c)/2/abs(a);
```

Next, function "ptInTriangle" determines if a given point lies in the interior of a given triangle. As shown in Fig. 6.27, the positions (x- and y-coordinates) of the three vertices of the triangle and the given point are denoted as $\mathbf{p}_1$, $\mathbf{p}_2$, $\mathbf{p}_3$, and $\mathbf{p}$ respectively. For notation convenience, we denote the position vector of $\mathbf{p}_1$ as $\mathbf{v}_0$ and let $\mathbf{v}_1$ and $\mathbf{v}_2$ be the vectors from $\mathbf{v}_0$ to the other two vertices. Then, the position vector of point $\mathbf{p}$, denoted as $\mathbf{v}$ can be written as

$$\mathbf{v} = \mathbf{v}_0 + a\mathbf{v}_1 + b\mathbf{v}_2 \tag{6.7}$$

where

$$a = \frac{det(\mathbf{v}\mathbf{v}_2) - det(\mathbf{v}_0\mathbf{v}_2)}{det(\mathbf{v}_1\mathbf{v}_2)} \tag{6.8}$$

$$b = -\frac{det(\mathbf{v}\mathbf{v}_1) - det(\mathbf{v}_0\mathbf{v}_1)}{det(\mathbf{v}_1\mathbf{v}_2)} \tag{6.9}$$

and

$$det(\mathbf{uv}) = \begin{vmatrix} u_x & u_y \\ v_x & v_y \end{vmatrix}. \tag{6.10}$$

Then it can be shown that the point $\mathbf{v}$ lies in the interior of the triangle if $a, b > 0$ and $a + b < 1$.

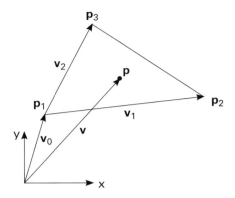

Figure 6.27 Determine if a point is in a triangle.

```
1  function [re,a,b]=PtInTriangle (p1, p2, p3, p)
2  global TOL;
3  v0=p1;
4  v1=p2−p1;
5  v2=p3−p1;
6  detpv1=p(1)*v1(2) − p(2)*v1(1);
7  detpv2=p(1)*v2(2) − p(2)*v2(1);
8  detv0v1=v0(1)*v1(2) − v0(2)*v1(1);
9  detv0v2=v0(1)*v2(2) − v0(2)*v2(1);
10 detv1v2=v1(1)*v2(2) − v1(2)*v2(1);
11 a=(detpv2 − detv0v2)/detv1v2;     % v=v_0 + a*v_1 + b*v_2
12 b=−(detpv1 − detv0v1)/detv1v2;    % cofficients  a and b
13
14 % if−block: determine return value based on different situations
15 if abs(a)<TOL && b<=1 && b>=0
16    re=0;                          % on triangle
17 elseif abs(b)<TOL && a<=1 && a>=0
18    re=0;                          % on triangle
19 elseif abs(a+b−1)<TOL && a<=1 && a>=0 && b<=1 && b>=0
20    re=0;                          % on triangle
21 elseif a>0 && b>0 && a+b<1
22    re=1;                          % in triangle
23 else
24    re=−1;                         % outside of triangle
25 end
```

In the Delaunay triangulation, when we delete all the edges that lie outside of the domain to be meshed, we need to determine whether or not a point is in the domain. A simple algorithm for doing this is to cast a downward vertical ray from that point and count how many times it crosses the boundary edges. If the count is odd, the point is inside, otherwise the point is outside. The codes are listed as the "IsInShape" and "IntersectVertSW" functions. It should be noted that one must pay attention to the special cases such as when the vertical ray is collinear with a boundary edge or is passing through an end node shared by two boundary edges, as shown in the functions.

```
1   function re=IsInShape(edge_in)
2   global edges nodes last_edge_row;
3
4   x1=nodes(edges(edge_in,1),1);
5   y1=nodes(edges(edge_in,1),2);
6   x2=nodes(edges(edge_in,2),1);
7   y2=nodes(edges(edge_in,2),2);
8   x=(x1+x2)/2; y=(y1+y2)/2;
9   count=0;
10  for i=1:last_edge_row
11      if edges(i,3)~=2, continue, end;
12      count=count + IntersectVertSW(i,x,y);  % intersection count
13      if mod(count,2)==0
14          re=0;                               % if count is even: outside
15      else
16          re=1;                               % if odd: inside
17      end
18  end
```

```
1   function re=IntersectVertSW(edge_in,ix,iy)
2   global edges nodes TOL;
3   % next 4 lines: get x- and y- coordinates of the input edge's endpoints
4   x1=nodes(edges(edge_in,1),1);
5   y1=nodes(edges(edge_in,1),2);
6   x2=nodes(edges(edge_in,2),1);
7   y2=nodes(edges(edge_in,2),2);
8
9   if (x1<=ix-TOL && x2<=ix-TOL) || (x1>=ix+TOL && x2>=ix+TOL)
10      re=0; % input point is outside of the x-range of the edge
11  elseif abs(x1-ix)<TOL && abs(x2-ix)<TOL % if edge is vertical
12      if (iy-y1)*(iy-y2)<TOL, re=-1;      % on the edge: re=-1
13      else re=2; % out of the edge = intersect twice
14      end
15  elseif (abs(x1-ix)<TOL && x2-ix>TOL) ||...
16         (abs(x2-ix)<TOL && x1-ix>TOL)
17      re=0; % vertical ray intersects left end of the edge
18  else
19      oy=(ix-x1)/(x2-x1)*(y2-y1)+y1;
20      if oy>iy+TOL, re=0; % intersection point is above the input point
21      elseif oy>iy-TOL, re=-1;            % on the edge: re=-1
22      else re=1; % intersection point is below the input point
23      end
24  end
```

In addition to the auxiliary computational geometry functions, a few utility functions are also created to simplify common tasks and avoid duplication. For a given interval (from "s" to "e"), the function "GetNext" shown below returns the element next to the current element in a circular fashion. Similarly, function "GetPrev" returns the previous element.

```
1 function current=GetNext(s, e, current)
2 current=current+1;
3 if current>e; current=s; end;
```

```
1 function current=GetPrev(s, e, current)
2 current=current−1;
3 if current<s; current=e; end;
```

6.4.3 Meshing Functions

In the following, we show the code of the meshing generator. In the first box is the main program that calls the rest of the functions for producing a mesh for the shape shown in Fig. 6.6. The two main meshing steps are the initial Delaunay triangulation (function "BuildDelaunay") and the Delaunay refinement (function "DelaunayRefinement"). In the following, we first summarize the functions with brief description of their tasks. Then the detailed codes are listed.

1. main code: read in the PSLG file of the shape to be meshed, create and display the mesh.
2. **function** [le,re]=BuildDelaunay(sid,eid): a recursive function for building the initial Delaunay triangulation. "sid" and "eid" are the left-most and right-most nodes of the node set to be triangulated.
3. **function** bpoints=SetBoundaryEdges(shape, bpos): with the input shape and boundary nodes, the function sets the flags for the boundary edges.
4. **function** DeleteOutsideEdges(): delete all the edges outside of the shape to be meshed.
5. **function** SetupTriangles(): set up the data structures for the triangles (matrix "tris") and update entries of matrix "edges."
6. **function** DelaunayRefinement(): perform Delaunay refinement to satisfy the requirements of mesh quality and element size.
7. **function** [Lp,Rp]=CompLCT(left_s,left_e, right_s,right_e): compute and returns two end nodes "Lp" and "Rp" of the lower common tangent of two sets of nodes. "left_s" and "left_e" define the left-most and right-most nodes of the left node set, and "right_s" and "right_e" are the left-most and right-most nodes of the right node set.
8. **function** Merge(Lp, Rp, left_s,left_e, right_s,right_e): merge two sets of nodes. The input variables have the same meaning as described in function "CompLCT."

9 **function** [left_s,left_e, right_s,right_e]=DivideInX(sid,eid): divide the node set defined by the left-most ("sid") and right-most ("eid") nodes into two halves. The left-most and right-most nodes of each half are returned by the function.
10 **function** nid=EncroachNode(eid): find the node "nid" that encroaches edge "eid."
11 **function** [tid, a, b]=FindBdngTriByPt(pt): find the triangle "tid" that encloses the point "pt." "a" and "b" are defined in Eqs. (6.7–6.9)
12 **function** FindNextLargeEdge(): find the next edge that is longer than the maximum length allowed in the mesh and enqueue the triangles associated with the edge for refinement.
13 **function** FindNextSmallAngle(): find the next triangle that has a vertex angle smaller than the minimum angle allowed and enqueue the triangle for refinement.
14 **function** Insert(pt): insert a point ("pt") into the triangulation.
15 **function** re=IsEdgeEncrByPt(eid,pt): determines whether or not the edge "eid" is encroached by a point "pt."
16 **function** [edge_out, node_out]=LeftMostCCWNext(node_in): for a given node "node_in," in the counterclockwise direction, find the first edge (and its outer end node) whose angle is larger than 180 degrees.
17 **function** [edge_out, node_out]=RightMostCWNext(node_in): similar to the function above, but from the 0 degree position and in the clockwise direction.
18 **function** LocalSwap(): test all the edges in the swap queue ("sw_queue"), swap if an edge fails the empty circumcircle test.
19 **function** Swap(a): perform swap operation on an edge "a."
20 **function** SetNodeCorner(nid, eid): bookkeeping function for setting up triangles.
21 **function** Split(eid, pt): split an edge "eid" with a point "pt."
22 **function** re=BdryEdgeIntNd(eid): get the interior node opposite the boundary edge "eid."
23 **function** [angle_a, angle_b]=TwoNodeRayAngles(node_a, node_b): obtain the ray angles of the edge connecting "node_a" and "node_b" with respect to the two nodes.
24 **function** PlotEdges(): display the mesh.

```
1  clear all ;
2  % next 10 lines : global   variables
3  global  edges nodes ray_angles ;
4  global  n_nodes last_node_row n_edges n_max_edges last_edge_row dflag;
5  global  tris  tris_data  n_tris  n_max_tris  last_tri_row  nd_corners;
6  global  TOL MAX_LENGTH;
7  global  ec_queue ec_head  ec_tail ;
8  global  sm_queue sm_head sm_tail sm_nodes;
9  global  sw_queue sw_head sw_tail;
10 global  test_node   test_edge ;
11 global  no_bad_node no_bad_edge;
12 global  angle_test_done  edge_test_done;
```

```
13  % next 5 lines : initilization of some of the global variables
14  dflag=0;                        % a test flag
15  TOL=1e-5;                       % tolerance
16  MAX_LENGTH=1.5;                 % maxium edge length allowed
17  angle_test_done=0;              % an angle test flag
18  edge_test_done=0;               % an edge test flag
19  load shape.dat;                 % load the PSLG file
20  vertices=shape(shape(1,1)+2:end,:); % get the vertices of the shape
21  % next 13 lines : initialization of the data structures
22  n_nodes=size( vertices ,1);
23  n_edges=0;
24  last_edge_row=0;
25  n_max_edges=10;
26  edges=zeros(n_max_edges,10);
27  nodes=zeros(n_nodes,23);
28  nodes (:,1:2) = vertices ;
29  nodes (:,3) =(linspace (1,n_nodes,n_nodes))';
30  ray_angles=zeros(n_nodes,20);
31  nodes=sortrows(nodes,[1 2]);
32  bnode_pos=nodes(:,3);
33  nodes (:,3) =0;
34  last_node_row=size(nodes,1);
35  % next 5 lines : main triangulation and refinement functions
36  [le,re]=BuildDelaunay(1,n_nodes);
37  bpoints=SetBoundaryEdges(shape, bnode_pos);
38  DeleteOutsideEdges();
39  SetupTriangles ();
40  DelaunayRefinement();
41  % next 2 lines : plot the mesh
42  figure (1);
43  clf ; PlotEdges();
```

```
1   function [le,re]=BuildDelaunay(sid,eid)
2   global nodes last_edge_row dflag ;
3   n_node_ids=eid-sid+1;
4   p1=sid; p2=sid+1;
5   if n_node_ids==2               % if number of nodes==2
6     Connect(p1,p2);              % connect the two nodes to form an edge
7     le=last_edge_row; re=le;
8   elseif n_node_ids==3           % if number of nodes==3
9     p3=eid;
10    Connect(p1,p2); a=last_edge_row;  % connect three nodes to form
11    Connect(p2,p3); b=last_edge_row;  % a triangle
12    Connect(p3,p1); c=last_edge_row;
13    if CCW(p1,p2,p3)>0
14      le=a; re=b;
15    else
16      le=c; re=c;
```

```
17    end
18  else
19    [ left_s , left_e , right_s , right_e ]=DivideInX(sid,eid);  % divide  points
20    [ ldo , ldi ]=BuildDelaunay( left_s ,  left_e );     % recursive   call  ( left )
21    [ rdi ,rdo]=BuildDelaunay(right_s ,  right_e );    % recursive   call  ( right )
22    le=ldo;  re=rdo;
23    [Lp,Rp]=CompLCT(left_s,left_e,  right_s , right_e );  % get  lowest
24                                                         % common tangent
25    Merge(Lp,Rp, left_s , left_e ,  right_s , right_e );    % merge  two  halves
26    if  dflag==1, return, end;
27  end
```

```
1   function  bpoints=SetBoundaryEdges(shape, bpos)
2   global  nodes n_nodes edges;
3
4   n_shapes=shape(1,1);
5    n_interior_shapes =n_shapes−1;
6   bpoints=zeros(n_nodes,1);
7
8   n_exterior_boundary_nodes=shape(2,1);
9   for  i=1:n_nodes % sequential  boundary node  list
10     bpoints(bpos(i,1),1)=i; % in  the  new  sorted  nodes  matrix
11  end
12  % next 5  lines : test  the   exterior   boundary  edges
13  for  i=1:n_exterior_boundary_nodes
14     j=GetNext(1,n_exterior_boundary_nodes,i);
15     edge=IsExistingEdge( bpoints(i), bpoints(j));
16     edges(edge,3)=2;
17  end
18  % next 11  lines : test  the   interior   boundary  edges
19  s=i+1;  e=s−1;
20  for  k=1: n_interior_shapes
21     n_interior_shape_nodes =shape(k+2,1);
22     e=e+ n_interior_shape_nodes;
23     for  i=s:e
24        j=GetNext(s,e,i);
25        edge=IsExistingEdge( bpoints(i), bpoints(j));
26        edges(edge,3)=2;
27     end
28     s=e+1;
29  end
```

```
1   function  DeleteOutsideEdges()
2   global  edges nodes last_edge_row;
3   for  i=1:last_edge_row                  % loop  over  the  edges
4     if  edges(i,3)~=1; continue; end;    % if  it's boundary edge, continue
5     if  ~IsInShape(i)                     % if  the  edge  is  outside  the  shape
```

```
6       DeleteEdge(i);              % delete it
7     end
8   end
```

```
1   function SetupTriangles ()
2   global edges nodes ray_angles;
3   global n_nodes n_edges n_max_edges last_edge_row;
4   global tris  n_tris n_max_tris  last_tri_row  nd_corners;
5   % next 5 lines : set up empty triangular matrix and  initialization
6   nd_corners=nodes;
7   n_tris =0;
8   n_max_tris=10;
9   tris =zeros(n_max_tris,6);
10  last_tri_row =0;
11
12  % set up the  triangles  based on  the  results  of  initial   triangulation
13  for i=1:n_nodes                 % loop over the nodes
14    n_corners=nd_corners(i,3);
15    p1=i;                         % first  point
16    for j=1:n_corners             % loop over the edges connected
17      if nd_corners(i,j+3) ~=−1   % to the  center node
18        p2=GetRayEndNode(i,nodes(i,j+3));  % second point
19        next_edge_id=nodes(i, 4+mod(j,n_corners));
20        p3=GetRayEndNode(i,next_edge_id); % third point
21        [sedge, tp]=GetNbrRayEndNode(p2, i, −1);
22        [eedge, tp]=GetNbrRayEndNode(p3, i, 1);
23        if sedge==eedge
24          row=last_tri_row +1;
25          if row>n_max_tris   % if tris  matrix is  full : double its size
26            n_max_tris=n_max_tris*2;
27            tmp=tris ;
28            tris =zeros(n_max_tris,6);
29            tris (1:n_max_tris /2,:) =tmp;
30          end;
31          tris (row,1:3)=[p1 p2 p3];
32          tris (row,4:6)=[nodes(i,j+3),eedge,next_edge_id];
33          nd_corners(i,j+3)=−1;
34          SetNodeCorner(p2,eedge);
35          SetNodeCorner(p3,next_edge_id);
36          last_tri_row =row;
37          n_tris =n_tris +1;
38        end
39      end
40    end
41  end
42
43  % for−loop: update the  relevant  entries  in "edge" matrix
44  for i=1: last_tri_row
```

```
45      for j=4:6
46         e_id= tris (i,j);
47         if edges(e_id,9)==0; edges(e_id,9)=i;
48         elseif edges(e_id,10)==0; edges(e_id,10)=i;
49         else fprintp ('Setup tris error\n');
50         end
51      end
52   end
```

```
1   function DelaunayRefinement()
2   global edges last_edge_row angle_test_done edge_test_done;
3   global ec_queue ec_head ec_tail ;
4   global sm_queue sm_head sm_tail sm_nodes;
5   global sw_queue sw_head sw_tail;
6   global test_node test_edge no_bad_node no_bad_edge;
7   % next line : create queue of encroached edges
8   [ec_queue, ec_head, ec_tail ] = QueueCreate(2);
9   % next line : create queue of candidate swap edges
10  [sw_queue, sw_head, sw_tail ] = QueueCreate(1);
11
12  while 1 % continue until exit from inside
13     % for-loop: enqueue all the encroached edges
14     for i=1:last_edge_row
15        n_id=EncroachNode(i);
16        if n_id>0
17           [ec_queue, ec_head, ec_tail ]= ...
18              QueueAdd(ec_queue, ec_head, ec_tail, [i n_id]);
19        end
20     end
21     if ec_head>ec_tail; break; end; % exit while-loop if queue is empty
22     % while-loop: split all the encroached edges
23     while ec_head<=ec_tail
24        [ec_queue, ec_head, ec_tail , ec_edge] ...
25           = QueueRemove(ec_queue, ec_head, ec_tail );
26        Split (ec_edge(1),ec_edge(2));
27     end
28  end
29
30  % next line : create queue of small angle and large size triangles
31  [sm_queue, sm_head, sm_tail ] = QueueCreate(3);
32  test_node=1;   test_edge=1;        % set up flags
33  no_bad_node=1; no_bad_edge=1;      % set up flags
34
35  while 1 % continue until exit from inside
36     % if-block: find next small angle or large triangle , add to sm_queue
37     if angle_test_done==0
38        FindNextSmallAngle();
39     elseif edge_test_done==0
```

```
40        FindNextLargeEdge();
41      else
42        break;
43    end
44    while sm_head<=sm_tail  % continue until sm_queue is empty
45      [sm_queue, sm_head, sm_tail, sm_nodes] ...
46        = QueueRemove(sm_queue, sm_head, sm_tail);
47      % get the next triangle in the sm_queue
48      [x,y,r]=CompCircumcircle(sm_nodes(1),sm_nodes(2), sm_nodes(3));
49      flag=0;
50      % for-loop: insert a node at its circumcenter
51      for i=1:last_edge_row
52        if edges(i,3)==2 && IsEdgeEncrByPt(i,[x y])
53          flag=1; % if a boundary edge is encroached: flag =1
54          nid=BdryEdgeIntNd(i);
55          Split(i,nid);
56          break;
57        end
58      end
59      if flag==0;  Insert([x y]); end;
60    end
61  end
```

```
1   function [Lp,Rp]=CompLCT(left_s,left_e, right_s, right_e)
2   global edges;
3   Lp=left_e; Rp=right_s; % initial Lp and Rp nodes
4   [redge,p1]=LeftMostCCWNext(right_s); % get next(CCW) node (right side)
5   [ledge,p2]=RightMostCWNext(left_e);  % get next(CW) node (left side)
6
7   while 1
8     if CCW(p1,Rp,Lp)>0
9       Rp=p1;
10      if edges(redge,1)==Rp
11        redge=edges(redge,5);
12        p1=GetRayEndNode(p1,redge);
13      else
14        redge=edges(redge,7);
15        p1=GetRayEndNode(p1,redge);
16      end
17    else
18      if CCW(p2,Rp,Lp)>0
19        Lp=p2;
20        if edges(ledge,1)==Lp
21          ledge=edges(ledge,4);
22          p2=GetRayEndNode(p2,ledge);
23        else
24          ledge=edges(ledge,6);
25          p2=GetRayEndNode(p2,ledge);
```

```
26         end
27      else
28         return;
29      end
30   end
31 end
```

```
1  function Merge(Lp, Rp, left_s, left_e, right_s, right_e)
2  [Lp, Rp]=CheckColinear(Lp, Rp);   % get any collinear node of Lp or Rp
3  Connect(Rp,Lp);   % connect the two points
4
5  while 1 % continue until exit from inside
6     % next 17 lines : get the candidate right side node for edge Lp-Rp
7     [re1,rp1]=GetNbrRayEndNode(Rp,Lp,-1);
8     if CCW(rp1,Lp,Rp)>0
9        invalid_rp1 =0;
10       [re2,rp2]=GetNbrRayEndNode(Rp,rp1,-1);
11       while InCircle (Rp, rp2, rp1, Lp)
12          if rp2>right_e || rp2<right_s
13             break;
14          end
15          bad_edge=re1;
16          re1=re2;
17          rp1=rp2;
18          [re2,rp2]=GetNbrRayEndNode(Rp,rp1,-1);
19          DeleteEdge(bad_edge);
20       end
21    else
22       invalid_rp1 =1;
23    end
24    % next 17 lines : get the candidate left side node for edge Lp-Rp
25    [le1,lp1]=GetNbrRayEndNode(Lp,Rp,1);
26    if CCW(lp1,Lp,Rp)>0
27       invalid_lp1 =0;
28       [le2,lp2]=GetNbrRayEndNode(Lp,lp1,1);
29       while InCircle (Lp, lp1, lp2, Rp)
30          if lp2>left_e || lp2<left_s
31             break;
32          end
33          bad_edge=le1;
34          le1=le2;
35          lp1=lp2;
36          [le2,lp2]=GetNbrRayEndNode(Lp,lp1,-1);
37          DeleteEdge(bad_edge);
38       end
39    else
40       invalid_lp1 =1;
41    end
```

```
42  % next line: if no candidate node for either side, exit
43  if invalid_rp1 && invalid_lp1; return; end;
44  % if-block: create the new Lp-Rp edge
45  if invalid_lp1 ==1 || (~invalid_rp1 && InCircle(lp1,Lp,Rp,rp1))
46    Connect(Lp,rp1);
47    Rp=rp1;
48  else
49    Connect(Rp,lp1);
50    Lp=lp1;
51  end
52 end
```

```
1  function [left_s,left_e,right_s,right_e]=DivideInX(sid,eid)
2  n_input_nodes=eid-sid+1;
3  d=round(n_input_nodes/2);
4  left_s =sid; left_e =sid+d-1;
5  right_s =sid+d; right_e =eid;
```

```
1  function nid=EncroachNode(eid)
2  global nodes edges TOL;
3  nid=0;
4  if edges(eid,3)~=2 return; end;
5  nid=BdryEdgeIntNd(eid);
6  if ~IsEdgeEncrByPt(eid,[nodes(nid,1) nodes(nid,2)])
7    nid=0;
8  end
```

```
1  function [tid, a, b]=FindBdngTriByPt(pt)
2  global tris nodes last_tri_row ;
3  re=-1;
4  tid=0;
5  for i=1: last_tri_row
6    if tris(i,6)==0; continue; end;
7    p1=nodes(tris(i,1),1:2)';
8    p2=nodes(tris(i,2),1:2)';
9    p3=nodes(tris(i,3),1:2)';
10   [re, a, b]=PtInTriangle(p1, p2, p3, pt);
11   if re==1 || re==0
12     tid=i;
13     return;
14   end
15  end
16  if re<0; fprintf("findBdngTriByPt error\n"); end;
```

```
1  function FindNextLargeEdge()
2  global edges last_edge_row sm_queue sm_head sm_tail;
3  global MAX_LENGTH test_edge no_bad_edge edge_test_done;
4
5  if test_edge>last_edge_row
6    if no_bad_edge==1
7      edge_test_done=1;
8      return;
9    else
10     test_edge=1;
11     no_bad_edge=1;
12   end
13 end
14
15 if edges(test_edge,3)==0;
16   test_edge=test_edge+1;
17   return;
18 end;
19
20 if edges(test_edge,8)>MAX_LENGTH   % if edge length > MAX_LENGTH
21   [eids,nids]=NeighborEdgesNodes(test_edge);
22   [sm_queue, sm_head, sm_tail]= ...   % enqueue the triangle
23   QueueAdd(sm_queue, sm_head, sm_tail, nids(1:3,1));
24   no_bad_edge=0;
25 end
26 test_edge=test_edge+1;
```

```
1  function FindNextSmallAngle()
2  global nodes ray_angles last_node_row;
3  global sm_queue sm_head sm_tail;
4  global test_node no_bad_node angle_test_done;
5
6  if test_node>last_node_row
7    if no_bad_node==1
8      angle_test_done=1;
9      return;
10   else
11     test_node=1;
12     no_bad_node=1;
13   end
14 end
15
16 for j=1:nodes(test_node,3)
17   e=j+1;
18   if j~=nodes(test_node,3)
19     dtheta=ray_angles(test_node,e)−ray_angles(test_node,j);
20   else
```

```
21      e=1;
22      dtheta=ray_angles(test_node, e)+2*pi−ray_angles(test_node, j);
23    end
24    if dtheta<0.3613671239 % if angle < 20.7 degrees
25      a=GetRayEndNode(test_node, nodes(test_node, 3+j));
26      b=GetRayEndNode(test_node, nodes(test_node, 3+e));
27      nids=[test_node a b];
28      [sm_queue, sm_head, sm_tail]= ...    % enqueue the triangle
29        QueueAdd(sm_queue, sm_head, sm_tail, nids);
30      no_bad_node=0;
31    end
32  end
33  test_node=test_node+1;
```

```
1  function Insert(pt)
2  global edges TOL last_node_row tris last_edge_row;
3  global sw_queue sw_head sw_tail;
4
5  [tid,a,b]=FindBdngTriByPt(pt);
6  if abs(a)<TOL
7    if abs(b)<TOL || abs(b−1)<TOL; return;
8    else Split( tris(tid,6), pt); return;
9    end
10  elseif abs(b)<TOL
11    if abs(a−1)<TOL; return;
12    else Split( tris(tid,4), pt); return;
13    end
14  elseif abs(a+b−1)<TOL
15    Split( tris(tid,5), pt);
16    return;
17  end
18
19  nids=tris(tid,1:3)';
20  eids=tris(tid,4:6)';
21  AddNode(pt(1), pt(2));
22  for i=1:size(nids,1)
23    Connect(last_node_row,nids(i));
24  end
25
26  % delete the old triangle
27  for i=1:size(eids,1)
28    for k=9:10
29      if edges(eids(i),k)==tid;
30        edges(eids(i),k)=0;
31      end
32    end
33  end
34  tris(tid,6)=0;
```

```
35
36  % create new triangles
37  tnids =[last_node_row  nids(1)  nids(2) ];
38  teids =[last_edge_row-2 eids(1) last_edge_row-1];
39  AddTriangle(tnids,teids);
40
41  tnids =[last_node_row  nids(2)  nids(3) ];
42  teids =[last_edge_row-1 eids(2) last_edge_row ];
43  AddTriangle(tnids,teids);
44
45  tnids =[last_node_row  nids(3)  nids(1) ];
46  teids =[last_edge_row  eids(3) last_edge_row-2];
47  AddTriangle(tnids,teids);
48
49  % enqueue all the edges opposite to the new node
50  for i=1:size(eids,1)
51    if edges(eids(i),3)~=2
52      [sw_queue, sw_head, sw_tail]= ...
53      QueueAdd(sw_queue, sw_head, sw_tail, eids(i));
54    end
55  end
56  LocalSwap();
```

```
1  function re=IsEdgeEncrByPt(eid,pt)
2  global edges nodes TOL;
3
4  o=edges(eid,1);
5  d=edges(eid,2);
6  x=(nodes(o,1)+nodes(d,1))/2.0;
7  y=(nodes(o,2)+nodes(d,2))/2.0;
8
9  dist =norm([pt(1)-x pt(2)-y]);
10 if dist -TOL>edges(eid,8)/2.0
11   re=0;
12 else
13   re=1;
14 end
```

```
1  function [edge_out, node_out]=LeftMostCCWNext(node_in)
2  global nodes ray_angles TOL;
3
4  n_rays=nodes(node_in,3);
5  if n_rays==0
6    node_out=node_in;
7    edge_out=0;
8    return;
```

```
 9    end
10    edge_out=nodes(node_in,4);
11    node_out=GetRayEndNode(node_in,edge_out);
12    dtheta=mod(ray_angles(node_in,1)+TOL−pi*1.5,pi*2);
13
14    for i=2:n_rays
15      t=mod(ray_angles(node_in,i)+TOL−pi*1.5,pi*2);
16      if t<dtheta
17        edge_out=nodes(node_in,3+i);
18        node_out=GetRayEndNode(node_in,edge_out);
19        dtheta=t;
20      end
21    end
```

```
 1    function [edge_out, node_out]=RightMostCWNext(node_in)
 2    global nodes ray_angles TOL;
 3
 4    n_rays=nodes(node_in,3);
 5    if n_rays==0
 6      node_out=node_in;
 7      edge_out=0;
 8      return;
 9    end
10    edge_out=nodes(node_in,4);
11    node_out=GetRayEndNode(node_in,edge_out);
12    dtheta=1.5*pi− ray_angles(node_in,1);
13
14    for i=2:n_rays
15      t=1.5*pi− ray_angles(node_in,i);
16      if t<dtheta
17        edge_out=nodes(node_in,3+i);
18        node_out=GetRayEndNode(node_in,edge_out);
19        dtheta=t;
20      end
21    end
```

```
 1    function LocalSwap()
 2    global edges last_node_row last_edge_row ;
 3    global tris last_tri_row ;
 4    global sw_queue sw_head sw_tail;
 5
 6    while sw_head<=sw_tail  % continue if the swap queue is not empty
 7      [sw_queue, sw_head, sw_tail , sw_edge] ...
 8          = QueueRemove(sw_queue, sw_head, sw_tail);
 9      [egs,nds]=NeighborEdgesNodes(sw_edge);
10      if edges(sw_edge,3)==2; continue; end;
11      if InCircle(nds(1), nds(2), nds(3), nds(4))>0  % if in circle
```

Mesh Generation

```
12      for j=1:size(egs,1)          % for all the neighbor edges
13          if edges(egs(j),1)~=last_node_row ...  % if edge not
14          && edges(egs(j),2)~=last_node_row ...  % connected
15          && edges(egs(j),3)~=2              % to the center
16              [sw_queue, sw_head, sw_tail]= ...  % add to sw_queue
17              QueueAdd(sw_queue, sw_head, sw_tail, egs(j));
18          end
19      end % end for
20      DeleteTriByEdge(sw_edge);      % delete the old tringles
21      Swap(sw_edge);                 % swap
22      tnids=[nds(1) nds(2) nds(4)];  % create a new triangle
23      teids=[egs(1) last_edge_row egs(4)];
24      AddTriangle(tnids, teids);     % add the new triangle
25      tnids=[nds(2) nds(3) nds(4)];  % create a new triangle
26      teids=[egs(2) egs(3) last_edge_row];
27      AddTriangle(tnids, teids);     % add the new triangle
28    end % end if
29 end
```

```
1  function Swap(a)
2  global edges;
3  if edges(a,3)==0 || edges(a,3)==2, return,  % skip if not an interior
4  end;                                        % edge
5  [oprev, onext, dprev, dnext]=NeighborEdges(a); % get neighbor edges
6  orig=edges(a,1);
7  dest=edges(a,2);
8
9  % if-block: assign correct node1
10 if edges(oprev,1)==orig;  node1=edges(oprev,2);
11 else  node1=edges(oprev,1);
12 end
13 % if-block: assign correct node2
14 if edges(onext,1)==orig;  node2=edges(onext,2);
15 else  node2=edges(onext,1);
16 end
17
18 % if-block: connect the new node1 and node2, delete the old edge
19 if node1~=node2
20    DeleteEdge(a);
21    Connect(node1, node2);
22 else  fprintf("Swap error\n");
23 end
```

```
1  function SetNodeCorner(nid, eid)
2  global nodes nd_corners;
3
4  for i=1:nodes(nid,3)
```

```
5    if nodes(nid,3+i)==eid
6      nd_corners(nid,3+i)=-1;
7    end
8  end
```

```
1   function Split(eid, pt)
2   global nodes edges last_node_row last_edge_row;
3   global sw_queue sw_head sw_tail;
4   global tris;
5   o=edges(eid,1);            % get the origin node of edge "eid"
6   d=edges(eid,2);            % get the destination node of edge "eid"
7   [eids, nids]=NeighborEdgesNodes(eid);  % get neighbor edges and nodes
8   DeleteTriByEdge(eid);      % delete triangles associated with eid
9   DeleteEdge(eid);           % delete the edge
10
11  if size(pt,2)==1
12      x=(nodes(o,1)+nodes(d,1)) /2.0;
13      y=(nodes(o,2)+nodes(d,2)) /2.0;
14  else
15      x=pt(1); y=pt(2);
16  end
17  AddNode(x,y);
18
19  for i=1:size(nids,1)
20      Connect(last_node_row,nids(i));
21      if size(pt,2)==1 && (nids(i)==o || nids(i)==d)
22          edges(last_edge_row,3)=2;
23      end
24  end
25  % if-block: create new triangles
26  if size(nids,1)==3
27      if edges(eids(1),1)==nids(1) || edges(eids(1),2)==nids(1)
28          tnids=[nids(1) nids(2) last_node_row ]';
29          teids=[eids(1) last_edge_row-1 last_edge_row-2]';
30          AddTriangle(tnids,teids);
31          tnids=[last_node_row nids(2) nids(3) ]';
32          teids=[last_edge_row-1 eids(2) last_edge_row ]';
33          AddTriangle(tnids,teids);
34      else
35          tnids=[nids(1) last_node_row nids(3) ]';
36          teids=[last_edge_row-2 last_edge_row eids(2) ]';
37          AddTriangle(tnids,teids);
38          tnids=[last_node_row nids(2) nids(3) ]';
39          teids=[last_edge_row-1 eids(1) last_edge_row ]';
40          AddTriangle(tnids,teids);
41      end
```

```
42  else
43        tnids=[nids(1)  nids(2)  last_node_row];
44        teids=[eids(1)  last_edge_row-2 last_edge_row-3];
45        AddTriangle(tnids,teids);
46        tnids=[nids(2)  nids(3)  last_node_row];
47        teids=[eids(2)  last_edge_row-1 last_edge_row-2];
48        AddTriangle(tnids,teids);
49        tnids=[nids(3)  nids(4)  last_node_row];
50        teids=[eids(3)  last_edge_row  last_edge_row-1];
51        AddTriangle(tnids,teids);
52        tnids=[nids(4)  nids(1)  last_node_row];
53        teids=[eids(4)  last_edge_row-3 last_edge_row];
54        AddTriangle(tnids,teids);
55  end
56
57  % queue all the edges opposite to the new node
58  for i=1:size(eids,1)
59     [sw_queue, sw_head, sw_tail]= ...
60         QueueAdd(sw_queue, sw_head, sw_tail, eids(i));
61  end
62  LocalSwap();   % swap the edges if necessary
```

```
1   function re=BdryEdgeIntNd(eid)
2   global edges;
3   if edges(eid,3)~=2; fprintf("BdryEdgeIntNd error\n"); end;
4   [oprev,onext,dprev,dnext]=NeighborEdges(eid); % get neighbor edges
5   o=edges(eid,1);              % origin node
6   d=edges(eid,2);              % destination node
7   nop=GetRayEndNode(o,oprev);  % get the end nodes
8   non=GetRayEndNode(o,onext);  % of the neighbor edges
9   ndp=GetRayEndNode(d,dprev);
10  ndn=GetRayEndNode(d,dnext);
11
12  % if-block: identify the correct interior node
13  if nop==ndn && non~=ndp
14     re=nop;
15  elseif non==ndp && nop~=ndn
16     re=non;
17  else
18     fprintf("BdryEdgeIntNd error\n");
19  end
```

```
1   function [angle_a, angle_b]=TwoNodeRayAngles(node_a, node_b)
2   global nodes;
3   angle_a= atan2(nodes(node_b,2)-nodes(node_a,2), ...    % atan(dy/dx)
4                  nodes(node_b,1)-nodes(node_a,1));
```

```
 5    if angle_a<0
 6       angle_a= angle_a+2*pi;     % make sure angle is 0-2*pi
 7    end
 8    angle_b= mod(angle_a +pi, 2*pi);   % the other angle is pi apart
```

```
 1  function PlotEdges()
 2  global edges nodes ray_angles last_edge_row;
 3
 4  hold on;
 5  for i=1:last_edge_row
 6     if edges(i,3)>0
 7        orig=edges(i,1);
 8        dest=edges(i,2);
 9        plot ([nodes(orig,1) nodes(dest,1) ],...
10             [nodes(orig,2) nodes(dest,2)],'-k','LineWidth',edges(i,3));
11     end
12  end
13  hold off
```

6.5 Summary

Upon completion of this chapter, you should be able to:

- understand the main ideas of various solid modeling approaches
- understand the main ideas of various meshing methods
- understand the details of the Delaunay triangulation and refinement method
- know how to manually construct the Voronoi diagram for a set of scattered nodes
- know how to manually perform Delaunay triangulation of a set of nodes by using the divide-and-conquer algorithm
- know how to manually eliminate an encroached edge in a Delaunay triangulation
- know how to manually insert a node into a Delaunay triangulation
- understand the criteria for identifying a "bad" triangle
- know how to manually eliminate a "bad" triangle from a Delaunay triangulation
- know how to implement Delaunay triangulation and Delaunay refinement algorithms to generate 2-D unstructured meshes of triangular elements
- employ the MATLAB program developed in this chapter to generate 2-D triangular meshes for arbitrary 2-D shapes.

6.6 Problems

6.1 Write a MATLAB program to generate a structured mesh for the shape shown in Fig. 6.28. The dimensions of the shape and the number of element divisions along the edges shown in the figure are the input parameters.

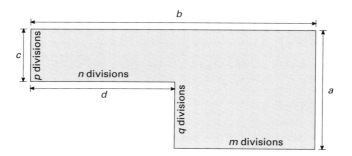

Figure 6.28 Structured mesh for Problem 6.1.

6.2 Write a MATLAB program to generate structured meshes for the two shapes shown in Fig. 6.29. The program should take the outer radius R, m, and n as the input parameters for the half circle (a) and an additional parameter, inner radius r, for the half annulus (b). Note: use straight lines to approximate the curve segments.

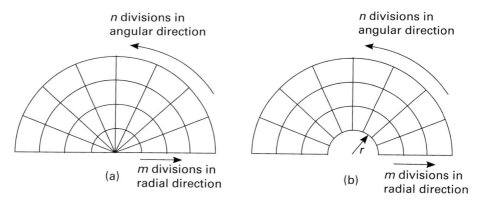

Figure 6.29 Structured meshes for Problem 6.2.

6.3 A quadrilateral shape is called strictly convex if all its four interior angles are less than 180 degrees. Consider a strictly convex quadrilateral with its four vertices not on a circle. Show that the quadrilateral can be partitioned into two triangles in two possible ways, and one of the triangulations is a Delaunay triangulation. Show that the Delaunay triangulation gives a larger minimum angle of the triangles.

6.4 Assuming the four vertices of the strictly convex quadrilateral in Problem 6.3 are on a circle (co-circular), show that the two possible triangulations of the quadrilateral are both Delaunay. Show that both triangulations give the same minimum angle of the triangles.

6.5 Let T be a Delaunay triangulation bounded by boundary edges. Show that if T has no encroached boundary edge, then the circumcenter of any triangle of T lies in T.

6.6 Show that if a corner of a PSLG has a very small angle as shown in Fig. 6.30, the Delaunay refinement algorithm described in Section 6.3.2 may not terminate (will loop forever).

Figure 6.30 PSLG for Problem 6.6.

6.7 Manually create Delaunay triangulation for the two sets of nodes (a) and (b), respectively, in Fig. 6.31, by using the divide-and-conquer algorithm discussed in the text.

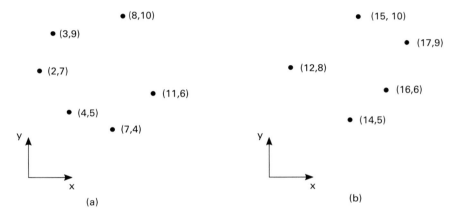

Figure 6.31 Nodes for Problem 6.7.

6.8 Merge the two Delaunay triangulations generated in Problem 6.7 by using the merge algorithm discussed in the text. Show the merge steps in drawings.

6.9 By using the programs discussed in this chapter, generate a 2-D mesh of triangular elements for the square domain shown in Fig. 6.32. Show the following results: (1) The PSLG file, (2) the initial Delaunay triangulation of the PSLG, (3) the initial Delaunay triangulation with no encroached boundary edges, (4) the mesh with maximum edge lengths of 1.0, (5) the mesh with maximum edge lengths of 0.2, and (6) the mesh with maximum edge lengths of 0.05.

6.10 Repeat the tasks described in Problem 6.9, but with the domain shape shown in Fig. 6.33. Note that the PSLG of the circular hole should be an approximation.

6.11 Repeat the tasks described in Problem 6.9, but with the domain shape shown in Fig. 6.28. Generate a coarse and a fine mesh (say >5000 elements) for the shape.

6.12 Repeat the tasks described in Problem 6.9, but with the domain shape shown in Fig. 6.29. Generate a coarse and a fine mesh (>5000 elements) for the shape.

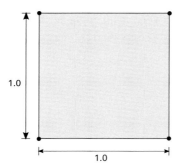
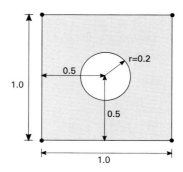

Figure 6.32 Square domain for Problem 6.9.   **Figure 6.33** Domain for Problem 6.10.

6.13 Show that if the function "InCircle(a,b,c,d)" listed in Section 6.4.2 returns 1 (i.e., "true"), then (a) "InCircle(c,d,a,b)" returns 1, and (b) "InCircle(b,c,d,a)" and "InCircle(d,a,b,c)" returns 0.

References

Baumgart, B. G. (1975), A polyhedron representation for computer vision, in *Proceedings of the May 19–22, 1975, national computer conference and exposition*, ACM, pp. 589–596.

Blazek, J. (2015), *Computational fluid dynamics: principles and applications*, Butterworth-Heinemann.

Chew, L. P. (1993), Guaranteed-quality mesh generation for curved surfaces, in *Proceedings of the ninth annual symposium on computational geometry*, ACM, pp. 274–280.

Dwyer, R. A. (1987), "A faster divide-and-conquer algorithm for constructing Delaunay triangulations," *Algorithmica* **2**(1–4), 137–151.

Guibas, L. & Stolfi, J. (1985), "Primitives for the manipulation of general subdivisions and the computation of Voronoi," *ACM Transactions on Graphics (TOG)* **4**(2), 74–123.

Lawson, C. L. (1977), Software for c1 surface interpolation, in *Mathematical software*, Elsevier, pp. 161–194.

Lee, D.-T. & Lin, A. K. (1986), "Generalized Delaunay triangulation for planar graphs," *Discrete & Computational Geometry* **1**(3), 201–217.

Lee, D.-T. & Schachter, B. J. (1980), "Two algorithms for constructing a Delaunay triangulation," *International Journal of Computer & Information Sciences* **9**(3), 219–242.

Pedoe, D. (1995), *Circles: a mathematical view*, Cambridge University Press.

Ruppert, J. (1995), "A Delaunay refinement algorithm for quality 2-dimensional mesh generation," *Journal of Algorithms* **18**(3), 548–585.

Shewchuk, J. R. (1996), Triangle: Engineering a 2D quality mesh generator and Delaunay triangulator, in *Applied computational geometry towards geometric engineering*, Springer, pp. 203–222.

Shewchuk, J. R. (2002), "Delaunay refinement algorithms for triangular mesh generation," *Computational Geometry* **22**(1–3), 21–74.

7 FEA for Multi-Dimensional Vector Field Problems

7.1 Overview

In this chapter, we discuss the general finite element analysis procedure for linear vector field problems. A vector field problem is a problem whose primary unknown physical quantity is a vector quantity at any spatial location in the domain. For example, 2-D or 3-D structural mechanics problems are vector field problems, since the displacement is the primary physical quantity in a typical finite element analysis, and at each spatial point in the structure of interest, the displacement is a vector having components in x, y (2-D) or x, y, z (3-D) directions. In other words, there are multiple degrees of freedom at each spatial point in a vector field problem. While this increases the total degrees of freedom of the finite element model and expands the dimension of element level matrices and vectors, it does not complicate the analysis procedure itself. The discretization or meshing process remain the same. The nodes and elements can be the same as well. The difference is that each node now has an unknown vector instead of an unknown scalar variable in the scalar field case. The finite element approximation also remains the same and is applied to each component of the unknown vector.

As solid mechanics is representative of vector field problems, this chapter focuses on a set of solid mechanics problems. The chapter contains four sections. The first section briefly reviews the theory of linear elasticity. The second section introduces the FEA procedure for structural analysis of a 2-D elasticity problem. The third section discusses a 3-D elasticity problem and illustrates the FEA steps. The fourth section discusses the finite element analysis procedure for 2-D steady state incompressible viscous flow problems. At the end of each section, MATLAB codes for solving these problems are presented.

7.2 2-D Elasticity

In this section, the theory of elasticity is briefly reviewed before a finite element analysis of a 2-D structure is presented. The derivation of equations of equilibrium and motion is described for a better understanding of its relation to the finite element formulation of various types of structures. The weak form of 2-D elasticity is derived by using the method of weighted residual, the principle of virtual work, and the variational method. Detailed analysis steps with intermediate numerical results are included.

7.2.1 Linear Elasticity: A Brief Review

Consider the three dimensional region Ω occupied by a linear elastic solid in Fig. 7.1. The origin of a fixed right-handed Cartesian coordinate system (x, y, z) is located at point O. We will refer to this system as the global system. The displacement of any point in Ω, such as P, has components (u, v, w) measured with respect to the global reference axes. The boundary of Ω is denoted as Γ and it consists of two parts: Γ_u, the portion on which displacements are specified, and Γ_t, the portion on which surface tractions are specified. The vector $\mathbf{n}$ is the unit outward boundary normal to Ω and has components (n_x, n_y, n_z) with respect to the global system.

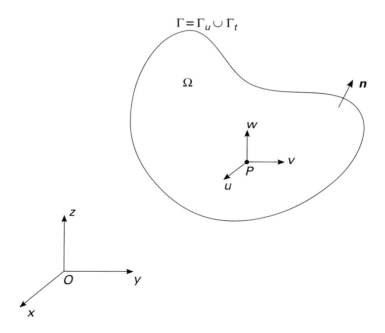

Figure 7.1 Three-dimensional problem domain.

Stress in Three Dimensions

If we cut the body Ω into two parts using an arbitrary plane, as shown in Fig. 7.2, to satisfy equilibrium for each half of Ω, a distributed force must act over the area A created by the cut. The distributed force per unit area acting on A is the traction, $\mathbf{t}(x, y, z)$. The total force acting on A is $\int_A \mathbf{t} dA$.

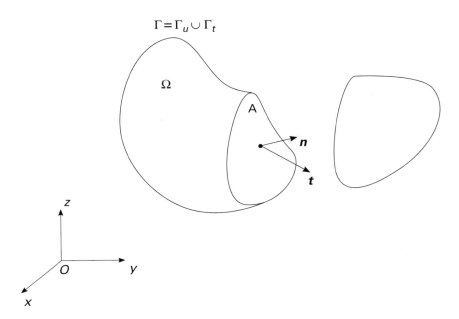

Figure 7.2 Internal forces.

Consider a point P lying on A. Cauchy's stress principle defines a tensor quantity, stress, that relates the traction, $\mathbf{t}$, to the unit normal to A at P, $\mathbf{n}$, i.e.,

$$\mathbf{t} = \sigma^T \mathbf{n} \tag{7.1}$$

where

$$\mathbf{t} = \begin{Bmatrix} t_x \\ t_y \\ t_z \end{Bmatrix} \quad \sigma = \begin{bmatrix} \sigma_{xx} & \sigma_{xy} & \sigma_{xz} \\ \sigma_{yx} & \sigma_{yy} & \sigma_{yz} \\ \sigma_{zx} & \sigma_{zy} & \sigma_{zz} \end{bmatrix} \quad \mathbf{n} = \begin{Bmatrix} n_x \\ n_y \\ n_z \end{Bmatrix} \tag{7.2}$$

or in component form

$$\begin{aligned} t_x &= \sigma_{xx} n_x + \sigma_{yx} n_y + \sigma_{zx} n_z \\ t_y &= \sigma_{xy} n_x + \sigma_{yy} n_y + \sigma_{zy} n_z \\ t_z &= \sigma_{xz} n_x + \sigma_{yz} n_y + \sigma_{zz} n_z. \end{aligned} \tag{7.3}$$

For linear elasticity, the stress tensor is symmetric, i.e.,

$$\begin{aligned} \sigma_{xy} &= \sigma_{yx} \\ \sigma_{xz} &= \sigma_{zx} \\ \sigma_{yz} &= \sigma_{zy}. \end{aligned} \tag{7.4}$$

Cut off an infinitesimal cube of material from Ω, with the edges of the cube parallel to the x-, y- and z-axes. The components of the stress tensor are the tractions acting on the faces of this cube as shown in Fig. 7.3. Hence, σ_{ij} is the j-th component of the traction acting on the plane with the normal in the i-th direction. Note that it is important to clearly distinguish between tractions (force per unit area acting on a surface) and stress (a mathematical concept that can be physically related to tractions when combined with a surface).

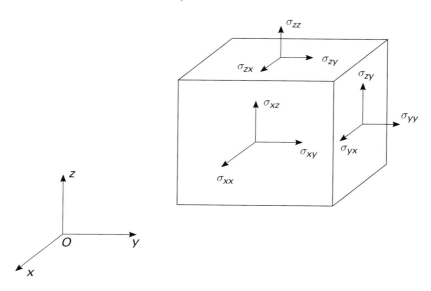

Figure 7.3 Physical interpretation of the stress tensor.

Differential Equations of Equilibrium

The equations of static equilibrium may be derived by considering a balance of forces on an infinitesimal element, as shown in Fig. 7.4. To prevent crowding of the figure, only forces in the x-direction are shown. Similar forces act in the y- and z-directions. Here b_x is the x-component of the body force (per unit volume). A balance of forces in the x-direction yields

$$\frac{\partial \sigma_{xx}}{\partial x} + \frac{\partial \sigma_{yx}}{\partial y} + \frac{\partial \sigma_{zx}}{\partial z} + b_x = 0. \qquad (7.5)$$

Similarly for the other directions,

$$\frac{\partial \sigma_{xy}}{\partial x} + \frac{\partial \sigma_{yy}}{\partial y} + \frac{\partial \sigma_{zy}}{\partial z} + b_y = 0 \qquad (7.6)$$

$$\frac{\partial \sigma_{xz}}{\partial x} + \frac{\partial \sigma_{yz}}{\partial y} + \frac{\partial \sigma_{zz}}{\partial z} + b_z = 0 \qquad (7.7)$$

where $\mathbf{b} = (b_x, b_y, b_z)$ is the body force per unit volume. These equations (together with the boundary conditions) are called the strong form of the equilibrium equations.

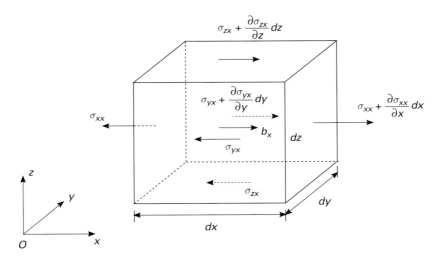

Figure 7.4 Force balance in the x-direction on an infinitesimal element.

Differential Equations of Motion

If the infinitesimal element is not in static equilibrium, the net force acting on the element results in acceleration in the direction of the net force. The equations of equilibrium become the equations of motion, i.e.,

$$\frac{\partial \sigma_{xx}}{\partial x} + \frac{\partial \sigma_{yx}}{\partial y} + \frac{\partial \sigma_{zx}}{\partial z} + b_x = \rho \frac{\partial^2 u}{\partial t^2} \quad (7.8)$$

$$\frac{\partial \sigma_{xy}}{\partial x} + \frac{\partial \sigma_{yy}}{\partial y} + \frac{\partial \sigma_{zy}}{\partial z} + b_y = \rho \frac{\partial^2 v}{\partial t^2} \quad (7.9)$$

$$\frac{\partial \sigma_{xz}}{\partial x} + \frac{\partial \sigma_{yz}}{\partial y} + \frac{\partial \sigma_{zz}}{\partial z} + b_z = \rho \frac{\partial^2 w}{\partial t^2} \quad (7.10)$$

where ρ is the mass density of the material. The equations of motion are the governing equations of the dynamic analysis, which will be discussed in Section 9.3.

Strain in Three Dimensions

For infinitesimal deformations, we define the following strains to measure the deformation (the strain–displacement relations):

$$\begin{aligned} \epsilon_{xx} &= \frac{\partial u}{\partial x} & \epsilon_{xy} &= \frac{\partial u}{\partial y} + \frac{\partial v}{\partial x} \\ \epsilon_{yy} &= \frac{\partial v}{\partial y} & \epsilon_{yz} &= \frac{\partial v}{\partial z} + \frac{\partial w}{\partial y} \\ \epsilon_{zz} &= \frac{\partial w}{\partial z} & \epsilon_{xz} &= \frac{\partial w}{\partial x} + \frac{\partial u}{\partial z}. \end{aligned} \quad (7.11)$$

Note that here we are using the engineering definition of strain. Consider a deformation experienced by the two infinitesimal segments shown in Fig. 7.5 that are initially parallel to the x- and y-axes, respectively. The normal strain measures the change in length of a line initially parallel to a coordinate axis, for example,

$$\epsilon_{xx} = \frac{ds - ds_0}{ds_0}. \tag{7.12}$$

The shear strain measures the change in angle between two initially perpendicular lines, for example,

$$\epsilon_{xy} = \frac{\pi}{2} - \phi \tag{7.13}$$

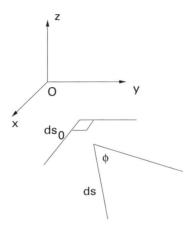

Figure 7.5 Physical interpretation of the strains.

The Constitutive Law for Linear Elastic Solids

A constitutive law gives the relationship between stresses and strains. The constitutive law of homogeneous linear elastic materials can be written in the form of Eq. (7.14), which is referred to as generalized Hooke's law. The six strains are related to the six stresses by

$$\begin{Bmatrix} \sigma_{xx} \\ \sigma_{yy} \\ \sigma_{zz} \\ \sigma_{xy} \\ \sigma_{yz} \\ \sigma_{xz} \end{Bmatrix} = \begin{bmatrix} C_{11} & C_{12} & C_{13} & C_{14} & C_{15} & C_{16} \\ C_{21} & C_{22} & C_{23} & C_{24} & C_{25} & C_{26} \\ C_{31} & C_{32} & C_{33} & C_{34} & C_{35} & C_{36} \\ C_{41} & C_{42} & C_{43} & C_{44} & C_{45} & C_{46} \\ C_{51} & C_{52} & C_{53} & C_{54} & C_{55} & C_{56} \\ C_{61} & C_{62} & C_{63} & C_{64} & C_{65} & C_{66} \end{bmatrix} \begin{Bmatrix} \epsilon_{xx} \\ \epsilon_{yy} \\ \epsilon_{zz} \\ \epsilon_{xy} \\ \epsilon_{yz} \\ \epsilon_{xz} \end{Bmatrix}. \tag{7.14}$$

In short form, Eq.(7.14) can be written as

$$\sigma = C\epsilon. \tag{7.15}$$

The entries of the **C** matrix are called elastic constants of the material. As the **C** matrix is symmetric, the total number of independent elastic constants is 21 for a general elastic material. It can be shown that for an isotropic elastic material,

$$\begin{Bmatrix} \sigma_{xx} \\ \sigma_{yy} \\ \sigma_{zz} \\ \sigma_{xy} \\ \sigma_{yz} \\ \sigma_{xz} \end{Bmatrix} = \frac{E}{(1+\nu)(1-2\nu)} \begin{bmatrix} (1-\nu) & \nu & \nu & 0 & 0 & 0 \\ \nu & (1-\nu) & \nu & 0 & 0 & 0 \\ \nu & \nu & (1-\nu) & 0 & 0 & 0 \\ 0 & 0 & 0 & \frac{1-2\nu}{2} & 0 & 0 \\ 0 & 0 & 0 & 0 & \frac{1-2\nu}{2} & 0 \\ 0 & 0 & 0 & 0 & 0 & \frac{1-2\nu}{2} \end{bmatrix} \begin{Bmatrix} \epsilon_{xx} \\ \epsilon_{yy} \\ \epsilon_{zz} \\ \epsilon_{xy} \\ \epsilon_{yz} \\ \epsilon_{xz} \end{Bmatrix} \quad (7.16)$$

where the two independent elastic constants are Young's modulus, E, and Poisson's ratio, ν. We can define other physically meaningful elastic constants using E and ν, such as the shear modulus G and bulk modulus K

$$G = \frac{E}{2(1+\nu)} \qquad K = \frac{E}{3(1-2\nu)}. \quad (7.17)$$

For different classes of anisotropic materials, there are more than two independent elastic constants and some of the zeros in Eq. (7.16) become nonzero. One important type of anisotropic material is called orthotropic material. This type of material has three orthogonal planes of symmetry. The lines of intersection of the symmetric planes form a rectangular Cartesian coordinate system. The axes are called principal axes of the orthotropic material. Without losing generality, we assume x-, y- and z-axes are the three principal axes, and the constitutive relation is then given by

$$\begin{Bmatrix} \sigma_{xx} \\ \sigma_{yy} \\ \sigma_{zz} \\ \sigma_{xy} \\ \sigma_{yz} \\ \sigma_{xz} \end{Bmatrix} = \begin{bmatrix} \frac{1-\nu_{yz}\nu_{zy}}{E_y E_z \Delta} & \frac{\nu_{yx}+\nu_{zx}\nu_{yz}}{E_y E_z \Delta} & \frac{\nu_{zx}+\nu_{yx}\nu_{zy}}{E_y E_z \Delta} & 0 & 0 & 0 \\ \frac{\nu_{xy}+\nu_{xz}\nu_{zy}}{E_z E_x \Delta} & \frac{1-\nu_{zx}\nu_{xz}}{E_z E_x \Delta} & \frac{\nu_{zy}+\nu_{zx}\nu_{xy}}{E_z E_x \Delta} & 0 & 0 & 0 \\ \frac{\nu_{xz}+\nu_{xy}\nu_{yz}}{E_x E_y \Delta} & \frac{\nu_{yz}+\nu_{xz}\nu_{yx}}{E_x E_y \Delta} & \frac{1-\nu_{xy}\nu_{yx}}{E_x E_y \Delta} & 0 & 0 & 0 \\ 0 & 0 & 0 & G_{xy} & 0 & 0 \\ 0 & 0 & 0 & 0 & G_{yz} & 0 \\ 0 & 0 & 0 & 0 & 0 & G_{xz} \end{bmatrix} \begin{Bmatrix} \epsilon_{xx} \\ \epsilon_{yy} \\ \epsilon_{zz} \\ \epsilon_{xy} \\ \epsilon_{yz} \\ \epsilon_{xz} \end{Bmatrix} \quad (7.18)$$

where

$$\Delta = \frac{1 - \nu_{xy}\nu_{yx} - \nu_{yz}\nu_{zy} - \nu_{zx}\nu_{xz} - 2\nu_{xy}\nu_{yz}\nu_{zx}}{E_x E_y E_z}. \quad (7.19)$$

In Eq. (7.18), there are nine independent parameters: three normal moduli (E_x, E_y, E_z), three Poisson's ratios (ν_{xy}, ν_{yz}, ν_{zx}), and three shear moduli (G_{xy}, G_{yz}, G_{zx}). Note that, although $\nu_{xy} \neq \nu_{yx}$, $\nu_{yz} \neq \nu_{zy}$, and $\nu_{zx} \neq \nu_{xz}$, these Poisson's ratios can be obtained from the symmetry of the **C** matrix as

$$\frac{\nu_{yz}}{E_y} = \frac{\nu_{zy}}{E_z} \qquad \frac{\nu_{zx}}{E_z} = \frac{\nu_{xz}}{E_x} \qquad \frac{\nu_{xy}}{E_x} = \frac{\nu_{yx}}{E_y}. \quad (7.20)$$

Plane Stress and Plane Strain

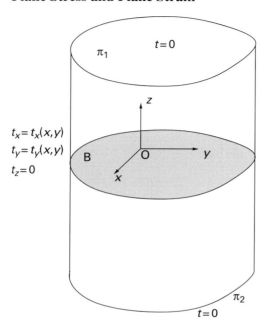

Figure 7.6 The plane problem.

There is an important class of three-dimensional problems (known as the plane problem) which can be approximated by two-dimensional solutions. Consider the prismatic cylinder shown in Fig. 7.6. The plane problem of elasticity consists of finding the displacements, strains, and stresses in the cylinder when the following boundary conditions are applied

$$t_x = t_y = t_z = 0 \qquad \text{on the end faces } \pi_1 \text{ and } \pi_2 \qquad (7.21)$$

and

$$\left.\begin{array}{l} t_x = t_x(x, y) \\ t_y = t_y(x, y) \\ t_z = 0 \end{array}\right\} \text{ on B, the lateral surface.} \qquad (7.22)$$

Two typical examples of plane problem are shown in Figs. 7.7 and 7.8: a thin plate with a hole under tension, and a gravity dam. A crucial feature of the plane problem is that the applied surface tractions are independent of z, and parallel to the xy-plane. This motivates the study of two-dimensional versions of this problem.

Plane Strain in Isotropic Materials

As the name implies, plane strain means the strains in the third direction are zero and the nonzero strains are two dimensional. We make the kinematic assumption

$$u = u(x, y) \qquad (7.23)$$

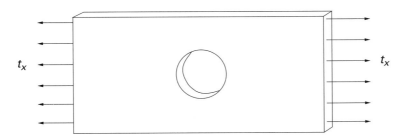

Figure 7.7 A thin plate with a hole.

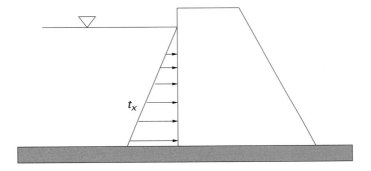

Figure 7.8 A gravity dam.

$$v = v(x, y) \tag{7.24}$$
$$w = 0. \tag{7.25}$$

Using these strain–displacement relations

$$\epsilon_{xz} = \frac{\partial w}{\partial x} + \frac{\partial u}{\partial z} = 0 \tag{7.26}$$

$$\epsilon_{yz} = \frac{\partial v}{\partial z} + \frac{\partial w}{\partial y} = 0 \tag{7.27}$$

$$\epsilon_{zz} = \frac{\partial w}{\partial z} = 0. \tag{7.28}$$

Using the constitutive law:

$$\sigma_{xz} = \sigma_{yz} = 0$$
$$\sigma_{zz} = \nu(\sigma_{xx} + \sigma_{yy}) \tag{7.29}$$
$$\left\{ \begin{array}{c} \sigma_{xx} \\ \sigma_{yy} \\ \sigma_{xy} \end{array} \right\} = \frac{E}{(1+\nu)(1-2\nu)} \begin{bmatrix} 1-\nu & \nu & 0 \\ \nu & 1-\nu & 0 \\ 0 & 0 & \frac{1-2\nu}{2} \end{bmatrix} \left\{ \begin{array}{c} \epsilon_{xx} \\ \epsilon_{yy} \\ \epsilon_{xy} \end{array} \right\}.$$

We would expect the plane strain conditions to be present when we have thick sections. Physically, the material in the z-direction provides enough constraint to prevent displacements in that direction, for example, a gravity dam as shown in Fig. 7.9.

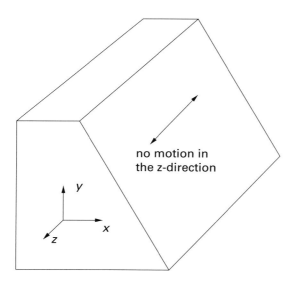

Figure 7.9 Plane strain: gravity dam.

Plane Stress in Isotropic Materials

We assume that the through-thickness stresses are negligible, i.e.,

$$\sigma_{xz} = \sigma_{yz} = \sigma_{zz} = 0 \tag{7.30}$$

From the three-dimensional constitutive law

$$\epsilon_{zz} = \frac{1}{E}\left[\sigma_{zz} - \nu(\sigma_{xx} + \sigma_{yy})\right]. \tag{7.31}$$

Hence, for plane stress

$$\epsilon_{zz} = -\frac{\nu}{E}(\sigma_{xx} + \sigma_{yy}) \tag{7.32}$$

and

$$\left\{\begin{array}{c}\sigma_{xx}\\ \sigma_{yy}\\ \sigma_{xy}\end{array}\right\} = \frac{E}{(1-\nu^2)}\begin{bmatrix}1 & \nu & 0\\ \nu & 1 & 0\\ 0 & 0 & \frac{1-\nu}{2}\end{bmatrix}\left\{\begin{array}{c}\epsilon_{xx}\\ \epsilon_{yy}\\ \epsilon_{xy}\end{array}\right\}. \tag{7.33}$$

Plane stress conditions are present in thin sections. The section is thin enough so that through-thickness stress cannot develop (recall π_1 and π_2 are traction free), for example, in a thin plate as shown in Fig. 7.10.

For both plane strain and plane stress conditions, the plane problem now reduces to finding displacements (u, v), stresses $(\sigma_{xx}, \sigma_{yy}, \sigma_{xy})$, and strains $(\epsilon_{xx}, \epsilon_{yy}, \epsilon_{xy})$ in a cross section of the cylinder that satisfies the applied boundary conditions. We have reduced the three-dimensional problem to a two-dimensional problem. We can write the relationship between in-plane stresses and strains as

7.2 2-D Elasticity

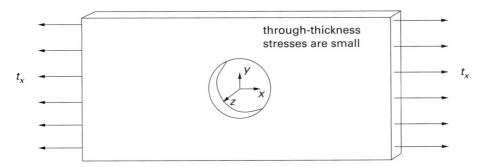

Figure 7.10 Plane stress: thin plate with a hole.

$$\begin{Bmatrix} \sigma_{xx} \\ \sigma_{yy} \\ \sigma_{xy} \end{Bmatrix} = \frac{E^*}{(1-(v^*)^2)} \begin{bmatrix} 1 & v^* & 0 \\ v^* & 1 & 0 \\ 0 & 0 & \frac{1-v^*}{2} \end{bmatrix} \begin{Bmatrix} \epsilon_{xx} \\ \epsilon_{yy} \\ \epsilon_{xy} \end{Bmatrix} \quad (7.34)$$

where

$$E^* = E, \quad v^* = v \quad \text{for plane stress} \quad (7.35)$$

and

$$E^* = \frac{E}{1-v^2}, \quad v^* = \frac{v}{1-v} \quad \text{for plane strain.} \quad (7.36)$$

This generalized formula can be used to simplify programming.

Plane Strain in Orthotropic Materials

Given $\epsilon_{xz} = \epsilon_{yz} = \epsilon_{zz} = 0$, Eq. (7.18) can be reduced to its 2-D version as

$$\begin{Bmatrix} \sigma_{xx} \\ \sigma_{yy} \\ \sigma_{xy} \end{Bmatrix} = \begin{bmatrix} \frac{1-v_{yz}v_{zy}}{E_y E_z \Delta} & \frac{v_{yx}+v_{zx}v_{yz}}{E_y E_z \Delta} & 0 \\ \frac{v_{xy}+v_{xz}v_{zy}}{E_z E_x \Delta} & \frac{1-v_{zx}v_{xz}}{E_z E_x \Delta} & 0 \\ 0 & 0 & G_{xy} \end{bmatrix} \begin{Bmatrix} \epsilon_{xx} \\ \epsilon_{yy} \\ \epsilon_{xy} \end{Bmatrix}. \quad (7.37)$$

Plane Stress in Orthotropic Materials

Different from the plane strain case, as discussed for isotropic materials, all the stresses related to the z-direction are zero, $\sigma_{xz} = \sigma_{yz} = \sigma_{zz} = 0$. However, ϵ_{zz} is not zero. To obtain the plane stress constitutive relation, we start with

$$\begin{Bmatrix} \sigma_{xx} \\ \sigma_{yy} \\ \sigma_{zz} \\ \sigma_{xy} \end{Bmatrix} = \begin{bmatrix} C_{11} & C_{12} & C_{13} & C_{14} \\ C_{21} & C_{22} & C_{23} & C_{24} \\ C_{31} & C_{32} & C_{33} & C_{34} \\ C_{41} & C_{42} & C_{43} & C_{44} \end{bmatrix} \begin{Bmatrix} \epsilon_{xx} \\ \epsilon_{yy} \\ \epsilon_{zz} \\ \epsilon_{xy} \end{Bmatrix}. \quad (7.38)$$

Since
$$\sigma_{zz} = C_{31}\epsilon_{xx} + C_{32}\epsilon_{yy} + C_{33}\epsilon_{zz} + C_{34}\epsilon_{xy} = 0 \qquad (7.39)$$
we obtain
$$\epsilon_{zz} = -\frac{C_{31}\epsilon_{xx} + C_{32}\epsilon_{yy} + C_{34}\epsilon_{xy}}{C_{33}}. \qquad (7.40)$$
Substituting Eq. (7.40) into Eq. (7.38), we obtain
$$\begin{Bmatrix} \sigma_{xx} \\ \sigma_{yy} \\ \sigma_{xy} \end{Bmatrix} = \begin{bmatrix} \overline{C}_{11} & \overline{C}_{12} & \overline{C}_{14} \\ \overline{C}_{21} & \overline{C}_{22} & \overline{C}_{24} \\ \overline{C}_{41} & \overline{C}_{42} & \overline{C}_{44} \end{bmatrix} \begin{Bmatrix} \epsilon_{xx} \\ \epsilon_{yy} \\ \epsilon_{xy} \end{Bmatrix} \qquad (7.41)$$
where
$$\overline{C}_{ij} = C_{ij} - \frac{C_{i3}C_{3j}}{C_{33}}. \qquad (7.42)$$

7.2.2 2-D Elasticity: Thin Plate Subject to External Loads

Step 1: Physical Problem
The FEA procedure for the static FEA analysis of 2-D elasticity problems is illustrated in this section by using an example, as shown in Fig. 7.11. In this example problem, a thin plate is subjected to a linearly varying pressure on the top and a point force on the right edge. The left edge of the structure is attached to the wall by a pin connector and a set of rollers as shown in the figure. Young's modulus and Poisson's ratio of the material are 10 MPa and 0.3, respectively. The structural dimensions are shown in the figure. The goal is to find the equilibrium deformation of the structure due to the applied load. Such elasticity problems under the condition of equilibrium are also referred to as elastostatic problems.

Step 2: Mathematical Model
Since the plate is thin, as discussed in Section 7.2.1, this problem can be regarded as a plane stress problem, i.e. $\sigma_{xz} = \sigma_{yz} = \sigma_{zz} = 0$. The 3-D governing partial differential equations of the linear elastostatics is given by
$$\frac{\partial \sigma_{xx}}{\partial x} + \frac{\partial \sigma_{yx}}{\partial y} + \frac{\partial \sigma_{zx}}{\partial z} + b_x = 0 \qquad (7.43)$$
$$\frac{\partial \sigma_{xy}}{\partial x} + \frac{\partial \sigma_{yy}}{\partial y} + \frac{\partial \sigma_{zy}}{\partial z} + b_y = 0 \qquad (7.44)$$
$$\frac{\partial \sigma_{xz}}{\partial x} + \frac{\partial \sigma_{yz}}{\partial y} + \frac{\partial \sigma_{zz}}{\partial z} + b_z = 0. \qquad (7.45)$$
Since $\sigma_{xz} = \sigma_{yz} = \sigma_{zz} = 0$, and due to the symmetric property of the Cauchy stress,
$$\sigma_{xz} = \sigma_{zx} = \sigma_{yz} = \sigma_{zy} = \sigma_{zz} = 0 \qquad (7.46)$$

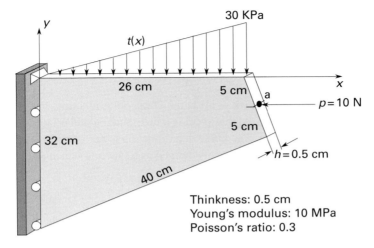

Figure 7.11 A 2-D elasticity problem.

and

$$\sigma_{yx} = \sigma_{xy} \tag{7.47}$$

the governing partial differential equations are reduced to

$$\frac{\partial \sigma_{xx}}{\partial x} + \frac{\partial \sigma_{xy}}{\partial y} + b_x = 0 \tag{7.48}$$

$$\frac{\partial \sigma_{xy}}{\partial x} + \frac{\partial \sigma_{yy}}{\partial y} + b_y = 0. \tag{7.49}$$

The boundary conditions of the problem can be obtained as

$$\text{Displacement}: \quad u = 0 \quad v = 0 \text{ at } x = 0 \text{ and } y = 0 \tag{7.50}$$

$$u = 0 \text{ at } x = 0. \tag{7.51}$$

$$\text{Surface traction}: \quad \mathbf{t} = \left\{ \begin{array}{c} 0 \\ -\dfrac{30}{0.26}x \end{array} \right\} \text{ (KPa) at } y = 0. \tag{7.52}$$

$$\text{Point force}: \quad \mathbf{p} = \left\{ \begin{array}{c} -10 \\ 0 \end{array} \right\} \text{ (N) at point } \mathbf{a}. \tag{7.53}$$

Point Force

In mechanical analysis, in addition to the body force and surface traction, another type of externally applied force is a point force. An example is shown in Fig. 7.12. Physically, the point force p can be viewed as a special case of distributed load (i.e., surface traction), which is distributed over a very small surface area. Mathematically, this concentrated distributed load at **a** can be represented by a Dirac delta function as

$$\mathbf{p}'(\mathbf{x}) = \mathbf{p}\Delta(|\mathbf{x} - \mathbf{a}|) \tag{7.54}$$

where **p** is the point force, $\Delta(|\mathbf{x} - \mathbf{a}|)$ is the Dirac delta function, and $|\mathbf{x} - \mathbf{a}|$ is the distance between **x** and **a**. Note that $\mathbf{p}'(\mathbf{x})$ is a distributed load/surface traction whose unit is Newton per unit area. The Dirac delta function is not a function in the conventional sense. However, it can be loosely defined as a function which goes to infinity at $x = 0$ and is zero everywhere else, i.e.,

$$\Delta(x) = \begin{cases} \infty & x = 0 \\ 0 & x \neq 0 \end{cases} \tag{7.55}$$

and it is also constrained by

$$\int_{-\infty}^{\infty} \Delta(x)dx = 1 \tag{7.56}$$

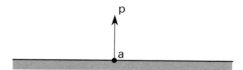

Figure 7.12 A point force.

An important property of the Dirac delta function is

$$\int_{-\infty}^{\infty} g(x)\Delta(x - a)dx = g(a). \tag{7.57}$$

The integral assigns to each continuous function $g(x)$ the value of the function at the point $x = 0$. With the mathematical representation given in Eq. (7.54), the point force can be treated as a surface traction in the derivation of the weak form. For example, the point force **p** applied at a point **a** on the surface of the plate in Fig. 7.11 can be written as

$$\mathbf{p}(\mathbf{a}) = \int_{S_a} \mathbf{p}\Delta(|\mathbf{x} - \mathbf{a}|)dS \tag{7.58}$$

where S_a is an area enclosing point **a**. Note that the integration is over a piece of surface area on the boundary of a 3-D volume.

Step 3: Weak Form

In this section, we derive the weak form by using three methods: the Galerkin weighted residual method, principle of virtual work, and the variational method.

7.2 2-D Elasticity

Galerkin Weighted Residual Method

As discussed in Chapter 4, there are several ways to obtain the weak form for linear elasticity problems. In this section, we first employ the Galerkin weighted residual method to derive the weak form from the strong form given in Eqs. (7.48–7.53).

We define the residuals of the two differential equations in the strong form as

$$R_x = \frac{\partial \sigma_{xx}}{\partial x} + \frac{\partial \sigma_{xy}}{\partial y} + b_x \qquad (7.59)$$

$$R_y = \frac{\partial \sigma_{xy}}{\partial x} + \frac{\partial \sigma_{yy}}{\partial y} + b_y. \qquad (7.60)$$

Let the weight functions for the two residuals be δu and δv, respectively, i.e., the variations of the u and v displacements. Note that, although other weight functions can be used, δu and δv are chosen to be the weight function for two reasons: (1) the choice is convenient and has apparent physical meaning, and (2) it makes more sense to approximate the displacements and their variations using the same shape functions. When the same shape functions are used for the displacements and their variations, the weighted residual method becomes the Galerkin weighted residual method. With the residuals defined, the weighted residual method requires that the products of the weight functions and the residuals are integrated over the domain. As we have shown that the elasticity problem shown in Fig. 7.11 can be treated as a 2-D problem, the residuals are first integrated over the 3-D volume, i.e.,

$$\int_V \delta u R_x dV = 0 \qquad (7.61)$$

$$\int_V \delta v R_y dV = 0. \qquad (7.62)$$

Note that, we will use Ω to denote the 2-D domain; here V is used to denote the 3-D volume. Substituting the residuals, we obtain

$$\int_V \delta u \left(\frac{\partial \sigma_{xx}}{\partial x} + \frac{\partial \sigma_{xy}}{\partial y} + b_x \right) dV = 0 \qquad (7.63)$$

$$\int_V \delta v \left(\frac{\partial \sigma_{xy}}{\partial x} + \frac{\partial \sigma_{yy}}{\partial y} + b_y \right) dV = 0. \qquad (7.64)$$

Equations (7.63, 7.64) can be written as

$$\int_V \nabla \cdot \left\{ \begin{array}{c} \sigma_{xx} \\ \sigma_{xy} \end{array} \right\} \delta u dV + \int_V \delta u b_x dV = 0 \qquad (7.65)$$

$$\int_V \nabla \cdot \left\{ \begin{array}{c} \sigma_{xy} \\ \sigma_{yy} \end{array} \right\} \delta v dV + \int_V \delta v b_y dV = 0. \qquad (7.66)$$

By using the relations

$$\nabla \cdot \left(\begin{Bmatrix} \sigma_{xx} \\ \sigma_{xy} \end{Bmatrix} \delta u \right) = \nabla \cdot \begin{Bmatrix} \sigma_{xx} \\ \sigma_{xy} \end{Bmatrix} \delta u + \nabla \delta u \cdot \begin{Bmatrix} \sigma_{xx} \\ \sigma_{xy} \end{Bmatrix} \quad (7.67)$$

$$\nabla \cdot \left(\begin{Bmatrix} \sigma_{xy} \\ \sigma_{yy} \end{Bmatrix} \delta v \right) = \nabla \cdot \begin{Bmatrix} \sigma_{xy} \\ \sigma_{yy} \end{Bmatrix} \delta v + \nabla \delta v \cdot \begin{Bmatrix} \sigma_{xy} \\ \sigma_{yy} \end{Bmatrix}. \quad (7.68)$$

Equations (7.65, 7.66) can be written as

$$\int_V \nabla \cdot \left(\begin{Bmatrix} \sigma_{xx} \\ \sigma_{xy} \end{Bmatrix} \delta u \right) dV - \int_V \nabla \delta u \cdot \begin{Bmatrix} \sigma_{xx} \\ \sigma_{xy} \end{Bmatrix} dV + \int_V \delta u b_x dV = 0 \quad (7.69)$$

$$\int_V \nabla \cdot \left(\begin{Bmatrix} \sigma_{xy} \\ \sigma_{yy} \end{Bmatrix} \delta v \right) dV - \int_V \nabla \delta v \cdot \begin{Bmatrix} \sigma_{xy} \\ \sigma_{yy} \end{Bmatrix} dV + \int_V \delta v b_y dV = 0. \quad (7.70)$$

Applying the divergence theorem to the first terms in Eqs. (7.69, 7.70), we obtain

$$\int_S \begin{Bmatrix} \sigma_{xx} \\ \sigma_{xy} \end{Bmatrix} \delta u \cdot \begin{Bmatrix} n_x \\ n_y \end{Bmatrix} dS - \int_V \nabla \delta u \cdot \begin{Bmatrix} \sigma_{xx} \\ \sigma_{xy} \end{Bmatrix} dV + \int_V \delta u b_x dV = 0 \quad (7.71)$$

$$\int_S \begin{Bmatrix} \sigma_{xy} \\ \sigma_{yy} \end{Bmatrix} \delta v \cdot \begin{Bmatrix} n_x \\ n_y \end{Bmatrix} dS - \int_V \nabla \delta v \cdot \begin{Bmatrix} \sigma_{xy} \\ \sigma_{yy} \end{Bmatrix} dV + \int_V \delta v b_y dV = 0 \quad (7.72)$$

where S is the surface area of the volume, n_x and n_y are the x and y components of the outward unit normal vector $\mathbf{n}$. Since the surface traction $\mathbf{t}$ can be expressed as

$$\mathbf{t} = \begin{Bmatrix} t_x \\ t_y \end{Bmatrix} = \begin{bmatrix} \sigma_{xx} & \sigma_{xy} \\ \sigma_{xy} & \sigma_{yy} \end{bmatrix} \begin{Bmatrix} n_x \\ n_y \end{Bmatrix}. \quad (7.73)$$

Equations (7.71, 7.72) can be further written as

$$\int_S \delta u t_x dS - \int_V \nabla \delta u \cdot \begin{Bmatrix} \sigma_{xx} \\ \sigma_{xy} \end{Bmatrix} dV + \int_V \delta u b_x dV = 0 \quad (7.74)$$

$$\int_S \delta v t_y dS - \int_V \nabla \delta v \cdot \begin{Bmatrix} \sigma_{xy} \\ \sigma_{yy} \end{Bmatrix} dV + \int_V \delta v b_y dV = 0. \quad (7.75)$$

Note that S denotes the boundary of the volume. However, as shown in Fig. 7.11, the plate is subjected to boundary forces only on the top, left, and right edges. Recall that the point force $\mathbf{p}$ on the right edge is treated as a special surface traction represented by the Dirac delta function. In addition, on the left edge with roller support, for a given direction (x- or y-direction) either the surface traction component or the variation of the displacement component is zero. Therefore, the first integrals in Eqs. (7.74, 7.75) are nonzero only on the top and right edges, where the surface traction and the point force are explicitly applied. Denoting the top edge as S_t and the right edge as S_p, Eqs. (7.74, 7.75) can be rewritten as

$$\int_{S_t} \delta u t_x dS + \int_{S_p} \delta u p_x \Delta(|\mathbf{x} - \mathbf{a}|) dS - \int_V \nabla \delta u \cdot \begin{Bmatrix} \sigma_{xx} \\ \sigma_{xy} \end{Bmatrix} dV + \int_V \delta u b_x dV = 0 \tag{7.76}$$

$$\int_{S_t} \delta v t_y dS + \int_{S_p} \delta v p_y \Delta(|\mathbf{x} - \mathbf{a}|) dS - \int_V \nabla \delta v \cdot \begin{Bmatrix} \sigma_{xy} \\ \sigma_{yy} \end{Bmatrix} dV + \int_V \delta v b_y dV = 0. \tag{7.77}$$

By using Eq. (7.58), Eqs. (7.76, 7.77) can be rewritten as

$$\int_{S_t} \delta u t_x dS + \delta u(\mathbf{a}) p_x - \int_V \nabla \delta u \cdot \begin{Bmatrix} \sigma_{xx} \\ \sigma_{xy} \end{Bmatrix} dV + \int_V \delta u b_x dV = 0 \tag{7.78}$$

$$\int_{S_t} \delta v t_y dS + \delta v(\mathbf{a}) p_y - \int_V \nabla \delta v \cdot \begin{Bmatrix} \sigma_{xy} \\ \sigma_{yy} \end{Bmatrix} dV + \int_V \delta v b_y dV = 0. \tag{7.79}$$

The summation of Eqs. (7.78, 7.79) gives

$$\int_{S_t} \left(\delta u t_x + \delta v t_y \right) dS + [\delta u(\mathbf{a}) \ \delta v(\mathbf{a})] \begin{Bmatrix} p_x \\ p_y \end{Bmatrix}$$
$$- \int_V \nabla \delta u \cdot \begin{Bmatrix} \sigma_{xx} \\ \sigma_{xy} \end{Bmatrix} dV - \int_V \nabla \delta v \cdot \begin{Bmatrix} \sigma_{xy} \\ \sigma_{yy} \end{Bmatrix} dV + \int_V \left(\delta u b_x + \delta v b_y \right) dV = 0. \tag{7.80}$$

Equation (7.80) can be rewritten as

$$\int_{S_t} [\delta u \ \delta v] \begin{Bmatrix} t_x \\ t_y \end{Bmatrix} dS + [\delta u(\mathbf{a}) \ \delta v(\mathbf{a})] \begin{Bmatrix} p_x \\ p_y \end{Bmatrix}$$
$$- \int_V \left(\nabla \delta u \cdot \begin{Bmatrix} \sigma_{xx} \\ \sigma_{xy} \end{Bmatrix} + \nabla \delta v \cdot \begin{Bmatrix} \sigma_{xy} \\ \sigma_{yy} \end{Bmatrix} \right) dV + \int_V [\delta u \ \delta v] \begin{Bmatrix} b_x \\ b_y \end{Bmatrix} dV = 0. \tag{7.81}$$

For the plane stress problem shown in Fig. 7.11, as the displacement, stress, and surface traction remain the same in the direction perpendicular to the plate, the domain can be considered as a 2-D domain by taking $dV = h d\Omega$ and $dS = h d\Gamma$, where h is the thickness of the plate, Ω is the 2-D domain (i.e., the surface of the plate in the xy-plane) and Γ is the boundary of the 2-D domain. Therefore, Eq. (7.81) can be rewritten in 2-D form as

$$h \int_{\Gamma_t} [\delta u \ \delta v] \begin{Bmatrix} t_x \\ t_y \end{Bmatrix} d\Gamma + [\delta u(\mathbf{a}) \ \delta v(\mathbf{a})] \begin{Bmatrix} p_x \\ p_y \end{Bmatrix}$$

$$- h \int_{\Omega} \left(\nabla \delta u \cdot \begin{Bmatrix} \sigma_{xx} \\ \sigma_{xy} \end{Bmatrix} + \nabla \delta v \cdot \begin{Bmatrix} \sigma_{xy} \\ \sigma_{yy} \end{Bmatrix} \right) d\Omega + h \int_{\Omega} [\delta u \ \delta v] \begin{Bmatrix} b_x \\ b_y \end{Bmatrix} d\Omega = 0.$$

(7.82)

Dividing all the terms by h, Eq. (7.82) can be rewritten as

$$\int_{\Gamma_t} [\delta u \ \delta v] \begin{Bmatrix} t_x \\ t_y \end{Bmatrix} d\Gamma + [\delta u(\mathbf{a}) \ \delta v(\mathbf{a})] \begin{Bmatrix} p_x/h \\ p_y/h \end{Bmatrix}$$

$$- \int_{\Omega} \left(\nabla \delta u \cdot \begin{Bmatrix} \sigma_{xx} \\ \sigma_{xy} \end{Bmatrix} + \nabla \delta v \cdot \begin{Bmatrix} \sigma_{xy} \\ \sigma_{yy} \end{Bmatrix} \right) d\Omega + \int_{\Omega} [\delta u \ \delta v] \begin{Bmatrix} b_x \\ b_y \end{Bmatrix} d\Omega = 0.$$

(7.83)

Next we combine the two terms in the second integral of Eq. (7.83) by expanding the vectors as follows.

$$\nabla \delta u \cdot \begin{Bmatrix} \sigma_{xx} \\ \sigma_{xy} \end{Bmatrix} + \nabla \delta v \cdot \begin{Bmatrix} \sigma_{xy} \\ \sigma_{yy} \end{Bmatrix}$$

$$= \begin{bmatrix} \dfrac{\partial \delta u}{\partial x} & \dfrac{\partial \delta u}{\partial y} \end{bmatrix} \begin{Bmatrix} \sigma_{xx} \\ \sigma_{xy} \end{Bmatrix} + \begin{bmatrix} \dfrac{\partial \delta v}{\partial x} & \dfrac{\partial \delta v}{\partial y} \end{bmatrix} \begin{Bmatrix} \sigma_{xy} \\ \sigma_{yy} \end{Bmatrix}$$

$$= \begin{bmatrix} \dfrac{\partial \delta u}{\partial x} & 0 & \dfrac{\partial \delta u}{\partial y} \end{bmatrix} \begin{Bmatrix} \sigma_{xx} \\ \sigma_{yy} \\ \sigma_{xy} \end{Bmatrix} + \begin{bmatrix} 0 & \dfrac{\partial \delta v}{\partial y} & \dfrac{\partial \delta v}{\partial x} \end{bmatrix} \begin{Bmatrix} \sigma_{xx} \\ \sigma_{yy} \\ \sigma_{xy} \end{Bmatrix}$$

$$= \begin{bmatrix} \dfrac{\partial \delta u}{\partial x} & \dfrac{\partial \delta v}{\partial y} & \dfrac{\partial \delta u}{\partial y} + \dfrac{\partial \delta v}{\partial x} \end{bmatrix} \begin{Bmatrix} \sigma_{xx} \\ \sigma_{yy} \\ \sigma_{xy} \end{Bmatrix}.$$

(7.84)

Therefore, Eq. (7.83) can be rewritten as

$$\int_{\Gamma_t} [\delta u \ \delta v] \begin{Bmatrix} t_x \\ t_y \end{Bmatrix} d\Gamma + [\delta u(\mathbf{a}) \ \delta v(\mathbf{a})] \begin{Bmatrix} p_x/h \\ p_y/h \end{Bmatrix}$$

$$- \int_{\Omega} \begin{bmatrix} \dfrac{\partial \delta u}{\partial x} & \dfrac{\partial \delta v}{\partial y} & \dfrac{\partial \delta u}{\partial y} + \dfrac{\partial \delta v}{\partial x} \end{bmatrix} \begin{Bmatrix} \sigma_{xx} \\ \sigma_{yy} \\ \sigma_{xy} \end{Bmatrix} d\Omega + \int_{\Omega} [\delta u \ \delta v] \begin{Bmatrix} b_x \\ b_y \end{Bmatrix} d\Omega = 0.$$

(7.85)

7.2 2-D Elasticity

Equation (7.85) is the weak form for the strong form governing equations given in Eqs. (7.48–7.53). Note that the partial derivatives of the displacement variations can be expressed as the strain variations, i.e.,

$$\frac{\partial \delta u}{\partial x} = \delta \frac{\partial u}{\partial x} = \delta \epsilon_{xx} \tag{7.86}$$

$$\frac{\partial \delta v}{\partial y} = \delta \frac{\partial v}{\partial y} = \delta \epsilon_{yy} \tag{7.87}$$

$$\frac{\partial \delta u}{\partial y} + \frac{\partial \delta v}{\partial x} = \delta \left(\frac{\partial u}{\partial y} + \frac{\partial v}{\partial x} \right) = \delta \epsilon_{xy}. \tag{7.88}$$

In addition, from Hooke's law,

$$\begin{Bmatrix} \sigma_{xx} \\ \sigma_{yy} \\ \sigma_{xy} \end{Bmatrix} = \mathbf{C} \begin{Bmatrix} \epsilon_{xx} \\ \epsilon_{yy} \\ \epsilon_{xy} \end{Bmatrix} \tag{7.89}$$

where $\mathbf{C}$ is the material matrix given by

$$\mathbf{C} = \frac{E}{(1+\nu)(1-2\nu)} \begin{bmatrix} (1-\nu) & \nu & 0 \\ \nu & (1-\nu) & 0 \\ 0 & 0 & \frac{1-2\nu}{2} \end{bmatrix} \quad \text{for plane strain} \tag{7.90}$$

and

$$\mathbf{C} = \frac{E}{(1-\nu^2)} \begin{bmatrix} 1 & \nu & 0 \\ \nu & 1 & 0 \\ 0 & 0 & \frac{1-\nu}{2} \end{bmatrix} \quad \text{for plane stress.} \tag{7.91}$$

By using these relations, Eq. (7.85) can be further written as

$$\int_{\Gamma_t} [\delta u \ \delta v] \begin{Bmatrix} t_x \\ t_y \end{Bmatrix} d\Gamma + [\delta u(\mathbf{a}) \ \delta v(\mathbf{a})] \begin{Bmatrix} p_x/h \\ p_y/h \end{Bmatrix}$$

$$- \int_\Omega [\delta \epsilon_{xx} \ \delta \epsilon_{yy} \ \delta \epsilon_{xy}] \mathbf{C} \begin{Bmatrix} \epsilon_{xx} \\ \epsilon_{yy} \\ \epsilon_{xy} \end{Bmatrix} d\Omega + \int_\Omega [\delta u \ \delta v] \begin{Bmatrix} b_x \\ b_y \end{Bmatrix} d\Omega = 0 \tag{7.92}$$

or

$$\int_{\Gamma_t} \begin{Bmatrix} \delta u \\ \delta v \end{Bmatrix} \cdot \begin{Bmatrix} t_x \\ t_y \end{Bmatrix} d\Gamma + \begin{Bmatrix} \delta u(\mathbf{a}) \\ \delta v(\mathbf{a}) \end{Bmatrix} \cdot \begin{Bmatrix} p_x/h \\ p_y/h \end{Bmatrix}$$

$$- \int_\Omega \begin{Bmatrix} \delta \epsilon_{xx} \\ \delta \epsilon_{yy} \\ \delta \epsilon_{xy} \end{Bmatrix} \cdot \mathbf{C} \begin{Bmatrix} \epsilon_{xx} \\ \epsilon_{yy} \\ \epsilon_{xy} \end{Bmatrix} d\Omega + \int_\Omega \begin{Bmatrix} \delta u \\ \delta v \end{Bmatrix} \cdot \begin{Bmatrix} b_x \\ b_y \end{Bmatrix} d\Omega = 0. \tag{7.93}$$

In short form, we have

$$\int_{\Gamma_t} \delta\mathbf{u} \cdot \mathbf{t} d\Gamma + \frac{\delta\mathbf{u}(\mathbf{a}) \cdot \mathbf{p}}{h} - \int_{\Omega} \delta\boldsymbol{\epsilon} \cdot \mathbf{C}\boldsymbol{\epsilon} d\Omega + \int_{\Omega} \delta\mathbf{u} \cdot \mathbf{b} d\Omega = 0 \qquad (7.94)$$

where

$$\delta\mathbf{u} = \left\{ \begin{array}{c} \delta u \\ \delta v \end{array} \right\} \quad \mathbf{t} = \left\{ \begin{array}{c} t_x \\ t_y \end{array} \right\} \quad \mathbf{p} = \left\{ \begin{array}{c} p_x \\ p_y \end{array} \right\} \quad \mathbf{b} = \left\{ \begin{array}{c} b_x \\ b_y \end{array} \right\} \qquad (7.95)$$

$$\delta\boldsymbol{\epsilon} = \left\{ \begin{array}{c} \delta\epsilon_{xx} \\ \delta\epsilon_{yy} \\ \delta\epsilon_{xy} \end{array} \right\} \quad \boldsymbol{\epsilon} = \left\{ \begin{array}{c} \epsilon_{xx} \\ \epsilon_{yy} \\ \epsilon_{xy} \end{array} \right\}. \qquad (7.96)$$

Principle of Virtual Work

As discussed in Chapter 4, one can use another approach to obtain the weak form referred to as the principle of virtual work. The principle of virtual work states that

$$\text{total internal virtual work } W_I = \text{total external virtual work } W_E. \qquad (7.97)$$

The total internal virtual work is the work done by the internal stresses over a virtual strain, i.e.,

$$W_I = h \int_{\Omega} \delta\boldsymbol{\epsilon} \cdot \boldsymbol{\sigma} d\Omega = h \int_{\Omega} \delta\boldsymbol{\epsilon} \cdot \mathbf{C}\boldsymbol{\epsilon} d\Omega. \qquad (7.98)$$

The total external virtual work is the work done by various externally applied forces. In this example, the externally applied forces include the body force, the surface traction, and the point force. The external virtual work done by the body force over a virtual displacement is

$$W_E(\mathbf{b}) = h \int_{\Omega} \delta\mathbf{u} \cdot \mathbf{b} d\Omega \qquad (7.99)$$

the external virtual work done by the surface traction is

$$W_E(\mathbf{t}) = h \int_{\Gamma_t} \delta\mathbf{u} \cdot \mathbf{t} d\Gamma \qquad (7.100)$$

and the external virtual work done by the point force is

$$W_E(\mathbf{p}) = \delta\mathbf{u}(\mathbf{a}) \cdot \mathbf{p}. \qquad (7.101)$$

Therefore, from the principle of virtual work

$$W_I = W_E(\mathbf{b}) + W_E(\mathbf{t}) + W_E(\mathbf{p})$$
$$\Rightarrow h \int_{\Gamma_t} \delta\mathbf{u} \cdot \mathbf{t} d\Gamma + \delta\mathbf{u}(\mathbf{a}) \cdot \mathbf{p} + h \int_{\Omega} \delta\mathbf{u} \cdot \mathbf{b} d\Omega - h \int_{\Omega} \delta\boldsymbol{\epsilon} \cdot \mathbf{C}\boldsymbol{\epsilon} d\Omega = 0.$$

$$(7.102)$$

Equation (7.102) is the weak form of the problem, which is the same as Eq. (7.94).

Variational Method
Another approach to obtain the weak form of the elastostatic problem is the variational approach. For elasticity problems, the functional is the total potential energy of the system given by

$$\Pi = \frac{1}{2}h\int_\Omega \epsilon \cdot \sigma d\Omega - h\int_\Omega \mathbf{u}\cdot\mathbf{b}d\Omega - h\int_{\Gamma_t}\mathbf{u}\cdot\mathbf{t}d\Gamma - \mathbf{u}\cdot\mathbf{p} \quad (7.103)$$

where the first integral on the right hand side is the strain energy of the structure, the second integral is the potential energy lost due to the body force, the third integral is the potential energy lost due to the surface traction, and the last integral is the potential energy lost due to the point force. In the variational calculus, an optimal solution is found when the first variation of the functional is zero, i.e.,

$$\delta\Pi = 0. \quad (7.104)$$

Substituting Eq. (7.103) into Eq. (7.104), we obtain

$$\delta\left[\frac{1}{2}h\int_\Omega \epsilon\cdot\sigma d\Omega - h\int_\Omega \mathbf{u}\cdot\mathbf{b}d\Omega - h\int_{\Gamma_t}\mathbf{u}\cdot\mathbf{t}d\Gamma - \mathbf{u}\cdot\mathbf{p}\right] = 0$$

$$\Rightarrow \frac{1}{2}h\delta\int_\Omega \epsilon\cdot\mathbf{C}\epsilon d\Omega - h\delta\int_\Omega \mathbf{u}\cdot\mathbf{b}d\Omega - h\delta\int_{\Gamma_t}\mathbf{u}\cdot\mathbf{t}d\Gamma - \delta\mathbf{u}\cdot\mathbf{p} = 0$$

$$\Rightarrow \frac{1}{2}h\int_\Omega 2\delta\epsilon\cdot\mathbf{C}\epsilon d\Omega - h\int_\Omega \delta\mathbf{u}\cdot\mathbf{b}d\Omega - h\int_{\Gamma_t}\delta\mathbf{u}\cdot\mathbf{t}d\Gamma - \delta\mathbf{u}\cdot\mathbf{p} = 0$$

$$\Rightarrow h\int_\Omega \delta\epsilon\cdot\mathbf{C}\epsilon d\Omega - h\int_\Omega \delta\mathbf{u}\cdot\mathbf{b}d\Omega - h\int_{\Gamma_t}\delta\mathbf{u}\cdot\mathbf{t}d\Gamma - \delta\mathbf{u}\cdot\mathbf{p} = 0. \quad (7.105)$$

Equation (7.105) is the weak form of the problem. Note that all the approaches lead to the same weak form of the 2-D elastostatic problem.

Step 4: Discretization

In this step, the 2-D plate domain is discretized into elements. The meshing process is the same as for the 2-D scalar problems. For illustration purposes, Fig. 7.13 shows a sample mesh containing four coarse 4-node quadrilateral elements and a total of nine nodes.

For the mesh shown in Fig. 7.13, node and element files can be generated as shown below the figure.

The first column of "nodes.dat" is the list of nodal index, the second and third columns are the x- and y-coordinates of the nodes, respectively. In "elements.dat", the first column is the list of element index, and the following four columns are the four nodes of the quadrilateral elements. Note that the node numbers are counted counterclockwise for each element, which is consistent with the node arrangement of the 4-node square master element. However, which node is the starting node is not important. For example, element 1 can also be listed, equivalently, as "1 4 5 2 1."

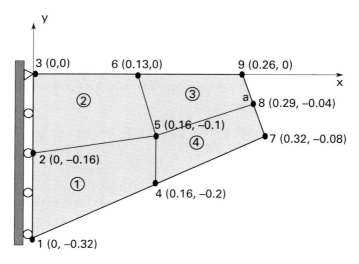

Figure 7.13 Discretization of the plate (meshing). The dimension unit is meter.

nodes.dat

1	0.0	−0.32
2	0.0	−0.16
3	0.0	0.0
4	0.16	−0.20
5	0.16	−0.10
6	0.13	0.0
7	0.32	−0.08
8	0.29	−0.04
9	0.26	0.0

elements.dat

1	1	4	5	2
2	5	6	3	2
3	5	8	9	6
4	4	7	8	5

In addition to the nodes and element files, we can create several other files to store the material properties, essential boundary conditions, and applied surface tractions and forces. For example,

materials.dat

$1e+7$	← Young's modulus (Pa)
0.3	← Poisson's ratio
0.005	← thickness of the plate (m)

bcsdisp.dat

1	1	0.0
2	1	0.0
3	1	0.0
3	2	0.0.

The material properties are listed in "materials.dat" as shown above. In "bcsdisp.dat," displacement boundary conditions are listed. The first column is the index of the nodes on which the displacement boundary conditions are specified. The second column is the direction of the displacement; "1" represents x-direction and "2" is for y-direction. The third column is the magnitude of the displacement.

The surface traction and force applied are stored separately in two files: "bcstraction.dat" and "bcsforce.dat,"

bcstraction.dat	bcsforce.dat
2 2 6 3 0.0 −15000.0 0.0 0.0	8 −10.0 0.0.
3 3 9 6 0.0 −30000.0 0.0 −15000.0	

In "bcstraction.dat," the columns are: element index, edge number, starting node of the edge, ending node of the edge, x-component of the traction on the start node, y-component of the traction on the start node, x-component of the traction on the end node, and y-component of the traction on the end node. For example, the first line states: element 2, on the second edge whose starting node is node 6 and end node is node 3, the x- and y-components of the traction on node 6 are 0 Pa and -15000 Pa, respectively, and the x- and y-components of the traction on node 3 are 0 Pa and 0 Pa, respectively. In "bcsforce.dat" the columns are: node index, x-component of the force on the node, and y-component of the force on the node.

Step 5: Approximation
After the meshing process, the unknown physical quantities are approximated by using the finite element shape functions. As the elements have irregular shapes, an isoparametric mapping is necessary. For the 4-node quadrilateral elements shown in Fig. 7.13, the finite element approximation for the displacements and their derivatives are then given by

$$u = \sum_{i=1}^{4} N_i(\xi, \eta) u_i \qquad v = \sum_{i=1}^{4} N_i(\xi, \eta) v_i \qquad (7.106)$$

$$\frac{\partial u}{\partial x} = \sum_{i=1}^{4} \frac{\partial N_i(\xi, \eta)}{\partial x} u_i \qquad \frac{\partial v}{\partial x} = \sum_{i=1}^{4} \frac{\partial N_i(\xi, \eta)}{\partial x} v_i \qquad (7.107)$$

$$\frac{\partial u}{\partial y} = \sum_{i=1}^{4} \frac{\partial N_i(\xi, \eta)}{\partial y} u_i \qquad \frac{\partial v}{\partial y} = \sum_{i=1}^{4} \frac{\partial N_i(\xi, \eta)}{\partial y} v_i. \qquad (7.108)$$

The same shape functions are used for the displacement variations (or virtual displacements).

Step 6: Element Matrices and Vectors
The element matrices and vectors can be obtained by using the element weak form. The global weak form obtained in the previous sections is repeated here

$$\int_\Omega \delta\boldsymbol{\epsilon} \cdot \mathbf{C}\boldsymbol{\epsilon} d\Omega - \int_\Omega \delta\mathbf{u} \cdot \mathbf{b} d\Omega - \int_{\Gamma_t} \delta\mathbf{u} \cdot \mathbf{t} d\Gamma - \frac{\delta\mathbf{u}(\mathbf{a}) \cdot \mathbf{p}}{h} = 0. \qquad (7.109)$$

Note that the displacement variation in the point force term is for the point **a** only. This requires that the point **a** has to be a node since the final global linear system is in terms of nodal displacements and nodal forces. Unlike the virtual displacement vector $\delta\mathbf{u}$ in the integrals of the weak form, the virtual displacement $\delta\mathbf{u}(\mathbf{a})$ is directly corresponding to the global node **a**. Therefore, the point force term is indeed global, i.e., it does not belong to an individual element and will be directly assembled to the global force vector. Therefore, the force term is omitted when the global weak form is rewritten for individual elements as

$$\int_\Omega^{(e)} \delta\boldsymbol{\epsilon} \cdot \mathbf{C}\boldsymbol{\epsilon} d\Omega - \int_\Omega^{(e)} \delta\mathbf{u} \cdot \mathbf{b} d\Omega - \int_{\Gamma_t}^{(e)} \delta\mathbf{u} \cdot \mathbf{t} d\Gamma = 0. \qquad (7.110)$$

For the sake of convenient derivation, the element weak form is rewritten in vector form as

$$\int_\Omega^{(e)} \delta\boldsymbol{\epsilon}^T \mathbf{C}\boldsymbol{\epsilon} d\Omega - \int_\Omega^{(e)} \delta\mathbf{u}^T \mathbf{b} d\Omega - \int_{\Gamma_t}^{(e)} \delta\mathbf{u}^T \mathbf{t} d\Gamma = 0. \qquad (7.111)$$

The element matrices and vectors can then be obtained by substituting the finite element approximations of the physical quantities in the weak form. The finite element approximation of the strain vector can be written as

$$\boldsymbol{\epsilon} = \left\{ \begin{array}{c} \dfrac{\partial u}{\partial x} \\[6pt] \dfrac{\partial v}{\partial y} \\[6pt] \dfrac{\partial u}{\partial y} + \dfrac{\partial v}{\partial x} \end{array} \right\}$$

$$= \left\{ \begin{array}{c} \dfrac{\partial N_1}{\partial x}u_1 + \dfrac{\partial N_2}{\partial x}u_2 + \dfrac{\partial N_3}{\partial x}u_3 + \dfrac{\partial N_4}{\partial x}u_4 \\[6pt] \dfrac{\partial N_1}{\partial y}v_1 + \dfrac{\partial N_2}{\partial y}v_2 + \dfrac{\partial N_3}{\partial y}v_3 + \dfrac{\partial N_4}{\partial y}v_4 \\[6pt] \dfrac{\partial N_1}{\partial y}u_1 + \dfrac{\partial N_1}{\partial x}v_1 + \dfrac{\partial N_2}{\partial y}u_2 + \dfrac{\partial N_2}{\partial x}v_2 + \dfrac{\partial N_3}{\partial y}u_3 + \dfrac{\partial N_3}{\partial x}v_3 + \dfrac{\partial N_4}{\partial y}u_4 + \dfrac{\partial N_4}{\partial x}v_4 \end{array} \right\}$$

$$= \begin{bmatrix} \dfrac{\partial N_1}{\partial x} & 0 & \dfrac{\partial N_2}{\partial x} & 0 & \dfrac{\partial N_3}{\partial x} & 0 & \dfrac{\partial N_4}{\partial x} & 0 \\ 0 & \dfrac{\partial N_1}{\partial y} & 0 & \dfrac{\partial N_2}{\partial y} & 0 & \dfrac{\partial N_3}{\partial y} & 0 & \dfrac{\partial N_4}{\partial y} \\ \dfrac{\partial N_1}{\partial y} & \dfrac{\partial N_1}{\partial x} & \dfrac{\partial N_2}{\partial y} & \dfrac{\partial N_2}{\partial x} & \dfrac{\partial N_3}{\partial y} & \dfrac{\partial N_3}{\partial x} & \dfrac{\partial N_4}{\partial y} & \dfrac{\partial N_4}{\partial x} \end{bmatrix} \begin{Bmatrix} u_1 \\ v_1 \\ u_2 \\ v_2 \\ u_3 \\ v_3 \\ u_4 \\ v_4 \end{Bmatrix}$$

$$= \mathbf{Bd}. \tag{7.112}$$

where $\mathbf{B}$ is referred to as the strain–displacement matrix and $\mathbf{d}$ is the nodal displacement vector. In short form, the strain is given by

$$\boldsymbol{\epsilon}_{3\times 1} = \mathbf{B}_{3\times 8}\mathbf{d}_{8\times 1}. \tag{7.113}$$

Similarly,

$$\delta\boldsymbol{\epsilon}_{3\times 1} = \mathbf{B}_{3\times 8}\delta\mathbf{d}_{8\times 1} \tag{7.114}$$

or

$$\delta\boldsymbol{\epsilon}^T_{1\times 3} = \delta\mathbf{d}^T_{1\times 8}\mathbf{B}^T_{8\times 3} \tag{7.115}$$

where $\delta\mathbf{d}$ is the virtual nodal displacement vector (or variation of the nodal displacement vector). Substituting the strain approximations into the first integral (strain energy) in Eq. (7.111), we obtain

$$\int_{\Omega}^{(e)} \delta\boldsymbol{\epsilon}^T \mathbf{C}\boldsymbol{\epsilon}\, d\Omega = \int_{\Omega}^{(e)} \delta\mathbf{d}^T_{1\times 8}\mathbf{B}^T_{8\times 3}\mathbf{C}_{3\times 3}\mathbf{B}_{3\times 8}\mathbf{d}_{8\times 1}\, d\Omega$$

$$= \delta\mathbf{d}^T_{1\times 8}\left(\int_{\Omega}^{(e)} \mathbf{B}^T_{8\times 3}\mathbf{C}_{3\times 3}\mathbf{B}_{3\times 8}\, d\Omega\right) \mathbf{d}_{8\times 1}. \tag{7.116}$$

By using the isoparametric mapping, the integral over the element can be mapped to the master element with

$$\int_{\Omega}^{(e)} \rightarrow \int_{-1}^{1}\int_{-1}^{1} \quad \text{and} \quad d\Omega = det(\mathbf{J})d\xi\, d\eta. \tag{7.117}$$

The material matrix $\mathbf{C}$ is a constant matrix and remains unchanged in mapping. For the plate problem, the plane stress $\mathbf{C}$ is obtained as

$$\mathbf{C} = \begin{bmatrix} 1.098901 & 0.329670 & 0 \\ 0.329670 & 1.098901 & 0 \\ 0 & 0 & 0.384615 \end{bmatrix} 10^7 (Pa). \tag{7.118}$$

However, the strain–displacement matrix $\mathbf{B}$ contains the derivatives of the shape functions. As discussed in the heat transfer analysis, these x- and y-derivatives of the shape functions need to be converted to the ξ- and η-derivatives of the shape functions defined in the master element. Typically, the relation

$$\left\{ \begin{array}{c} \dfrac{\partial}{\partial x} \\ \dfrac{\partial}{\partial y} \end{array} \right\} = \mathbf{J}^{-1} \left\{ \begin{array}{c} \dfrac{\partial}{\partial \xi} \\ \dfrac{\partial}{\partial \eta} \end{array} \right\} \tag{7.119}$$

is used to perform the transformation. Two approaches can be used to compute the $\mathbf{B}$ matrix. In the first approach, the x- and y-derivatives of the shape functions are directly calculated by using Eq. (7.119), i.e.,

$$\frac{\partial N_i}{\partial x} = \mathbf{J}^{-1}(1,1) \frac{\partial N_i}{\partial \xi} + \mathbf{J}^{-1}(1,2) \frac{\partial N_i}{\partial \eta} \qquad i = 1,2,3,4 \tag{7.120}$$

$$\frac{\partial N_i}{\partial y} = \mathbf{J}^{-1}(2,1) \frac{\partial N_i}{\partial \xi} + \mathbf{J}^{-1}(2,2) \frac{\partial N_i}{\partial \eta} \qquad i = 1,2,3,4. \tag{7.121}$$

The calculated $\dfrac{\partial N_i}{\partial x}$ and $\dfrac{\partial N_i}{\partial y}$, $i = 1,2,3,4$, are then directly substituted into the $\mathbf{B}$ matrix.

In the second approach, we rewrite the $\mathbf{B}$ matrix as

$$\mathbf{B} = \begin{bmatrix} \dfrac{\partial N_1}{\partial x} & 0 & \dfrac{\partial N_2}{\partial x} & 0 & \dfrac{\partial N_3}{\partial x} & 0 & \dfrac{\partial N_4}{\partial x} & 0 \\ 0 & \dfrac{\partial N_1}{\partial y} & 0 & \dfrac{\partial N_2}{\partial y} & 0 & \dfrac{\partial N_3}{\partial y} & 0 & \dfrac{\partial N_4}{\partial y} \\ \dfrac{\partial N_1}{\partial y} & \dfrac{\partial N_1}{\partial x} & \dfrac{\partial N_2}{\partial y} & \dfrac{\partial N_2}{\partial x} & \dfrac{\partial N_3}{\partial y} & \dfrac{\partial N_3}{\partial x} & \dfrac{\partial N_4}{\partial y} & \dfrac{\partial N_4}{\partial x} \end{bmatrix}$$

7.2 2-D Elasticity

$$= \begin{bmatrix} 1 & 0 & 0 & 0 \\ 0 & 0 & 0 & 1 \\ 0 & 1 & 1 & 0 \end{bmatrix} \begin{bmatrix} \frac{\partial N_1}{\partial x} & 0 & \frac{\partial N_2}{\partial x} & 0 & \frac{\partial N_3}{\partial x} & 0 & \frac{\partial N_4}{\partial x} & 0 \\ \frac{\partial N_1}{\partial y} & 0 & \frac{\partial N_2}{\partial y} & 0 & \frac{\partial N_3}{\partial y} & 0 & \frac{\partial N_4}{\partial y} & 0 \\ 0 & \frac{\partial N_1}{\partial x} & 0 & \frac{\partial N_2}{\partial x} & 0 & \frac{\partial N_3}{\partial x} & 0 & \frac{\partial N_4}{\partial x} \\ 0 & \frac{\partial N_1}{\partial y} & 0 & \frac{\partial N_2}{\partial y} & 0 & \frac{\partial N_3}{\partial y} & 0 & \frac{\partial N_4}{\partial y} \end{bmatrix}.$$
(7.122)

Since

$$\left\{ \begin{array}{c} \frac{\partial N_i}{\partial x} \\ \frac{\partial N_i}{\partial y} \end{array} \right\} = \mathbf{J}^{-1} \left\{ \begin{array}{c} \frac{\partial N_i}{\partial \xi} \\ \frac{\partial N_i}{\partial \eta} \end{array} \right\} \quad (7.123)$$

we can obtain

$$\begin{bmatrix} \frac{\partial N_i}{\partial x} & 0 \\ \frac{\partial N_i}{\partial y} & 0 \\ 0 & \frac{\partial N_i}{\partial x} \\ 0 & \frac{\partial N_i}{\partial y} \end{bmatrix}_{4\times 2} = \begin{bmatrix} \mathbf{J}^{-1} & 0 \\ 0 & \mathbf{J}^{-1} \end{bmatrix}_{4\times 4} \begin{bmatrix} \frac{\partial N_i}{\partial \xi} & 0 \\ \frac{\partial N_i}{\partial \eta} & 0 \\ 0 & \frac{\partial N_i}{\partial \xi} \\ 0 & \frac{\partial N_i}{\partial \eta} \end{bmatrix}_{4\times 2}.$$
(7.124)

Therefore, the **B** matrix can be rewritten as

$$\mathbf{B} = \begin{bmatrix} 1 & 0 & 0 & 0 \\ 0 & 0 & 0 & 1 \\ 0 & 1 & 1 & 0 \end{bmatrix} \begin{bmatrix} \frac{\partial N_1}{\partial x} & 0 & \frac{\partial N_2}{\partial x} & 0 & \frac{\partial N_3}{\partial x} & 0 & \frac{\partial N_4}{\partial x} & 0 \\ \frac{\partial N_1}{\partial y} & 0 & \frac{\partial N_2}{\partial y} & 0 & \frac{\partial N_3}{\partial y} & 0 & \frac{\partial N_4}{\partial y} & 0 \\ 0 & \frac{\partial N_1}{\partial x} & 0 & \frac{\partial N_2}{\partial x} & 0 & \frac{\partial N_3}{\partial x} & 0 & \frac{\partial N_4}{\partial x} \\ 0 & \frac{\partial N_1}{\partial y} & 0 & \frac{\partial N_2}{\partial y} & 0 & \frac{\partial N_3}{\partial y} & 0 & \frac{\partial N_4}{\partial y} \end{bmatrix}$$

$$= \begin{bmatrix} 1 & 0 & 0 & 0 \\ 0 & 0 & 0 & 1 \\ 0 & 1 & 1 & 0 \end{bmatrix} \times \begin{bmatrix} \mathbf{J}^{-1} & 0 \\ 0 & \mathbf{J}^{-1} \end{bmatrix}$$

$$\times \begin{bmatrix} \dfrac{\partial N_1}{\partial \xi} & 0 & \dfrac{\partial N_2}{\partial \xi} & 0 & \dfrac{\partial N_3}{\partial \xi} & 0 & \dfrac{\partial N_4}{\partial \xi} & 0 \\ \dfrac{\partial N_1}{\partial \eta} & 0 & \dfrac{\partial N_2}{\partial \eta} & 0 & \dfrac{\partial N_3}{\partial \eta} & 0 & \dfrac{\partial N_4}{\partial \eta} & 0 \\ 0 & \dfrac{\partial N_1}{\partial \xi} & 0 & \dfrac{\partial N_2}{\partial \xi} & 0 & \dfrac{\partial N_3}{\partial \xi} & 0 & \dfrac{\partial N_4}{\partial \xi} \\ 0 & \dfrac{\partial N_1}{\partial \eta} & 0 & \dfrac{\partial N_2}{\partial \eta} & 0 & \dfrac{\partial N_3}{\partial \eta} & 0 & \dfrac{\partial N_4}{\partial \eta} \end{bmatrix}.$$

(7.125)

In short form

$$\mathbf{B} = \mathbf{H}_{3\times 4} \bar{\mathbf{J}}_{4\times 4} \mathbf{N}'_{4\times 8} \tag{7.126}$$

where the three matrices in Eq. (7.125) are denoted as $\mathbf{H}$, $\bar{\mathbf{J}}$, and $\mathbf{N}'$, respectively. The advantage of rewriting the $\mathbf{B}$ matrix in the form of Eq. (7.126) is that $\mathbf{H}$ and $\mathbf{N}'$ are all constant. They only need to be set up once. Substituting Eqs. (7.126, 7.117) into Eq. (7.116), the strain energy integral is transformed to the master element as

$$\int_{\Omega}^{(e)} \delta \boldsymbol{\epsilon}^T \mathbf{C} \boldsymbol{\epsilon} \, d\Omega$$

$$= \delta \mathbf{d}_{1\times 8}^T \left(\int_{\Omega}^{(e)} \mathbf{B}_{8\times 3}^T \mathbf{C}_{3\times 3} \mathbf{B}_{3\times 8} \, d\Omega \right) \mathbf{d}_{8\times 1}$$

$$= \delta \mathbf{d}_{1\times 8}^T \left(\int_{-1}^{1} \int_{-1}^{1} \mathbf{N}'^T_{8\times 4} \bar{\mathbf{J}}^T_{4\times 4} \mathbf{H}^T_{4\times 3} \mathbf{C}_{3\times 3} \mathbf{H}_{3\times 4} \bar{\mathbf{J}}_{4\times 4} \mathbf{N}'_{4\times 8} \det(\mathbf{J}) \, d\xi \, d\eta \right) \mathbf{d}_{8\times 1}$$

$$= \delta \mathbf{d}_{1\times 8}^T \mathbf{K}_{8\times 8}^{(e)} \mathbf{d}_{8\times 1} \tag{7.127}$$

where $\mathbf{K}^{(e)}$ is the element stiffness matrix. The dimension of the element matrix $\mathbf{K}^{(e)}$ depends on the type of the element. For example, for 8-node quadratic quadrilateral elements, the dimension of the $\mathbf{K}^{(e)}$ matrix is 16×16. In practice, quadratic elements are preferred for stress analysis due to their accuracy and flexibility in modeling complex geometry such as curved boundaries. The element stiffness matrix can be computed by using numerical integration over the master element (2-D integration). Details of 2-D Gauss quadrature have been discussed in Chapter 5. For the 4-node quadrilateral elements in the plate problem, 2×2 Gauss quadrature is used. Here we illustrate the calculation of the element matrix of element 1 in the 2-D plate problem. In Eq. (7.138), the strain–displacement matrix $\mathbf{B}$ needs to be computed at four Gauss points in the master element. For Gauss point $(-1/\sqrt{3}, 1/\sqrt{3})$,

$$\mathbf{N}' = \begin{bmatrix} 0.39434 & 0.00000 & -0.39434 & 0.00000 & -0.10566 & 0.00000 & 0.10566 & 0.00000 \\ 0.10566 & 0.00000 & 0.39434 & 0.00000 & -0.39434 & 0.00000 & -0.10566 & 0.00000 \\ 0.00000 & 0.39434 & 0.00000 & -0.39434 & 0.00000 & -0.10566 & 0.00000 & 0.10566 \\ 0.00000 & 0.10566 & 0.00000 & 0.39434 & 0.00000 & -0.39434 & 0.00000 & -0.10566 \end{bmatrix} \quad (7.128)$$

$$\bar{\mathbf{J}} = \begin{bmatrix} -12.50000 & 11.90551 & 0.00000 & 0.00000 \\ 0.00000 & -17.74946 & 0.00000 & 0.00000 \\ 0.00000 & 0.00000 & -12.50000 & 11.90551 \\ 0.00000 & 0.00000 & 0.00000 & -17.74946 \end{bmatrix}. \quad (7.129)$$

For Gauss point $(1/\sqrt{3}, 1/\sqrt{3})$,

$$\mathbf{N}' = \begin{bmatrix} 0.39434 & 0.00000 & -0.39434 & 0.00000 & -0.10566 & 0.00000 & 0.10566 & 0.00000 \\ 0.39434 & 0.00000 & 0.10566 & 0.00000 & -0.10566 & 0.00000 & -0.39434 & 0.00000 \\ 0.00000 & 0.39434 & 0.00000 & -0.39434 & 0.00000 & -0.10566 & 0.0000 & 0.10566 \\ 0.00000 & 0.39434 & 0.00000 & 0.10566 & 0.00000 & -0.10566 & 0.0000 & -0.39434 \end{bmatrix} \quad (7.130)$$

$$\bar{\mathbf{J}} = \begin{bmatrix} -12.50000 & 9.10604 & 0.00000 & 0.00000 \\ 0.00000 & -13.57584 & 0.00000 & 0.00000 \\ 0.00000 & 0.00000 & -12.50000 & 9.10604 \\ 0.00000 & 0.00000 & 0.00000 & -13.57584 \end{bmatrix}. \quad (7.131)$$

For Gauss point $(-1/\sqrt{3}, -1/\sqrt{3})$,

$$\mathbf{N}' = \begin{bmatrix} 0.10566 & 0.00000 & -0.10566 & 0.00000 & -0.39434 & 0.00000 & 0.39434 & 0.00000 \\ 0.10566 & 0.00000 & 0.39434 & 0.00000 & -0.39434 & 0.00000 & -0.10566 & 0.00000 \\ 0.00000 & 0.10566 & 0.00000 & -0.10566 & 0.00000 & -0.39434 & 0.00000 & 0.39434 \\ 0.00000 & 0.10566 & 0.00000 & 0.39434 & 0.00000 & -0.39434 & 0.00000 & -0.10566 \end{bmatrix} \quad (7.132)$$

$$\bar{\mathbf{J}} = \begin{bmatrix} -12.50000 & 8.06264 & 0.00000 & 0.00000 \\ 0.00000 & -17.74946 & 0.00000 & 0.00000 \\ 0.00000 & 0.00000 & -12.50000 & 8.06264 \\ 0.00000 & 0.00000 & 0.00000 & -17.74946 \end{bmatrix}. \quad (7.133)$$

For Gauss point $(1/\sqrt{3}, -1/\sqrt{3})$,

$$\mathbf{N}' = \begin{bmatrix} 0.10566 & 0.00000 & -0.10566 & 0.00000 & -0.39434 & 0.00000 & 0.39434 & 0.00000 \\ 0.39434 & 0.00000 & 0.10566 & 0.00000 & -0.10566 & 0.00000 & -0.39434 & 0.00000 \\ 0.00000 & 0.10566 & 0.00000 & -0.10566 & 0.00000 & -0.39434 & 0.00000 & 0.39434 \\ 0.00000 & 0.39434 & 0.00000 & 0.10566 & 0.00000 & -0.10566 & 0.00000 & -0.39434 \end{bmatrix}$$
(7.134)

$$\bar{\mathbf{J}} = \begin{bmatrix} -12.50000 & 6.16678 & 0.00000 & 0.00000 \\ 0.00000 & -13.57584 & 0.00000 & 0.00000 \\ 0.00000 & 0.00000 & -12.50000 & 6.16678 \\ 0.00000 & 0.00000 & 0.00000 & -13.57584 \end{bmatrix}.$$
(7.135)

After numerical integration using the Gauss quadrature, the element stiffness matrix for the first element is then computed to be

$$\mathbf{K}^{\textcircled{1}} = \begin{bmatrix} 2.29545 & 0.30120 & -1.41997 & -0.97643 & 0.04634 & -0.94664 & -0.92182 & 1.62187 \\ 0.30120 & 4.36624 & -0.70171 & 1.51678 & -0.94664 & -1.99755 & 1.34715 & -3.88546 \\ -1.41997 & -0.70171 & 9.88184 & -3.65749 & -2.32689 & 1.73441 & -6.13498 & 2.62478 \\ -0.97643 & 1.51678 & -3.65749 & 7.95090 & 2.00914 & -5.30667 & 2.62478 & -4.16101 \\ 0.04634 & -0.94664 & -2.32689 & 2.00914 & 3.70053 & -0.08606 & -1.41997 & -0.97643 \\ -0.94664 & -1.99755 & 1.73441 & -5.30667 & -0.08606 & 5.78744 & -0.70171 & 1.51678 \\ -0.92183 & 1.34715 & -6.13498 & 2.62478 & -1.41997 & -0.70171 & 8.47676 & -3.27022 \\ 1.62187 & -3.88546 & 2.62478 & -4.16101 & -0.97643 & 1.51678 & -3.27022 & 6.52969 \end{bmatrix} \times 10^6 \frac{N}{m^2}$$
(7.136)

The second integral in the weak form, Eq. (7.111), is the virtual work done by the body force. Typically, the body force is given as externally applied loading. With the finite element approximation of the virtual displacement (or the variation of the displacement) and the isoparametric mapping, the body force virtual work integral can be rewritten as

$$\int_{\Omega}^{e} \delta \mathbf{u}^T \mathbf{b} d\Omega$$

$$= \begin{bmatrix} \delta u_1 & \delta v_1 & \delta u_2 & \delta v_2 & \delta u_3 & \delta v_3 & \delta u_4 & \delta v_4 \end{bmatrix} \int_{-1}^{1} \int_{-1}^{1} \begin{bmatrix} N_1 & 0 \\ 0 & N_1 \\ N_2 & 0 \\ 0 & N_2 \\ N_3 & 0 \\ 0 & N_3 \\ N_4 & 0 \\ 0 & N_4 \end{bmatrix} \begin{Bmatrix} b_x \\ b_y \end{Bmatrix} det(\mathbf{J}) d\xi d\eta$$

7.2 2-D Elasticity

$$= \delta\mathbf{d}^T \left[\int_{-1}^{1} \int_{-1}^{1} \begin{bmatrix} N_1 b_x \\ N_1 b_y \\ N_2 b_x \\ N_2 b_y \\ N_3 b_x \\ N_3 b_y \\ N_4 b_x \\ N_4 b_y \end{bmatrix} det(\mathbf{J}) d\xi\, d\eta \right] = \delta\mathbf{d}^T \mathbf{f}_b^{(e)}. \tag{7.137}$$

The vector in the brackets $\mathbf{f}_b^{(e)}$ is the element body force vector. In the 2-D plate example, there is no body force applied. Therefore, the body force vector is zero. The work done by the surface traction is given by

$$\int_{\Gamma_t}^{(e)} \delta\mathbf{u}^T \mathbf{t}\, d\Gamma = \begin{bmatrix} \delta u_1 & \delta v_1 & \delta u_2 & \delta v_2 & \delta u_3 & \delta v_3 & \delta u_4 & \delta v_4 \end{bmatrix} \int_{\Gamma_t}^{(e)} \begin{bmatrix} N_1 & 0 \\ 0 & N_1 \\ N_2 & 0 \\ 0 & N_2 \\ N_3 & 0 \\ 0 & N_3 \\ N_4 & 0 \\ 0 & N_4 \end{bmatrix} \begin{Bmatrix} t_x \\ t_y \end{Bmatrix} d\Gamma$$

$$= \delta\mathbf{d}^T \int_{\Gamma_t}^{(e)} \begin{bmatrix} N_1 t_x \\ N_1 t_y \\ N_2 t_x \\ N_2 t_y \\ N_3 t_x \\ N_3 t_y \\ N_4 t_x \\ N_4 t_y \end{bmatrix} d\Gamma . \tag{7.138}$$

In the isoparametric mapping, the boundary integration (line integral) is required to be transformed to the boundary of the master element, i.e., the boundary Γ_t is required to be mapped to the boundary of the master element. In Section 5.2, the mapping of an infinitesimal line segment between the physical and the master elements has been derived, i.e.,

$$d\Gamma = \begin{cases} \left[\sqrt{\left(\dfrac{\partial x}{\partial \xi}\right)^2 + \left(\dfrac{\partial y}{\partial \xi}\right)^2}\right] d\xi & \text{if } d\Gamma \text{ is corresponding to } d\xi \\[2ex] \left[\sqrt{\left(\dfrac{\partial x}{\partial \eta}\right)^2 + \left(\dfrac{\partial y}{\partial \eta}\right)^2}\right] d\eta & \text{if } d\Gamma \text{ is corresponding to } d\eta. \end{cases}$$

$$\tag{7.139}$$

As shown in Fig. 7.14, the surface traction on an edge of the physical element can be corresponding to either a horizontal or a vertical edge of the master element. Denoting $\sqrt{\left(\dfrac{\partial x}{\partial \xi}\right)^2 + \left(\dfrac{\partial y}{\partial \xi}\right)^2}$ and $\sqrt{\left(\dfrac{\partial x}{\partial \eta}\right)^2 + \left(\dfrac{\partial y}{\partial \eta}\right)^2}$ as L_h and L_v, respectively, the surface virtual work, Eq. (7.138), can be transformed to the master element edge as

$$\int_{\Gamma_t}^{(e)} \delta \mathbf{u}^T \mathbf{t} d\Gamma = \delta \mathbf{d}^T \int_{-1}^{1} \begin{bmatrix} N_1 t_x \\ N_1 t_y \\ N_2 t_x \\ N_2 t_y \\ N_3 t_x \\ N_3 t_y \\ N_4 t_x \\ N_4 t_y \end{bmatrix} L_h d\xi = \delta \mathbf{d}^T \mathbf{f}_t^{(e)} \quad (7.140)$$

when the surface traction edge of the physical element is mapped to a horizontal edge of the master element, and

$$\int_{\Gamma_t}^{(e)} \delta \mathbf{u}^T \mathbf{t} d\Gamma = \delta \mathbf{d}^T \int_{-1}^{1} \begin{bmatrix} N_1 t_x \\ N_1 t_y \\ N_2 t_x \\ N_2 t_y \\ N_3 t_x \\ N_3 t_y \\ N_4 t_x \\ N_4 t_y \end{bmatrix} L_v d\eta = \delta \mathbf{d}^T \mathbf{f}_t^{(e)} \quad (7.141)$$

when the surface traction edge of the physical element is mapped to a vertical edge of the master element. Note that the vector $\mathbf{f}_t^{(e)}$ is the element surface traction vector.

The surface traction is integrated over an edge of the master element. The procedure can be illustrated by using element 2 in our 2-D plate problem as shown in Fig. 7.15. In the figure, the physical element edge where the surface traction is applied is mapped to the left edge (i.e., edge 2, as listed in "tractions.dat") of the master element.

Since the second edge of the master element is a vertical edge, Eq. (7.141) should be used for the calculation of the element surface traction vector, i.e.,

$$\mathbf{f}_t^{(e)} = \int_{-1}^{1} \begin{bmatrix} N_1 & 0 \\ 0 & N_1 \\ N_2 & 0 \\ 0 & N_2 \\ N_3 & 0 \\ 0 & N_3 \\ N_4 & 0 \\ 0 & N_4 \end{bmatrix} \begin{Bmatrix} t_x \\ t_y \end{Bmatrix} L_v d\eta = \int_{-1}^{1} \begin{bmatrix} N_1 t_x \\ N_1 t_y \\ N_2 t_x \\ N_2 t_y \\ N_3 t_x \\ N_3 t_y \\ N_4 t_x \\ N_4 t_y \end{bmatrix} L_v d\eta \quad (7.142)$$

where $L_v = \sqrt{\left(\dfrac{\partial x}{\partial \eta}\right)^2 + \left(\dfrac{\partial y}{\partial \eta}\right)^2}$. For the sake of clarity, we take the expression in the middle to calculate $\mathbf{f}_t^{(e)}$. Note that the shape functions N_i, $i = 1, 2, 3, 4$, are the 2-D shape functions of the master element, i.e.,

7.2 2-D Elasticity

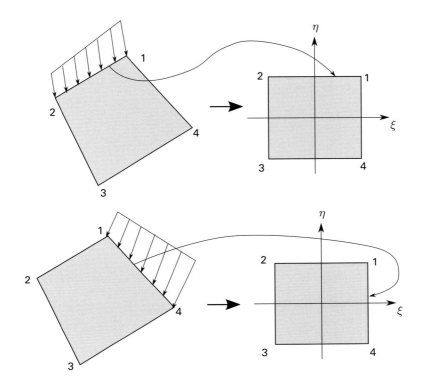

Figure 7.14 Surface traction edges.

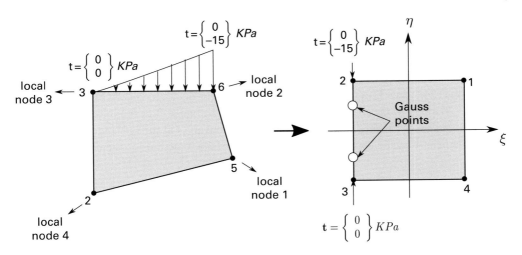

Figure 7.15 Surface traction integration.

$$N_1(\xi, \eta) = \frac{(1+\xi)(1+\eta)}{4}, \qquad N_2(\xi, \eta) = \cdots . \qquad (7.143)$$

As shown in Fig. 7.15, we use 2-point 1-D Gauss quadrature to integrate the surface traction on the edge, i.e.,

$$\mathbf{f}_t^e = \int_{-1}^{1} \begin{bmatrix} N_1 & 0 \\ 0 & N_1 \\ N_2 & 0 \\ 0 & N_2 \\ N_3 & 0 \\ 0 & N_3 \\ N_4 & 0 \\ 0 & N_4 \end{bmatrix} \begin{Bmatrix} t_x \\ t_y \end{Bmatrix} L_v d\eta$$

$$= \begin{bmatrix} N_1(-1,\eta_1) & 0 \\ 0 & N_1(-1,\eta_1) \\ N_2(-1,\eta_1) & 0 \\ 0 & N_2(-1,\eta_1) \\ N_3(-1,\eta_1) & 0 \\ 0 & N_3(-1,\eta_1) \\ N_4(-1,\eta_1) & 0 \\ 0 & N_4(-1,\eta_1) \end{bmatrix} \begin{Bmatrix} t_x(-1,\eta_1) \\ t_y(-1,\eta_1) \end{Bmatrix} L_v(-1,\eta_1) w_1$$

$$+ \begin{bmatrix} N_1(-1,\eta_2) & 0 \\ 0 & N_1(-1,\eta_2) \\ N_2(-1,\eta_2) & 0 \\ 0 & N_2(-1,\eta_2) \\ N_3(-1,\eta_2) & 0 \\ 0 & N_3(-1,\eta_2) \\ N_4(-1,\eta_2) & 0 \\ 0 & N_4(-1,\eta_2) \end{bmatrix} \begin{Bmatrix} t_x(-1,\eta_2) \\ t_y(-1,\eta_2) \end{Bmatrix} L_v(-1,\eta_2) w_2 \quad (7.144)$$

where the location of the two edge Gauss points are $(-1,\eta_1) = (-1, 1/\sqrt{3})$ and $(-1,\eta_2) = (-1, -1/\sqrt{3})$, and the weight for both Gauss points is 1. As shown in the figure, the surface traction $\mathbf{t}$ is not constant over the physical element edge. Therefore, $t_x(-1,\eta_1)$, $t_y(-1,\eta_1)$, $t_x(-1,\eta_2)$, and $t_y(-1,\eta_2)$ in Eq. (7.144) need to be computed.

Mapped to the master element, $\mathbf{t}$ at node 2 and node 3 in the master element are $\mathbf{t} = \{0 \ -15KPa\}^T$ and $\mathbf{t} = \{0 \ 0\}^T$, respectively. On the left edge of the master element, $\mathbf{t}$ can be approximated by using 1-D shape functions on the edge,

$$\mathbf{t} = N_1(\eta) \mathbf{t}_{(node\ 2)} + N_2(\eta) \mathbf{t}_{(node\ 3)} \quad (7.145)$$

where the shape functions in Eq. (7.145) are 1-D shape functions defined on the left edge of the master element

$$N_1(\eta) = \frac{1+\eta}{2} \qquad N_2(\eta) = \frac{1-\eta}{2}. \quad (7.146)$$

Therefore, the traction vector in Eq. (7.144)

$$\begin{Bmatrix} t_x(-1,\eta_j) \\ t_y(-1,\eta_j) \end{Bmatrix} = \begin{Bmatrix} N_1(\eta_j) \times 0.0 + N_2(\eta_j) \times 0.0 \\ N_1(\eta_j) \times -15000.0 + N_2(\eta_j) \times 0.0 \end{Bmatrix} \quad j=1,2. \quad (7.147)$$

As discussed in Section 5.2, an alternative approach to compute **t** on the edge Gauss points is to utilize the 2-D shape functions that are already defined on the 2-D master element, i.e.,

$$\mathbf{t}(-1,\eta) = N_2(-1,\eta)\mathbf{t}_{(node\ 2)} + N_3(-1,\eta)\mathbf{t}_{(node\ 3)}$$
$$= \sum_{i=2}^{3} N_i(-1,\eta)\mathbf{t}_{(node\ i)} \qquad (7.148)$$

where N_i are 2-D shape functions. Note that, in this case, using the 2-D shape functions to approximate **t** on the edges of the master element is equivalent to the 1-D approximation given in Eq. (7.145). However, it is generally not true for higher order elements. For those elements, 1-D approximation is preferred. For element 2 in the 2-D plate problem, the element surface traction vector is computed to be

$$\mathbf{f}_t^{(2)} = \begin{Bmatrix} 0.0 \\ 0.0 \\ 0.0 \\ -650.0 \\ 0.0 \\ -325.0 \\ 0.0 \\ 0.0 \end{Bmatrix} \frac{N}{m}. \qquad (7.149)$$

Note that, since typically only a small number of element edges are on the surface of the structure, it is not necessary to loop over all the element edges to obtain the element surface traction vectors. A more efficient way is to loop over the rows of the "traction.dat." An element force vector can be computed from each row of the "traction.dat" and then assembled into the global force vector.

Combining the equations above, for an individual element, the element discretized weak form without including the global point force vector can be written in short form as

$$\delta\mathbf{d}^T\mathbf{K}^{(e)}\mathbf{d} - \delta\mathbf{d}^T\mathbf{f}_b^{(e)} - \delta\mathbf{d}^T\mathbf{f}_t^{(e)} = 0. \qquad (7.150)$$

The displacement variations can be canceled and the discretized element weak form can be rewritten as a linear system as

$$\mathbf{K}^{(e)}\mathbf{d} = \mathbf{f}_b^{(e)} + \mathbf{f}_t^{(e)} \qquad (7.151)$$

where the right hand vectors are the applied external force vectors on the element.

For the sake of completeness, the work done by the point force **p** is provided here. It can be simply approximated as

$$\frac{\delta\mathbf{u}(\mathbf{a})^T\mathbf{p}}{h} = \begin{bmatrix} \delta u(\mathbf{a}) & \delta v(\mathbf{a}) \end{bmatrix} \begin{Bmatrix} p_x/h \\ p_y/h \end{Bmatrix}. \qquad (7.152)$$

Note that it is not an element vector, i.e., it does not belong to an individual element and is directly assembled to the global force vector.

Step 7: Global Matrix Assembly

The global matrix assembly procedure is similar to the 2-D heat transfer except that there are two degrees of freedom (two nodal displacements) for each node in the 2-D elasticity problem. For example, as shown in Fig. 3.16, the global nodes of the element 2 are 5, 6, 3, and 2, corresponding to the local nodes 1, 2, 3, and 4, respectively. The 8 × 8 element stiffness matrix can be seen as a matrix with 4 × 4 blocks. Each block has the dimension of 2 × 2. The row and column positions of the blocks in the element matrix are corresponding to the nodes' local IDs, and their positions in the global matrix are corresponding to the nodes' global IDs. Instead of assembling component by component as in the heat transfer problem, the global matrix of the elasticity problem is assembled block by block. The global right hand side force vector is assembled in a similar fashion.

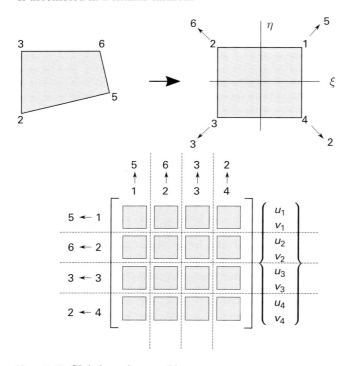

Figure 7.16 Global matrix assembly.

After assembly, the global system is obtained as

$$\mathbf{Ku} = \mathbf{F} \qquad (7.153)$$

where $\mathbf{K}$ is the global stiffness matrix, $\mathbf{u}$ is the global nodal displacement vector, and $\mathbf{F}$ is the global force vector. The 2-D force vector is assembled from the element force vectors as shown in Eq. (7.149). The point force vector $\mathbf{p}/h = \{-10/h\ 0\}^T$ given in Eq. (7.152) is directly assembled into the global force vector. As the point force is applied on node 8, the 2 × 1 point force vector is added to row 15 and row 16 of $\mathbf{F}$. The assembled global system for the 2-D plate problem is shown in Eq. (7.154).

$$\begin{bmatrix} 2.29545 & 0.30120 & -0.92182 & 1.62187 & 0 & -1.41997 & -0.97643 & 0.04634 & -0.94664 & 0 & 0 & 0 & 0 & 0 & 0 & 0 & 0 & 0 \\ 0.30120 & 4.36624 & 1.34715 & -3.88546 & 0 & -0.70171 & 1.51678 & -0.94664 & -1.99755 & 0 & 0 & 0 & 0 & 0 & 0 & 0 & 0 & 0 \\ -0.92183 & 1.34715 & 11.7271 & -1.81959 & 0.48612 & -6.13498 & 2.62478 & -3.54361 & -1.37437 & -1.61283 & -1.53898 & 0 & 0 & 0 & 0 & 0 & 0 & 0 \\ 1.62187 & -3.88546 & -1.81959 & 11.7271 & 0.48628 & -2.99020 & -4.16101 & -1.37437 & 2.71455 & -1.53898 & -2.95347 & 0 & 0 & 0 & 0 & 0 & 0 & 0 \\ 0 & 0 & 0.48612 & 0.48628 & 5.51994 & -1.98028 & 0 & -3.27779 & 1.85932 & -2.72827 & -2.95347 & 0 & 0 & 0 & 0 & 0 & 0 & 0 \\ 0 & 0 & 0.76100 & -2.99020 & -1.98028 & 4.81279 & 0 & 1.85932 & -2.35853 & -0.64004 & 0.53595 & 0 & 0 & 0 & 0 & 0 & 0 & 0 \\ -1.41997 & -0.70171 & -6.13498 & 2.62478 & 0 & 0 & 12.73092 & -4.38730 & -5.70628 & 4.39396 & 0 & 1.36504 & -1.89424 & -0.83473 & -0.03549 & 0 & 0 & 0 \\ -0.97643 & 1.51678 & 2.62478 & -4.16101 & 0 & 0 & -4.38730 & 14.8838 & 4.39396 & -11.6629 & 0 & -1.61952 & 3.78087 & -0.03549 & -4.35753 & 0 & 0 & 0 \\ 0.04634 & -0.94664 & -3.54361 & -1.37437 & -3.27779 & 1.85932 & -5.70628 & 4.39396 & 21.5932 & -4.67621 & 1.78466 & -5.90217 & 2.61187 & 0.15647 & -2.43238 & -2.00760 & -1.22020 & 0 \\ -0.94664 & -1.99755 & -1.37437 & 2.71455 & 1.85932 & -2.35853 & 4.39396 & -11.6629 & -4.67621 & 26.5012 & -8.78088 & 2.61187 & -4.09827 & -2.43238 & 5.38029 & -1.22020 & -5.69787 & 0 \\ 0 & 0 & -1.61283 & -1.53898 & -2.72827 & -0.64004 & 0 & 0 & -1.35856 & 1.78466 & 9.11517 & -4.09827 & -8.78088 & -2.65229 & 1.69426 & -0.76322 & -0.82995 & 0 \\ 0 & 0 & -1.53898 & -2.95347 & -0.36531 & 0.53595 & 0 & 0 & 1.78466 & -8.78088 & -0.46996 & 11.1400 & -0.46996 & 1.69426 & -1.94822 & -1.10467 & 2.00665 & 0 \\ 0 & 0 & 0 & 0 & 0 & 0 & 1.36504 & -1.61952 & -5.90217 & 2.61187 & -4.09827 & 0 & 9.64312 & -4.03315 & -5.10598 & 3.04079 & 0 & -1950 \\ 0 & 0 & 0 & 0 & 0 & 0 & -1.89424 & 3.78087 & 2.61187 & -4.09827 & -8.78088 & 2.61187 & -4.03315 & 8.88105 & 3.31552 & -8.56366 & 0 & 0 \\ 0 & 0 & 0 & 0 & 0 & 0 & -0.83473 & -0.03549 & 0.15647 & -2.43238 & -2.65229 & 0.15647 & -5.10598 & 3.31552 & 11.1098 & -3.66934 & -2.67330 & 1.12743 \\ 0 & 0 & 0 & 0 & 0 & 0 & -0.03549 & -4.35753 & -2.43238 & 5.38029 & 1.69426 & -2.43238 & 3.04079 & -8.56366 & -3.66934 & 17.2225 & 1.40216 & -7.73334 \\ 0 & 0 & 0 & 0 & 0 & 0 & 0 & 0 & -2.00760 & -1.22020 & -0.76322 & -2.00760 & 0 & 0 & -2.67330 & 1.40216 & 5.44413 & 0.92272 \\ 0 & 0 & 0 & 0 & 0 & 0 & 0 & 0 & -1.22020 & -5.69787 & -0.82995 & -1.22020 & 0 & 0 & 1.12743 & -7.73334 & 0.92272 & 11.4246 \end{bmatrix} \times 10^6 \begin{Bmatrix} u_1 \\ v_1 \\ u_2 \\ v_2 \\ u_3 \\ v_3 \\ u_4 \\ v_4 \\ u_5 \\ v_5 \\ u_6 \\ v_6 \\ u_7 \\ v_7 \\ u_8 \\ v_8 \\ u_9 \\ v_9 \end{Bmatrix} = \begin{Bmatrix} 0 \\ 0 \\ 0 \\ 0 \\ 0 \\ -325 \\ 0 \\ 0 \\ 0 \\ 0 \\ 0 \\ -1950 \\ 0 \\ 0 \\ -2000 \\ 0 \\ 0 \\ -1625 \end{Bmatrix}$$

(7.154)

Step 8: Boundary Conditions and Solution

The natural (surface traction) boundary condition is already included in the weak form. For a structure containing N nodes, the total degrees of freedom is $2N$, i.e., the dimension of the global matrix and the right hand side vector is $2N \times 2N$ and $2N \times 1$. The application of the essential (displacement) boundary condition can be implemented by using the direct substitution method and the penalty method discussed in previous chapters. These methods are equally applicable in 2-D elastostatic analysis.

In this section, we introduce another method of applying the essential boundary conditions: the Lagrange multiplier method. We start with the functional statement of the problem, given in Eq. (7.103) which is rewritten below

$$\Pi = \frac{1}{2} h \int_\Omega \boldsymbol{\epsilon} \cdot \mathbf{C}\boldsymbol{\epsilon} d\Omega - h \int_\Omega \mathbf{u} \cdot \mathbf{b} d\Omega - h \int_{\Gamma_t} \mathbf{u} \cdot \mathbf{t} d\Omega - \mathbf{u} \cdot \mathbf{p}. \tag{7.155}$$

If we directly discretize the global integrals over the elements of the entire domain and substitute the finite approximations into each of the integrals, we can obtain a fully expanded matrix-vector expression of the functional that looks similar to Eq. (4.79), except that the row vectors in front of the matrices contain nodal displacements, not their variations. Adding all the terms together leads to a globally discretized functional in the form of

$$\Pi = \frac{1}{2} [u_1 \ v_1 \ u_2 \ v_2 \cdots v_N] \begin{bmatrix} K_{11} & K_{12} & \cdots & \cdots \\ K_{21} & K_{22} & \cdots & \cdots \\ \cdots & \cdots & \cdots & \cdots \\ \cdots & \cdots & \cdots & \cdots \end{bmatrix} \begin{Bmatrix} u_1 \\ v_1 \\ u_2 \\ v_2 \\ \vdots \\ v_N \end{Bmatrix}$$

$$+ [u_1 \ v_1 \ u_2 \ v_2 \cdots v_N] \begin{Bmatrix} f_1 \\ f_2 \\ f_3 \\ \vdots \\ f_{2N} \end{Bmatrix}. \tag{7.156}$$

In short form, Eq. (7.156) can be written as

$$\Pi = \frac{1}{2} \mathbf{d}^T \mathbf{K} \mathbf{d} + \mathbf{d}^T \mathbf{F}. \tag{7.157}$$

In the Lagrange multiplier method, the functional is modified to include the displacement boundary conditions as

$$\Pi = \frac{1}{2} \mathbf{d}^T \mathbf{K} \mathbf{d} + \mathbf{d}^T \mathbf{F} + \sum_{j=1}^{nb} \lambda_j (d_j - \overline{d}_j) \tag{7.158}$$

where $\overline{d}_j$ (u or v) is the prescribed value of the nodal displacement d_j, λ_i denotes the Lagrange multiplier of d_j, and nb is the number of constrained nodal displacements.

Note that, the modified functional is the same as the original one when the displacement boundary conditions are satisfied, i.e., $d_j = \bar{d}_j$, $j = 1, 2, \ldots$. Invoking the stationary condition of the modified functional, we have

$$\delta \Pi = 0$$

$$\rightarrow \delta \mathbf{d}^T \mathbf{K} \mathbf{d} + \delta \mathbf{d}^T \mathbf{F} + \sum_{j=1}^{nb} \delta \lambda_j (d_j - \bar{d}_j) + \sum_{j=1}^{nb} \lambda_j \delta d_j = 0. \quad (7.159)$$

To make the illustration clear, without loss of generality, we assume that we have only one prescribed x-displacement of node j, $u_j = \bar{u}_j$. Equation (7.159) can be rewritten as

$$\delta \mathbf{u}^T \left(\mathbf{K} \mathbf{u} - \mathbf{F} + \begin{Bmatrix} 0 \\ 0 \\ \vdots \\ \lambda \\ \vdots \\ 0 \end{Bmatrix} \right) + \delta \lambda (u_j - \bar{u}_j) = 0. \quad (7.160)$$

For a 2-D elasticity problem, the position of λ in the column vector is $2j - 1$. Since both $\delta \mathbf{u}$ and $\delta \lambda$ are arbitrary, Eq. (7.160) gives

$$\mathbf{K} \mathbf{u} - \mathbf{F} + \begin{Bmatrix} 0 \\ 0 \\ \vdots \\ \lambda \\ \vdots \\ 0 \end{Bmatrix} = 0. \quad (7.161)$$

$$u_j = \bar{u}_j \quad (7.162)$$

In matrix form,

$$\begin{bmatrix} & & & 0 \\ & & & \vdots \\ & \mathbf{K} & & 1 \\ & & & \vdots \\ 0 & \cdots & 1 & \cdots & 0 \end{bmatrix} \begin{Bmatrix} u_1 \\ v_1 \\ u_2 \\ \vdots \\ u_i \\ \vdots \\ u_N \\ v_N \\ \lambda \end{Bmatrix} = \begin{Bmatrix} f_1 \\ f_2 \\ f_3 \\ \vdots \\ f_{2N} \\ \bar{u}_i \end{Bmatrix}. \quad (7.163)$$

The positions of the "1"s in the right column and bottom row of the matrix are $(2j-1)$-th row and column, respectively. By solving the global system Eq. (7.163), the unknown displacements are computed, the boundary condition is satisfied and the Lagrange multiplier is obtained too.

Step 9: Solution of the Global System

Once the essential boundary conditions are applied, the final linear system can be solved directly in MATLAB, the unknown nodal displacements can be computed. The displacement solution is given by

Table 7.1 Nodal displacement solution

Node	u (m)	v (m)
1	0.000000000	−0.0008142999
2	0.000000000	−0.0006691279
3	0.000000000	−0.000000000
4	−0.0002251297	−0.0010788983
5	−0.0002036185	−0.0011220109
6	0.0001132064	−0.0011588491
7	−0.0002783309	−0.0016931126
8	−0.0002387272	−0.0016688928
9	0.0000415561	−0.0016212858

Step 10: Post-processing

Once the nodal displacements are computed by solving the global linear system, the deformed position of the structure compared to its undeformed position can be visualized as shown in Fig. 7.17.

In the post-processing stage, typically, the strain and stress distributions in the domain are calculated by using the nodal displacements at the nodes by using Eqs. (7.181, 7.125). For each element (4-node linear quadrilateral element in this case), the strains are computed by

$$\left\{\begin{array}{c} \epsilon_{xx} \\ \epsilon_{yy} \\ \epsilon_{xy} \end{array}\right\} = \begin{bmatrix} 1 & 0 & 0 & 0 \\ 0 & 0 & 0 & 1 \\ 0 & 1 & 1 & 0 \end{bmatrix} \begin{bmatrix} \mathbf{J}^{-1} & 0 \\ 0 & \mathbf{J}^{-1} \end{bmatrix}$$

$$\times \begin{bmatrix} \frac{\partial N_1}{\partial \xi} & 0 & \frac{\partial N_2}{\partial \xi} & 0 & \frac{\partial N_3}{\partial \xi} & 0 & \frac{\partial N_4}{\partial \xi} & 0 \\ \frac{\partial N_1}{\partial \eta} & 0 & \frac{\partial N_2}{\partial \eta} & 0 & \frac{\partial N_3}{\partial \eta} & 0 & \frac{\partial N_4}{\partial \eta} & 0 \\ 0 & \frac{\partial N_1}{\partial \xi} & 0 & \frac{\partial N_2}{\partial \xi} & 0 & \frac{\partial N_3}{\partial \xi} & 0 & \frac{\partial N_4}{\partial \xi} \\ 0 & \frac{\partial N_1}{\partial \eta} & 0 & \frac{\partial N_2}{\partial \eta} & 0 & \frac{\partial N_3}{\partial \eta} & 0 & \frac{\partial N_4}{\partial \eta} \end{bmatrix} \left\{\begin{array}{c} u_1 \\ v_1 \\ u_2 \\ v_2 \\ u_3 \\ v_3 \\ u_4 \\ v_4 \end{array}\right\}.$$

(7.164)

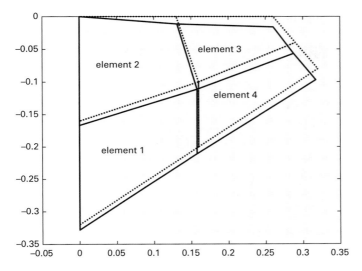

Figure 7.17 Deformed (solid line) and undeformed (dotted line) shapes of the plate.

The stresses can then be obtained by using Eq. (7.89)

$$\left\{ \begin{array}{c} \sigma_{xx} \\ \sigma_{yy} \\ \sigma_{xy} \end{array} \right\} = \mathbf{C} \left\{ \begin{array}{c} \epsilon_{xx} \\ \epsilon_{yy} \\ \epsilon_{xy} \end{array} \right\}. \tag{7.165}$$

Note that, in general, the stresses will be discontinuous across the elements. To compute a stress at a node, one can compute the nodal stress from all the elements sharing this node and then average the values obtained. For a coarse mesh, the stresses at a node obtained from the neighboring elements could be quite different, which indicates a mesh refinement is needed. As the element size becomes smaller, the difference in the nodal stresses reduces. While averaging stresses at the nodes is convenient, there are cases where the stresses should not be averaged at the nodes. For example, at the interface of two different materials, stresses are typically discontinuous. Numerically averaging the element stresses at the interface nodes would distort the actual physical characteristics of the materials.

An alternative approach can be employed to compute the nodal stresses more accurately. This approach is based on the observation that the stresses are often more accurate at the Gauss points of a quadrature rule one order less than that required for exact integration of the element stiffness matrix. The idea is then to compute the stress at the Gauss points within each element and extrapolate the stress onto the nodes. Next, the nodal stresses obtained from all the elements sharing the node are averaged. Although this approach typically gives a more accurate nodal solution of the stresses, it is more involved due to extrapolation.

7.2.3 Computer Implementation

In this section, by using the plate problem as an example, we present the complete MATLAB code for the finite element analysis.

```
1  clear all ;
2  % next 3 lines : read the input files
3  filenames = {'nodes.dat','elements.dat','materials.dat', ...
4               'bcstraction.dat','bcsforce.dat', 'bcsdisp.dat'};
5  for i = 1:numel(filenames); load(filenames{i}); end;
6
7  % next 5 lines : bookkeeping
8  n_nodes=size(nodes,1);
9  n_elements=size(elements,1);
10 n_bcstraction =size(bcstraction,1);
11 n_bcsforce=size(bcsforce,1);
12 n_bcsdisp=size(bcsdisp,1);
13
14 K=CompK(nodes, elements, materials);   % compute global K matrix and
15 F=CompF(nodes, elements, materials, bcstraction, bcsforce); % F vector
16
17 % next 7 lines : apply displacement boundary condition
18 coeff=abs(max(K(bcsdisp(1,1)*2−1,:)))*1e8;   % penalty number
19 for i=1:n_bcsdisp
20    node_id=bcsdisp(i,1);
21    direction =bcsdisp(i,2);
22    K(2*node_id−2 + direction , 2*node_id−2 + direction )=coeff;
23    F(2*node_id−2 + direction , 1) = bcsdisp(i,3)*coeff;
24 end
25
26 u=K\F;  % solve the global linear system
27
28 % next 6 lines : save the displacement results in file
29 U=zeros(n_nodes,5);
30 for n=1:n_nodes
31    U(n,:)=[n nodes(n,2:3)  u(2*n−1:2*n,1)'];
32 end
33 save −ascii −double feU.dat U;
```

```
1  % Compute K matrix
2  % Input: nodes, elements, material properties ( materials ),
3  % Output: K matrix
4  function K=CompK(nodes, elements, materials)
5  n_nodes = size(nodes,1);
6  n_elements = size(elements,1);
7  n_nodes_per_element = size(elements,2)−1;
8  K=zeros(n_nodes*2,n_nodes*2);
9  C=CompCPlaneStress(materials); % compute material matrix
10 H=CompH();                      % set up H matrix
```

```
11  [gauss_points, gauss_weights]=GetQuadGauss(2,2);
12  n_gauss_points=size(gauss_points,1);
13  [N,Nx,Ny]=CompNDNatPointsQuad4(gauss_points(:,1), gauss_points(:,2));
14
15  % compute K matrix : loop over all the elements
16  for e=1:n_elements
17    ke=zeros(n_nodes_per_element*2, n_nodes_per_element*2);
18    [element_nodes, node_id_map]= SetElementNodes(e, nodes, elements);
19    % next 11 lines : compute element stiffness matrix ke
20    for g=1:n_gauss_points
21      J=CompJacobian2DatPoint(element_nodes, Nx(:,g), Ny(:,g));
22      detJ=det(J);
23
24      Jinv=inv(J);
25      Jb(1:2,1:2) =Jinv;
26      Jb(3:4,3:4) =Jinv;
27      B=CompB4x8Quad4atPoint(Nx(:,g), Ny(:,g));
28      HJB=(H*Jb*B);
29      ke=ke+HJB'*C*HJB*detJ*gauss_weights(g);
30    end
31    % assemble ke into global K
32    K= AssembleGlobalMatrix(K,ke,node_id_map,2);
33  end
```

```
1   % Compute F vector
2   % Input: nodes, elements, materials, bcstraction, bcsforce
3   % Output: F vector
4   function F=CompF(nodes, elements, materials, bcstraction, bcsforce)
5   n_nodes = size(nodes,1);
6   n_elements = size(elements,1);
7   n_traction_edges = size(bcstraction,1);
8   n_force_nodes = size(bcsforce,1);
9   n_nodes_per_element = size(elements,2)-1;
10   traction_start =zeros(2,1);
11   traction_end =zeros(2,1);
12   traction =zeros(2,1);
13   F=zeros(n_nodes*2,1);
14   n_edge_gauss_points=2;
15   [gauss_points, gauss_weights]=GetQuadEdgeGauss(n_edge_gauss_points);
16   edge_gauss=-gauss_points(1:n_edge_gauss_points,1)
17   [N,Nx,Ny]=CompNDNatPointsQuad4(gauss_points(:,1),gauss_points(:,2));
18
19  % for-loop: loop over the number of edges affected by traction
20  % to apply surface traction
21  for t=1: n_traction_edges
22    fe=zeros(n_nodes_per_element*2, 1);
23    eid= bcstraction (t,1);
24    [element_nodes, node_id_map]= SetElementNodes(eid,nodes,elements);
```

```
25      edge=bcstraction(t,2);
26      traction_start (1:2,1) =bcstraction(t,5:6)
27      traction_end (1:2,1) =bcstraction(t,7:8)
28      % loop over the edge Gauss points
29      for g=1:n_edge_gauss_points
30        gid=n_edge_gauss_points*(edge-1)+g;   % Gauss point ID
31        J=CompJacobian2DatPoint(element_nodes,Nx(:,gid),Ny(:,gid));
32        if (edge==1) | (edge==3)
33          lengthJ=sqrt(J(1,1)^2+J(1,2)^2);
34        else
35          lengthJ=sqrt(J(2,1)^2+J(2,2)^2);
36        end
37        fxy=zeros(8,2);
38        for k=1:n_nodes_per_element
39          fxy(2*k-1,1)=N(k, gid);
40          fxy(2*k,2)=N(k, gid);
41        end
42        traction (1,1)=(1-edge_gauss(g))/2* traction_start (1,1) ...
43                      +(1+edge_gauss(g))/2* traction_end (1,1);
44        traction (2,1)=(1-edge_gauss(g))/2* traction_start (2,1) ...
45                      +(1+edge_gauss(g))/2* traction_end (2,1);
46
47        fe =fe+ fxy* traction *lengthJ *gauss_weights(gid);
48      end
49      F= AssembleGlobalVector(F, fe, node_id_map, 2); % assemble F
50    end
51    % for loop: apply point forces
52    for i=1:n_force_nodes
53      row=2*bcsforce(i,1)-1;
54      F(row,1) = F(row,1)+ bcsforce(i,2) / materials (3); % materials (3) is
55      F(row+1,1) = F(row+1,1)+ bcsforce(i,3) / materials (3); % thickness
56    end
```

```
1   % Compute material matrix (plane stress)
2   function C= CompCPlaneStress(materials)
3   C=zeros(3,3);      % set up 3x3 empty matrix
4   E=materials(1);    % Young's modulus
5   nu=materials(2);   % Poisson's ratio
6   t = E/(1-(nu*nu));
7   C(1,1)=t;
8   C(2,2)=t;
9   C(1,2)=t*nu;
10  C(2,1)=t*nu;
11  C(3,3)=t*((1-nu)/2);
```

```
1   % Set up H matrix
2   function H= CompH()
```

```
3  H=zeros(3,4);
4  H(1,1)=1;  H(2,4)=1;
5  H(3,2)=1;  H(3,3)=1;
```

```
1  % Compute strain − displacement matrix
2  function B= CompB4x8Quad4atPoint(Nxi_vec, Neta_vec)
3  B=zeros(4,8);
4  for i=1:4
5      B(1,2*i−1)=Nxi_vec(i);
6      B(3,2*i)= Nxi_vec(i);
7      B(2,2*i−1)=Neta_vec(i);
8      B(4,2*i)= Neta_vec(i);
9  end
```

In the MATLAB programs listed above, the functions which have already been discussed previously are not repeated here. They are listed below for your reference:
function [gauss_points, gauss_weights]=GetQuadGauss(rows, cols): Section 5.2.1.
function [gauss_points, gauss_weights]=GetQuadEdgeGauss(neg): Section 5.2.1.
function [N,Nx,Ny]=CompNDNatPointsQuad4(xi_vector, eta_vector): Section 5.2.1.
function [element_nodes, node_id_map]= SetElementNodes(ele, nodes, elements): Section 5.2.1.
function J=CompJacobian2DatPoint(element_nodes, Nxi, Neta): Section 5.2.1.
function K=AssembleGlobalMatrix(K, ke, node_id_map, ndDOF): Section 5.2.1.
function F=AssembleGlobalVector(F, fe, node_id_map, ndDOF): Section 5.2.1.

7.3   3-D Elasticity

In this section, the finite element analysis of the 2-D elastic structure demonstrated in Section 7.2.2 is extended into 3-D. A 3-D elastic structure is solved step by step in this section. It is shown that extension from 2-D to 3-D cases is straightforward. A MATLAB code for 3-D analysis is presented.

7.3.1   3-D Elastic Structure Subjected to External Loads

Step 1: Physical Problem

A T-shaped elastic structure is fixed at the bottom and subjected to point forces as shown in Fig. 7.18. The structure is symmetric about the yz-plane. The elastic material has a Young's modulus of 200 GPa and a Poisson's ratio of 0.3. There is no body force and there is no surface traction acting on the surface of the structure. The dimensions of the structure are shown in the figure. The applied forces are: $\mathbf{f} = \{0 -10\,\text{kN} -10\,\text{kN}\}^T$ at (42.0, 45.0, 0), and $\mathbf{f} = \{0 -10\,\text{kN} -10\,\text{kN}\}^T$ at (−42.0,

45.0, 8.0), as shown in the figure. The deformation of the structure are to be determined by using the finite element method.

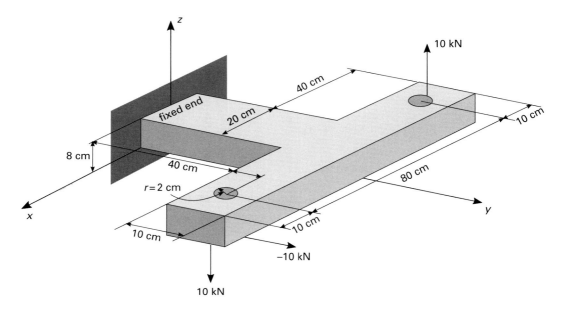

Figure 7.18 T-beam subjected to external forces.

Step 2: Mathematical Model
The 3-D governing partial differential equations of linear elastostatics are given by

$$\frac{\partial \sigma_{xx}}{\partial x} + \frac{\partial \sigma_{xy}}{\partial y} + \frac{\partial \sigma_{xz}}{\partial z} + b_x = 0 \tag{7.166}$$

$$\frac{\partial \sigma_{xy}}{\partial x} + \frac{\partial \sigma_{yy}}{\partial y} + \frac{\partial \sigma_{yz}}{\partial z} + b_y = 0 \tag{7.167}$$

$$\frac{\partial \sigma_{xz}}{\partial x} + \frac{\partial \sigma_{yz}}{\partial y} + \frac{\partial \sigma_{zz}}{\partial z} + b_z = 0. \tag{7.168}$$

For our example problem, the body force is zero. The boundary conditions are

$$\mathbf{u} = \left\{ \begin{array}{c} u \\ v \\ w \end{array} \right\} = \mathbf{0} \qquad \text{at } y = 0 \tag{7.169}$$

and

$$\mathbf{f}(42, 45, 0) = \left\{ \begin{array}{c} 0 \\ -10 \\ -10 \end{array} \right\} \text{ kN}, \qquad \mathbf{f}(-42, 45, 8) = \left\{ \begin{array}{c} 0 \\ 0 \\ 10 \end{array} \right\} \text{ kN}. \tag{7.170}$$

Equations (7.166–7.170) establish the strong form of the 3-D elastostatic problem.

Step 3: Weak Form

The methods used to derive the weak form for 2-D elasticity problems are equally applicable to 3-D problems. The only difference is the one extra dimension in the z-direction. In this section, we simply use the principle of virtual work to obtain the weak form directly. For deriving the weak form using other methods, see Section 7.2.2.

The principle of virtual work gives the following equation

$$W_I = W_E(\mathbf{b}) + W_E(\mathbf{t}) + W_E(\mathbf{p}) \tag{7.171}$$

where W_I is the total internal virtual work which, in static analysis of elasticity problems, is the virtual work done by the stresses, $W_E(\mathbf{b})$ is the external virtual work done by the body force over a virtual displacement, $W_E(\mathbf{t})$ is the external virtual work done by the surface traction, and $W_E(\mathbf{p})$ is the external virtual work done by the point forces, $\mathbf{p}_1, \mathbf{p}_2, \ldots$, acting at locations $\mathbf{a}_1, \mathbf{a}_2, \ldots$. The expressions of these virtual works are given by

$$W_E(\mathbf{b}) = \int_\Omega \delta\mathbf{u} \cdot \mathbf{b} d\Omega, \tag{7.172}$$

$$W_E(\mathbf{t}) = \int_{\Gamma_t} \delta\mathbf{u} \cdot \mathbf{t} d\Gamma \tag{7.173}$$

$$W_E(\mathbf{p}) = \sum_j \delta\mathbf{u}(\mathbf{a}_j) \cdot \mathbf{p}_j \qquad j = 1, 2, \ldots \tag{7.174}$$

Therefore, from the principle of virtual work

$$W_I = W_E(\mathbf{b}) + W_E(\mathbf{t}) + W_E(\mathbf{p})$$
$$\Rightarrow \int_{\Gamma_t} \delta\mathbf{u} \cdot \mathbf{t} d\Gamma + \sum_j \delta\mathbf{u}(\mathbf{a}_j) \cdot \mathbf{p}_j + \int_\Omega \delta\mathbf{u} \cdot \mathbf{b} d\Omega - \int_\Omega \delta\boldsymbol{\epsilon} \cdot \mathbf{C}\boldsymbol{\epsilon} d\Omega = 0. \tag{7.175}$$

Equation (7.175) is the weak form of 3-D elastostatics. Note that while the form of the weak form is essentially the same as that in 2-D, all the vectors are now corresponding to the physical quantities in 3-D and contain the z-components. Furthermore, although there is no body force and surface traction in the example problem, we include them in the weak form for the sake of illustration.

Step 4: Discretization

For this example, we discretize the 3-D volume of the T-beam into hexahedral elements. Figure 7.19 shows a mesh for the T-beam structure. This particular mesh contains 1,108 elements. Eight-node linear hexahedral elements are used in the example and there are a total of 1730 nodes in the mesh. Creating 3-D unstructured mesh is beyond the scope of this book. The mesh shown in Fig. 7.19 was generated by using commercial software.

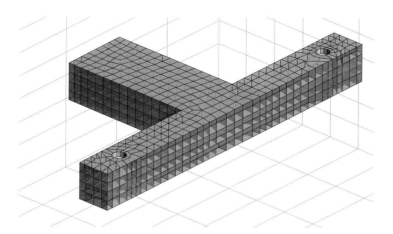

Figure 7.19 T-beam mesh.

Step 5: Finite Element Approximation

Finite element approximation over 3-D elements has been discussed in detail in Chapter 5.3.1. The unknown physical quantities and their derivatives are approximated over the master elements in the reference coordinate system. Then the approximations are mapped to the physical elements through isoparametric mapping. For 3-D elasticity problems, each node has three unknown quantities: three components of the displacement vector, u, v, and w. Each of the components is approximated the same way as the temperature is approximated in the heat transfer problem. For the 8-node hexahedral elements in the mesh shown in Fig. 7.19, the finite element approximation for the displacements and their derivatives are then given by

$$u = \sum_{i=1}^{8} N_i(\xi, \eta) u_i, \qquad v = \sum_{i=1}^{8} N_i(\xi, \eta) v_i, \qquad w = \sum_{i=1}^{8} N_i(\xi, \eta) w_i \qquad (7.176)$$

$$\frac{\partial u}{\partial x} = \sum_{i=1}^{8} \frac{\partial N_i(\xi, \eta)}{\partial x} u_i \qquad \frac{\partial v}{\partial x} = \sum_{i=1}^{8} \frac{\partial N_i(\xi, \eta)}{\partial x} v_i \qquad \frac{\partial w}{\partial x} = \sum_{i=1}^{8} \frac{\partial N_i(\xi, \eta)}{\partial x} w_i$$

$$\frac{\partial u}{\partial y} = \sum_{i=1}^{8} \frac{\partial N_i(\xi, \eta)}{\partial y} u_i \qquad \frac{\partial v}{\partial y} = \sum_{i=1}^{8} \frac{\partial N_i(\xi, \eta)}{\partial y} v_i \qquad \frac{\partial w}{\partial y} = \sum_{i=1}^{8} \frac{\partial N_i(\xi, \eta)}{\partial y} w_i$$

$$\frac{\partial u}{\partial z} = \sum_{i=1}^{8} \frac{\partial N_i(\xi, \eta)}{\partial z} u_i \qquad \frac{\partial v}{\partial z} = \sum_{i=1}^{8} \frac{\partial N_i(\xi, \eta)}{\partial z} v_i \qquad \frac{\partial w}{\partial z} = \sum_{i=1}^{8} \frac{\partial N_i(\xi, \eta)}{\partial z} w_i.$$

$$(7.177)$$

The same shape functions are used for the displacement variations (or virtual displacements).

Step 6: Element Matrices and Vectors

The element matrices and vectors can obtained by using the element weak form, which is the element version of Eq. (7.175) by omitting the global force vector term, i.e.

$$\int_\Omega^{(e)} \delta\epsilon^T \mathbf{C}\epsilon\, d\Omega - \int_\Omega^{(e)} \delta\mathbf{u}^T \mathbf{b}\, d\Omega - \int_{\Gamma_t}^{(e)} \delta\mathbf{u}^T \mathbf{t}\, d\Gamma = 0. \tag{7.178}$$

The element matrices and vectors can then be obtained by substituting the finite element approximations of the displacements and strains in the weak form. In 3-D, the finite element approximation of the strain vector can be written as

$$\epsilon = \begin{Bmatrix} \epsilon_{xx} \\ \epsilon_{yy} \\ \epsilon_{zz} \\ \epsilon_{xy} \\ \epsilon_{yz} \\ \epsilon_{xz} \end{Bmatrix} = \begin{Bmatrix} \dfrac{\partial u}{\partial x} \\ \dfrac{\partial v}{\partial y} \\ \dfrac{\partial w}{\partial z} \\ \dfrac{\partial u}{\partial y} + \dfrac{\partial v}{\partial x} \\ \dfrac{\partial v}{\partial z} + \dfrac{\partial w}{\partial y} \\ \dfrac{\partial w}{\partial x} + \dfrac{\partial u}{\partial z} \end{Bmatrix} = [\mathbf{B}_1\ \mathbf{B}_2\ \mathbf{B}_3\ \cdots\ \mathbf{B}_8]_{6\times 24} \begin{Bmatrix} u_1 \\ v_1 \\ w_1 \\ u_2 \\ v_2 \\ \vdots \\ u_8 \\ v_8 \\ w_8 \end{Bmatrix}_{24\times 1}$$

$$= \mathbf{B}_{6\times 24}\mathbf{d}_{24\times 1}. \tag{7.179}$$

where $\mathbf{B}$ is the strain–displacement matrix and $\mathbf{d}$ is the nodal displacement vector. For the 8-node hexahedral element, each node has three nodal displacements u, v, and w. Therefore, there are a total of 24 nodal displacements in the $\mathbf{d}$ vector, listed as shown in Eq. (7.179). The strain–displacement matrix $\mathbf{B}$ can be considered as a vector of eight smaller $\mathbf{B}_i$, $i = 1, \ldots, 8$, matrices with a repeated pattern of matrix entries, which are

$$\mathbf{B}_i = \begin{Bmatrix} \frac{\partial N_i}{\partial x} & 0 & 0 \\ 0 & \frac{\partial N_i}{\partial y} & 0 \\ 0 & 0 & \frac{\partial N_i}{\partial z} \\ \frac{\partial N_i}{\partial x} & \frac{\partial N_i}{\partial y} & 0 \\ 0 & \frac{\partial N_i}{\partial y} & \frac{\partial N_i}{\partial z} \\ \frac{\partial N_i}{\partial x} & 0 & \frac{\partial N_i}{\partial z} \end{Bmatrix} \quad i = 1, \ldots, 8. \tag{7.180}$$

In short form, the strain vector is given by

$$\boldsymbol{\epsilon}_{6 \times 1} = \mathbf{B}_{6 \times 24} \mathbf{d}_{24 \times 1}. \tag{7.181}$$

Similarly,

$$\delta\boldsymbol{\epsilon}_{6 \times 1} = \mathbf{B}_{6 \times 24} \delta\mathbf{d}_{24 \times 1}. \tag{7.182}$$

It is shown in Eq. (7.180) that the **B** matrix contains the derivatives of the shape functions with respect to x, y, and z. However, the shape functions are defined in the master element in terms of ξ, η, and ζ. As discussed in previous chapters, the transformation is done via the Jacobian matrix as

$$\begin{Bmatrix} \frac{\partial N_i}{\partial x} \\ \frac{\partial N_i}{\partial y} \\ \frac{\partial N_i}{\partial z} \end{Bmatrix} = \mathbf{J}^{-1} \begin{Bmatrix} \frac{\partial N_i}{\partial \xi} \\ \frac{\partial N_i}{\partial \eta} \\ \frac{\partial N_i}{\partial \zeta} \end{Bmatrix} \quad i = 1, \ldots, 8 \tag{7.183}$$

where as shown in Eq. (5.166) **J** is given by

$$\mathbf{J} = \begin{bmatrix} \frac{\partial x}{\partial \xi} & \frac{\partial y}{\partial \xi} & \frac{\partial z}{\partial \xi} \\ \frac{\partial x}{\partial \eta} & \frac{\partial y}{\partial \eta} & \frac{\partial z}{\partial \eta} \\ \frac{\partial x}{\partial \zeta} & \frac{\partial y}{\partial \zeta} & \frac{\partial z}{\partial \zeta} \end{bmatrix}. \tag{7.184}$$

By using the relation

$$x = \sum_{i=1}^{8} N_i(\xi, \eta, \zeta) x_i \quad y = \sum_{i=1}^{8} N_i(\xi, \eta, \zeta) y_i \quad z = \sum_{i=1}^{8} N_i(\xi, \eta, \zeta) z_i \tag{7.185}$$

the Jacobian matrix can be calculated by

$$\mathbf{J} = \begin{bmatrix} \dfrac{\partial N_1}{\partial \xi} & \dfrac{\partial N_2}{\partial \xi} & \dfrac{\partial N_3}{\partial \xi} & \cdots & \dfrac{\partial N_8}{\partial \xi} \\ \dfrac{\partial N_1}{\partial \eta} & \dfrac{\partial N_2}{\partial \eta} & \dfrac{\partial N_3}{\partial \eta} & \cdots & \dfrac{\partial N_8}{\partial \eta} \\ \dfrac{\partial N_1}{\partial \zeta} & \dfrac{\partial N_2}{\partial \zeta} & \dfrac{\partial N_3}{\partial \zeta} & \cdots & \dfrac{\partial N_8}{\partial \zeta} \end{bmatrix} \begin{bmatrix} x_1 & y_1 & z_1 \\ x_2 & y_2 & z_2 \\ x_3 & y_3 & z_3 \\ \vdots & \vdots & \vdots \\ x_8 & y_8 & z_8 \end{bmatrix} \quad (7.186)$$

where x_i, y_i, and z_i are the coordinates of the i-th node of the physical element.

Substituting the strain approximations into the first integral in the element weak form, we obtain

$$\int_{\Omega^{(e)}} \delta \boldsymbol{\epsilon}^T \mathbf{C} \boldsymbol{\epsilon} \, d\Omega = \int_{\Omega^{(e)}} \delta \mathbf{d}_{1 \times 24}^T \mathbf{B}_{24 \times 6}^T \mathbf{C}_{6 \times 6} \mathbf{B}_{6 \times 24} \mathbf{d}_{24 \times 1} \, d\Omega$$
$$= \delta \mathbf{d}_{1 \times 24}^T \left(\int_{\Omega^{(e)}} \mathbf{B}_{24 \times 6}^T \mathbf{C}_{6 \times 6} \mathbf{B}_{6 \times 24} \, d\Omega \right) \mathbf{d}_{24 \times 1}. \quad (7.187)$$

The integral in the parenthesis is the element stiffness matrix

$$\mathbf{K}^{(e)} = \int_{\Omega^{(e)}} \mathbf{B}_{24 \times 6}^T \mathbf{C}_{6 \times 6} \mathbf{B}_{6 \times 24} \, d\Omega. \quad (7.188)$$

By using the isoparametric mapping, the integral over the element can be mapped to the master element with

$$\int_{\Omega^{(e)}} \rightarrow \int_{-1}^{1} \int_{-1}^{1} \int_{-1}^{1} \quad \text{and} \quad d\Omega = det(\mathbf{J}) d\xi \, d\eta \, d\zeta. \quad (7.189)$$

The numerical integration of the $\mathbf{K}^{(e)}$ matrix is performed by using Gauss quadrature. The numerical integration can be written as

$$\mathbf{K}^{(e)} = \sum_{g=1}^{ng} \mathbf{B}^T(\xi_g, \eta_g, \zeta_g) \mathbf{C} \mathbf{B}(\xi_g, \eta_g, \zeta_g) det(\mathbf{J}(\xi_g, \eta_g, \zeta_g)) w_g \quad (7.190)$$

where ng is the number of integration points, and w_g is the weight for the g-th integration point.

The second integral in the weak form, Eq. (7.178), is the virtual work done by the body force. Substituting the finite element approximation of the virtual displacement and by following the derivation shown in Section 7.2.2, the body force virtual work integral can be rewritten as

$$\int_{\Omega^{(e)}} \delta \mathbf{u}^T \mathbf{b} \, d\Omega$$

$$= \delta \mathbf{d}^T \left[\int_{-1}^{1} \int_{-1}^{1} \int_{-1}^{1} \begin{bmatrix} N_1 b_x \\ N_1 b_y \\ N_1 b_z \\ N_2 b_x \\ N_2 b_y \\ N_2 b_z \\ \vdots \\ N_8 b_x \\ N_8 b_y \\ N_8 b_z \end{bmatrix} det(\mathbf{J}) d\xi \, d\eta \, d\zeta \right] = \delta \mathbf{d}^T \mathbf{f}_b^{(e)}. \quad (7.191)$$

The vector in the brackets $\mathbf{f}_b^{(e)}$ is the element body force vector. In our T-beam example, there is no body force applied. Therefore, the body force vector is zero.

By following the same procedure, the third term in the element weak form, which is the virtual work done by the surface traction can be rewritten as

$$\int_{\Gamma_t^{(e)}} \delta \mathbf{u}^T \mathbf{t} \, d\Omega$$

$$= \delta \mathbf{d}^T \left[\int_{\Gamma_t^{(e)}} \begin{bmatrix} N_1 t_x \\ N_1 t_y \\ N_1 t_z \\ N_2 t_x \\ N_2 t_y \\ N_2 t_z \\ \vdots \\ N_8 t_x \\ N_8 t_y \\ N_8 t_z \end{bmatrix} det(\mathbf{J}) d\xi \, d\eta \, d\zeta \right] = \delta \mathbf{d}^T \mathbf{f}_t^{(e)}. \quad (7.192)$$

Since the displacement variation vector in each term can be canceled, the discretized element weak form can be rewritten as a linear system as

$$\mathbf{K}^{(e)} \mathbf{d} = \mathbf{f}_b^{(e)} + \mathbf{f}_t^{(e)} \quad (7.193)$$

where the right hand vectors are the applied external force vectors on the element. Note that for the T-beam problem, both force vectors in Eq. (7.193) are zero.

Step 7: Global Matrix and Vector Assembly

The global matrix and vector assembly procedure is mostly the same as in 2-D elastostatic analysis. The process illustrated in Fig. 7.16 is applicable to the 3-D case, except that the dimension of the element stiffness matrix is 24×24, and each primitive assembling block is 3×3.

The nodal point forces, which were omitted in writing out the element weak form, are now to be assembled into the global force vector $\mathbf{F}$. The location in $\mathbf{F}$ a point force is assembled at depends on the node on which it is acting. For example, if a point force $\mathbf{f} = \{f_x, f_y, f_z\}$ is acting on node m, then the force components should be added to $\mathbf{F}$ at rows $3m - 2$, $3m - 1$, and $3m$, respectively. The final linear system is obtained as

$$\mathbf{K}_{3N \times 3N} \mathbf{u}_{3N \times 1} = \mathbf{F}_{3N \times 1} \qquad (7.194)$$

where N is the total number of nodes.

Step 8: Apply the Essential Boundary Condition

The application of the essential (displacement) boundary condition can be implemented by using the direct substitution method, the penalty method, and the Lagrange multiplier method discussed in previous chapters. These methods are equally applicable in 3-D elastostatic analysis.

Step 9: Solve the Global System

After the essential boundary conditions are applied, the final linear system can be solved in MATLAB. The unknown nodal displacement vector $\mathbf{u}$ can be computed.

Step 10: Post-processing

We first visualize the displacement results. A simple but effective way is plotting the deformed shape of the structure. The deformed positions of the nodes are obtained by adding the displacements to the coordinates of undeformed nodes. Note that, in many engineering applications, the displacements can be quite small so that the deformed shape looks not very different from the undeformed shape. In this case, a scaling factor is useful to artificially enlarge the displacement for visualization. Figure 7.20 shows the deformed shape of the T-beam. The MATLAB code that does the plotting is shown below.

In addition, by using the displacement results, we can compute secondary physical quantities. For static elasticity analysis, these physical quantities of interest are typically strains and stresses. Methods for strain and stress calculation in the post-processing stage are introduced in Section 7.2.2. By definition, the strains are the derivatives of the displacements, and the stresses can be obtained using Hooke's law. Here we leave the actual calculations of the stresses to the reader as an exercise.

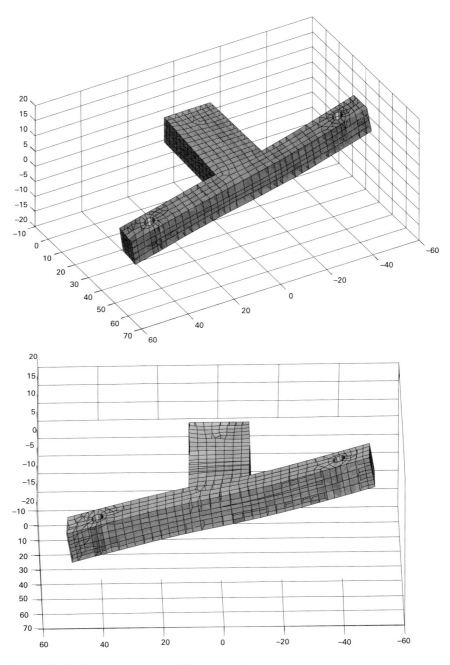

Figure 7.20 Deformed shape of the T-beam (viewing from two different angles).

```matlab
1  % next 3 lines load files
2  load elements.dat;
3  load nodes.dat;
4  load feU.dat;      % results file obtained from FEA
5   scaling_factor =100;  % set scaling factor
6  nodes=nodes(:,2:4)+feU(:,5:7)*scaling_factor ; % deformed nodes
7
8  % next 16 lines : create patches of the element faces
9  p=patch('Vertices',nodes,'Faces',elements(:,2:5) );
10 set(p,'facecolor','[.9 .9 .9]','edgecolor','black');
11 s1=elements(:,[2 3 7 6]);
12 p=patch('Vertices',nodes,'Faces',s1);
13 set(p,'facecolor','[.9 .9 .9]','edgecolor','black');
14 s1=elements(:,[3 4 8 7]);
15 p=patch('Vertices',nodes,'Faces',s1);
16 set(p,'facecolor','[.9 .9 .9]','edgecolor','black');
17 s1=elements(:,[4 5 9 8]);
18 p=patch('Vertices',nodes,'Faces',s1);
19 set(p,'facecolor ','[.9 .9 .9]','edgecolor','black');
20 s1=elements(:,[5 2 6 9]);
21 p=patch('Vertices',nodes,'Faces',s1);
22 set(p,'facecolor','[.9 .9 .9]','edgecolor','black');
23 p=patch('Vertices',nodes,'Faces',elements(:,6:9) );
24 set(p,'facecolor','[.9 .9 .9]','edgecolor','black');
25
26 % next 6 lines : set 3-D view
27 daspect([1 1 0.7]);
28 view(150,35);
29 grid on;
30 camlight; lighting gouraud;
31 alpha(.75)
32 axis([-60 60 -10 70 -20 20]);
```

7.3.2 Computer Implementation

In this section, by using the T-beam problem as an example, we present the data files and the complete MATLAB code for the finite element analysis.

Data Files

The input data files are shown in the tables below. The nodes, elements, material properties, loads, boundary conditions, and other model options are stored in forms of arrays and matrices. There are 1108 elements and 1730 nodes in the mesh shown in Fig. 7.19. Only the first five lines of the "nodes.dat" and "elements.dat" are shown in the tables. For the same reason, "bcsdisp.dat" only shows the first five lines in the table.

MATLAB Code

The first code is the main program of the 3-D elastostatics analysis:

Table 7.2 Data files for analysis

nodes.dat

1	−0.3562	9.0424	8.0000
2	1.3138	16.3431	8.0000
3	1.2822	19.0760	8.0000
4	1.3340	13.6104	8.0000
5	1.3126	21.8556	8.0000
⋮	⋮	⋮	⋮

elements.dat

1	541	417	414	368	583	1102	1106	1280
2	583	1102	1106	1280	584	1103	1108	1486
3	584	1103	1108	1486	585	1104	1110	1692
4	585	1104	1110	1692	608	1101	1111	45
5	369	420	416	361	1281	1090	1094	1119
⋮	⋮	⋮	⋮	⋮	⋮	⋮	⋮	⋮

bcsdisp.dat			bcsforce.dat				materials.dat
451	1	0	420	0	−10000	−10000	% Young's modulus (Pa) 200e9
451	2	0	1025	0	0	10000	
451	3	0					% Poisson's ratio 0.3
452	1	0					
452	2	0					
⋮	⋮						

```
1  clear all ;
2  % next 3 lines : read the input files
3  filenames = { 'nodes.dat' , 'elements.dat' , 'materials. dat' ,   ...
```

```
4                   'bcsforce.dat', 'bcsdisp.dat'};
5   for i = 1:numel(filenames); load(filenames{i}); end;
6
7   % next 4 lines : bookkeeping
8   n_nodes=size(nodes,1);
9   n_elements=size(elements,1);
10  n_bcsforce=size(bcsforce,1);
11  n_bcsdisp=size(bcsdisp,1);
12
13  K=CompK(nodes, elements, materials);   % compute global K matrix
14  F=CompF(nodes, elements, bcsforce);    % compute global F vector
15
16  % next 7 lines : apply displacement boundary condition
17  coeff=abs(max(K(bcsdisp(1,1)*3−2,:)))*1e7;   % penalty number
18  for i=1:n_bcsdisp
19      node_id=bcsdisp(i,1);
20      direction =bcsdisp(i,2);
21      K(3*node_id−3 + direction, 3*node_id−3 + direction )=coeff;
22      F(3*node_id−3 + direction, 1) = bcsdisp(i,3)*coeff;
23  end
24
25  u=K\F;  % solve the global linear system
26
27  % next 5 lines : save the displacement results in file
28  U=zeros(n_nodes,7);
29  for n=1:n_nodes
30      U(n,1:7)=[n nodes(n,2:4) u(3*n−2:3*n,1)'];
31  end
32  save −ascii −double feU.dat U
```

The function to compute the element stiffness matrices and assemble them into the global matrix:

```
1   function K=CompK(nodes, elements, materials)
2   n_nodes = size(nodes,1);
3   n_elements = size(elements,1);
4   n_nodes_per_element = size(elements,2) −1;
5   K=zeros(n_nodes*3,n_nodes*3);
6   C=CompC3D(materials);
7   [gauss_points, gauss_weights]=GetHexaGauss(2,2,2);
8   n_gauss_points=size(gauss_points,1)
9   [N,Nx,Ny,Nz]=CompNDNatPointsHexa8(gauss_points(:,1), ...
10                  gauss_points (:,2), gauss_points (:,3) );
11
12  % compute K matrix : loop over all the elements
13  for e=1:n_elements
14      ke=zeros(n_nodes_per_element*3, n_nodes_per_element*3);
15      [element_nodes, node_id_map]= SetElementNodes(e, nodes, elements);
```

```
16   % next 7 lines : compute element stiffness matrix ke
17   for g=1:n_gauss_points
18       J=CompJacobian3DatPoint(element_nodes, Nx(:,g), Ny(:,g), Nz(:,g));
19       detJ=det(J);
20       Jinv=inv(J);
21       B=CompB6x24Hexa8atPoint(Nx(:,g), Ny(:,g), Nz(:,g), Jinv);
22       ke=ke+B'*C*B*detJ*gauss_weights(g);
23   end
24   % assemble ke into global K
25   K= AssembleGlobalMatrix(K,ke,node_id_map,3);
26   end
```

The function to compute the element force vectors and assemble them into the global right hand side vector:

```
1  function F=CompF(nodes, elements, bcsforce)
2  n_nodes = size(nodes,1);
3  n_force_nodes = size(bcsforce,1);
4  F=zeros(n_nodes*3,1);
5
6  % for-loop: apply point forces
7  for i=1:n_force_nodes
8      row=3*bcsforce(i,1)-2;
9      F(row,1) = F(row,1)+ bcsforce(i,2);
10     F(row+1,1) = F(row+1,1)+ bcsforce(i,3);
11     F(row+2,1) = F(row+2,1)+ bcsforce(i,4);
12 end
```

The function to compute the 3×3 Jacobian matrix at a given integration point:

```
1  % Compute Jacobian matrix at (xi, eta, zeta) in a 3-D master element
2  % Input: element_nodes: the physical coordinates of a 3-D element
3  %        in the format of [x1, y1, z1; x2 y2 z2; x3, y3 z3; ...]
4  % Input: Nxi, Neta, Nzeta: dN/dxi, dN/deta, dN/dzeta vectors at point
5  % Output: Jacobian matrix at the point
6  function J= CompJacobian3DatPoint(element_nodes, Nxi, Neta, Nzeta)
7  J=zeros(3,3);
8  for j=1:3
9      J(1,j) = Nxi' * element_nodes(:,j);
10     J(2,j) = Neta' * element_nodes(:,j);
11     J(3,j) = Nzeta' * element_nodes(:,j);
12 end
```

The function to compute the strain–displacement matrix:

```
1  % Compute strain - displacement matrix
2  function B= CompB6x24Hexa8atPoint(Nx_v, Ny_v, Nz_v, Jinv)
3  B=zeros(6,24);
```

```
4   dNmaster=zeros(3,1);
5   for i=1:8
6     dNmaster(1,1)=Nx_v(i);
7     dNmaster(2,1)=Ny_v(i);
8     dNmaster(3,1)=Nz_v(i);
9     dN=Jinv*dNmaster;
10    B(1,3*i-2)=dN(1,1);
11    B(4,3*i-1)=dN(1,1);
12    B(6,3*i)  = dN(1,1);
13    B(2,3*i-1)=dN(2,1);
14    B(4,3*i-2)=dN(2,1);
15    B(5,3*i)  = dN(2,1);
16    B(3,3*i)=dN(3,1);
17    B(5,3*i-1)=dN(3,1);
18    B(6,3*i-2) = dN(3,1);
19  end
```

The function to compute the 3-D material matrix:

```
1   % Compute material matrix (3-D)
2   function C= CompC3D(materials)
3   C=zeros(6,6);
4   E=materials(1);   % Young's modulus
5   nu=materials(2);  % Poisson's ratio
6   t = E/((1+nu)*(1-2*nu));
7   c11=t*(1-nu);
8   c12=t*nu;
9   c44=t*((1-2*nu)/2);
10  C=[c11 c12 c12   0   0   0
11     c12 c11 c12   0   0   0
12     c12 c12 c11   0   0   0
13      0   0   0  c44   0   0
14      0   0   0   0  c44   0
15      0   0   0   0   0  c44];
```

In the MATLAB programs listed above, the functions which have already been discussed previously are not repeated here. They are listed below for your reference:

function [gauss_points, gauss_weights]=GetHexaGauss(n_xi,n_eta,n_zeta): Section 5.3.1.

function [N,Nx,Ny]=CompNDNatPointsHexa8(xi_v, eta_v, zeta_v): Section 5.3.1.

function [element_nodes, node_id_map]= SetElementNodes(ele, nodes, elements): Section 5.2.1.

function K=AssembleGlobalMatrix(K, e, node_id_map, ndDOF): Section 5.2.1.

7.4 2-D Steady State Incompressible Viscous Flow

Fluid flow in multi-dimensions is another example of multi-dimensional vector field problems. While fluid flow problems can be classified into many sub-categories as shown in Fig. 4.25, generally a fluid flow is represented by its velocity and pressure profiles in the prescribed spatial and temporal domains. Therefore, there are multiple degrees of freedom (velocity and pressure) at any spatial location of the fluid domain, making it a vector field problem. In this section, we use a classical cavity driven flow problem to illustrate the finite analysis steps for steady state incompressible viscous fluid flow.

7.4.1 2-D Steady State Cavity Driven Flow

Step 1: Physical Problem

We consider a 2-D square shaped cavity as shown in Fig. 7.21. The cavity is filled with some type of fluid. The top lid of the cavity is moving at a unit speed to the right. Assume the fluid is incompressible and has a unit viscosity. The flow is laminar due to its relatively low velocity. The dimensions of the cavity are shown in the figure. In this example, we calculate the steady state velocity profile along the vertical centerline of the cavity.

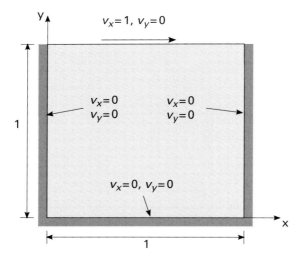

Figure 7.21 Cavity driven flow.

Step 2: Mathematical Model

The equations of motion derived for solids and shown in Eqs. (7.8–7.10) are essentially Newton's second law $\mathbf{F} = m\mathbf{a}$. Note that $m\mathbf{a}$ is simply $d(m\mathbf{v})/dt$ for a constant mass. $m\mathbf{v}$ is the momentum of the object. Therefore, Newton's second law is equivalent to stating that the net force equals the rate of change of momentum. This statement is called conservation of linear momentum. For fluids, the law of conservation of linear momentum gives

$$\frac{\partial \sigma_{xx}}{\partial x} + \frac{\partial \sigma_{xy}}{\partial y} + \frac{\partial \sigma_{xz}}{\partial z} + b_x = \rho \frac{\partial v_x}{\partial t} + \rho \left(v_x \frac{\partial v_x}{\partial x} + v_y \frac{\partial v_x}{\partial y} + v_z \frac{\partial v_x}{\partial z} \right) \quad (7.195)$$

$$\frac{\partial \sigma_{xy}}{\partial x} + \frac{\partial \sigma_{yy}}{\partial y} + \frac{\partial \sigma_{yz}}{\partial z} + b_y = \rho \frac{\partial v_y}{\partial t} + \rho \left(v_x \frac{\partial v_y}{\partial x} + v_y \frac{\partial v_y}{\partial y} + v_z \frac{\partial v_y}{\partial z} \right) \quad (7.196)$$

$$\frac{\partial \sigma_{xz}}{\partial x} + \frac{\partial \sigma_{yz}}{\partial y} + \frac{\partial \sigma_{zz}}{\partial z} + b_z = \rho \frac{\partial v_z}{\partial t} + \rho \left(v_x \frac{\partial v_z}{\partial x} + v_y \frac{\partial v_z}{\partial y} + v_z \frac{\partial v_z}{\partial z} \right) \quad (7.197)$$

where ρ is the mass desnity, v_x, v_y, and v_z are the velocity components in the x-, y- and z-directions, respectively. Comparing Eqs. (7.8–7.10) and Eqs. (7.195–7.197), there are several differences on the right hand side of the equations. First is that the primary physical quantities are the velocities in the fluid case, and displacements in the solid case. Second, there are extra terms in the parenthesis in the fluid case. Both of these differences are due to the different reference coordinate systems used to describe the deformation and motion of the two type of materials. For solids, it is easy and natural to keep track of the positions of material particles in an object when it undergoes a deformation. Therefore, it is natural to use the initial positions of the material particles in the initial coordinate system as references to describe their deformed positions as a function of time. The coordinates of the material particles in the initial coordinate system are referred to as the material coordinates. The linear elasticity equations of solids are all defined using material coordinates. For fluids, however, the initial positions of the fluid material particles (e.g. water molecules) are usually unknown, and it is also difficult to follow the fluid particles and keep track of their deformed positions as they typically travel over very large distances with complex trajectories. It is more convenient to describe the flow field by using a setting much like a camera frame. A fluid flow is described with the flow velocity and pressure at all the locations inside the camera frame at any given instant of time. That is, snapshots of flow velocity and pressure fields inside the camera frame are taken as time proceeds. These snapshots together provide a description of the transient behavior of the fluid flow. The snapshot is an instantaneous description of the flow at a given instant of time. In this description, the individual fluid particles are not being tracked. Instead, we keep track of the velocity and pressure of the fluid flow at given spatial locations at given instants of time. The coordinates of the fluid particle inside the camera frame are referred to as the current, instant, or spatial coordinates. This important difference between the kinematics descriptions of solids and fluids leads to different expressions of the same conservation law of linear momentum. For more information, the reader is referred to a continuum mechanics textbook (Holzapfel 2002, Chandrasekharaiah and Debnath 2014).

In the cavity driven flow problem, the terms in the parenthesis can be neglected since the velocity and velocity gradient are both small due to the low Reynolds number of the flow. In addition, the problem is a 2-D steady state problem, meaning time derivatives are equal to zero and all physical quantities in the z-direction are either zero or a constant. The linear momentum equations shown in Eqs. (7.195–7.197) are then reduced to

$$\frac{\partial \sigma_{xx}}{\partial x} + \frac{\partial \sigma_{xy}}{\partial y} + b_x = 0 \tag{7.198}$$

$$\frac{\partial \sigma_{xy}}{\partial x} + \frac{\partial \sigma_{yy}}{\partial y} + b_y = 0. \tag{7.199}$$

It is clear that the steady state linear momentum equations, Equations (7.198, 7.199), are simply the equations of equilibrium. In addition to the conservation of linear momentum, a conservation law of mass must also be obeyed in the fluid analysis. The mass is automatically conserved in the linear elasticity since material coordinates are used. In the fluid case, however, conservation law of mass must be explicitly enforced within a given volume in the instant coordinate system. The conservation of mass equation is given by (Chandrasekharaiah and Debnath 2014, Holzapfel 2002)

$$\frac{\partial \rho}{\partial t} + \mathbf{v} \cdot \nabla \rho + \rho \nabla \cdot \mathbf{v} = 0 \tag{7.200}$$

where $\mathbf{v}$ is the velocity vector. For incompressible flow, the mass density ρ is constant and $\nabla \rho$ is zero. For steady state flow, $\frac{\partial \rho}{\partial t} = 0$. Therefore, for the cavity driven flow problem, the conservation of mass equation becomes

$$\nabla \cdot \mathbf{v} = 0 \tag{7.201}$$

or

$$\frac{\partial v_x}{\partial x} + \frac{\partial v_y}{\partial y} = 0. \tag{7.202}$$

Equation (7.202) is also called the continuity equation. Note that, the momentum equations, Eqs. (7.198, 7.199), are in terms of stresses. They can be rewritten in terms of the velocity and pressure using the constitutive relations. For 2-D incompressible flow, the relations are (Chandrasekharaiah and Debnath 2014)

$$\begin{bmatrix} \sigma_{xx} & \sigma_{xy} \\ \sigma_{xy} & \sigma_{yy} \end{bmatrix} = -p \begin{bmatrix} 1 & 0 \\ 0 & 1 \end{bmatrix} + \mu \begin{bmatrix} 2\frac{\partial v_x}{\partial x} & \frac{\partial v_x}{\partial y} + \frac{\partial v_y}{\partial x} \\ \frac{\partial v_x}{\partial y} + \frac{\partial v_y}{\partial x} & 2\frac{\partial v_y}{\partial y} \end{bmatrix} \tag{7.203}$$

where p is the pressure and μ is the viscosity. Substituting Eq. (7.203) into Eqs. (7.198, 7.199), and using the continuity equation Eq. (7.198), we obtain

$$\mu \left(\frac{\partial^2 v_x}{\partial x^2} + \frac{\partial^2 v_x}{\partial y^2} \right) - \frac{\partial p}{\partial x} + b_x = 0 \tag{7.204}$$

$$\mu \left(\frac{\partial^2 v_y}{\partial x^2} + \frac{\partial^2 v_y}{\partial y^2} \right) - \frac{\partial p}{\partial y} + b_y = 0. \tag{7.205}$$

The momentum equations in terms of velocity and pressure, Eqs. (7.204, 7.205), together with the continuity equation, Eq. (7.202), are the governing equations of the

2-D incompressible viscous flow. The laminar incompressible viscous flow is also referred to as the Stokes flow.

From Fig. 7.21, the velocity boundary conditions are given by

$$v_x = 1, \quad v_y = 0 \quad \text{at} \quad y = 1, \quad (7.206)$$

$$v_x = 0, \quad v_y = 0 \quad \text{at} \quad x = 0, \ x = 1, \text{ and } y = 0. \quad (7.207)$$

While there is no explicitly applied pressure boundary condition, it is necessary to define a reference pressure at a location of the computational domain. In this example, we set

$$p = 0 \quad \text{at} \quad x = 0.5, \ y = 1. \quad (7.208)$$

The boundary conditions combined with the governing equations, Eqs. (7.204, 7.205) and Eq. (7.202), represent the strong form of the 2-D incompressible viscous cavity driven flow problem.

Step 3: Weak Form

In this section, we derive the weak form of the 2-D incompressible viscous flow problem by using the Galerkin weighted residual method. Taking the residual of the three PDEs of the strong form, multiplying them by the variations of the three physical quantities v_x, v_x, and p, and then integrating the product over the domain, we have (Reddy 1993)

$$\int_\Omega \delta v_x \left[\mu \left(\frac{\partial^2 v_x}{\partial x^2} + \frac{\partial^2 v_x}{\partial y^2} \right) - \frac{\partial p}{\partial x} + b_x \right] d\Omega = 0 \quad (7.209)$$

$$\int_\Omega \delta v_y \left[\mu \left(\frac{\partial^2 v_y}{\partial x^2} + \frac{\partial^2 v_y}{\partial y^2} \right) - \frac{\partial p}{\partial y} + b_y \right] d\Omega = 0 \quad (7.210)$$

$$\int_\Omega -\delta p \left[\frac{\partial v_x}{\partial x} + \frac{\partial v_y}{\partial y} \right] d\Omega = 0. \quad (7.211)$$

It should be pointed out that the terms in the square brackets in Eqs. (7.209, 7.210) have the physical meaning of force per unit volume. The variation of the velocities multiplied by the force per unit volume can be interpreted as virtual power density. In the third equation, the divergence of velocity term can be interpreted as the volume change per unit time per unit undeformed volume. Multiplying that by the pressure variation also represents a virtual power (hydrostatic in this case). Therefore, Eqs. (7.209–7.211) imply the application of the principle of virtual power. Note that the minus sign in front of δp is used to make the element matrices symmetric. It can be observed from Eqs. (7.209, 7.210) that the variations of the velocities are multiplied by the Laplacian of the velocity components, which is similar to the form of Eq. (5.19) in heat transfer. Therefore, for these terms we can derive further by using Green's first identity and the procedure illustrated for the 2-D heat transfer problem following Eq. (5.19). For the terms containing only first derivatives, there is no need to change the order of the derivatives. Therefore, Eqs. (7.209–7.211) can be rewritten as

$$\int_\Gamma \delta v_x \mu \frac{\partial v_x}{\partial \mathbf{n}} d\Gamma - \int_\Omega \mu \left(\nabla \delta v_x\right)^T \nabla v_x d\Omega - \int_\Omega \delta v_x \frac{\partial p}{\partial x} d\Omega + \int_\Omega \delta v_x b_x d\Omega = 0 \quad (7.212)$$

$$\int_\Gamma \delta v_y \mu \frac{\partial v_y}{\partial \mathbf{n}} d\Gamma - \int_\Omega \mu \left(\nabla \delta v_y\right)^T \nabla v_y d\Omega - \int_\Omega \delta v_y \frac{\partial p}{\partial y} d\Omega + \int_\Omega \delta v_y b_y d\Omega = 0 \quad (7.213)$$

$$\int_\Omega -\delta p \left[\frac{\partial v_x}{\partial x} + \frac{\partial v_y}{\partial y}\right] d\Omega = 0. \quad (7.214)$$

Adding Eq. (7.212) to Eq. (7.213) gives

$$\int_\Gamma \mu \begin{Bmatrix} \delta v_x \\ \delta v_y \end{Bmatrix} \cdot \begin{Bmatrix} \frac{\partial v_x}{\partial \mathbf{n}} \\ \frac{\partial v_y}{\partial \mathbf{n}} \end{Bmatrix} - \int_\Omega \mu \begin{Bmatrix} \frac{\partial \delta v_x}{\partial x} \\ \frac{\partial \delta v_x}{\partial y} \\ \frac{\partial \delta v_y}{\partial x} \\ \frac{\partial \delta v_y}{\partial y} \end{Bmatrix} \cdot \begin{Bmatrix} \frac{\partial v_x}{\partial x} \\ \frac{\partial v_x}{\partial y} \\ \frac{\partial v_y}{\partial x} \\ \frac{\partial v_y}{\partial y} \end{Bmatrix} d\Omega$$

$$-\int_\Omega \begin{Bmatrix} \delta v_x \\ \delta v_y \end{Bmatrix} \cdot \nabla p d\Omega + \int_\Omega \begin{Bmatrix} \delta v_x \\ \delta v_y \end{Bmatrix} \cdot \begin{Bmatrix} b_x \\ b_y \end{Bmatrix} d\Omega = 0. \quad (7.215)$$

Applying Green's first identity, Eq. (5.33), to the third term (pressure gradient term) on the left hand side, Eq. (7.215) can be re-written as

$$\int_\Gamma \mu \begin{Bmatrix} \delta v_x \\ \delta v_y \end{Bmatrix} \cdot \begin{Bmatrix} \frac{\partial v_x}{\partial \mathbf{n}} \\ \frac{\partial v_y}{\partial \mathbf{n}} \end{Bmatrix} - \int_\Omega \mu \begin{Bmatrix} \frac{\partial \delta v_x}{\partial x} \\ \frac{\partial \delta v_x}{\partial y} \\ \frac{\partial \delta v_y}{\partial x} \\ \frac{\partial \delta v_y}{\partial y} \end{Bmatrix} \cdot \begin{Bmatrix} \frac{\partial v_x}{\partial x} \\ \frac{\partial v_x}{\partial y} \\ \frac{\partial v_y}{\partial x} \\ \frac{\partial v_y}{\partial y} \end{Bmatrix} d\Omega$$

$$+ \int_\Omega p \left(\frac{\partial \delta v_x}{\partial x} + \frac{\partial \delta v_y}{\partial y}\right) d\Omega - \int_\Gamma p \left(\delta v_x n_x + \delta v_y n_y\right) d\Gamma$$

$$+ \int_\Omega \begin{Bmatrix} \delta v_x \\ \delta v_y \end{Bmatrix} \cdot \begin{Bmatrix} b_x \\ b_y \end{Bmatrix} d\Omega = 0. \quad (7.216)$$

Equation (7.216), together with Eq. (7.214), which is repeated below

$$\int_\Omega -\delta p \left[\frac{\partial v_x}{\partial x} + \frac{\partial v_y}{\partial y} \right] d\Omega = 0 \quad (7.217)$$

is the weak form governing equation of the Stokes flow.

Step 4: Discretization

For the square domain of the cavity driven flow, the meshing process is particularly simple. We generate a uniform mesh of square elements for the computational domain by using the uniform mesh generating code shown in Chapter 6.

Step 5: Finite Element Approximation

The finite element approximation of the unknown physical quantities, velocities and pressures, is essentially the same as those discussed for other types of problems. However, there is a special situation in the weak form of the Stokes flow: the order of the derivative of the velocities is different from that of the pressure. Equations (7.216, 7.217) contain the first derivatives of the velocities, but no derivative of the pressure. Therefore, the continuity requirement for the velocities is an order higher than that for the pressure. The situation is shown in Fig. 7.22. If a linear quadrilateral element is used for v_x and v_y as shown in Fig. 7.22(a), the pressure is only required to be approximated as a constant over the element which can be represented by using a single node (typically center of the element). If v_x and v_y are approximated by using an 8-node quadratic element, then p can be approximated by a linear element, as shown in Fig. 7.22(b). In the latter case, the velocities and pressure share the same set of nodes. The velocities use all the eight nodes, and the pressure only uses the four corner nodes.

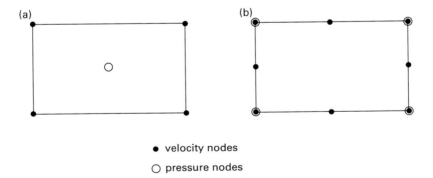

- velocity nodes
- ○ pressure nodes

Figure 7.22 Elements with different orders of approximation for velocity and pressure: (a) linear velocity and constant pressure; (b) quadratic velocity and linear pressure.

Assuming we use the quadratic/linear element shown in Fig. 7.22 (right) for the approximation, the velocities and pressure are then expressed as

$$v_x(x,y) = \sum_{i=1}^{8} N_i(x,y) v_{xi} \qquad v_y(x,y) = \sum_{i=1}^{8} N_i(x,y) v_{yi} \quad (7.218)$$

and

$$p(x, y) = \sum_{i=1}^{4} \tilde{N}_i(x, y) p_i \tag{7.219}$$

where the linear shape functions are denoted as $\tilde{N}$ to differentiate from the quadratic shape functions. While Eqs. (7.218, 7.219) are expressed in terms of x and y, as we discussed before, isoparametric mapping is typically employed to enable the use of general quadrilateral elements. The master element corresponding to Fig. 7.22 (right) is the familiar square master element in the $\xi\eta$-coordinate system, but with the different order of approximation for v_x, v_y and for p, that is 8-node for v_x, v_y and 4-node for p.

Step 6: Element Matrices and Vectors

For the cavity driven flow problem, the weak form can be further simplified due to the boundary conditions at the walls. The velocity components are specified everywhere on the boundary, which implies that the variations of the velocity components, δv_x and δv_y are zero everywhere on the boundary. Therefore, the boundary integrals in the weak form vanish. In addition, since there is no body force applied in the fluid, the body force terms are zero. The weak form given in Eqs. (7.216, 7.217) then becomes

$$\int_{\Omega} \mu \left\{ \begin{array}{c} \frac{\partial \delta v_x}{\partial x} \\ \frac{\partial \delta v_x}{\partial y} \\ \frac{\partial \delta v_y}{\partial x} \\ \frac{\partial \delta v_y}{\partial y} \end{array} \right\} \cdot \left\{ \begin{array}{c} \frac{\partial v_x}{\partial x} \\ \frac{\partial v_x}{\partial y} \\ \frac{\partial v_y}{\partial x} \\ \frac{\partial v_y}{\partial y} \end{array} \right\} d\Omega - \int_{\Omega} p \left(\frac{\partial \delta v_x}{\partial x} + \frac{\partial \delta v_y}{\partial y} \right) d\Omega = 0 \tag{7.220}$$

$$\int_{\Omega} -\delta p \left[\frac{\partial v_x}{\partial x} + \frac{\partial v_y}{\partial y} \right] d\Omega = 0. \tag{7.221}$$

Equations (7.220, 7.221) become element weak form when the integrals are over an element "e." Substituting the approximations into the element weak form, we obtain

$$\begin{bmatrix} \delta v_{x1} & \delta v_{y1} & \cdots & \delta v_{x8} & \delta v_{y8} \end{bmatrix} \times$$

$$\left(\int_{\Omega}^{(e)} \begin{bmatrix} \frac{\partial N_1}{\partial x} & \frac{\partial N_1}{\partial y} & 0 & 0 \\ 0 & 0 & \frac{\partial N_1}{\partial x} & \frac{\partial N_1}{\partial y} \\ \vdots & \vdots & \vdots & \vdots \\ \frac{\partial N_8}{\partial x} & \frac{\partial N_8}{\partial y} & 0 & 0 \\ 0 & 0 & \frac{\partial N_8}{\partial x} & \frac{\partial N_8}{\partial y} \end{bmatrix} \mu \begin{bmatrix} \frac{\partial N_1}{\partial x} & 0 & \cdots & \frac{\partial N_8}{\partial x} & 0 \\ \frac{\partial N_1}{\partial y} & 0 & \cdots & \frac{\partial N_8}{\partial y} & 0 \\ 0 & \frac{\partial N_1}{\partial x} & \cdots & 0 & \frac{\partial N_8}{\partial x} \\ 0 & \frac{\partial N_1}{\partial y} & \cdots & 0 & \frac{\partial N_8}{\partial y} \end{bmatrix} d\Omega \right) \begin{Bmatrix} v_{x1} \\ v_{y1} \\ \vdots \\ v_{x8} \\ v_{y8} \end{Bmatrix}$$

$$-\begin{bmatrix} \delta v_{x1} & \delta v_{y1} & \cdots & \delta v_{x8} & \delta v_{y8} \end{bmatrix} \times$$

$$\left(\int_\Omega^{(e)} \begin{bmatrix} \frac{\partial N_1}{\partial x} \\ \frac{\partial N_1}{\partial y} \\ \vdots \\ \frac{\partial N_8}{\partial x} \\ \frac{\partial N_8}{\partial y} \end{bmatrix} \begin{bmatrix} \tilde{N}_1 & \tilde{N}_2 & \tilde{N}_3 & \tilde{N}_4 \end{bmatrix} d\Omega \right) \begin{Bmatrix} p_1 \\ p_2 \\ p_3 \\ p_4 \end{Bmatrix} = 0 \qquad (7.222)$$

$$-\begin{bmatrix} \delta p_1 & \delta p_2 & \cdots & \delta p_4 \end{bmatrix} \left(\int_\Omega^{(e)} \begin{bmatrix} \tilde{N}_1 \\ \tilde{N}_2 \\ \tilde{N}_3 \\ \tilde{N}_4 \end{bmatrix} \begin{bmatrix} \frac{\partial N_1}{\partial x} & \frac{\partial N_1}{\partial y} & \cdots & \frac{\partial N_8}{\partial x} & \frac{\partial N_8}{\partial y} \end{bmatrix} d\Omega \right) \begin{Bmatrix} v_{x1} \\ v_{y1} \\ \vdots \\ v_{x8} \\ v_{y8} \end{Bmatrix} = 0.$$

$$(7.223)$$

Canceling out the vector of variations, Eqs. (7.222, 7.223) can be written in short form as

$$\mathbf{K}_{vv}^{(e)} \mathbf{v} - \mathbf{K}_{vp}^{(e)} \mathbf{p} = \mathbf{0} \qquad (7.224)$$

$$-\mathbf{K}_{pv}^{(e)} \mathbf{v} + \mathbf{0} = \mathbf{0} \qquad (7.225)$$

where $\mathbf{K}_{vv}^{(e)}$ and $\mathbf{K}_{vp}^{(e)}$ are the integrals in the first and second parentheses in Eq. (7.222), $\mathbf{K}_{pv}^{(e)}$ is the integral in the parenthesis in Eq. (7.223). Combining Eqs. (7.224, 7.225), one can write

$$\begin{bmatrix} \mathbf{K}_{vv}^{(e)} & -\mathbf{K}_{vp}^{(e)} \\ -\mathbf{K}_{pv}^{(e)} & \mathbf{0} \end{bmatrix} \begin{Bmatrix} \mathbf{v}^{(e)} \\ \mathbf{p}^{(e)} \end{Bmatrix} = \begin{Bmatrix} \mathbf{0} \\ \mathbf{0} \end{Bmatrix}. \qquad (7.226)$$

It is now clear that with the minus sign in Eq. (7.223), and $\mathbf{K}_{vp}^{(e)} = \left(\mathbf{K}_{pv}^{(e)} \right)^T$, symmetry of the coefficient matrix is ensured. Once again, evaluating the coefficient matrices for general shaped quadrilateral or triangular elements requires isoparametric mapping. The procedure of isoparametric mapping between the actual and the master elements is the same as discussed in previous chapters (e.g. 2-D heat transfer). It is not repeated here.

Step 7: Global Matrix and Vector Assembly
The global matrix and vector assembly process follows the same procedure described in previous chapters. The only difference in the fluid flow problem is that the nodal

velocities and pressure are listed separately in the right hand side vectors. This arrangement is for convenience. It is also straightforward to put the velocity and pressure degrees of freedom together for each node. As shown in Fig. 7.23, the element matrix contains four main blocks, each nonzero block corresponding to an integral term in Eqs. (7.222, 7.223). Since the approximation order is different for the velocity and pressure, the nodal velocity and pressure degrees of freedom are defined on different sets of nodes as shown in Fig. 7.22. For the velocity, there are two degrees of freedom (one in x and one in y) for each of the eight nodes in an element. Therefore the dimension of $\mathbf{K}_{vv}^{(e)}$ is 16×16. The scalar pressure is approximated using four nodes, making the element pressure degrees of freedom equal to four. Thus the dimensions of $\mathbf{K}_{vp}^{(e)}$ and $\mathbf{K}_{pv}^{(e)}$ are 16×4 and 4×16, respectively.

The global system is partitioned in the same manner as shown in Fig. 7.23. In the assembly process, the element matrix and vector entries are assembled into their corresponding vv, vp, and pv blocks. As described before, the global assembling locations of the elements are determined by using the mapping between the global and local nodal indices/IDs.

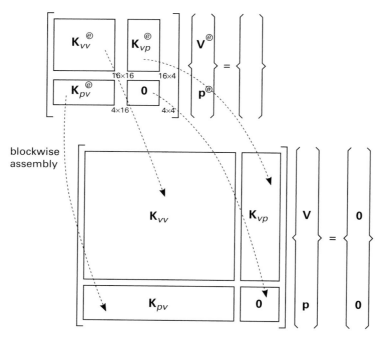

Figure 7.23 Assembling the global system.

Steps 8 and 9: Apply Boundary Conditions and Solve the Global System

In the cavity driven flow problem, there is no derivative (natural) boundary condition. The essential (velocity and pressure) boundary conditions can be applied by using the methods described in Section 5.1. After the essential boundary conditions are applied, the final linear system can be solved directly in MATLAB. The global velocity-and-pressure vector $\{\mathbf{v}\ \mathbf{p}\}^T$ can be computed.

Step 10: Post-processing

For the cavity driven flow problem, we plot the velocity field in the cavity. The velocity field is a vector field with magnitude and direction of the velocity at all spatial locations in the computational domain. A natural way to visualize the velocity field is plotting the velocity vectors at the nodes. In MATLAB, such a function is readily available. A quiver plot displays velocity vectors as arrows with components (v_x, v_y) at the points. Note that the velocity components and positions are all vectors. A sample code for plotting the velocity results of the current problem is shown below.

```
1  load sol.dat;   % load solution
2  figure(1)
3  quiver(sol(:,2), sol(:,3), sol(:,4), sol(:,5),2.0,...
4          'Color',[0 0 0],'LineWidth',2);   % plotting
5  axis([0 1 0 1.1]);          % set axis range
6  xlabel('x-coordinate');  % set xlabel
7  ylabel('y-coordinate');  % set ylabel
8  set(gca,'fontsize',16);  % set font size
```

Note that "sol.dat" is the results file generated by the main simulation code. The text file contains five columns. The first column contains the node indices. The second and third columns are the x- and y-coordinates of the nodes, respectively. The fourth and fifth columns are the x- and y-components of the velocity, respectively. Figure 7.24 shows the velocity plot produced by the code.

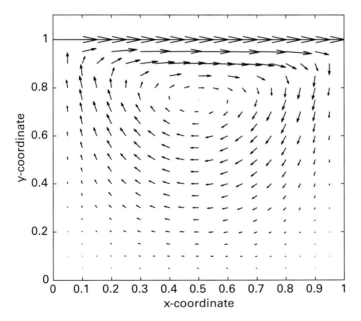

Figure 7.24 Velocity vector plot.

A benchmark result for the well-known cavity driven flow problem is the horizontal velocity along the vertical center line of the cavity. Figure 7.25 shows the convergence of the result for different meshes. It is shown that while the 10 × 10 gives a smoother curve, the mesh of the 4 × 4 elements has already provided a result with reasonable accuracy.

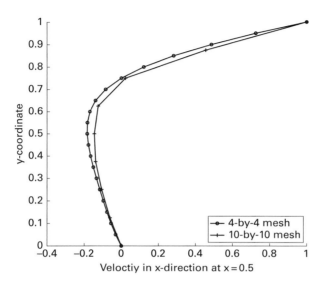

Figure 7.25 Horizontal velocity along the vertical center line of the cavity.

7.4.2 Computer Implementation

In this section, by using the cavity driven flow problem as an example, we present the data files and the complete MATLAB code for the finite element analysis.

Data Files

The input data files are shown in the tables below. The nodes, elements, material properties, boundary conditions, and other model options are stored in forms of arrays and matrices. Note that since quadratic elements are used for the velocity, and the pressure is approximated linearly within the quadratic elements, velocity and pressure have different nodes and elements files. In this case, the "nodes.dat" and "elements.dat" are for the velocity, and "pelements.dat" is for the pressure. The "pnodes.dat" file provides the relationship between the pressure nodes and the velocity nodes. For example, the first three lines of the "pnodes.dat" are "1 1," "2 0," and "3 2." The meaning of the lines are: velocity node 1 is corresponding to pressure node 1, velocity node 2 is corresponding to no pressure node, velocity node 3 is pressure node 2, respectively. There are 65 nodes in the mesh even though only 4 × 4 elements are used. Therefore, only the first five lines of the nodes and elements files are shown in the tables. The boundary condition file "bcsvp.dat" stores the velocity and pressure boundary conditions. The first column of the file has the node numbers. The second column stores the boundary condition type: 1 represents the x-component of the velocity, 2 is the y-component of

7.4 2-D Steady State Incompressible Viscous Flow

Table 7.3 Data files for analysis

nodes.dat

1	0.000	0.000
2	0.125	0.000
3	0.250	0.000
4	0.375	0.000
5	0.500	0.000
⋮	⋮	⋮

elements.dat

1	1	3	17	15	2	11	16	10
2	3	5	19	17	4	12	18	11
3	5	7	21	19	6	13	20	12
4	7	9	23	21	8	14	22	13
5	15	17	31	29	16	25	30	24
⋮	⋮	⋮	⋮	⋮	⋮	⋮	⋮	⋮

pnodes.dat

1	1
2	0
3	2
4	0
5	3
⋮	⋮

pelements.dat

1	1	3	17	15
2	3	5	19	17
3	5	7	21	19
4	7	9	23	21
5	15	17	31	29
⋮	⋮	⋮	⋮	⋮

bcsvp.dat

1	1	0.0
1	2	0.0
2	1	0.0
2	2	0.0
⋮	⋮	⋮

materials.dat

% Viscosity
1

the velocity, and 3 is the pressure. The third column contains the value of the boundary conditions.

MATLAB Code
The first code is the main program of the cavity driven flow analysis:

```matlab
clear all;
UniformMeshQuad8n4(0,0,1,1,10,10); % create uniform mesh
% next 3 lines : read the input files
filenames = {'nodes.dat','elements.dat','materials.dat', ...
             'pnodes.dat','pelements.dat'};
for i = 1:numel(filenames); load(filenames{i}); end;
SetBCs(nodes);  % set up boundary conditions
load bcsvp.dat;
n_nodes=size(nodes,1);
n_bcsvp=size(bcsvp,1);
n_pnodes=max(pnodes(:,2));

% next two lines : compute global K & F
K=CompK(nodes, pnodes, elements, pelements, materials);
F=zeros(n_nodes*2+n_pnodes,1);

% next 14 lines : apply v & P boundary conditions
coeff=abs(max(K(bcsvp(1,1)*2-1,:)))*1e7;  % penalty number
prow=2*n_nodes;
for i=1:n_bcsvp
    node_id=bcsvp(i,1);
    component=bcsvp(i,2);
    if component<=2
       K(2*node_id-2 + component, 2*node_id-2 + component)=coeff;
       F(2*node_id-2 + component, 1) = bcsvp(i,3)*coeff;
    else
       pnode_id=pnodes(node_id,2);
       K(prow + pnode_id, prow + pnode_id)=coeff;
       F(prow + pnode_id, 1) = bcsvp(i,3)*coeff;
    end
end

u=K\F;  % solve the global linear system

% next 15 lines : save the displacement results in file
U=zeros(n_nodes,6);
k=1;
for i=1:n_nodes
    U(i,1:5)=[nodes(i,1:3) u(2*i-1:2*i,1)'];
    pnode_id=pnodes(i,2);
    U(i,6)=u(prow + pnode_id,1);
    if nodes(i,2)==0.5
      vxline(k,1:2)=U(i,3:4);
```

```
44        k=k+1;
45      end
46  end
47  vxline
48  save -ascii -double sol.dat U
49  save -ascii -double vx.dat vxline
50  disp('Velocity and pressure results stored in sol.dat');
51  PlotQuiver;   % plot the velocity results
```

The function to set up the boundary conditions for this example problem:

```
1  function SetBCs(nodes)
2  fid=fopen('bcsvp.dat','w+');  % open file to write to
3  for i=1:size(nodes,1)
4    x=nodes(i,2);
5    y=nodes(i,3);
6    if y==1.0         % top edge: velocity boundary condition
7      fprintf( fid ,'%d %d %.10f\n', i, 1, 1);
8      fprintf( fid ,'%d %d %.10f\n', i, 2, 0);
9      if x==0.5       % top edge center: zero pressure
10       fprintf( fid ,'%d %d %.10f\n',i, 3, 0);
11     end
12   elseif x==0.0 || y==0.0 || x==1.0  % the other 3 edges
13     fprintf( fid ,'%d %d %.10f\n', i, 1, 0); % x-velocity =0
14     fprintf( fid ,'%d %d %.10f\n', i, 2, 0); % y-velocity =0
15   end
16  end
17  fclose( fid );
```

The function to compute the element matrices and assemble them into the global matrix:

```
1  function K=CompK(nodes, pnodes, elements, pelements, materials )
2  n_nodes = size(nodes,1);
3  n_pnodes=max(pnodes(:,2));
4  n_elements = size(elements,1);
5  n_nodes_per_element = size(elements,2)-1;
6  K=zeros(n_nodes*2+n_pnodes,n_nodes*2+n_pnodes);
7  C=materials(1)*[2 0 0; 0 2 0; 0 0 1];
8  H=CompH();
9  [gauss_points, gauss_weights]=GetQuadGauss(3,3);
10 n_gauss_points=size( gauss_points ,1);
11 [N,Nx,Ny]=CompNDNatPointsQuad8(gauss_points(:,1),gauss_points(:,2));
12
13 % for-loop: compute K11 block of K
14 for e=1:n_elements       % loop over all the elements
15   ke=zeros(n_nodes_per_element*2, n_nodes_per_element*2);
```

```
16    [element_nodes, node_id_map]= SetElementNodes(e, nodes, elements);
17    % for-loop: compute element stiffness matrix ke
18    for g=1:n_gauss_points
19      J=CompJacobian2DatPoint(element_nodes,Nx(:,g),Ny(:,g));
20      detJ=det(J);
21      Jinv=inv(J);
22      Jb (1:2,1:2) =Jinv;
23      Jb (3:4,3:4) =Jinv;
24      B=CompBQuadGaussatPoint(n_nodes_per_element, Nx(:,g), Ny(:,g));
25      HJB=(H*Jb*B);
26      ke=ke+HJB'*C*HJB*detJ*gauss_weights(g);
27    end
28    K= AssembleGlobalMatrix(K,ke,node_id_map,2); % assemble global K
29  end
30
31  % setup for K12 block
32  n_pnodes_per_element = size(pelements,2) -1;
33  K12=zeros(n_nodes*2,n_pnodes);
34  [NP,NPx,NPy]=CompNDNatPointsQuad4(gauss_points(:,1), gauss_points(:,2));
35  % for-loop: compute K12 block of K
36  for e=1:n_elements
37    ke=zeros(n_nodes_per_element*2, n_pnodes_per_element);
38    [element_nodes, node_id_map]=SetElementNodes(e, nodes, elements);
39    [pelement_nodes, pnode_id_map]=SetElementNodes(e,nodes,pelements);
40    % for-loop: compute element stiffness matrix ke
41    for g=1:n_gauss_points
42      J=CompJacobian2DatPoint(element_nodes,Nx(:,g),Ny(:,g));
43      detJ=det(J);
44      Jinv=inv(J);
45      dN=zeros(2*n_nodes_per_element,1);
46      for j=1: n_nodes_per_element
47        dN(2*j-1:2*j,1)= Jinv *[Nx(j,g) Ny(j,g) ]';
48      end
49      ke=ke+dN*(NP(:,g))'*detJ*gauss_weights(g);
50    end
51    % for-loop: assemble ke into global KP
52    for i = 1:n_nodes_per_element
53      row_node = node_id_map(i,1);
54      row=2*row_node - 1;
55      for j = 1:n_pnodes_per_element
56        col_node = pnode_id_map(j,1);
57        pnode_id=pnodes(col_node,2);
58        col=pnode_id;
59        K12(row:row+1, col)=K12(row:row+1, col) + ke(2*i-1:i*2, j);
60      end
61    end
62  end
63  % next 2 lines: fill in the symmetric part of K
```

```
64  K(1:2*n_nodes, 2*n_nodes+1:2*n_nodes+n_pnodes)=-K12;
65  K(2*n_nodes+1:2*n_nodes+n_pnodes, 1:2*n_nodes)=-K12';
```

The function to compute the strain–displacement matrix:

```
1  function B= CompBQuadGaussatPoint(n_elnodes, Nxi_vec, Neta_vec)
2  B=zeros(4,2*n_elnodes);
3  for i=1:n_elnodes
4      B(1,2*i-1)=Nxi_vec(i);
5      B(3,2*i)= Nxi_vec(i);
6      B(2,2*i-1)=Neta_vec(i);
7      B(4,2*i)= Neta_vec(i);
8  end
```

The function to compute the shape functions and their derivatives of the 8-node square master element at given set of points:

```
1   % Compute the shape functions and their first derivatives at a set
2   % of points defined by ( xi_vector , eta_vector ) in a 2-D 8-node square
3   % master element. Input is a single point when the length of
4   % xi _vector and eta _vector is one
5   % Input : xi_vector , eta_vector : coordinates of the input points
6   % Output : N: matrix storing shape function values with the format
7   %          [N1(xi_1, eta_1)  N1(xi_2, eta_2)  N1(xi_3, eta_3) ...
8   %           N2(xi_1, eta_1)  N2(xi_2, eta_2)  N2(xi_3, eta_3) ...
9   %           N3(xi_1, eta_1)  N3(xi_2, eta_2)  N3(xi_3, eta_3) ...
10  %           ...              ...              ...            ...]
11  % Output : Nx, Ny: matrices of dNi/dxi(xi, eta) and dNi/deta(xi, eta)
12  %          respectively , format is the same as N
13  function [N,Nx,Ny]=CompNDNatPointsQuad8(xi_vector, eta_vector)
14  np=size( xi_vector ,1);
15  N=zeros(8,np); Nx=zeros(8,np); Ny=zeros(8,np); % set up empty matrices
16  master_nodes=[1 1; -1 1; -1 -1; 1 -1; ...    % coordinates of the nodes
17               0 1; -1 0; 0 -1; 1 0];          % of the master element
18  % for loop : compute N, Nx, Ny
19  for j=1:np                                   % columns for point 1,2 ...
20      xi=xi_vector(j);                         % xi- coordinate of point j
21      eta=eta_vector(j);                       % eta- coordinate of point j
22      N(5,j)=1/2*(1-xi^2)*(1+eta);
23      N(6,j)=1/2*(1-xi)*(1-eta^2);
24      N(7,j)=1/2*(1-xi^2)*(1-eta);
25      N(8,j)=1/2*(1+xi)*(1-eta^2);
26      Nx(5,j)=-xi*(1+eta);
27      Nx(6,j)=-1/2*(1-eta^2);
28      Nx(7,j)=-xi*(1-eta);
29      Nx(8,j)=1/2*(1-eta^2);
30      Ny(5,j)=1/2*(1-xi^2);
```

```
31    Ny(6,j)=(1−xi)*(−eta);
32    Ny(7,j)=−1/2*(1−xi^2);
33    Ny(8,j)=(1+xi)*(−eta);
34    for i=1:4
35       nx=master_nodes(i,1);
36       ny=master_nodes(i,2);
37       if i==1; jj=8; else jj=i+3; end
38       k=i+4;
39       N(i,j)=(1.0 + nx*xi)*(1.0 + ny*eta)/4.0  − 1/2*(N(jj,j)+N(k,j));
40       Nx(i,j)= nx*(1.0 + ny*eta)/4.0  − 1/2*(Nx(jj,j)+Nx(k,j));
41       Ny(i,j)= ny*(1.0 + nx*xi)/4.0  − 1/2*(Ny(jj,j)+Ny(k,j));
42    end
43 end
```

In the MATLAB programs listed above, the functions which have already been discussed previously are not repeated here. They are listed below for your reference:
function [gauss_points, gauss_weights]=GetQuadGauss(rows, cols): Section 5.2.1.
function [N,Nx,Ny]=CompNDNatPointsQuad4(xi_vector, eta_vector): Section 5.2.1.
function [element_nodes, node_id_map]= SetElementNodes(ele, nodes, elements): Section 5.2.1.
function J=CompJacobian2DatPoint(element_nodes, Nxi, Neta): Section 5.2.1.
function K=AssembleGlobalMatrix(K, ke, node_id_map, ndDOF): Section 5.2.1.
function function H= CompH(): Section 7.2.3.

7.5 Summary

Upon completion of this chapter, you should be able to:

- understand the fundamental theory of 2-D and 3-D elasticity and derive the differential governing equations
- understand the difference between plane stress and plane strain conditions
- derive the elasticity weak forms by using the Galerkin weighted residual method, principle of virtual work, and minimization of potential energy
- derive the expressions of element matrices and vectors from a weak form for 2-D and 3-D elasticity problems
- know how to assemble element matrices and vectors into the global matrix and vector for 2-D and 3-D elasticity problems
- know how to apply the essential boundary conditions by using the direct substitution, the penalty, and the Lagrange multiplier methods for 2-D and 3-D elasticity problems
- know how to calculate the strains and stresses after the nodal displacements are obtained
- implement MATLAB functions to perform the post-processing steps for 2-D and 3-D elasticity problems

- implement a complete MATLAB program to perform FEA of 2-D and 3-D elasticity problems
- understand the fundamental theory of 2-D steady state Stokes flow and derive the differential governing equations
- derive the 2-D steady state Stokes flow weak form by using the Galerkin weighted residual method and mixed approximations
- derive the expressions of element matrices and vectors from a weak form for 2-D Stoke flow problems
- know how to assemble element matrices and vectors into the global matrix and vector for 2-D steady state Stokes flow problems
- know how to apply the essential boundary conditions by using the penalty methods for 2-D steady state Stokes flow problems
- implement a complete MATLAB program to perform FEA of 2-D steady state Stokes flow problems.

7.6 Problems

7.1 In the plane stress case, we assume that the through-thickness stresses are zero, i.e.,

$$\sigma_{xz} = \sigma_{yz} = \sigma_{zz} = 0. \tag{7.227}$$

Show that

(a)
$$\epsilon_{zz} = \frac{1}{E}\left[\sigma_{zz} - \nu(\sigma_{xx} + \sigma_{yy})\right] \tag{7.228}$$

(b)
$$\epsilon_{zz} = -\frac{\nu}{E}(\sigma_{xx} + \sigma_{yy}) \tag{7.229}$$

(c)
$$\begin{Bmatrix} \sigma_{xx} \\ \sigma_{yy} \\ \sigma_{xy} \end{Bmatrix} = \frac{E}{(1-\nu^2)} \begin{bmatrix} 1 & \nu & 0 \\ \nu & 1 & 0 \\ 0 & 0 & \frac{1-\nu}{2} \end{bmatrix} \begin{Bmatrix} \epsilon_{xx} \\ \epsilon_{yy} \\ \epsilon_{xy} \end{Bmatrix}. \tag{7.230}$$

7.2 Consider a 2-D plate as shown in Fig. 7.26(a). The displacements (u, v) of the corner points 1, 2, 3, and 4 are $(1, 1)$, $(2, 1)$, $(1, 2)$, and $(2, 2)$, respectively. The unit is cm for the dimensions and is mm for the displacements.

(a) Discretize the plate into two linear triangular elements as shown in Fig. 7.26(b). Calculate the strains $(\epsilon_{xx}, \epsilon_{yy}, \epsilon_{xy})$ for each element and show that the strains are constant within each element.

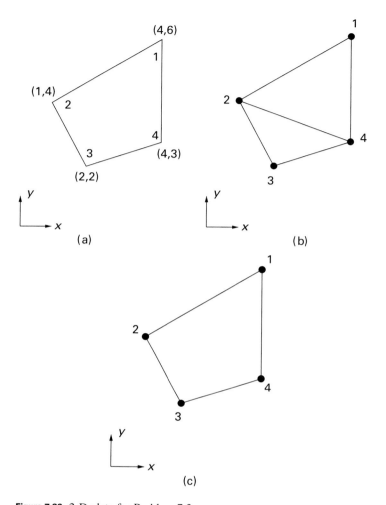

Figure 7.26 2-D plate for Problem 7.2.

(b) Now represent the plate by using a 4-node quadrilateral element as shown in Fig. 7.26(c). Calculate the strains at node 1 and node 2. Are the strains constant within the quadrilateral element?

7.3 Continuing with Problem 7.2, calculate the stresses ($\sigma_{xx}, \sigma_{yy}, \sigma_{xy}$) in the two elements shown in Fig. 7.26(b)
(a) when the elements are plane stress.
(b) when the elements are plane strain.
Let $E = 5 \times 10^9$ Pa and $\nu = 0.4$ and use unit thickness for the elements.

7.4 Describe at least two scenarios in which the strains and stresses are constant in a linear quadrilateral element. Demonstrate the cases by calculating the strains in the element.

7.5 Consider the mapped 4-node quadrilateral element shown in Fig. 7.27. used in a mesh to model a structure subjected to a plane stress deformation. The coordinates and the displacements of the nodes are summarized in the table. Assuming Young's modulus $E = 4$ and Poisson's ratio $\nu = 0.25$, compute:
(a) the (x, y) location of the 1×1 Gauss point,
(b) the stresses at the 1×1 Gauss point.
Note we neglect the units here for simplicity.

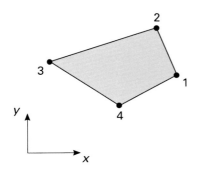

node	x	y	u	v
1	18	6	2	2
2	16	12	1	1
3	0	8	2	0
4	10	2	1	2

Figure 7.27 Quadrilateral element for Problem 7.5.

7.6 A rectangular element is subjected to a surface traction as shown in Fig. 7.28. Compute the element vector **f** for the rectangular element.

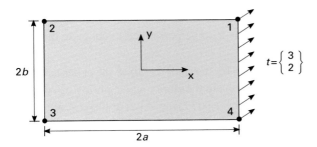

Figure 7.28 Rectangular element for Problem 7.6.

7.7 A rectangular element is subjected to a surface traction as shown in Fig. 7.29. Compute the element vector **f** for the rectangular element.

7.8 A triangular element is subjected to a surface traction as shown in Fig. 7.30. Compute the element vector **f** for the element.

7.9 A force is applied to the end of a wrench to turn a nut as shown in Fig. 7.31. Assuming that the nut is rigid, there is no relative movement between the wrench and the nut at the contact points a and b, and the wrench–nut system is in equilibrium, a

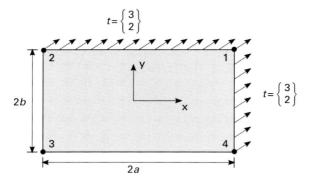

Figure 7.29 Rectangular element for Problem 7.7.

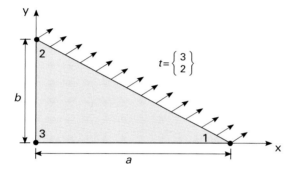

Figure 7.30 Triangular element for Problem 7.8.

2-D finite element static analysis is to be performed to compute the stress and strain in the wrench.

(a) Draw a mesh on the wrench for the analysis. Show the nodes clearly (do not make the elements too small, maintain the clarity and legibility of your drawing).
(b) What material law should be used? Plane stress or plane strain?
(c) List the boundary conditions you need to apply for this problem.

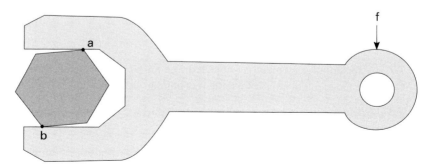

Figure 7.31 Wrench for Problem 7.9.

7.10 Consider the two-dimensional linear elastic plate with a square hole subjected to the loading shown in Fig. 7.32(a):

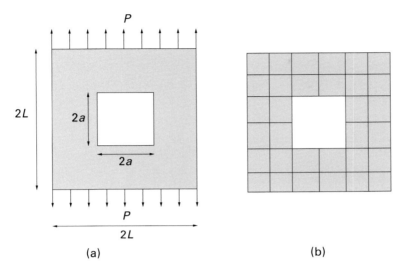

Figure 7.32 Plate with square hole for Problem 7.10.

(a) If the plate is discretized into a coarse mesh shown in Fig. 7.32(b) by using 4-node quadrilateral elements (32 elements total), what is the dimension of the element stiffness matrix? What is the dimension of the global stiffness matrix?

(b) If 8-node quadrilateral elements are used, what is the dimension of the element stiffness matrix? What is the dimension of the global stiffness matrix?

(c) For the loading condition shown in Fig. 7.32(a), indicate the region that you would mesh so as to take full advantage of the symmetry of the loading and geometry. Clearly state the boundary conditions you would apply.

7.11 Consider the two-dimensional linear elastic plate subjected to a point force F as shown in Fig. 7.33. The plate is discretized into four 3-node linear triangular elements. The element list (i.e., "elements.dat") is given as

element index	node indices		
1	1	2	4
2	2	1	3
3	2	5	4
4	5	2	3

The global element and node indices are shown in the figure.

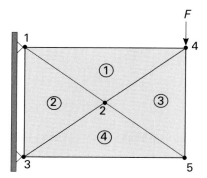

Figure 7.33 Plate for Problem 7.11.

(a) What is the dimension of the global stiffness matrix?
(b) Assuming the element stiffness matrices of elements ① and ② are obtained as

$$\mathbf{k}^{①} = \begin{bmatrix} a_{11} & a_{12} & a_{13} & a_{14} & a_{15} & a_{16} \\ a_{21} & a_{22} & a_{23} & a_{24} & a_{25} & a_{26} \\ a_{31} & a_{32} & a_{33} & a_{34} & a_{35} & a_{36} \\ a_{41} & a_{42} & a_{43} & a_{44} & a_{45} & a_{46} \\ a_{51} & a_{52} & a_{53} & a_{54} & a_{55} & a_{56} \\ a_{61} & a_{62} & a_{63} & a_{64} & a_{65} & a_{66} \end{bmatrix} \quad (7.231)$$

$$\mathbf{k}^{②} = \begin{bmatrix} b_{11} & b_{12} & b_{13} & b_{14} & b_{15} & b_{16} \\ b_{21} & b_{22} & b_{23} & b_{24} & b_{25} & b_{26} \\ b_{31} & b_{32} & b_{33} & b_{34} & b_{35} & b_{36} \\ b_{41} & b_{42} & b_{43} & b_{44} & b_{45} & b_{46} \\ b_{51} & b_{52} & b_{53} & b_{54} & b_{55} & b_{56} \\ b_{61} & b_{62} & b_{63} & b_{64} & b_{65} & b_{66} \end{bmatrix}$$

write down the third row of the global stiffness matrix right after assembling the two element matrices. Note that, at this point, the element stiffness matrices of elements ③ and ④ have not been assembled yet.

7.12 A thin plate with 1.0 in thickness is subjected to loading as shown in Fig. 7.34. Young's modulus of the plate material is 30×10^6 psi and Poisson's ratio is 0.3. Assuming the plate is discretized into two triangular elements, determine the nodal displacements and element stresses.

(a) Perform the FEA procedure manually and obtain the results.
(b) Implement an FEA program to compute the deformation and stresses. Check with your manual calculation.

(c) Refine your mesh and compare the results obtained from different meshes. Describe the convergence behavior of the largest displacements.

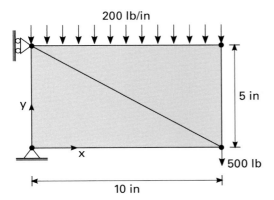

Figure 7.34 Plate for Problem 7.12.

7.13 A thin plate with a hole at the center is subjected to a pressure loading as shown in Fig. 7.35. The geometry is shown in the figure. Young's modulus of the plate is 1×10^5 Pa and Poisson's ratio is 0.3. Write FEA code to compute the deformation and stress distributions of the plate. Refine your mesh and compare the results obtained from different meshes, and discuss the physical meaning of your FEA results.

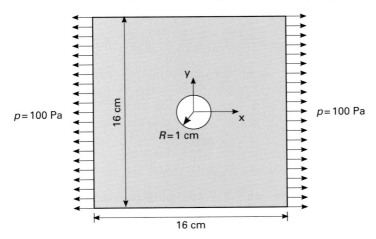

Figure 7.35 Problem 7.13: a plate with a hole.

7.14 A steel ($E = 200$ GPa, $\nu = 0.3$) curved beam structure is subjected to a vertical force at point C as shown in Fig. 7.36. The thickness of the beam is 0.01 m. Write suitable code and perform a static analysis of the structure using linear quadrilateral elements. Use a 2-D model to represent the structure. In the analysis you are required to perform a convergence check by adjusting the mesh size until a converged solution is obtained. That is, you need to refine the mesh and perform the simulation repeatedly until the maximum displacement of the structure does not change by more than 1% when another finer mesh is used.

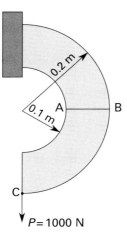

Figure 7.36 Curved beam for Problem 7.14.

The following results are expected:
(a) A plot showing the deformed shape of the structure.
(b) A plot showing the shear stress profile of the deformed shape of the curved beam with your final mesh.

7.15 For the element shown in Fig. 7.37, the displacements of the nodes are:

node	u (mm)	v (mm)	w (mm)
1	0.1	−0.2	−0.2
2	−0.3	0.05	0
3	−0.1	−0.1	−0.1
4	0.04	0	0.3.

Determine the strains at nodes 3 and 4.

7.16 For the element shown in Fig. 7.38, the displacements of the nodes are:

node	u (mm)	v (mm)	w (mm)
1	−0.01	0.02	0.1
2	0.03	0.05	0
3	0.15	−0.1	−0.1
4	0.4	0.3	−0.1
5	0.15	−0.05	0.25
6	0.1	0.25	−0.3
7	0.3	0.15	−0.2
8	−0.04	0.3	0.

Determine the strains at the center of the top and bottom faces of the element.

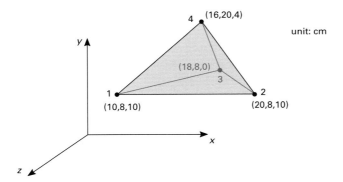

Figure 7.37 Element for Problem 7.15.

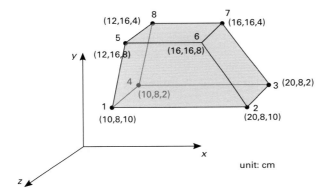

Figure 7.38 Elements for Problem 7.16.

7.17 The element shown in Fig. 7.39 is subjected to a concentrated force. Let $E = 200$ GPa, $\nu = 0.3$. Write a MATLAB program and compute the displacements and strains at the nodes.

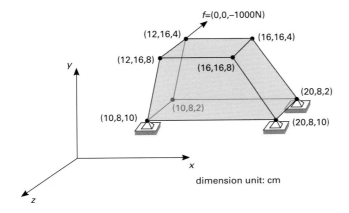

Figure 7.39 Element for Problem 7.17.

References

Chandrasekharaiah, D. & Debnath, L. (2014), *Continuum mechanics*, Elsevier.

Holzapfel, G. A. (2002), "Nonlinear solid mechanics: a continuum approach for engineering science," *Meccanica* **37**(4), 489–490.

Reddy, J. N. (1993), *An introduction to the finite element method*, McGraw-Hill.

8 Structural Elements

8.1 Overview

In this chapter, we discuss several special types of structures: trusses, beams, frames, and plates. These types of structures are defined based on their geometric characteristics and mechanical behavior. A truss is an assembly of slender elastic bars that are connected with each other by pins or joints allowing rotation. The slender elastic bars in a truss are only subjected to axial forces. A beam is also a slender elastic bar or rod, but with a general loading condition. Frames are complex structures formed by beams rigidly connected to each other at their ends. Plates are thin flat structures subjected to transverse and/or in-plane loads. From a geometric point of view, these structures can be viewed as structures with lower dimensions, i.e., line segments and 2-D surfaces in the 3-D space. In finite element analysis, special types of elements were devised to model the low-dimensional structures on both geometric and mechanics aspects. In this regard, the mathematical models of these structural elements can be considered as reduced-order 3-D elasticity elements. The finite element formulations are derived based on such model reductions or simplifications. In the three sections of this chapter, we discuss the finite element analysis procedures for these types of structures. At the end of each section, a MATLAB code for solving the section's example problem is presented.

8.2 Trusses

In engineering, a truss or a truss system is an assembly of elastic bars or rods that forms a load bearing structure. The bars in a truss system are connected to their neighbors by pin or ball joints. These pin/ball joints do not resist any moment about them. The elastic bars are free to rotate about these joints. For this reason, the bars/rods in the truss system are "two force members," that is, the forces acting on the two ends of a bar/rod through the pin/ball joints must be equal in magnitude, opposite in direction, and collinear. If the bar/rod is straight, the forces must be along the axial direction (i.e., axial forces). In this section, finite element analysis of truss systems is illustrated by demonstrating the step-by-step procedure of solving a 3-D truss problem.

8.2.1 3-D Truss System Subjected to External Loads

Step 1: Physical Problem

The truss system contains three elastic rods. The rods are connected to other rods and the rigid walls via ball joints as shown in the Fig. 8.1. The plane in which rods AC and BC reside is parallel to the yz-plane. Rods BC and CD are in the xy-plane. The lengths of the rods are shown in the figure. The rods are made of steel (Young's modulus = 200 GPa) and the cross sectional area of all rods is 1.0 cm^2. Three forces, 200 N, 50 N, and 100 N, are applied at the ball joint C in x-, y-, and z-directions, respectively. From the finite element analysis, we would like to compute the displacement of joint C and the axial (normal) stresses in the rods.

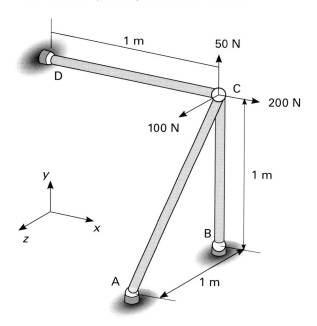

Figure 8.1 A 3-D truss system subjected to external loads.

Step 2: Mathematical Model

As described previously, a truss rod/bar with pin or ball joint connections at both ends can only be subjected to axial load from its joints. Such structures are referred to as axially loaded members. Another simple elastic structure that can only be axially loaded is a spring. In fact, axially loaded rods/bars are nothing but solid springs. Therefore, it is straightforward to obtain the equations of equilibrium of the axially loaded rods/bars from those of springs.

Assuming the longitudinal direction of a spring is along the x-axis, then its equation of equilibrium is simply

$$f = k\Delta x \tag{8.1}$$

8.2 Trusses

where f is the tensile (or compressive) force experienced by the spring in its longitudinal direction, k is the spring constant, and Δx is the elongation (or contraction) of the spring. For the spring shown in Fig. 8.2(a), from equilibrium, it is obvious that the longitudinal forces acting at the two ends should be equal in magnitude and opposite in direction, that is, $f_{l2} = -f_{l1}$. Therefore, the tensile force experienced by the spring is $f = f_{l2} = -f_{l1}$. With the longitudinal displacements of the two ends of the spring defined as u_1 and u_2, the elongation of the spring is then $\Delta x = u_2 - u_1$. Thus, Eq. (8.1) can be rewritten for this spring as

$$f_{l2} = -f_{l1} = k(u_2 - u_1). \tag{8.2}$$

The axially loaded rod shown in Fig. 8.2(b) behaves the same way as the spring. The only difference is that the spring constant for the rod can now be expressed in terms of Young's modulus of the rod material, E, the cross sectional area, A, of the rod, and the length of the rod, L:

$$k = \frac{AE}{L}. \tag{8.3}$$

Then Eq. (8.2) becomes

$$f_{l2} = -f_{l1} = \frac{AE}{L}(u_2 - u_1). \tag{8.4}$$

In matrix form, it can be rewritten as

$$\begin{Bmatrix} f_{l1} \\ f_{l2} \end{Bmatrix} = \frac{AE}{L} \begin{bmatrix} 1 & -1 \\ -1 & 1 \end{bmatrix} \begin{Bmatrix} u_1 \\ u_2 \end{Bmatrix}. \tag{8.5}$$

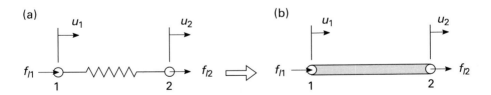

Figure 8.2 (a) Spring and (b) 1-D bar element.

Equation (8.5) is the matrix form of the equation of equilibrium for the 1-D axially loaded rod shown in Fig. 8.2. Figure 8.3(a) shows an axially loaded rod in 2-D with forces applied at the ends. Since the rod is now in a 2-D plane, the displacements of the ends are now vectors containing horizontal u and vertical v components. Similarly, the forces have horizontal f_x and vertical f_y components too. It should be noted that, as shown in Fig. 8.3(b), the vector sum of the horizontal and vertical force components results in the net force vector which is directed in the longitudinal (axial) direction of the rod. There is no net transverse force at the ends. Equation (8.5) can then be rewritten in 2-D as

$$\begin{Bmatrix} f_{l1} \\ f_{t1} \\ f_{l2} \\ f_{t2} \end{Bmatrix} = \frac{AE}{L} \begin{bmatrix} 1 & 0 & -1 & 0 \\ 0 & 0 & 0 & 0 \\ -1 & 0 & 1 & 0 \\ 0 & 0 & 0 & 0 \end{bmatrix} \begin{Bmatrix} u_{l1} \\ u_{t1} \\ u_{l2} \\ u_{t2} \end{Bmatrix} \quad (8.6)$$

where the subscripts l and t denote longitudinal and transverse, respectively.

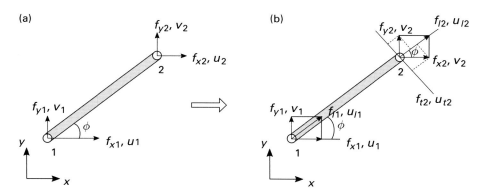

Figure 8.3 2-D bar element.

Assuming the angle between the longitudinal axis of the rod and the x-axis is ϕ, then we have the following geometric relations at end 2:

$$f_{l2} = f_{x2}\cos\phi + f_{y2}\sin\phi \quad (8.7)$$
$$f_{t2} = -f_{x2}\sin\phi + f_{y2}\cos\phi \quad (8.8)$$
$$u_{l2} = u_2\cos\phi + v_2\sin\phi \quad (8.9)$$
$$u_{t2} = -u_2\sin\phi + v_2\cos\phi. \quad (8.10)$$

The same set of relations holds for the forces and displacements at end 1. Together, we have

$$\begin{Bmatrix} f_{l1} \\ f_{t1} \\ f_{l2} \\ f_{t2} \end{Bmatrix} = \frac{AE}{L} \begin{bmatrix} \cos\phi & \sin\phi & 0 & 0 \\ -\sin\phi & \cos\phi & 0 & 0 \\ 0 & 0 & \cos\phi & \sin\phi \\ 0 & 0 & -\sin\phi & \cos\phi \end{bmatrix} \begin{Bmatrix} f_{x1} \\ f_{y1} \\ f_{x2} \\ f_{y2} \end{Bmatrix} = \mathbf{R} \begin{Bmatrix} f_{x1} \\ f_{y1} \\ f_{x2} \\ f_{y2} \end{Bmatrix}$$
$$(8.11)$$

where $\mathbf{R}$ is the rotation matrix, and similarly

$$\begin{Bmatrix} u_{l1} \\ u_{t1} \\ u_{l2} \\ u_{t2} \end{Bmatrix} = \frac{AE}{L} \begin{bmatrix} \cos\phi & \sin\phi & 0 & 0 \\ -\sin\phi & \cos\phi & 0 & 0 \\ 0 & 0 & \cos\phi & \sin\phi \\ 0 & 0 & -\sin\phi & \cos\phi \end{bmatrix} \begin{Bmatrix} u_1 \\ v_1 \\ u_2 \\ v_2 \end{Bmatrix} = \mathbf{R} \begin{Bmatrix} u_1 \\ v_1 \\ u_2 \\ v_2 \end{Bmatrix}.$$
$$(8.12)$$

Substituting Eqs. (8.12, 8.11) into Eq. (8.6), we have

$$\mathbf{R} \begin{Bmatrix} f_{x1} \\ f_{y1} \\ f_{x2} \\ f_{y2} \end{Bmatrix} = \frac{AE}{L} \begin{bmatrix} 1 & 0 & -1 & 0 \\ 0 & 0 & 0 & 0 \\ -1 & 0 & 1 & 0 \\ 0 & 0 & 0 & 0 \end{bmatrix} \mathbf{R} \begin{Bmatrix} u_1 \\ v_1 \\ u_2 \\ v_2 \end{Bmatrix}. \qquad (8.13)$$

Multiplying $\mathbf{R}^{-1}$ to both sides, we have

$$\begin{Bmatrix} f_{x1} \\ f_{y1} \\ f_{x2} \\ f_{y2} \end{Bmatrix} = \frac{AE}{L} \mathbf{R}^{-1} \begin{bmatrix} 1 & 0 & -1 & 0 \\ 0 & 0 & 0 & 0 \\ -1 & 0 & 1 & 0 \\ 0 & 0 & 0 & 0 \end{bmatrix} \mathbf{R} \begin{Bmatrix} u_1 \\ v_1 \\ u_2 \\ v_2 \end{Bmatrix}. \qquad (8.14)$$

Explicitly,

$$\begin{Bmatrix} f_{x1} \\ f_{y1} \\ f_{x2} \\ f_{y2} \end{Bmatrix} = \frac{AE}{L} \begin{bmatrix} \cos^2\phi & \cos\phi\sin\phi & -\cos^2\phi & -\cos\phi\sin\phi \\ \cos\phi\sin\phi & \sin^2\phi & -\cos\phi\sin\phi & -\sin^2\phi \\ -\cos^2\phi & -\cos\phi\sin\phi & \cos^2\phi & \cos\phi\sin\phi \\ -\cos\phi\sin\phi & -\sin^2\phi & \cos\phi\sin\phi & \sin^2\phi \end{bmatrix} \begin{Bmatrix} u_1 \\ v_1 \\ u_2 \\ v_2 \end{Bmatrix}.$$
(8.15)

Equation (8.15) is the matrix form of the equation of equilibrium for the 2-D axially loaded rod shown in Fig. 8.3.

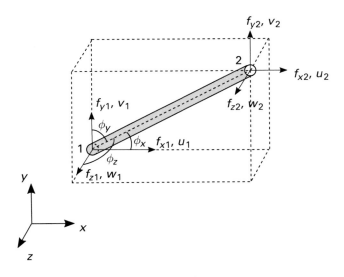

Figure 8.4 A 3-D axially loaded bar element.

Figure 8.4 shows an axially loaded rod in 3-D with forces applied at the ends. In this case, both the force and displacement are vectors with three components. By following the same procedure shown for the 2-D case, the matrix form of the equation of equilibrium for the 3-D axially loaded rod shown in Fig. 8.4 can be obtained as

$$\begin{Bmatrix} f_{x1} \\ f_{y1} \\ f_{z1} \\ f_{x2} \\ f_{y2} \\ f_{z2} \end{Bmatrix} = \frac{AE}{L} \begin{bmatrix} c_x^2 & c_x c_y & c_x c_z & -c_x^2 & -c_x c_y & -c_x c_z \\ c_x c_y & c_y^2 & c_y c_z & -c_x c_y & -c_y^2 & -c_y c_z \\ c_x c_z & c_y c_z & c_z^2 & -c_x c_z & -c_y c_z & -c_z^2 \\ -c_x^2 & -c_x c_y & -c_x c_z & c_x^2 & c_x c_y & c_x c_z \\ -c_x c_y & -c_y^2 & -c_y c_z & c_x c_y & c_y^2 & c_y c_z \\ -c_x c_z & -c_y c_z & -c_z^2 & c_x c_z & c_y c_z & c_z^2 \end{bmatrix} \begin{Bmatrix} u_1 \\ v_1 \\ w_1 \\ u_2 \\ v_2 \\ w_2 \end{Bmatrix}$$

(8.16)

where c_x, c_y, and c_z denote the cosine of the angles between the rod and the coordinate system axes, $\cos \phi_x$, $\cos \phi_y$, and $\cos \phi_z$, respectively, as shown in the figure.

Equations (8.5, 8.15, 8.16) are the mathematical models of an axially loaded elastic rod in one, two, and three dimensions. These axially loaded rods are the building blocks of a truss system.

Step 3: Weak Form

As illustrated in Step 2, with the conditions of regular geometry of the rods (straight, uniform cross section), pin/ball connections, and linear elastic material property, the equations of equilibrium, Eqs. (8.5, 8.15, 8.16), are obtained directly from the mechanics laws and already expressed as a set of algebraic equations in terms of the forces and displacements at the joints. There is no differentiation in these equations. Thus, the steps of converting differential equations (strong form) to integral equations (weak form) are no longer needed in this case. For this reason, the approach is also called the direct method.

Step 4: Discretization

Since the axially loaded rod/bar is only subjected to tensile or compressive force in its longitudinal direction and the rod/bar has uniform geometry and material property along its length, the deformation of the rod/bar is also uniform along its length. There is no variation of physical quantities within the rod/bar. Thus each rod/bar can be considered as one element. There is no need to further discretize a rod/bar into multiple elements since the deformation, stress, strain would be all the same in these elements even if the discretization was done.

For the example problem shown in Fig. 8.1, it is clear that the system can be represented by three axially loaded rod/bar elements (referred to as truss elements from this point forward). The two ends of a rod/bar are the two nodes of the element. The truss elements are connected by the shared nodes. Figure 8.5 shows the discretized model of the truss system. There are a total of three elements and four nodes in the system. Based on the dimensions shown in the figure, the nodes and elements data are given in Table 8.1.

8.2 Trusses

Table 8.1 Nodes and elements

nodes.dat				elements.dat		
1	1.0	0	0	1	1	3
2	1.0	0	1.0	2	2	3
3	1.0	1.0	0	3	3	4
4	0	1.0	0			

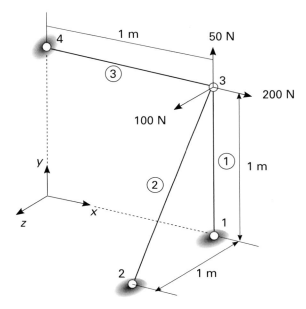

Figure 8.5 Elements of the truss system.

Step 5: Finite Element Approximation

Different from the elasticity problems introduced in the previous chapters, the equations of equilibrium of truss elements are obtained directly and the stress or strain in any given truss element is constant as

$$\epsilon_l = \frac{AE}{L}(u_{l2} - u_{l1}) \qquad \sigma_l = E\epsilon_l. \qquad (8.17)$$

The axial stress and strain expressions are exact for the entire truss element. Therefore, there is no need for numerical approximation in this case.

Step 6: Element Matrices and Vectors

The equations of equilibrium, Eqs. (8.5, 8.15, 8.16), obtained from 1-D, 2-D, and 3-D truss elements in Step 2 are already the discretized algebraic equations for the elements. The coefficient matrices of the displacement vector are the element stiffness matrices ready to be assembled into a global linear system. For the example truss system shown in Fig. 8.5, the element stiffness matrices can be obtained as follows.

For element 1, length is 1.0 m, cosine of the direction angles are 0, 1, 0, respectively. From Eq. (8.16)

$$\mathbf{k}^{\textcircled{1}} = \begin{bmatrix} 0 & 0 & 0 & 0 & 0 & 0 \\ 0 & 2 & 0 & 0 & -2 & 0 \\ 0 & 0 & 0 & 0 & 0 & 0 \\ 0 & 0 & 0 & 0 & 0 & 0 \\ 0 & -2 & 0 & 0 & 2 & 0 \\ 0 & 0 & 0 & 0 & 0 & 0 \end{bmatrix} \times 10^7 \text{ N/m}. \quad (8.18)$$

For element 2, length is $\sqrt{2}$ m, cosine of the direction angles are $0, 1/\sqrt{2}, -1/\sqrt{2}$, respectively, and

$$\mathbf{k}^{\textcircled{2}} = \begin{bmatrix} 0 & 0 & 0 & 0 & 0 & 0 \\ 0 & 0.7071 & -0.7071 & 0 & -0.7071 & 0.7071 \\ 0 & -0.7071 & 0.7071 & 0 & 0.7071 & -0.7071 \\ 0 & 0 & 0 & 0 & 0 & 0 \\ 0 & -0.7071 & 0.7071 & 0 & 0.7071 & -0.7071 \\ 0 & 0.7071 & -0.7071 & 0 & -0.7071 & 0.7071 \end{bmatrix} \times 10^7 \text{ N/m}. \quad (8.19)$$

For element 3, length is 1.0 m, cosine of the direction angles are $-1, 0, 0$, respectively, and

$$\mathbf{k}^{\textcircled{3}} = \begin{bmatrix} 2 & 0 & 0 & -2 & 0 & 0 \\ 0 & 0 & 0 & 0 & 0 & 0 \\ 0 & 0 & 0 & 0 & 0 & 0 \\ -2 & 0 & 0 & 2 & 0 & 0 \\ 0 & 0 & 0 & 0 & 0 & 0 \\ 0 & 0 & 0 & 0 & 0 & 0 \end{bmatrix} \times 10^7 \text{ N/m}. \quad (8.20)$$

Note that the element spring constant (i.e., stiffness) is AE/L which is dependent on the length of the element.

The force vectors in Eqs. (8.5, 8.15, 8.16), however, are unknown for each individual element since they are the forces experienced at the ends of the element. These forces include the forces exerted from the neighboring elements as well as any external force applied at the joint. The externally applied concentrated forces at the joints will be directly assembled into the global force vector at the location corresponding to the node and direction, in the same way as the point forces are applied in the continuum elasticity problems. The internal forces due to the other truss element, because of the force balance requirement of equilibrium, will be canceled by each other in the assembly process, giving zero net effect to the global force vector. Therefore, the element force vectors can be treated as zero vectors in this step.

Step 7: Assembly the Global System

The assembly process follows the same procedure described in previous chapters. The process illustrated in Fig. 7.16 is applicable to the truss system, except that the dimension of the element stiffness matrix is 6×6, and each primitive assembling block is 3×3.

The assembled stiffness matrix of the example problem is given below:

$$K = \begin{bmatrix} 0 & 0 & 0 & 0 & 0 & 0 & 0 & 0 & 0 & 0 & 0 & 0 \\ 0 & 2.0 & 0 & 0 & 0 & 0 & 0 & -2.0 & 0 & 0 & 0 & 0 \\ 0 & 0 & 0 & 0 & 0 & 0 & 0 & 0 & 0 & 0 & 0 & 0 \\ 0 & 0 & 0 & 0 & 0 & 0 & 0 & 0 & 0 & 0 & 0 & 0 \\ 0 & 0 & 0 & 0 & 0.707 & -0.707 & 0 & -0.707 & 0.707 & 0 & 0 & 0 \\ 0 & 0 & 0 & 0 & -0.707 & 0.707 & 0 & 0.707 & -0.707 & 0 & 0 & 0 \\ 0 & 0 & 0 & 0 & 0 & 0 & 2.0 & 0 & 0 & -2.0 & 0 & 0 \\ 0 & -2.0 & 0 & 0 & -0.707 & 0.707 & 0 & 2.707 & -0.707 & 0 & 0 & 0 \\ 0 & 0 & 0 & 0 & 0.707 & -0.707 & 0 & -0.707 & 0.707 & 0 & 0 & 0 \\ 0 & 0 & 0 & 0 & 0 & 0 & -2.0 & 0 & 0 & 2.0 & 0 & 0 \\ 0 & 0 & 0 & 0 & 0 & 0 & 0 & 0 & 0 & 0 & 0 & 0 \\ 0 & 0 & 0 & 0 & 0 & 0 & 0 & 0 & 0 & 0 & 0 & 0 \end{bmatrix}$$

$\times 10^7$ N/m. (8.21)

The global force vector is obtained as

$$F = \begin{Bmatrix} 0 \\ 0 \\ 0 \\ 0 \\ 0 \\ 0 \\ 200 \\ 50 \\ 100 \\ 0 \\ 0 \\ 0 \end{Bmatrix} \text{ N}. \qquad (8.22)$$

Step 8: Apply the Essential Boundary Condition
The application of the essential (displacement) boundary condition can be implemented by using the methods discussed in previous chapters. These methods are equally applicable here. The essential boundary condition for the truss system is

$$\mathbf{u} = \mathbf{0} \qquad \text{for nodes 1, 2, and 4.} \qquad (8.23)$$

Step 9: Solve the Global Linear System
After the essential boundary conditions are applied, the final linear system can be solved in MATLAB. The unknown nodal displacement vector **u** can be computed. For this example, the computed displacement vector is

$$\mathbf{u} = \begin{Bmatrix} 0 \\ 0 \\ 0 \\ 0 \\ 0 \\ 0 \\ 0.1 \\ 0.075 \\ 0.2164 \\ 0 \\ 0 \\ 0 \end{Bmatrix} 10^{-4} \text{m}. \qquad (8.24)$$

Note that in this example the only node that is deformable is node 3.

Step 10: Post-processing

The computed displacements are first visualized in this step. Figure 8.6 shows the truss system before (dashed lines) and after (solid lines) deformation.

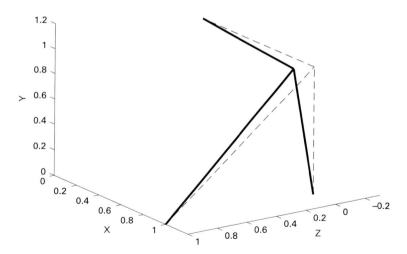

Figure 8.6 Deformed shape of the truss system.

Other than the truss deformation, two other physical quantities, support reaction forces and axial stresses, are often computed for engineering analysis purposes. The support reaction forces can be calculated by multiplying the stiffness matrix shown in Eq. (8.21) by the computed displacement vector:

$$\mathbf{F}_r = \mathbf{Ku} = \begin{Bmatrix} 0 \\ -150.0 \\ 0 \\ 0 \\ 100.0 \\ -100.0 \\ 200.0 \\ 50.0 \\ 100.0 \\ -200.0 \\ 0 \\ 0 \end{Bmatrix} \text{ N.} \qquad (8.25)$$

Note that the **K** matrix is the global stiffness matrix before the essential boundary conditions are applied. This is the reason why the $\mathbf{F}_r$ vector has nonzero values associated with the fixed nodes 1, 2, and 4. The resultant $\mathbf{F}_r$ vector gives the support reaction forces as

$$\mathbf{f}_r(\text{node 1}) = \begin{Bmatrix} 0 \\ -150.0 \\ 0 \end{Bmatrix} \text{ N} \qquad \mathbf{f}_r(\text{node 2}) = \begin{Bmatrix} 0 \\ 100.0 \\ -100.0 \end{Bmatrix} \text{ N} \qquad (8.26)$$

$$\mathbf{f}_r(\text{node 4}) = \begin{Bmatrix} -200.0 \\ 0 \\ 0 \end{Bmatrix} \text{ N.} \qquad (8.27)$$

The axial stress of a rod/bar element can be obtained by

$$\sigma = E\epsilon = E \frac{u_{l2} - u_{l1}}{L}. \qquad (8.28)$$

For a 3-D rod/bar element, it is easy to see that

$$\begin{Bmatrix} u_{l1} \\ u_{l2} \end{Bmatrix} = \begin{bmatrix} c_x & c_y & c_z & 0 & 0 & 0 \\ 0 & 0 & 0 & c_x & c_y & c_z \end{bmatrix} \begin{Bmatrix} u_1 \\ v_1 \\ w_1 \\ u_2 \\ v_2 \\ w_2 \end{Bmatrix}. \qquad (8.29)$$

Substituting Eq. (8.29) into Eq. (8.28), we have

$$\sigma = \frac{E}{L}\begin{bmatrix} -c_x & -c_y & -c_z & c_x & c_y & c_z \end{bmatrix} \begin{Bmatrix} u_1 \\ v_1 \\ w_1 \\ u_2 \\ v_2 \\ w_2 \end{Bmatrix}. \qquad (8.30)$$

For each element the nodal displacement vector can be extracted from the global displacement vector and the axial stress is obtained for the truss system as

$$\sigma^{\textcircled{1}} = 1.5 \times 10^3 \text{ KPa} \qquad \sigma^{\textcircled{2}} = -1.4142 \times 10^3 \text{ KPa} \qquad \sigma^{\textcircled{3}} = 2.0 \times 10^3 \text{ KPa}. \qquad (8.31)$$

8.2.2 Computer Implementation

In this section, we present the data files and the MATLAB code for the finite element analysis of the truss example.

Data Files

The input files "nodes.dat" and "elements.dat" are already shown in Table 8.1. Our simple truss system has three elements and four nodes as shown in Fig. 8.5. The material properties, loads, boundary conditions, and other model options are stored in the data files shown in the table below.

bcsdisp.dat	bcsforce.dat	materials.dat	options.dat
1 1 0	3 200 50 100	% Young's	% Dimensions
1 2 0		% modulus (Pa)	
1 3 0			3
2 1 0		200e9	
2 2 0			% Cross section
2 3 0			% area (m^2)
4 1 0			
4 2 0			1e-4
4 3 0			

MATLAB Code
The code of the truss analysis is listed below:

```matlab
clear all;
% next 3 lines: read the input files
filenames = {'nodes.dat','elements.dat','materials.dat', ...
             'options.dat','forces.dat', 'bcsdisp.dat'};
for i = 1:numel(filenames); load(filenames{i}); end;

% next 8 lines: set up constants and empty matrices
nNodes=size(nodes,1);
nElements=size(elements,1);
E=materials(1,1);
A=options(2,1);
K=zeros(3*nNodes,3*nNodes);
F=zeros(3*nNodes,1);
length=zeros(nElements,1);
direction_cos =zeros(nElements,3);

% for-loop: compute the global stiffness matrix
for e=1:nElements
    nids=elements(e,2:3);
    dv=nodes(nids(2),2:4)-nodes(nids(1),2:4);
    length(e)=norm(dv);
    direction_cos(e,:)=dv/length(e);
    cx=direction_cos(e,1);
    cy=direction_cos(e,2);
    cz=direction_cos(e,3);
    ke=[cx^2    cx*cy   cx*cz   -cx^2   -cx*cy  -cx*cz
        cx*cy   cy^2    cy*cz   -cx*cy  -cy^2   -cy*cz
        cx*cz   cy*cz   cz^2    -cx*cz  -cy*cz  -cz^2
        -cx^2   -cx*cy  -cx*cz  cx^2    cx*cy   cx*cz
        -cx*cy  -cy^2   -cy*cz  cx*cy   cy^2    cy*cz
        -cx*cz  -cy*cz  -cz^2   cx*cz   cy*cz   cz^2];
    ke=ke*A*E/length(e);
    % for-loop: assemble the element matrices into the global K
    for j=1:2           % loop over the row blocks
        for k=1:2       % loop over the column blocks
            K(3*nids(j)-2:3*nids(j), 3*nids(k)-2:3*nids(k)) ...
              =K(3*nids(j)-2:3*nids(j), 3*nids(k)-2:3*nids(k)) ...
              + ke(3*j-2:3*j,3*k-2:3*k);
        end
    end
end

% for-loop: set up the global force vector
for i=1:size(forces,1)
    node = forces(i,1);
    F(3*node-2:3*node)= forces(i,2:4);
```

```
47    end
48
49    % next 8 lines : apply the displacement BC using the penalty method
50    Kold=K
51    penalty=abs(max(max(K)))*1e7;
52    for i=1:size(bcsdisp,1)
53       node = bcsdisp(i,1);
54       direction = bcsdisp(i,2);
55       K(3*node + direction − 3, 3*node + direction − 3)= penalty;
56       F(3*node + direction − 3)= penalty*bcsdisp(i,3);
57    end
58
59    d=K\F  % solve global linear system
60
61    % next 32 lines : post-processing
62    reaction_force =Kold*d;  % compute support reactions
63    % next 10 lines : compute element stresses
64     stress =zeros(nElements,1);
65    for e=1:nElements
66       nids=elements(e,2:3);
67       cx=direction_cos(e,1);
68       cy=direction_cos(e,2);
69       cz=direction_cos(e,3);
70       d_vector= [d(nids(1)*3 −2) d(nids(1)*3−1) d(nids(1)*3) ...
71                  d(nids(2)*3 −2) d(nids(2)*3−1) d(nids(2)*3) ]';
72        stress (e)=E/length(e)*[−cx −cy −cz cx cy cz]*d_vector;
73    end
74    % next 19 lines : plotting
75    figure (1);
76    d=d*10000;  % scale the displacement for plotting
77    clf ;
78    hold on;
79    % for−loop: plot the deformed and undeformed truss system
80    for e=1:nElements
81       nd1=elements(e,2);
82       nd2=elements(e,3);
83       x1=nodes(nd1,2); y1=nodes(nd1,4); z1=nodes(nd1,3);
84       x2=nodes(nd2,2); y2=nodes(nd2,4); z2=nodes(nd2,3);
85       plot3([x1 x2], [y1 y2], [z1 z2],'k−−');
86       plot3([x1+d(3*nd1−2) x2+d(3*nd2−2)], [y1+d(3*nd1) y2+d(3*nd2)],...
87              [z1+d(3*nd1−1) z2+d(3*nd2−1)], 'k−','LineWidth',3);
88    end
89    axis([0 1.2 −0.3 1 −0 1.2]);
90    set(gca,'Ydir','reverse');  % reverse the y−axis for plotting
91    xlabel('X'); ylabel('Z'); zlabel('Y');
92    view(55,25);
93    hold off ;
```

8.3 Beams and Space Frames

Long slender straight bars or rods subjected to loads beyond axial force are often referred to as beams. Beams are different from axially loaded rods/bars as they can also undergo shear and bending deformations. In engineering, when the rods/bars are welded together at the ends, they are not two force members. They are beams. The assembly of beams is not a truss, but a space frame structure. In this section, finite element analysis of beams and space frames is illustrated by demonstrating the step-by-step procedure of solving a 3-D space frame structure as shown in Fig. 8.7.

8.3.1 3-D Space Frame Structure Subjected to External Loads

Step 1: Physical Problem

A space frame structure is composed of three straight segments as shown in Fig. 8.7. The frame is under the following combined load conditions (1) a load of 100 N is applied in the direction of $-y$ at B, and (2) a load of 100 N is applied in the direction of z at C. Take the material properties as $E = 2 \times 10^7$ N/cm^2 and $G = 0.8 \times 10^7$ N/cm^2. The unit for all the lengths is cm. Through a finite element analysis, we would like to compute the deformation of the structure.

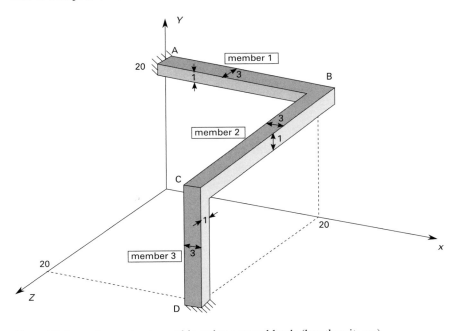

Figure 8.7 Space frame structure subjected to external loads (length unit: cm).

Step 2: Mathematical Model

In engineering applications, although a long beam may have stresses caused by bending, shear, axial, torsional internal loads, bending deformation and stresses often dominate. In this case, the transverse shear deformation can be neglected and the beam can be modeled as an Euler–Bernoulli beam. For a beam segment as shown in Fig. 8.8, the Euler–Bernoulli beam theory gives the relationship between beam deflection (vertical displacement of the beam) and its internal moment as

$$\frac{M}{EI} = \frac{d^2v}{dx^2}. \tag{8.32}$$

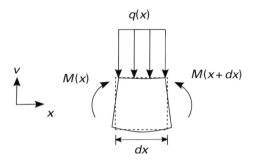

Figure 8.8 A beam segment.

In addition, it can be shown that

$$V = \frac{dM}{dx} \qquad q = -\frac{dV}{dx} \tag{8.33}$$

where V is the internal shear force of the beam. Combining Eqs. (8.32, 8.33), we obtain

$$\frac{d^2}{dx^2}\left(EI\frac{d^2v}{dx^2}\right) = -q(x). \tag{8.34}$$

Equation (8.34) is a fourth order differential equation of the 1-D Euler–Bernoulli beam. Note that it is referred to as 1-D for the reason that the differential equation of equilibrium is defined over a 1-D domain (in x-direction) and the beam deformation is a 1-D function (deflection v is only a function of x). Along with proper boundary conditions, Eq. (8.34) represents the strong form of the governing equation of the 1-D beam.

When the beam is also subjected to forces in y- and z-directions and/or moments about x- and y-axes, the beam deformation becomes three dimensional. In this case, the differential equations of equilibrium of an Euler–Bernoulli beam are given by (assuming cross section is symmetric about y- and z-axes)

$$\text{axial:} \qquad \frac{d}{dx}\left(EA\frac{du}{dx}\right) + q_x = 0 \tag{8.35}$$

$$\text{torsion:} \qquad \frac{d}{dx}\left(GJ\frac{d\theta_x}{dx}\right) + m(x) = 0 \tag{8.36}$$

$$\text{bending about y-axis:} \quad -\frac{d^2}{dx^2}\left(EI_y\frac{d^2w}{dx^2}\right) + q_z + \frac{dm_y}{dx} = 0 \tag{8.37}$$

bending about z-axis: $-\dfrac{d^2}{dx^2}\left(EI_z\dfrac{d^2v}{dx^2}\right) + q_y + \dfrac{dm_z}{dx} = 0$ (8.38)

where q_x, q_y, q_z are the distributed forces in x, y, and z directions, respectively, and m_x, m_y, m_z are the distributed moments about x, y, and z axes, respectively, as shown in Fig. 8.9. It should be noted that, m_x, m_y, m_z are zeros in most cases.

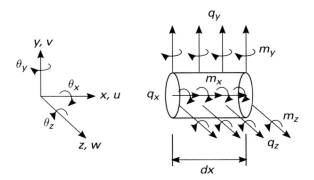

Figure 8.9 Beam segment subjected to 3-D loads.

Step 3: Weak Form

For the 1-D Euler–Bernoulli beam equation, Eq. (8.34), the weak form can be obtained straightforwardly by using the method of weighted residual. Multiplying the variation of the deflection to both sides of Eq. (8.34) and then integrating the resultant products over the computational domain, i.e.,

$$\int_0^L \delta v \left[\dfrac{d^2}{dx^2}\left(EI\dfrac{d^2v}{dx^2}\right) + q(x) \right] dx = 0. \quad (8.39)$$

Integrating by parts twice, we obtain

$$\left(EI\dfrac{d^3v}{dx^3}\delta v - EI\dfrac{d^2v}{dx^2}\dfrac{d\delta v}{dx}\right)\bigg|_0^L + \int_0^L \dfrac{d^2\delta v}{dx^2} EI \dfrac{d^2v}{dx^2} dx + \int_0^L \delta v q(x) dx = 0. \quad (8.40)$$

In light of Eqs. (8.32, 8.33), the integral equation can be rewritten as

$$\left(V\delta v - M\dfrac{d\delta v}{dx}\right)\bigg|_0^L + \int_0^L \dfrac{d^2\delta v}{dx^2} EI \dfrac{d^2v}{dx^2} dx + \int_0^L \delta v q(x) dx = 0. \quad (8.41)$$

Equation (8.41) is the weak form of the 1-D Euler–Bernoulli beam. Another approach to obtain the weak form is by minimizing the potential energy of the system. Assuming a pure bending deformation, the potential energy of the 1-D Euler–Bernoulli beam is

$$\Pi = \text{strain energy} - \text{work done by external loads}$$

$$= \int_0^L \dfrac{(M(x))^2}{2EI} dx + \int_0^L vq(x) dx + \left(Vv - M\dfrac{dv}{dx}\right)\bigg|_0^L. \quad (8.42)$$

Invoking the condition of minimizing the potential energy:

$$\delta \Pi = 0$$

$$\rightarrow \delta \left(\int_0^L \frac{1}{2EI} \left(EI \frac{\partial^2 v}{\partial x^2} \right)^2 dx + \int_0^L vq(x)dx + \left. \left(Vv - M\frac{dv}{dx} \right) \right|_0^L \right) = 0$$

$$\rightarrow \int_0^L EI \frac{d^2\delta v}{dx^2} \frac{d^2 v}{dx^2} dx + \int_0^L \delta v q(x) dx + \left. \left(V\delta v - M\frac{d\delta v}{dx} \right) \right|_0^L = 0. \quad (8.43)$$

The weak form obtained in Eq. (8.43) is identical to Eq. (8.41).

For the 3-D case shown in Fig. 8.9, the strong form shown in Eq. (8.38) can be converted to weak form by using the method of weighted residual. However, it is more straightforward to use the variational principle of potential energy. For the loading case shown in Fig. 8.9, the potential energy of a beam of length L can be written as

$$\Pi = \int_0^L \left(\frac{(N(x))^2}{2EA} + \frac{(M_z(x))^2}{2EI_z} + \frac{(M_y(x))^2}{2EI_y} + \frac{(T(x))^2}{2GJ} + \frac{k_y (V_y(x))^2}{2GA} + \frac{k_z (V_z(x))^2}{2GA} \right) dx$$

$$- \int_0^L \left(uq_x + vq_y + wq_z \right) dx - \int_0^L \left(\theta_x m_x + \frac{dw}{dx} m_y + \frac{dv}{dx} m_z \right) dx$$

$$- \left. \left(\bar{N}u + \bar{V}_y v + \bar{V}_z w + \bar{M}_z \frac{dv}{dx} + \bar{M}_y \frac{dw}{dx} + \bar{T}\theta_x \right) \right|_0^L \quad (8.44)$$

where $N(x)$, $V_y(x)$, $V_z(x)$, $M_z(x)$, $M_y(x)$, $T(x)$ are the internal section normal force, shear forces, bending moments, and torque, $\bar{N}$, $\bar{V}_y$, $\bar{V}_z$, $\bar{M}_z$, $\bar{M}_y$, $\bar{T}$ are the external normal force, shear forces in y- and z-directions, bending moments about y- and z-directions, and torque applied at the ends of the beam, respectively, k_y and k_z are the cross section shape factors for the transverse shear stress in y- and z-directions, respectively. Equation (8.44) represents the comprehensive version of the potential energy of the beam. For Euler–Bernoulli beams, the strain energy contribution from the transverse shear stresses are typically quite small compared to the bending energy. Therefore, the strain energy terms containing $V_y(x)$ and $V_z(x)$ are typically neglected. In addition, m_x, m_y, m_z are zeros in most cases. Thus a simplified version of Eq. (8.44) is given by

$$\Pi = \int_0^L \left(\frac{(N(x))^2}{2EA} + \frac{(M_z(x))^2}{2EI_z} + \frac{(M_y(x))^2}{2EI_y} + \frac{(T(x))^2}{2GJ} \right) dx$$

$$- \int_0^L \left(uq_x + vq_y + wq_z \right) dx - \left. \left(\bar{N}u + \bar{V}_y v + \bar{V}_z w + \bar{M}_z \frac{dv}{dx} + \bar{M}_y \frac{dw}{dx} + \bar{T}\theta_x \right) \right|_0^L.$$

$$(8.45)$$

From the variational principle, the first variation of the potential energy vanishes when potential energy is minimized.

Table 8.2 Nodes and elements

nodes.dat				elements.dat		
1	0	20	0	1	1	2
2	20	20	0	2	2	3
3	20	20	10	3	3	4
4	20	20	20	4	4	5
5	20	0	20			

$$\delta \Pi = \int_0^L \left(EA \frac{\partial u}{\partial x} \frac{\partial \delta u}{\partial x} + EI_z \frac{\partial^2 v}{\partial x^2} \frac{\partial^2 \delta v}{\partial x^2} + EI_y \frac{\partial^2 w}{\partial x^2} \frac{\partial^2 \delta w}{\partial x^2} + GJ \frac{\partial \theta_x}{\partial x} \frac{\partial \delta \theta_x}{\partial x} \right) dx$$

$$- \int_0^L \left(\delta u q_x + \delta v q_y + \delta w q_z \right) dx$$

$$- \left(\bar{N} \delta u + \bar{V}_y \delta v + \bar{V}_z \delta w + \bar{M}_z \frac{d \delta v}{dx} + \bar{M}_y \frac{d \delta w}{dx} + \bar{T} \delta \theta_x \right) \Bigg|_0^L .$$

$$= 0. \tag{8.46}$$

Equation (8.46) is the weak form of an Euler–Bernoulli beam subjected to 3-D loads.

Step 4: Discretization

Similar to the truss systems, the slender beams in space frame structures are represented by line elements. Two important differences between truss systems and space frames are: (1) the rods/bars in truss systems are only axially loaded due to the pin/ball joints at the ends while the beams in space frames can be loaded in all directions as shown in Fig. 8.9 because the beams are welded to each other at the ends; (2) the rods/bars in truss systems are only subjected to axial deformation (uniform deformation if material and cross section are uniform along the length) while space frame beams can simultaneously undergo bending, elongation/contraction, torsion, and more depending on the loading and beam geometry. Therefore, the deformation is usually non-uniform along the length of a segment of a space frame. Therefore, it is often necessary to model the segments in a space frame using multiple elements.

Figure 8.10 shows a discretization (mesh) of the space frame example problem. For illustration purposes, the entire structure is discretized into four elements. To demonstrate the non-uniform deformation of the straight beam segments, the center beam member is discretized into two elements. Therefore, there are a total of four elements and five nodes in the system. Based on the dimensions shown in the figure, the nodes and elements data are given in Table 8.2.

Step 5: Finite Element Approximation

As shown in the weak form of the 1-D Euler–Bernoulli beam, Eq. (8.41), the order of differentiation for the beam deflection v and its variation δv is two. The continuity condition requires the finite element approximation of v and δv to be first-order (C^1)

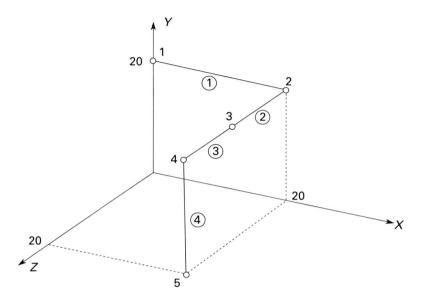

Figure 8.10 Discretization of the space frame structure.

continuous across the elements. That is, the approximation of v and its first derivative $\frac{dv}{dx}$ must be continuous across the elements, which means the approximated deflection function should be smooth at the interface of the elements. There are multiple approaches to achieve the C^1 continuity of the approximation function. In this text, we use the Hermite functions (see Chapter 3) for approximating the beam deflection function.

As shown in Fig. 8.11, the beam element of length L has two nodes and four degrees of freedom (DOF). At each node, the deflection v and its section angle of rotation $\theta = \frac{dv}{dx}$ are the unknown DOFs. Approximating the deflection function $v(x)$ for the element using a polynomial, we have

$$v(x) = a_1 x^3 + a_2 x^2 + a_3 x + a_4 \tag{8.47}$$

Note that, to ensure a unique solution of the unknown coefficients $a_1, a_2, \ldots$, the oder of the approximating polynomial is determined by the DOFs of the element. In this case, DOF=4 leads to the cubic polynomial.

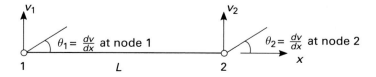

Figure 8.11 1-D beam element.

For the element shown in Fig. 8.11, assuming the left node is located at $x = 0$, then we have

$$v(x = 0) = v_1$$
$$\frac{dv}{dx}(x = 0) = \theta_1$$
$$v(x = L) = v_2$$
$$\frac{dv}{dx}(x = L) = \theta_2. \qquad (8.48)$$

Substituting Eq. (8.47) into the linear system Eq. (8.48), we can solve for the unknown coefficients of the polynomial:

$$a_1 = \frac{2}{L^3}(v_1 - v_2) + \frac{1}{L^2}(\theta_1 + \theta_2)$$
$$a_2 = -\frac{3}{L^2}(v_1 - v_2) - \frac{1}{L}(2\theta_1 + \theta_2)$$
$$a_3 = \theta_1$$
$$a_4 = v_1. \qquad (8.49)$$

Substituting the coefficients into Eq. (8.47), we obtain

$$v(x) = \left(\frac{2x^3 - 3x^2L + L^3}{L^3}\right)v_1 + \left(\frac{x^3 - 2x^2L + xL^2}{L^2}\right)\theta_1$$
$$+ \left(\frac{-2x^3 + 3x^2L}{L^3}\right)v_2 + \left(\frac{x^3 - x^2L}{L^2}\right)\theta_2$$
$$= \tilde{N}_1(x)v_1 + \tilde{N}_2(x)\theta_1 + \tilde{N}_3(x)v_2 + \tilde{N}_4(x)\theta_2$$
$$= \begin{bmatrix} \tilde{N}_1(x) & \tilde{N}_2(x) & \tilde{N}_3(x) & \tilde{N}_4(x) \end{bmatrix} \begin{Bmatrix} v_1 \\ \theta_1 \\ v_2 \\ \theta_2 \end{Bmatrix}$$
$$= \tilde{N}\mathbf{d}. \qquad (8.50)$$

Note that Eq. (8.50) can also be obtained from Eq. (3.27). Thus, the shape functions, $\tilde{N}_i(x)$, $i=1,2,3,4$, for the 1-D beam element are defined according to Eq. (8.50).

For 3-D beam elements, however, deflections in other directions and cross section rotations about other axes are allowed. More nodal DOFs are required to reflect extra deformation modes. Figure 8.12 shows a 3-D beam element with its longitudinal axis being the x-axis. In this element, each node (end) has six DOFs: three translational displacements and three rotational displacements. Approximation is required for all the six DOFs. The weak form of the 3-D beam, Eq. (8.46), shows that the integrals of the strain energy contain second derivatives of v and w, and only first derivatives of u and θ_x. Therefore, u and θ_x can be approximated by using the regular Lagrange interpolation function as described in previous chapters. w can be treated the same way as v is approximated in Eq. (8.50).

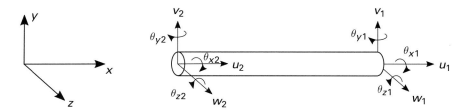

Figure 8.12 3-D beam element.

Therefore, in the 3-D beam scenario, the finite element approximations are

$$v(x) = \begin{bmatrix} \tilde{N}_1(x) & \tilde{N}_2(x) & \tilde{N}_3(x) & \tilde{N}_4(x) \end{bmatrix} \begin{Bmatrix} v_1 \\ \theta_{z1} \\ v_2 \\ \theta_{z2} \end{Bmatrix} \quad (8.51)$$

$$w(x) = \begin{bmatrix} \tilde{N}_1(x) & \tilde{N}_2(x) & \tilde{N}_3(x) & \tilde{N}_4(x) \end{bmatrix} \begin{Bmatrix} w_1 \\ \theta_{y1} \\ w_2 \\ \theta_{y2} \end{Bmatrix} \quad (8.52)$$

$$u(x) = \begin{bmatrix} N_1(x) & N_2(x) \end{bmatrix} \begin{Bmatrix} u_1 \\ u_2 \end{Bmatrix} \quad (8.53)$$

$$\theta_x(x) = \begin{bmatrix} N_1(x) & N_2(x) \end{bmatrix} \begin{Bmatrix} \theta_{x1} \\ \theta_{x2} \end{Bmatrix} \quad (8.54)$$

where

$$\tilde{N}_1(x) = \frac{2x^3 - 3x^2 L + L^3}{L^3}, \qquad \tilde{N}_2(x) = \frac{x^3 - 2x^2 L + xL^2}{L^2}$$

$$\tilde{N}_3(x) = \frac{-2x^3 + 3x^2 L}{L^3}, \qquad \tilde{N}_4(x) = \frac{x^3 - x^2 L}{L^2} \quad (8.55)$$

and

$$N_1(x) = \frac{L-x}{L}, \qquad N_2(x) = \frac{x}{L}. \quad (8.56)$$

Step 6: Element Matrices and Vectors

By substituting the approximations into the weak forms Eqs. (8.41, 8.46) for individual elements, the element matrices and vectors can be obtained. The derivation procedure for obtaining the expression of element matrices and vectors is the same as described in previous chapters and will not repeated here. The element stiffness matrix obtained from the integral of $\dfrac{\partial^2 v}{\partial x^2}$ in the weak form Eq. (8.41) is

$$\mathbf{k}^{(e)} = \frac{EI}{L^3} \begin{bmatrix} 12 & 6L & -12 & 6L \\ 6L & 4L^2 & -6L & 2L^2 \\ -12 & -6L & 12 & -6L \\ 6L & 2L^2 & -6L & 4L^2 \end{bmatrix}. \tag{8.57}$$

The displacement and force vectors are

$$\mathbf{d}^{(e)} = \begin{Bmatrix} v_1 \\ \theta_1 \\ v_2 \\ \theta_2 \end{Bmatrix} \qquad \mathbf{f}^{(e)} = \begin{Bmatrix} f_1 \\ \bar{m}_1 \\ f_2 \\ \bar{m}_2 \end{Bmatrix} \tag{8.58}$$

where f and $\bar{m}$ are force and moment applied to the nodes of the element.

For the 3-D beam weak form shown in Eq. (8.46), each of the terms in the strain energy integral generates a stiffness matrix. Following the order of the terms in the first integral of Eq. (8.46), the stiffness matrices and associated displacement vectors are:

$$\mathbf{k}_u^{(e)} = \frac{EA}{L} \begin{bmatrix} 1 & -1 \\ -1 & 1 \end{bmatrix} \qquad \mathbf{d}_u^{(e)} = \begin{Bmatrix} u_1 \\ u_2 \end{Bmatrix} \tag{8.59}$$

$$\mathbf{k}_v^{(e)} = \frac{EI_z}{L^3} \begin{bmatrix} 12 & 6L & -12 & 6L \\ 6L & 4L^2 & -6L & 2L^2 \\ -12 & -6L & 12 & -6L \\ 6L & 2L^2 & -6L & 4L^2 \end{bmatrix} \qquad \mathbf{d}_v^{(e)} = \begin{Bmatrix} v_1 \\ \theta_{z1} \\ v_2 \\ \theta_{z2} \end{Bmatrix} \tag{8.60}$$

$$\mathbf{k}_w^{(e)} = \frac{EI_y}{L^3} \begin{bmatrix} 12 & 6L & -12 & 6L \\ 6L & 4L^2 & -6L & 2L^2 \\ -12 & -6L & 12 & -6L \\ 6L & 2L^2 & -6L & 4L^2 \end{bmatrix} \qquad \mathbf{d}_w^{(e)} = \begin{Bmatrix} w_1 \\ \theta_{y1} \\ w_2 \\ \theta_{y2} \end{Bmatrix} \tag{8.61}$$

$$\mathbf{k}_\theta^{(e)} = \frac{GJ}{L} \begin{bmatrix} 1 & -1 \\ -1 & 1 \end{bmatrix} \qquad \mathbf{d}_\theta^{(e)} = \begin{Bmatrix} \theta_{x1} \\ \theta_{x2} \end{Bmatrix}. \tag{8.62}$$

The above stiffness matrices and vectors can be combined into a single element stiffness matrix and displacement vector:

$$\mathbf{k}^{(e)} = \begin{bmatrix} \frac{EA}{L} & 0 & 0 & 0 & 0 & 0 & \frac{-EA}{L} & 0 & 0 & 0 & 0 & 0 \\ 0 & \frac{12EI_z}{L^3} & 0 & 0 & 0 & \frac{6EI_z}{L^2} & 0 & \frac{-12EI_z}{L^3} & 0 & 0 & 0 & \frac{6EI_z}{L^2} \\ 0 & 0 & \frac{12EI_y}{L^3} & 0 & \frac{-6EI_y}{L^2} & 0 & 0 & 0 & \frac{-12EI_y}{L^3} & 0 & \frac{-6EI_y}{L^2} & 0 \\ 0 & 0 & 0 & \frac{GJ}{L} & 0 & 0 & 0 & 0 & 0 & \frac{-GJ}{L} & 0 & 0 \\ 0 & 0 & \frac{-6EI_y}{L^2} & 0 & \frac{4EI_y}{L} & 0 & 0 & 0 & \frac{6EI_y}{L^2} & 0 & \frac{2EI_y}{L} & 0 \\ 0 & \frac{6EI_z}{L^2} & 0 & 0 & 0 & \frac{4EI_z}{L} & 0 & \frac{-6EI_z}{L^2} & 0 & 0 & 0 & \frac{2EI_z}{L} \\ -\frac{EA}{L} & 0 & 0 & 0 & 0 & 0 & \frac{EA}{L} & 0 & 0 & 0 & 0 & 0 \\ 0 & \frac{-12EI_z}{L^3} & 0 & 0 & 0 & \frac{-6EI_z}{L^2} & 0 & \frac{12EI_z}{L^3} & 0 & 0 & 0 & \frac{-6EI_z}{L^2} \\ 0 & 0 & \frac{-12EI_y}{L^3} & 0 & \frac{6EI_y}{L^2} & 0 & 0 & 0 & \frac{12EI_y}{L^3} & 0 & \frac{6EI_y}{L^2} & 0 \\ 0 & 0 & 0 & \frac{-GJ}{L} & 0 & 0 & 0 & 0 & 0 & \frac{GJ}{L} & 0 & 0 \\ 0 & 0 & \frac{-6EI_y}{L^2} & 0 & \frac{2EI_y}{L} & 0 & 0 & 0 & \frac{6EI_y}{L^2} & 0 & \frac{4EI_y}{L} & 0 \\ 0 & \frac{6EI_z}{L^2} & 0 & 0 & 0 & \frac{2EI_z}{L} & 0 & \frac{-6EI_z}{L^2} & 0 & 0 & 0 & \frac{4EI_z}{L} \end{bmatrix}$$

(8.63)

$$\mathbf{d}^{(e)} = \{ u_1 \; v_1 \; w_1 \; \theta_{x1} \; \theta_{y1} \; \theta_{z1} \; u_2 \; v_2 \; w_2 \; \theta_{x2} \; \theta_{y2} \; \theta_{z2} \}^T. \quad (8.64)$$

The element force vector is then

$$\mathbf{f}^{(e)} = \{ f_{x1} \; f_{y1} \; f_{z1} \; \bar{m}_{x1} \; \bar{m}_{y1} \; \bar{m}_{z1} \; f_{x2} \; f_{y2} \; f_{z2} \; \bar{m}_{x2} \; \bar{m}_{y2} \; \bar{m}_{z2} \}^T. \quad (8.65)$$

The discretized element equations of equilibrium are

$$\mathbf{k}^{(e)} \mathbf{d}^{(e)} = \mathbf{f}^{(e)}. \quad (8.66)$$

The element matrices and vectors shown above are obtained for the 3-D element shown in Fig. 8.12 where the longitudinal axis coincides with the x-axis, and the y- and z-axes are the axes of symmetry of the beam's cross section. The coordinate system shown in Fig. 8.12 is referred to as the local coordinate system of the element. In general, however, a space frame structure containing many beam segments is placed in a global coordinate system. In such cases, a beam segment's local coordinate system is different from the global coordinate system. Therefore, the element matrices and vectors of the beam elements derived from their local coordinate systems need to be transformed into the global coordinate system.

The transformation between the representations of a vector (taking vector **u** as an example as shown in Fig. 8.13) in local and global coordinate systems is given in Eq. (8.67).

$$\left\{\begin{array}{c} u \\ v \\ w \end{array}\right\} = \left[\begin{array}{ccc} \cos(\widehat{x,X}) & \cos(\widehat{x,Y}) & \cos(\widehat{x,Z}) \\ \cos(\widehat{y,X}) & \cos(\widehat{y,Y}) & \cos(\widehat{y,Z}) \\ \cos(\widehat{z,X}) & \cos(\widehat{z,Y}) & \cos(\widehat{z,Z}) \end{array}\right] \left\{\begin{array}{c} U \\ V \\ W \end{array}\right\} \qquad (8.67)$$

where u, v, w are the displacement components in the local coordinate system, U, V, W denote the displacement components in the global coordinate system, and $\cos(\widehat{x,X})$ is the cosine of the angle between the x-axis of the local coordinate system and the X-axis of the global coordinate system.

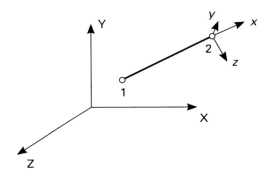

Figure 8.13 Beam element in global coordinate system.

It is easy to see that

$$\cos(\widehat{x,X}) = \frac{X_2 - X_1}{L} \qquad \cos(\widehat{x,Y}) = \frac{Y_2 - Y_1}{L} \qquad \cos(\widehat{x,Z}) = \frac{Z_2 - Z_1}{L}. \qquad (8.68)$$

However, it is not trivial to calculate the other cosines in the transformation matrix. Here we do it through two steps by using an intermediate coordinate system as shown in Fig. 8.14 (Rao 2017). The first step is to set up the intermediate coordinate system such that $\bar{x} = x$, and $\bar{z}$ is parallel to the X-Z plane. The second step is to rotate the intermediate coordinate system about the $\bar{x}$ axis to the position such that $\bar{z} = z$. Then the intermediate coordinate system $\bar{x}\bar{y}\bar{z}$ becomes the local coordinate system xyz.

Mathematically, the first step is described as

$$\left\{\begin{array}{c} \bar{x} \\ \bar{y} \\ \bar{z} \end{array}\right\} = \lambda_1 \left\{\begin{array}{c} X \\ Y \\ Z \end{array}\right\} \qquad (8.69)$$

and the second step is

$$\left\{\begin{array}{c} x \\ y \\ z \end{array}\right\} = \lambda_2 \left\{\begin{array}{c} \bar{x} \\ \bar{y} \\ \bar{z} \end{array}\right\}. \qquad (8.70)$$

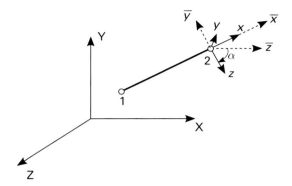

Figure 8.14 An intermediate coordinate system ($\bar{x}\bar{y}\bar{z}$) associated with the beam element.

Then we have

$$\begin{Bmatrix} x \\ y \\ z \end{Bmatrix} = \lambda_1 \lambda_2 \begin{Bmatrix} X \\ Y \\ Z \end{Bmatrix}. \tag{8.71}$$

For λ_1, it is observed that the unit vector in the $\bar{x}$-direction can be written as

$$\bar{\mathbf{x}} = \begin{Bmatrix} \dfrac{X_2 - X_1}{L} \\ \dfrac{Y_2 - Y_1}{L} \\ \dfrac{Z_2 - Z_1}{L} \end{Bmatrix} = \begin{Bmatrix} \cos(\widehat{\bar{x}, X}) \\ \cos(\widehat{\bar{x}, Y}) \\ \cos(\widehat{\bar{x}, Z}) \end{Bmatrix} \tag{8.72}$$

whose components are the coordinates of the unit vector $\bar{\mathbf{x}}$ in the global coordinate system, and also, by definition, the first row of the λ_1. In addition, the $\bar{z}$ directional vector can be obtained as

$$\bar{\mathbf{z}} = \bar{\mathbf{x}} \times \mathbf{Y}$$

$$= \begin{Bmatrix} \dfrac{X_2 - X_1}{L} \\ \dfrac{Y_2 - Y_1}{L} \\ \dfrac{Z_2 - Z_1}{L} \end{Bmatrix} \times \begin{Bmatrix} 0 \\ 1 \\ 0 \end{Bmatrix} = \begin{Bmatrix} -\dfrac{Z_2 - Z_1}{L} \\ 0 \\ \dfrac{X_2 - X_1}{L} \end{Bmatrix}. \tag{8.73}$$

8.3 Beams and Space Frames

Normalizing $\bar{\mathbf{z}}$ we have

$$\bar{\mathbf{z}} = \frac{1}{\sqrt{\left(\frac{Z_2-Z_1}{L}\right)^2 + \left(\frac{X_2-X_1}{L}\right)^2}} \left\{ \begin{array}{c} -\frac{Z_2-Z_1}{L} \\ 0 \\ \frac{X_2-X_1}{L} \end{array} \right\}. \qquad (8.74)$$

The components of the unit $\bar{\mathbf{z}}$ vector are the components of the third row of the λ_1. Finally,

$$\bar{\mathbf{y}} = \bar{\mathbf{z}} \times \bar{\mathbf{x}}$$

$$= \frac{1}{\sqrt{\left(\frac{Z_2-Z_1}{L}\right)^2 + \left(\frac{X_2-X_1}{L}\right)^2}} \left\{ \begin{array}{c} -\frac{Z_2-Z_1}{L} \\ 0 \\ \frac{X_2-X_1}{L} \end{array} \right\} \times \left\{ \begin{array}{c} \frac{X_2-X_1}{L} \\ \frac{Y_2-Y_1}{L} \\ \frac{Z_2-Z_1}{L} \end{array} \right\}$$

$$= \frac{1}{L\sqrt{(Z_2-Z_1)^2 + (X_2-X_1L)^2}} \left\{ \begin{array}{c} -(X_2-X_1)(Y_2-Y_1) \\ (X_2-X_1)^2 + (Z_2-Z_1)^2 \\ -(Y_2-Y_1)(Z_2-Z_1) \end{array} \right\}. \qquad (8.75)$$

Therefore, the transformation matrix λ_1 is

$$\lambda_1 = \left\{ \begin{array}{c} (\bar{\mathbf{x}}^T)_{1 \times 3} \\ (\bar{\mathbf{y}}^T)_{1 \times 3} \\ (\bar{\mathbf{z}}^T)_{1 \times 3} \end{array} \right\}. \qquad (8.76)$$

For the second step transformation, the intermediate coordinate system is rotated about its $\bar{\mathbf{x}}$ axis to move the $\bar{\mathbf{z}}$ vector to the $\mathbf{z}$ vector. The scenario is shown in Fig. 8.15 (a). Viewing the coordinate system from the direction of $\bar{\mathbf{x}}$, the rotation occurs entirely in the plane of $\mathbf{z}\bar{\mathbf{z}}$. Assuming the angle of rotation is α as shown in Fig. 8.15 (a), the transformation can written as

$$\left\{ \begin{array}{c} x \\ y \\ z \end{array} \right\} = \left[\begin{array}{ccc} 1 & 0 & 0 \\ 0 & \cos(\alpha) & \sin(\alpha) \\ 0 & -\sin(\alpha) & \cos(\alpha) \end{array} \right] \left\{ \begin{array}{c} \bar{x} \\ \bar{y} \\ \bar{z} \end{array} \right\} = \lambda_2 \left\{ \begin{array}{c} \bar{x} \\ \bar{y} \\ \bar{z} \end{array} \right\}. \qquad (8.77)$$

With λ_1 and λ_2 obtained, we can calculate λ by $\lambda = \lambda_1 \lambda_2$. While the two-step transformation is general and applicable to arbitrary vectors, there is a special scenario for which the two-step transformation becomes invalid and a different treatment is required. The special scenario is when **x** coincides with **Y**. In this case, there is no need of an intermediate coordinate system; the transformation between xyz and XYZ can be directly constructed as shown in Eq. (8.78). The transformation matrix is

$$\boldsymbol{\lambda} = \begin{bmatrix} 0 & 1 & 0 \\ -\cos(\alpha) & 0 & \sin(\alpha) \\ \sin(\alpha) & 0 & \cos(\alpha) \end{bmatrix}. \tag{8.78}$$

By using the two-step transformation, the displacement and force vectors defined for beam elements in their local coordinate systems can be transformed into the global coordinate system shared by all the beam elements for assembly. Equation (8.66) can be further written in the global coordinate system as

$$\mathbf{k}^{(e)} \begin{Bmatrix} u_1 \\ v_1 \\ w_1 \\ \theta_{x1} \\ \theta_{y1} \\ \theta_{z1} \\ u_2 \\ v_2 \\ w_2 \\ \theta_{x2} \\ \theta_{y2} \\ \theta_{z2} \end{Bmatrix} = \begin{Bmatrix} f_{x1} \\ f_{y1} \\ f_{z1} \\ \bar{m}_{x1} \\ \bar{m}_{y1} \\ \bar{m}_{z1} \\ f_{x2} \\ f_{y2} \\ f_{z2} \\ \bar{m}_{x2} \\ \bar{m}_{y2} \\ \bar{m}_{z2} \end{Bmatrix}$$

$$\rightarrow \mathbf{k}^{(e)} \begin{Bmatrix} \boldsymbol{\lambda} & & & \\ & \boldsymbol{\lambda} & & \\ & & \boldsymbol{\lambda} & \\ & & & \boldsymbol{\lambda} \end{Bmatrix} \begin{Bmatrix} U_1 \\ V_1 \\ W_1 \\ \theta_{X1} \\ \theta_{Y1} \\ \theta_{Z1} \\ U_2 \\ V_2 \\ W_2 \\ \theta_{X2} \\ \theta_{Y2} \\ \theta_{Z2} \end{Bmatrix} = \begin{Bmatrix} \boldsymbol{\lambda} & & & \\ & \boldsymbol{\lambda} & & \\ & & \boldsymbol{\lambda} & \\ & & & \boldsymbol{\lambda} \end{Bmatrix} \begin{Bmatrix} f_{X1} \\ f_{Y1} \\ f_{Z1} \\ \bar{m}_{X1} \\ \bar{m}_{Y1} \\ \bar{m}_{Z1} \\ f_{X2} \\ f_{Y2} \\ f_{Z2} \\ \bar{m}_{X2} \\ \bar{m}_{Y2} \\ \bar{m}_{Z2} \end{Bmatrix}.$$

(8.79)

Written in short form, Eq. (8.79) becomes

$$\mathbf{k}^{(e)} \bar{\boldsymbol{\lambda}} \mathbf{d}_g^{(e)} = \bar{\boldsymbol{\lambda}} \mathbf{f}_g^{(e)} \tag{8.80}$$

$$\rightarrow \bar{\boldsymbol{\lambda}}^{-1} \mathbf{k}^{(e)} \bar{\boldsymbol{\lambda}} \mathbf{d}_g^{(e)} = \mathbf{f}_g^{(e)} \tag{8.81}$$

$$\rightarrow \bar{\mathbf{k}}^{(e)} \mathbf{d}_g^{(e)} = \mathbf{f}_g^{(e)} \tag{8.82}$$

where $\bar{\mathbf{k}}^e$, $\mathbf{d}_g^e$, and $\mathbf{f}_g^e$ are the element stiffness matrix, displacement, and force vectors in the global coordinate system, respectively.

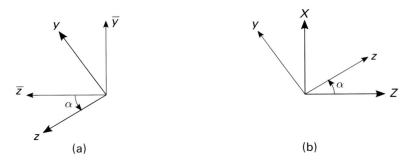

Figure 8.15 Scenarios of coordinate system transformation.

Step 7: Global Matrix and Vector Assembly

The global matrix and vector assembly procedure is the same as assembling the truss element matrices and vectors. Once again, the process illustrated in Fig. 7.16 is applicable to 3-D space frames, except that the dimension of the element stiffness matrix is 12 × 12, and each primitive assembling block is 6 × 6.

It should be noted that the element force vector in Eq. (8.82) is the force vector applied externally to the element which includes the forces and moments from external loads as well as those exerted by the beam elements connected to it. When assembled into the global force vector these beam–beam interactions are canceled as action is equal to reaction, and the net nodal forces and moments are the ones from external loads. Therefore, only distributed forces need to be calculated and added to the force vector for each element. The externally applied nodal forces and moments should be directly applied to the global force vector. The procedure is similar to applying external loads to truss systems. The final linear system is obtained as

$$\mathbf{K}_{6N \times 6N} \mathbf{u}_{6N \times 1} = \mathbf{F}_{6N \times 1} \tag{8.83}$$

where N is the total number of nodes.

Step 8: Apply the Essential Boundary Condition

The application of the essential (displacement) boundary condition can be implemented by using the methods discussed in previous chapters. These methods are equally applicable here. The essential boundary conditions for the 3-D space frame structure shown in Fig. 8.10 are

$$U, V, W = 0 \quad \text{for nodes 1 and 5} \tag{8.84}$$
$$\theta_X, \theta_Y, \theta_Z = 0 \quad \text{for nodes 1 and 5.} \tag{8.85}$$

Step 9: Solve the Global Linear System

After the essential boundary conditions are applied, the final linear system can be solved in MATLAB. The unknown nodal displacement vector **u**, which contains both translational and rotational displacements of the nodes, can be computed. For our example problem, the computed displacement vector is

$$\mathbf{u} = \{ \begin{matrix} 0.0000 & -0.0000 & 0.0000 & -0.0000 & -0.0000 & -0.0000 \\ 0.0008 & -1.5747 & 0.4316 & -0.0428 & -0.0260 & -0.0740 \\ -0.2042 & -0.8424 & 0.4336 & -0.0911 & -0.0159 & -0.0520 \\ -0.3335 & -0.0012 & 0.4355 & -0.0646 & -0.0109 & -0.0299 \\ -0.0000 & -0.0000 & -0.0000 & -0.0000 & -0.0000 & -0.0000 \end{matrix} \}^T \times 10^{-2} \text{cm}. \quad (8.86)$$

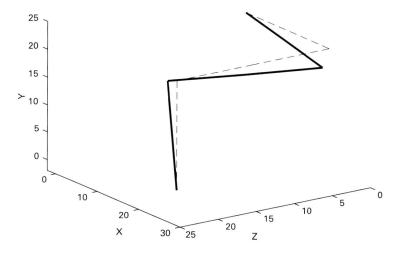

Figure 8.16 Deformed shape of the space frame.

Step 10: Post-processing

The computed displacements are visualized in this step. Figure 8.16 shows the space frame before (dashed lines) and after (solid lines) deformation. With the nodal displacements obtained, we can compute the end forces of the elements by using the following steps. First, we extract each element's nodal displacement and rotational angle vector $\mathbf{d}_g^{(e)}$ from the global nodal displacement vector. Note that the extracted nodal displacements and rotational angles are computed in the global coordinate system in Step 9. Then by using Eq. (8.80), the element force vector $\mathbf{f}^{(e)}$ is calculated in each element's local coordinate system as

$$\mathbf{k}^{(e)} \bar{\boldsymbol{\lambda}} \mathbf{d}_g^{(e)} = \mathbf{f}^{(e)}. \quad (8.87)$$

Once the element force vector $\mathbf{f}^{(e)}$ is obtained, the stresses in each element can be calculated by using formulas of classical mechanics of materials such as flexure, torsion, and shear formulas.

8.3.2 Computer Implementation

In this section, we present the data files and the complete MATLAB code for the finite element analysis of the space frame example.

Data Files

The input files "nodes.dat" and "elements.dat" are already shown in Table 8.2. The boundary conditions, loads, material properties, geometric properties, and model options are stored as shown in the tables below.

bcsdisp.dat	forces.dat	materials.dat
% node id, type, value	% node id, fx, fy, fz, mx, my, mz	% Young's modulus (N/cm^2)
1 1 0	2 0 −100 0 0 0 0	
1 2 0	4 0 0 100 0 0 0	2e7
1 3 0		
1 4 0		% Shear modulus (N/cm^2)
1 5 0		
1 6 0		0.8e7
5 1 0		
5 2 0		
5 3 0		
5 4 0		
5 5 0		
5 6 0		

geometry.dat	options.dat
% element id, alpha, A, Iy, Iz, J	% Dimensions
1 0 3 2.25 0.25 2.5	
2 0 3 2.25 0.25 2.5	3
3 0 3 2.25 0.25 2.5	
4 −90 3 2.25 0.25 2.5	

MATLAB Code
The code of the space frame analysis is listed below:

```matlab
clear all;
% next 3 lines: read the input files
filenames = {'nodes.dat','elements.dat','materials.dat', ...
             'options.dat','forces.dat', 'bcsdisp.dat',...
             'geometry.dat'};
for i = 1:numel(filenames); load(filenames{i}); end;

% next 13 lines: set up constants and empty matrices
nNodes=size(nodes,1);
nElements=size(elements,1);
E=materials(1,1);
G=materials(2,1);
alpha=geometry(:,2)/180*pi;
A=geometry(:,3);
Iy=geometry(:,4);
Iz=geometry(:,5);
J=geometry(:,6);
K=zeros(6*nNodes,6*nNodes);
F=zeros(6*nNodes,1);
length=zeros(nElements,1);
direction_cos =zeros(nElements,3);

% for-loop: compute the global stiffness matrix
for e=1:nElements
    node1=elements(e,2);
    node2=elements(e,3);
    dv=nodes(node2,2:4)-nodes(node1,2:4);
    length(e)=norm(dv);
    direction_cos (e ,:) =dv/length(e);
    % set up intermediate quantities
    EAL=E*A(e)/length(e);
    GJL=G*J(e)/length(e);
    EIzL=E*Iz(e)/(length(e));
    EIyL=E*Iy(e)/(length(e));
    EIzL2=EIzL/length(e);
    EIyL2=EIyL/length(e);
    EIzL3=EIzL2/length(e);
    EIyL3=EIyL2/length(e);
    % next 9 lines: element k matrix in local coordinate system
    k=zeros(12,12);
    k(1,1)=EAL; k(2,2)=12*EIzL3; k(3,3)=12*EIyL3; k(4,4)=GJL;
    k(5,3)=-6*EIyL2; k(5,5)=4*EIyL; k(6,2)=6*EIzL2; k(6,6)=4*EIzL;
    k(7,1)=-EAL; k(7,7)=EAL; k(8,2)=-12*EIzL3; k(8,6)=-6*EIzL2;
    k(8,8)=12*EIzL3; k(9,3)=-12*EIyL3; k(9,5)=6*EIyL2;
    k(9,9)=12*EIyL3; k(10,4)=-GJL; k(10,10)=GJL; k(11,3)=-6*EIyL2;
```

```
k(11,5)=2*EIyL; k(11,9)=6*EIyL2; k(11,11)=4*EIyL; k(12,2)=6*EIzL2;
k(12,6)=2*EIzL; k(12,8)=-6*EIzL2; k(12,12)=4*EIzL;
k=k+k'-eye(12).*diag(k);
% next 16 lines : calculate lambda
d=sqrt(dv(1)^2 + dv(3)^2);
if abs(d)>1e-10*length(e)
   lambda1=zeros(3,3);
   lambda1(1,:) = direction_cos(e,:);
   lambda1(2,:) =1/(d*length(e)) *...
       [-dv(1)*dv(2) dv(1)^2+dv(3)^2 -dv(2)*dv(3)];
   lambda1(3,:)=1/d*[-dv(3) 0 dv(1)];
   lambda2=[1 0    0
           0 cos(alpha(e)) sin(alpha(e))
           0 -sin(alpha(e)) cos(alpha(e))];
   lambda=lambda1*lambda2;
else
   lambda=[0 1 0
          -cos(alpha(e)) 0 sin(alpha(e))
           sin(alpha(e)) 0 cos(alpha(e))];
end
% for-loop: create the big lambda matrix
for i=0:3
   Lambda(i*3+1:i*3+3,i*3+1:i*3+3)=lambda;
end
k=inv(Lambda)*k*Lambda; % transform to global coordinate system
% assembly
K(6*node1-5:6*node1, 6*node1-5:6*node1) ...
       =K(6*node1-5:6*node1, 6*node1-5:6*node1) +k(1:6,1:6);
K(6*node1-5:6*node1, 6*node2-5:6*node2) ...
       =K(6*node1-5:6*node1, 6*node2-5:6*node2) +k(1:6,7:12);
K(6*node2-5:6*node2, 6*node1-5:6*node1) ...
       =K(6*node2-5:6*node2, 6*node1-5:6*node1) +k(7:12,1:6);
K(6*node2-5:6*node2, 6*node2-5:6*node2) ...
       =K(6*node2-5:6*node2, 6*node2-5:6*node2) +k(7:12,7:12);
end

% set up the global force vector
for i=1:size(forces,1)
   node = forces(i,1);
   F(6*node-5:6*node)= forces(i,2:7);
end

% next 7 lines : apply the displacement BC using the penalty method
penalty=abs(max(max(K)))*1e7;
for i=1:size(bcsdisp,1)
   node = bcsdisp(i,1);
   disp_type = bcsdisp(i,2);
   K(6*node + disp_type - 6, 6*node + disp_type - 6)= penalty;
   F(6*node + disp_type - 6)= penalty*bcsdisp(i,3);
```

```
95  end
96
97  d=K\F % solve global linear system
98
99  % next 20 lines : plotting
100 figure(1);
101 d=d*100;
102 clf;
103 hold on;
104 % for-loop: plot the deformed and undeformed space frame system
105 for e=1:nElements
106    nd1=elements(e,2);
107    nd2=elements(e,3);
108    x1=nodes(nd1,2); y1=nodes(nd1,4); z1=nodes(nd1,3);
109    x2=nodes(nd2,2); y2=nodes(nd2,4); z2=nodes(nd2,3);
110    plot3([x1 x2], [y1 y2], [z1 z2],'k--');
111    plot3([x1+d(6*nd1-5) x2+d(6*nd2-5)],[y1+d(6*nd1-3) ...
112         y2+d(6*nd2-3)], [z1+d(6*nd1-4) z2+d(6*nd2-4)],...
113         'k-','LineWidth',3);
114 end
115 axis([-2 30 0 25 -2 25]);
116 set(gca,'Ydir','reverse');   % reverse the y-axis for plotting
117 xlabel('X'); ylabel('Z'); zlabel('Y');
118 view(55,25);
119 hold off;
```

8.4 Plates

In solid mechanics, structures can be simplified with low-dimensional models if one or more dimensions is significantly smaller than the others. For instance, a thin structure can be treated as a surface if its thickness is small compared to the length and width. Flat surfaces can be approximated by plate models, and curved surfaces can be approximated by shell models. In this section, finite element analysis of 3-D thin plates is illustrated by demonstrating the step-by-step procedure of solving a triangular plate structure as shown in Fig. 8.17. Note that there are different plate theories, finite element formulations, and elements for treating different types of plates. In this section we only consider thin flat plates subjected to pure bending deformation.

8.4.1 Kirchhoff Plates Subjected to External Loads

Step 1: Physical Problem

A triangular plate with a hole is subjected to a transverse load of 500 N as shown in Fig. 8.17. The material properties are given by $E = 205$ GPa and $v = 0.33$. The thickness of the plate $t = 2$ mm. The dimensions are provided in the figure. The plate is symmetric about the y-axis. Note that there is no in-plane load. We perform a finite element analysis and calculate the deformation of the plate.

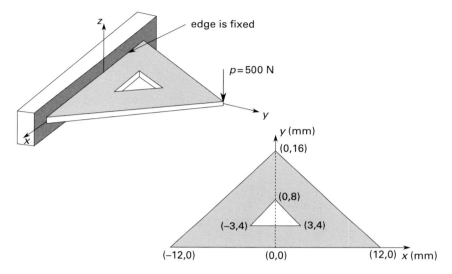

Figure 8.17 A triangular plate subjected to a load.

Step 2: Mathematical Model

For thin plates, bending is often considered to be the only out-of-plane deformation. In the classical pure-bending plate theory it is assumed that:

1. Straight lines that are normal to the plate's neutral surface (middle surface for homogeneous materials) remain straight after deformation.
2. These straight lines also remain normal to the neutral surface after deformation.
3. These straight lines also experience no normal strains after deformation.

This is called the Kirchhoff–Love assumption (Ventsel and Krauthammer 2001). The deformation field in Cartesian coordinates can be expressed as:

$$u(x, y, z) = u_0(x, y) - z\frac{\partial w_0(x, y)}{\partial x} \tag{8.88}$$

$$v(x, y, z) = v_0(x, y) - z\frac{\partial w_0(x, y)}{\partial y} \tag{8.89}$$

$$w(x, y, z) = w_0(x, y) \tag{8.90}$$

where u,v,w represent the displacements of point (x, y, z) in x-, y-, z-directions, respectively. u_0, v_0, w_0 are the corresponding values at the neutral surface.

The equilibrium equation of classical Kirchhoff–Love plate element shown in Fig. 8.18 is:

$$\frac{\partial N_x}{\partial x} + \frac{\partial N_{xy}}{\partial y} = -p_x(x, y) \tag{8.91}$$

$$\frac{\partial N_{xy}}{\partial x} + \frac{\partial N_y}{\partial y} = -p_y(x,y) \tag{8.92}$$

$$\frac{\partial^2 M_x}{\partial x^2} + 2\frac{\partial^2 M_{xy}}{\partial x \partial y} + \frac{\partial^2 M_y}{\partial y^2} = -q(x,y) \tag{8.93}$$

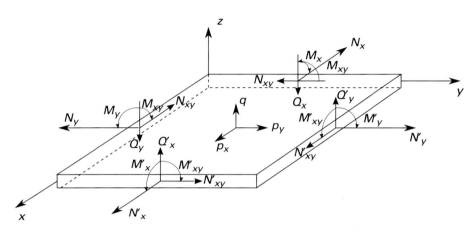

Figure 8.18 Internal loading on an infinitesimal plate element.

where the resultant forces N and moments M can be obtained by:

$$\begin{bmatrix} N_x \\ N_y \\ N_{xy} \\ M_x \\ M_y \\ M_{xy} \end{bmatrix} = \begin{bmatrix} \int_{-h/2}^{h/2} \sigma_{xx} dz \\ \int_{-h/2}^{h/2} \sigma_{yy} dz \\ \int_{-h/2}^{h/2} \sigma_{xy} dz \\ \int_{-h/2}^{h/2} \sigma_{xx} z dz \\ \int_{-h/2}^{h/2} \sigma_{yy} z dz \\ \int_{-h/2}^{h/2} \sigma_{xy} z dz \end{bmatrix} = \begin{bmatrix} \mathbf{A}_{3\times 3} & 0 \\ 0 & \mathbf{D}_{3\times 3} \end{bmatrix} \begin{bmatrix} \epsilon_{xx} \\ \epsilon_{yy} \\ \epsilon_{xy} \\ \kappa_{xx} \\ \kappa_{yy} \\ \kappa_{xy} \end{bmatrix} \tag{8.94}$$

$$\begin{bmatrix} \epsilon_{xx} \\ \epsilon_{yy} \\ \epsilon_{xy} \end{bmatrix} = \begin{bmatrix} \frac{\partial}{\partial x} & 0 \\ 0 & \frac{\partial}{\partial y} \\ \frac{\partial}{\partial y} & \frac{\partial}{\partial x} \end{bmatrix} \begin{bmatrix} u \\ v \end{bmatrix}, \quad \begin{bmatrix} \kappa_{xx} \\ \kappa_{yy} \\ \kappa_{xy} \end{bmatrix} = \begin{bmatrix} -\frac{\partial^2 w}{\partial x^2} \\ -\frac{\partial^2 w}{\partial y^2} \\ -2\frac{\partial^2 w}{\partial x \partial y} \end{bmatrix}$$

where k_{xx}, k_{yy} are the curvatures in the x- and y-directions, respectively, and k_{xy} is the twist curvature. We define matrix **A** and matrix **B** as the in-plane stiffness matrix and bending stiffness matrix, respectively. For traditional isotropic homogeneous materials, they can be expressed as:

$$\mathbf{A} = \begin{bmatrix} A & \nu A & 0 \\ \nu A & A & 0 \\ 0 & 0 & \frac{1-\nu}{2}A \end{bmatrix} \quad \mathbf{D} = \begin{bmatrix} D & \nu D & 0 \\ \nu D & D & 0 \\ 0 & 0 & \frac{1-\nu}{2}D \end{bmatrix} \quad (8.95)$$

where

$$A = \frac{Et}{1-\nu^2}, \quad D = \frac{Et^3}{12(1-\nu^2)}. \quad (8.96)$$

For pure bending (i.e., neglecting the in-plane displacements), substituting Eq. (8.94) into Eq. (8.93), we obtain the governing equation for the plate written in terms of deflection as

$$-D\left(\frac{\partial^4 w}{\partial x^4} + 2\frac{\partial^4 w}{\partial x^2 \partial y^2} + \frac{\partial^4 w}{\partial y^4}\right) + q = 0. \quad (8.97)$$

As shown in Fig. 8.19, possible boundary conditions on edges of a plate can be given by:

$$\begin{aligned}
w(x, y) &= w_0 & \text{on } \Gamma_w, & \quad w_{,n}(x, y) = \theta_0 & \text{on } \Gamma_\theta \\
u(x, y) &= u_0 & \text{on } \Gamma_u, & \quad v(x, y) = v_0 & \text{on } \Gamma_v \\
\mathbf{M}(x, y) &= \mathbf{M}_0 & \text{on } \Gamma_m, & \quad Q(x, y) = Q_0 & \text{on } \Gamma_q \\
\mathbf{T}(x, y) &= \mathbf{N}_0 & \text{on } \Gamma_t &
\end{aligned} \quad (8.98)$$

where Q is the transverse shear, **T** is the in-plane force vector, and **M** is the moment vector. The load components (normal and tangent to the edge) are shown in Fig. 8.19.

Step 3: Weak Form

In this section, we use the principle of virtual work to obtain the weak form of the plate bending equation. We assume deformation due to transverse shear stress is negligible. The virtual work of internal bending moment is:

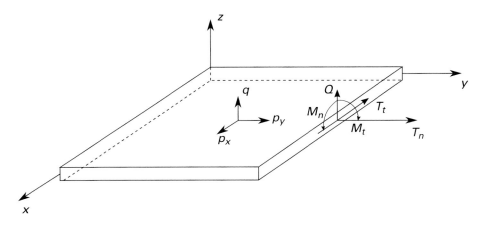

Figure 8.19 External loads on a plate.

$$\delta \Pi_{bint} = - \int_\Omega \left(M_x \frac{\partial^2 \delta w}{\partial x^2} + 2 M_{xy} \frac{\partial^2 \delta w}{\partial x \partial y} + M_y \frac{\partial^2 \delta w}{\partial y^2} \right) d\Omega. \qquad (8.99)$$

Then recall the expression of the virtual work of in-plane forces:

$$\delta \Pi_{mint} = t \int_\Omega \delta \epsilon_0^T \sigma_0 d\Omega = t \int_\Omega \delta \epsilon_0^T \mathbf{C} \epsilon_0 d\Omega \qquad (8.100)$$

where $\epsilon_0 = \{ \; \epsilon_{x0} \quad \epsilon_{y0} \quad \epsilon_{xy0} \; \}^T$ is the strain vector at mid-plane, and

$$\mathbf{C} = \frac{E}{(1-\nu^2)} \begin{bmatrix} 1 & \nu & 0 \\ \nu & 1 & 0 \\ 0 & 0 & \frac{1-\nu}{2} \end{bmatrix}. \qquad (8.101)$$

The external work can be calculated as follows:

$$\delta \Pi_{ext} = \int_\Omega \left(\delta u p_x + \delta v p_y + \delta w q \right) d\Omega + \int_{\Gamma_t} \left(\delta u T_x + \delta v T_y \right) d\Gamma$$
$$+ \int_{\Gamma_q} \delta w Q d\Gamma + \int_{\Gamma_m} \left(-\frac{\partial \delta w}{\partial \mathbf{n}} M_t + \frac{\partial \delta w}{\partial \mathbf{t}} M_n \right) d\Gamma \qquad (8.102)$$

where $\mathbf{n}, \mathbf{t}$ are normal and tangential directions of the boundary as shown in Fig. 8.19. T_x and T_y are the x- and y-components of the in-plane force vector $\mathbf{T}$, respectively. Next we apply the virtual work principle:

$$\delta \Pi_{int} = \delta \Pi_{ext}$$

$$\Rightarrow 0 = \int_\Omega -\left(M_x \frac{\partial^2 \delta w}{\partial x^2} + 2M_{xy} \frac{\partial^2 \delta w}{\partial x \partial y} + M_y \frac{\partial^2 \delta w}{\partial y^2}\right) d\Omega + \int_\Omega \delta\epsilon_0 \cdot \mathbf{C} \cdot \epsilon_0 d\Omega$$

$$- \int_\Omega (\delta u p_x + \delta v p_y + \delta w q) d\Omega - \int_{\Gamma_t} (\delta u T_x + \delta v T_y) d\Gamma - \int_{\Gamma_q} \delta w Q d\Gamma$$

$$+ \int_{\Gamma_m} \left(\frac{\partial \delta w}{\partial \mathbf{n}} M_t - \frac{\partial \delta w}{\partial \mathbf{t}} M_n\right) d\Gamma. \tag{8.103}$$

Consider the strain–displacement relation and moment–curvature relation:

$$\begin{Bmatrix} \epsilon_{x0} \\ \epsilon_{y0} \\ \epsilon_{xy0} \end{Bmatrix} = \begin{Bmatrix} \frac{\partial u}{\partial x} \\ \frac{\partial v}{\partial y} \\ \frac{\partial u}{\partial y} + \frac{\partial v}{\partial x} \end{Bmatrix} = \begin{bmatrix} \frac{\partial}{\partial x} & 0 \\ 0 & \frac{\partial}{\partial y} \\ \frac{\partial}{\partial y} & \frac{\partial}{\partial x} \end{bmatrix} \begin{Bmatrix} u \\ v \end{Bmatrix} = \mathcal{L}\mathbf{u} \tag{8.104}$$

$$\begin{Bmatrix} M_x \\ M_y \\ M_{xy} \end{Bmatrix} = -\begin{bmatrix} D & \nu D & 0 \\ \nu D & D & 0 \\ 0 & 0 & \frac{1}{2}D(1-\nu) \end{bmatrix} \begin{Bmatrix} \frac{\partial^2 w}{\partial x^2} \\ \frac{\partial^2 w}{\partial y^2} \\ 2\frac{\partial^2 w}{\partial x \partial y} \end{Bmatrix} = -\mathbf{D}(\mathcal{L}\nabla)w. \tag{8.105}$$

Then Eq. (8.103) can be rewritten as:

$$\int_\Omega (\mathcal{L}(\nabla \delta w))^T \mathbf{D} (\mathcal{L}(\nabla w)) d\Omega + \int_\Omega (\mathcal{L}\delta \mathbf{u})^T \mathbf{C}(\mathcal{L}\mathbf{u}) d\Omega$$

$$- \int_\Omega (\delta u p_x + \delta v p_y + \delta w q) d\Omega - \int_{\Gamma_t} (\delta u T_x + \delta v T_y) d\Gamma - \int_{\Gamma_q} \delta w Q d\Gamma$$

$$+ \int_{\Gamma_m} \left(\frac{\partial \delta w}{\partial \mathbf{n}} M_t - \frac{\partial \delta w}{\partial \mathbf{t}} M_n\right) d\Gamma = 0. \tag{8.106}$$

Note that the in-plane deformations and bending deformation are decoupled. Therefore Eq. (8.106) can be further rewritten as:

$$\int_\Omega [(\mathcal{L}(\nabla \delta w))^T \mathbf{D} (\mathcal{L}(\nabla w)) - \delta w q] d\Omega - \int_{\Gamma_q} \delta w Q d\Gamma$$

$$+ \int_{\Gamma_m} \left(\frac{\partial \delta w}{\partial \mathbf{n}} M_t - \frac{\partial \delta w}{\partial \mathbf{t}} M_n\right) d\Gamma = 0 \tag{8.107}$$

and

$$\int_\Omega [(\mathcal{L}\delta\mathbf{u})^T \mathbf{C}(\mathcal{L}\mathbf{u}) - \delta u p_x - \delta v p_y] d\Omega - \int_{\Gamma_t} (\delta u T_x + \delta v T_y) d\Gamma = 0. \quad (8.108)$$

Steps 4–5: Discretization and Approximation

For our plate problem, the 2-D plate is discretized into a set of 2-D elements, although these 2-D elements are located in the 3-D space. Similar to the truss and space frame cases, each of the 2-D elements are defined using their local coordinate systems. Then the quantities are transformed into the global 3-D coordinate system. As an example, Fig. 8.20 shows a mesh of the triangular plate.

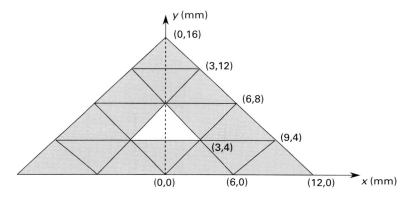

Figure 8.20 A mesh of the triangular plate.

Table 8.3 lists the nodes and elements for the mesh shown in Fig. 8.20. Note that the node list in the "elements.dat" file is following the counterclockwise direction.

In the local coordinate system of a plate element, we approximate u, v by using shape functions and nodal values

$$u = \sum_{i=1}^n N_i u_i = \mathbf{N}\mathbf{u}_d \qquad \delta u = \sum_{i=1}^n N_i \delta u_i = \mathbf{N}\delta\mathbf{u}_d$$

$$v = \sum_{i=1}^n N_i v_i = \mathbf{N}\mathbf{v}_d \qquad \delta v = \sum_{i=1}^n N_i \delta v_i = \mathbf{N}\delta\mathbf{v}_d \quad (8.109)$$

$$\mathbf{u} = \begin{Bmatrix} u \\ v \end{Bmatrix} = \begin{bmatrix} \mathbf{N} & 0 \\ 0 & \mathbf{N} \end{bmatrix} \begin{Bmatrix} \mathbf{u}_d \\ \mathbf{v}_d \end{Bmatrix} = \tilde{\mathbf{N}}\mathbf{d}_d$$

where $\mathbf{N} = \{N_1 N_2 \ldots N_n\}$. For in-plane displacement approximation in a triangular element

$$u = \sum_{i=1}^3 N_i(x, y) u_i \quad (8.110)$$

8.4 Plates

Table 8.3 Nodes and elements

nodes.dat				elements.dat			
1	−12	0	0	1	12	1	2
2	−6	0	0	2	12	2	13
3	0	0	0	3	12	13	11
4	6	0	0	4	13	15	11
5	12	0	0	5	10	11	15
6	9	4	0	6	10	15	8
7	6	8	0	7	10	8	9
8	3	12	0	8	8	15	7
9	0	16	0	9	6	7	14
10	−3	12	0	10	5	6	4
11	−6	8	0	11	14	4	6
12	−9	4	0	12	15	14	7
13	−3	4	0	13	3	14	13
14	3	4	0	14	3	13	2
15	0	8	0	15	14	3	4

$$v = \sum_{i=1}^{3} N_i(x,y) v_i \qquad (8.111)$$

where

$$N_1 = \frac{(x_2 y_3 - x_3 y_2) + (y_2 - y_3)x + (x_3 - x_2)y}{2A} \qquad (8.112)$$

$$N_2 = \frac{(x_3 y_1 - x_1 y_3) + (y_3 - y_1)x + (x_1 - x_3)y}{2A} \qquad (8.113)$$

$$N_3 = \frac{(x_1 y_2 - x_2 y_1) + (y_1 - y_2)x + (x_2 - x_1)y}{2A}. \qquad (8.114)$$

and A is the area of the triangular element, i.e.,

$$A = \frac{1}{2} \begin{vmatrix} 1 & x_1 & y_1 \\ 1 & x_2 & y_2 \\ 1 & x_3 & y_3 \end{vmatrix}. \qquad (8.115)$$

For bending deflection w, the fourth order derivatives in the governing equation, Eq. (8.97), require C^1 continuity across the elements. That is, the deflection, w, as well as the first derivatives of w should be continuous from one element to another to satisfy the compatibility condition. It is shown that, to meet such a requirement for rectangular and triangular elements, Hermite type or higher order polynomials are necessary to approximate the deflection, leading to more degrees of freedom for each element. To simply the problem and improve efficiency, non-conforming approximations were proposed for both rectangular and triangular elements. Practice shows that they give reasonable performance for most engineering problems. In this section, a simple non-conforming triangular element is employed for solving the given problem. For more information about the other non-conforming and conforming elements, the reader is referred to (Rao 2017).

In the non-conforming triangular element we employ here, the deflection function $w(x,y)$ is approximated as

$$w = a_1 + a_2 x + a_3 y + a_4 x^2 + a_5 xy + a_6 y^2 + a_7 x^3 + a_8 (x^2 y + xy^2) + a_9 y^3. \tag{8.116}$$

There are a total of nine unknown constants in the polynomial. The triangular element in its local coordinate system is shown in Fig. 8.21. The triangular element has three nodes, each having three degrees of freedom for bending: the deflection w, and its two first derivatives $w_x = \dfrac{\partial w}{\partial x}$ and $w_y = \dfrac{\partial w}{\partial y}$. The local coordinate system of the triangular element is defined as follows. The three nodes are numbered in the counterclockwise direction. Node 1 is set to be the origin of the local coordinate system. The y-axis is directed from node 1 to node 2. The x-axis points to the side on which node 3 lies, as shown in Fig. 8.21.

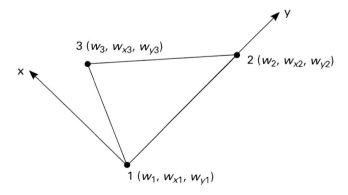

Figure 8.21 A pure bending triangular element and its degrees of freedom.

For the triangular element shown in Fig. 8.21, the element degrees of freedom vector (pure bending) can be written as

$$\begin{Bmatrix} w_1 \\ w_{x1} \\ w_{y1} \\ w_2 \\ w_{x2} \\ w_{y2} \\ w_3 \\ w_{x3} \\ w_{y3} \end{Bmatrix} = \begin{bmatrix} 1 & 0 & 0 & 0 & 0 & 0 & 0 & 0 & 0 \\ 0 & 1 & 0 & 0 & 0 & 0 & 0 & 0 & 0 \\ 0 & 0 & 1 & 0 & 0 & 0 & 0 & 0 & 0 \\ 1 & 0 & y_2 & 0 & 0 & y_2^2 & 0 & 0 & y_2^3 \\ 0 & 1 & 0 & 0 & y_2 & 0 & 0 & y_2^2 & 0 \\ 0 & 0 & 1 & 0 & 0 & 2y_2 & 0 & 0 & 3y_2^2 \\ 1 & x_3 & y_3 & x_3^2 & x_3 y_3 & y_3^2 & x_3^3 & x_3^2 y_3 + x_3 y_3^2 & y_3^3 \\ 0 & 1 & 0 & 2x_3 & y_3 & 0 & 3x_3^2 & y_3^2 + 2x_3 y_3 & 0 \\ 0 & 0 & 1 & 0 & x_3 & 2y_3 & 0 & 2x_3 y_3 + x_3^2 & 3y_3^2 \end{bmatrix} \begin{Bmatrix} a_1 \\ a_2 \\ a_3 \\ a_4 \\ a_5 \\ a_6 \\ a_7 \\ a_8 \\ a_9 \end{Bmatrix}$$

(8.117)

In short, it is written as

$$\mathbf{d}_w = \mathbf{Z}\mathbf{a} \tag{8.118}$$

where $\mathbf{d}_w$ is the bending degree of freedom vector, $\mathbf{a}$ is the coefficient vector, and $\mathbf{Z}$ is the coordinate matrix. The deflection function over the element can then be expressed as

$$w(x) = \begin{bmatrix} 1 & x & y & x^2 & xy & y^2 & x^3 & (x^2 y + xy^2) & y^3 \end{bmatrix} \begin{Bmatrix} a_1 \\ a_2 \\ a_3 \\ a_4 \\ a_5 \\ a_6 \\ a_7 \\ a_8 \\ a_9 \end{Bmatrix}$$

$$= \mathbf{b}^T \mathbf{a} = \mathbf{b}^T \mathbf{Z}^{-1} \mathbf{d}_w \tag{8.119}$$

where $\mathbf{b}$ is the basis function vector. The variation of w is approximated in the same way as

$$\delta w(x, y) = \mathbf{b}^T \mathbf{Z}^{-1} \delta \mathbf{d}_w \tag{8.120}$$

where $\delta \mathbf{d}_w$ is the degree of freedom variation vector. Note that, the deflection is approximated directly through Eq. (8.119), without converting to an expression in terms of shape functions. While for some of the conforming elements, the deflection function approximation is still expressed in terms of shape functions (Hermite functions), the approximation given in Eq. (8.119) is more convenient for the non-conforming triangular element.

From Eq. (8.105), Eq. (8.116), and Eq. (8.117), we have

$$(\mathcal{L}\nabla)w = \begin{Bmatrix} \dfrac{\partial^2 w}{\partial x^2} \\ \dfrac{\partial^2 w}{\partial y^2} \\ 2\dfrac{\partial^2 w}{\partial x \partial y} \end{Bmatrix} = \begin{bmatrix} 0 & 0 & 0 & 2 & 0 & 0 & 6x & 2y & 0 \\ 0 & 0 & 0 & 0 & 0 & 2 & 0 & 2x & 6y \\ 0 & 0 & 0 & 0 & 2 & 0 & 0 & 4(x+y) & 0 \end{bmatrix} \begin{Bmatrix} a_1 \\ a_2 \\ a_3 \\ a_4 \\ a_5 \\ a_6 \\ a_7 \\ a_8 \\ a_9 \end{Bmatrix}$$

$$= \overline{\mathbf{B}}\mathbf{a} = \overline{\mathbf{B}}\mathbf{Z}^{-1}\mathbf{d}_w \tag{8.121}$$

and

$$(\mathcal{L}\nabla)\delta w = \overline{\mathbf{B}}\mathbf{Z}^{-1}\delta \mathbf{d}_w. \tag{8.122}$$

Step 6: Element Matrices and Vectors

Assuming a general case in which the deflection w is expressed in terms of shape functions and its degrees of freedom vector $\mathbf{d}_w$ as $w = \mathbf{N}\mathbf{d}_w$, from Eq. (8.105), we obtain

$$(\mathcal{L}\nabla)w = (\mathcal{L}\nabla)\mathbf{N}\mathbf{d}_w = \mathcal{L}(\nabla \mathbf{N}\mathbf{w}_d) \tag{8.123}$$

Substituting Eq. (8.123) and Eq. (8.109) into Eq. (8.107) and Eq. (8.108), for an element we get

$$\int_\Omega^{(e)} [(\mathcal{L}(\nabla \mathbf{N}\delta\mathbf{w}_d))^T \mathbf{D} (\mathcal{L}(\nabla \mathbf{N}\mathbf{w}_d)) - \mathbf{N}\delta\mathbf{w}_d q] d\Omega$$

$$- \int_{\Gamma_q}^{(e)} \mathbf{N}\delta\mathbf{w}_d Q d\Gamma + \int_{\Gamma_m}^{(e)} \left(\dfrac{\partial(\mathbf{N}\delta\mathbf{w}_d)}{\partial \mathbf{n}} M_t - \dfrac{\partial(\mathbf{N}\delta\mathbf{w}_d)}{\partial \mathbf{t}} M_n \right) d\Gamma = 0 \tag{8.124}$$

and

$$\int_\Omega^{(e)} \left[(\mathcal{L}(\widetilde{\mathbf{N}}\delta\mathbf{d}_d))^T \mathbf{C} (\mathcal{L}(\widetilde{\mathbf{N}}\mathbf{d}_d)) - \mathbf{N}\delta\mathbf{u}_d p_x - \mathbf{N}\delta\mathbf{v}_d p_y \right] d\Omega$$

$$- \int_{\Gamma_t}^{(e)} \left(\mathbf{N}\delta\mathbf{u}_d T_x + \mathbf{N}\delta\mathbf{v}_d T_y \right) d\Gamma = 0 \tag{8.125}$$

We obtain the matrix form of the equations for bending and in-plane deformation

$$\mathbf{k}_b^{(e)} \mathbf{w}_d = \mathbf{f}_b^{(e)} \qquad (8.126)$$

$$\mathbf{k}_m^{(e)} \mathbf{d}_d = \mathbf{f}_m^{(e)} \qquad (8.127)$$

where

$$\begin{aligned}
\mathbf{k}_b^{(e)} &= \int_\Omega^{(e)} (\mathcal{L}(\nabla \mathbf{N}))^T \mathbf{D} \left(\mathcal{L}(\nabla \mathbf{N}) \right) d\Omega \\
\mathbf{f}_b^{(e)} &= \int_\Omega^{(e)} \mathbf{N}^T q d\Omega + \int_{\Gamma_q}^{(e)} \mathbf{N}^T Q d\Gamma - \int_{\Gamma_m}^{(e)} \left(\frac{\partial \mathbf{N}^T}{\partial \mathbf{n}} M_t - \frac{\partial \mathbf{N}^T}{\partial \mathbf{t}} M_n \right) d\Gamma \\
\mathbf{k}_m^{(e)} &= \int_\Omega^{(e)} (\mathcal{L}\tilde{\mathbf{N}})^T \mathbf{C} \left(\mathcal{L}\tilde{\mathbf{N}} \right) d\Omega \\
\mathbf{f}_m^{(e)} &= \int_\Omega^{(e)} \begin{Bmatrix} \mathbf{N}^T p_x \\ \mathbf{N}^T p_y \end{Bmatrix} d\Omega + \int_{\Gamma_t}^{(e)} \begin{Bmatrix} \mathbf{N}^T T_x \\ \mathbf{N}^T T_y \end{Bmatrix} d\Gamma,
\end{aligned} \qquad (8.128)$$

The above element matrices and vectors are defined in the local coordinate system. Similar to the space frame case, these matrices and vectors need to be transformed into the global coordinate system.

For the given example problem, the deformation of the plate is assumed to be pure bending due to the small thickness of the structure and the load type (transverse load). For the bending, we use the non-conforming triangular element shown in Fig. 8.21 and the approximations shown in Eqs. (8.119, 8.121). Substituting Eqs. (8.121, 8.122) into the first term of Eq. (8.107) and omitting the distributed load term since there is none applied, we obtain

$$\begin{aligned}
&\int_\Omega^{(e)} (\mathcal{L}(\nabla \delta w))^T \mathbf{D} \left(\mathcal{L}(\nabla w) \right) d\Omega \\
&= \int_\Omega^{(e)} \left(\overline{\mathbf{B}} \mathbf{Z}^{-1} \delta \mathbf{d}_w \right)^T \mathbf{D} \left(\overline{\mathbf{B}} \mathbf{Z}^{-1} \mathbf{d}_w \right) d\Omega = (\delta \mathbf{d}_w)^T \mathbf{Z}^{-T} \left(\int_\Omega^{(e)} \overline{\mathbf{B}}^T \mathbf{D} \overline{\mathbf{B}} d\Omega \right) \mathbf{Z}^{-1} \mathbf{d}_w \\
&= (\delta \mathbf{d}_w)^T \mathbf{Z}^{-T} \overline{\mathbf{K}}_b^{(e)} \mathbf{Z}^{-1} \mathbf{d}_w.
\end{aligned} \qquad (8.129)$$

Note that the terms which are brought out of the integral are not functions of x, y and z. Substituting Eq. (8.121) and $\mathbf{D}$ into the integral, we obtain

$$\begin{aligned}
\overline{\mathbf{K}}_b^{(e)} &= \int_\Omega^{(e)} \overline{\mathbf{B}}^T \mathbf{D} \overline{\mathbf{B}} d\Omega \\
&= \int_{area}^{(e)} \left(\int_{-t/2}^{t/2} \overline{\mathbf{B}}^T \mathbf{D} \overline{\mathbf{B}} dz \right) dA \\
&= \frac{Et^3}{12(1-\nu^2)} \int_{area}^{(e)} dx dy \times
\end{aligned}$$

$$\begin{bmatrix} 0 & & & & & & & & \\ 0 & 0 & & & & \text{symmetric} & & & \\ 0 & 0 & 0 & & & & & & \\ 0 & 0 & 0 & 4 & & & & & \\ 0 & 0 & 0 & 0 & 2(1-\nu) & & & & \\ 0 & 0 & 0 & 4\nu & 0 & 4 & & & \\ 0 & 0 & 0 & 12x & 0 & 12\nu x & 36x^2 & & \\ 0 & 0 & 0 & 4(\nu x+y) & 4(1-\nu)(x+y) & 4(x+\nu y) & 12x(\nu x+y) & [(12-8\nu)(x+y)^2 & \\ & & & & & & & -8(1-\nu)xy] & \\ 0 & 0 & 0 & 12\nu y & 0 & 12y & 36\nu xy & 12(x+\nu y)y & 36y^2 \end{bmatrix}.$$

(8.130)

The integrals (matrix elements) in Eq. (8.130) can be calculated analytically by using the following results for the triangular element shown in Fig. 8.21 (Rao 2017)

$$\int_{area}^{(e)} dxdy = \frac{1}{2}x_3 y_2 \tag{8.131}$$

$$\int_{area}^{(e)} xdxdy = \frac{1}{6}x_3^2 y_2 \tag{8.132}$$

$$\int_{area}^{(e)} ydxdy = \frac{1}{6}x_3 y_2 (y_2 + y_3) \tag{8.133}$$

$$\int_{area}^{(e)} x^2 dxdy = \frac{1}{12}x_3^3 y_2 \tag{8.134}$$

$$\int_{area}^{(e)} xydxdy = \frac{1}{24}x_3^2 y_2 (y_2 + 2y_3) \tag{8.135}$$

$$\int_{area}^{(e)} y^2 dxdy = \frac{1}{12}x_3 y_2 (y_2^2 + y_2 y_3 + y_3^2). \tag{8.136}$$

At this point, the matrices $\mathbf{Z}$ and $\overline{\mathbf{K}_b}^{(e)}$ in the discretized integral given in Eq. (8.129) can all be calculated for a given element. Note that, the element DOF and DOF variation vectors $\mathbf{d}_w$ and $\delta\mathbf{d}_w$ are all defined in the local coordinate system of the element. As discussed previously for the beam elements in space frames, the degrees of freedom defined in the local coordinate systems need to be transformed to the global coordinate system. In this example problem, we assume the local xy-coordinate system is also on the global xy-plane. Then we have

$$\left\{\begin{array}{c} w(x,y) \\ \frac{\partial w}{\partial x}(x,y) \\ \frac{\partial w}{\partial y}(x,y) \end{array}\right\} = \left[\begin{array}{ccc} 1 & 0 & 0 \\ 0 & \cos(\widehat{x,X}) & \cos(\widehat{x,Y}) \\ 0 & \cos(\widehat{y,X}) & \cos(\widehat{y,Y}) \end{array}\right] \left\{\begin{array}{c} w(X,Y,Z) \\ \frac{\partial w}{\partial X}(X,Y,Z) \\ \frac{\partial w}{\partial Y}(X,Y,Z) \end{array}\right\} \quad (8.137)$$

where $\cos(\widehat{x,X})$ is the direction cosine of the x-axis and X-axis. The other direction cosines are defined in the same way. Denoting the transformation matrix as λ, the bending DOF vector defined in the local coordinate system $\mathbf{d}_w$ can be written in terms of the DOF vector in the global coordinate system $\mathbf{d}_W$ as

$$\mathbf{d}_w = \left[\begin{array}{ccc} \lambda & \mathbf{0} & \mathbf{0} \\ \mathbf{0} & \lambda & \mathbf{0} \\ \mathbf{0} & \mathbf{0} & \lambda \end{array}\right]_{9\times 9} \mathbf{d}_W = \mathbf{\Lambda}\mathbf{d}_W \quad (8.138)$$

where the $\mathbf{0}$s are 3×3 zero matrices. The discretized integral in Eq. (8.129) for a given element e can then be rewritten as

$$\int_{\Omega}^{(e)} (\mathcal{L}(\nabla \delta w))^T = (\delta \mathbf{d}_w)^T (\mathbf{Z})^{-T} \overline{\mathbf{K}}_b^{(e)} (\mathbf{Z})^{-1} \mathbf{d}_w$$
$$= (\delta \mathbf{d}_W)^T (\mathbf{\Lambda})^T (\mathbf{Z})^{-T} \overline{\mathbf{K}}_b^{(e)} (\mathbf{Z})^{-1} \mathbf{\Lambda} \mathbf{d}_W$$
$$= (\delta \mathbf{d}_W)^T \mathbf{K}_b^{(e)} \mathbf{d}_W \quad (8.139)$$

where the element bending stiffness matrix $\mathbf{K}_b^{(e)}$ is obtained as

$$\mathbf{K}_b^{(e)} = (\mathbf{\Lambda})^T (\mathbf{Z})^{-T} \overline{\mathbf{K}}_b^{(e)} (\mathbf{Z})^{-1} \mathbf{\Lambda}. \quad (8.140)$$

The element force vectors due to the distributed transverse loads on the surface and edges of the element, and bending and twisting moments acting on the edges can be derived similarly. The derivation is left as an exercise for the reader. For the example problem, the vectors are all zero since there is no distributed loading or external moment. The concentrated force is applied directly on the node after the global system is assembled in Step 7.

Step 7: Global Matrix and Vector Assembly
The global matrix and vector assembly process is the same as that discussed previously. Figure 7.16 is applicable to the plate elements. For pure bending, the dimension of the element stiffness matrix $\mathbf{K}_b^{(e)}$ is 9×9, and each primitive assembling block is 3×3. The assembled system is written as

$$\mathbf{K}_{3N\times 3N}\mathbf{d}_{3N\times 1} = \mathbf{f}_{3N\times 1} \quad (8.141)$$

where N is the number of nodes. The external transverse concentrated force is directly added to the force vector entry corresponding to the deflection w degree of freedom of the node on which the force is applied.

Steps 8–9: Solution

Once the global matrix and vector is assembled. The deflection (w) and slope ($\frac{\partial w}{\partial x}$, $\frac{\partial w}{\partial y}$) boundary conditions can be applied by using the methods discussed in previous chapters. The boundary conditions for the plate shown in Fig. 8.17 are

$$w = 0 \quad \text{at y=0} \tag{8.142}$$

$$\frac{\partial w}{\partial x} = 0 \quad \text{and} \quad \frac{\partial w}{\partial y} = 0 \quad \text{at y=0.} \tag{8.143}$$

Step 10: Post-processing

The deformed plate is visualized in this step. Figure 8.22 shows the comparison of the deformed and undeformed shapes of the thin plate structure.

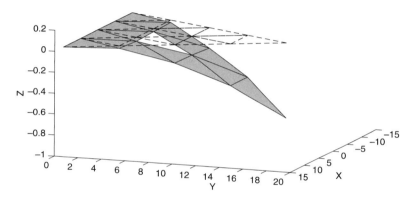

Figure 8.22 Deformed and undeformed shapes of the thin plate structure.

8.4.2 Computer Implementation

In this section, we present the data files and the MATLAB code for the finite element analysis of the plate bending problem.

Data Files

The input files "nodes.dat" and "elements.dat" are shown in Table 8.3. The boundary conditions, loads, material properties, geometric properties, and model options are shown in the table below.

MATLAB Code

The code of the plate analysis is listed below:

```
1  clear all ;
2  % next 3 lines : read the input files
3  filenames = {'nodes.dat','elements.dat','materials.dat', ...
```

bcsdisp.dat	forces.dat	materials.dat	options.dat
% node id, type, value	% node id, fz, mx, my	% Young's modulus	% Thickness
1 1 0	9 −60 0 0	% (N/cm^2)	% (cm)
1 2 0		2e7	0.2
1 3 0		% Poisson's ratio	
2 1 0		0.3	
2 2 0			
2 3 0			
3 1 0			
3 2 0			
3 3 0			
4 1 0			
4 2 0			
4 3 0			
5 1 0			
5 2 0			
5 3 0			

```
4              'options.dat','forces.dat','bcsdisp.dat'};
5   for i = 1:numel(filenames); load(filenames{i}); end;
6   % next 11 lines: set up constants and empty matrices
7   nNodes=size(nodes,1);
8   nElements=size(elements,1);
9   E=materials(1,1);
10  nu=materials(2,1);
11  thk=options(1,1);
12  K=zeros(3*nNodes,3*nNodes);
13  F=zeros(3*nNodes,1);
14  r=zeros(9,9);
15  [r(1,1),r(2,2),r(3,3),r(4,1),r(5,2),r(6,3)] =deal(1);
16  [r(7,1),r(8,2),r(9,3) =deal(1);
17  Lambda=zeros(9,9);
18
```

```
19  % for-loop: compute the global stiffness matrix
20  for e=1:nElements
21      elnodes=elements(e,2:4);
22      X=nodes(elnodes,2); Y=nodes(elnodes,3); Z=nodes(elnodes,4);
23      v12=[X(2)-X(1) Y(2)-Y(1) Z(2)-Z(1)]';
24      length_12=norm(v12);
25      oy=v12/length_12;
26      v13=[X(3)-X(1) Y(3)-Y(1) Z(3)-Z(1)]';
27      proj_13=v13'*oy;
28      ox=v13-proj_13*oy;
29      length_ox=norm(ox);
30      ox=ox/length_ox;
31      OX=[1 0 0]'; OY=[0 1 0]';
32      lox=ox'*OX; mox=ox'*OY;
33      loy=oy'*OX; moy=oy'*OY;
34      x=[0 0 length_ox ]';
35      y=[0 length_12 proj_13 ]';
36      % next 10 lines  Z matrix
37      r(4,3)=y(2); r(4,6)=y(2)^2; r(4,9)=y(2)^3;
38      r(5,5)=y(2); r(5,8)=y(2)^2;
39      r(6,6)=2*y(2); r(6,9)=3*y(2)^2;
40      r(7,2)=x(3); r(7,3)=y(3); r(7,4)=x(3)^2;
41      r(7,5)=x(3)*y(3); r(7,6)=y(3)^2; r(7,7)=x(3)^3;
42      r(7,8)=x(3)^2*y(3)+ x(3)*y(3)^2; r(7,9)=y(3)^3;
43      r(8,4)=2*x(3); r(8,5)=y(3);
44      r(8,7)=3*x(3)^2; r(8,8)=y(3)^2+2*x(3)8y(3)
45      r(9,5)=x(3); r(9,6)=2*y(3);
46      r(9,8)=2*x(3)*y(3) + x(3)^2; r(9,9)=3*y(3)^2
47      % next 6 lines : integrals
48      I=1/2*x(3)*y(2);
49      Ix=1/6*x(3)^2*y(2);
50      Iy=1/6*x(3)*y(2)*(y(2)+y(3));
51      Ix2=1/12*x(3)^3*y(2);
52      Ixy=1/24*x(3)^2*y(2)*(y(2)+2*y(3));
53      Iy2=1/12*x(3)*y(2)*(y(2)^2+y(2)*y(3)+y(3)^2);
54      % next 11 lines : set up element k matrix
55      k=zeros(9,9);
56      k(4,4)=4*I; k(5,5)=2*(1-nu)*I; k(6,4)=4*nu*I;
57      k(6,6)=4*I; k(7,4)=12*Ix; k(7,6)=12*nu*Ix;
58      k(7,7)=36*Ix2; k(8,4)=4*(nu*Ix + Iy);
59      k(8,5)=4*(1-nu)*(Ix + Iy); k(8,6)=4*(Ix+nu*Iy);
60      k(8,7)=12*nu*Ix2 + 12*Ixy;
61      k(8,8)=(12-8*nu)*(Ix2 + 2*Ixy + Iy2) - 8*(1-nu)*Ixy;
62      k(9,4)=12*nu*Iy; k(9,6)=12*Iy; k(9,7)=36*nu*Ixy;
63      k(9,8)=12*Ixy+ 12*nu*Iy2; k(9,9)=36*Iy2;
64      k=k+k'-eye(9).*diag(k);
65      k=k*(E*thk^3/(12*(1-nu^2)));
66      % next 4 lines : set up coordinate transformation matrix
67      lambda=[-1 0 0; 0 lox mox; 0 loy moy];
```

```matlab
   Lambda(1:3,1:3)=lambda;
   Lambda(4:6,4:6)=lambda;
   Lambda(7:9,7:9)=lambda;
   ir=inv(r);
   k=Lambda'*(ir'*k*ir)*Lambda;
   % next 9 lines : assemble  global  K matrix
   for i=1:3
     ni=elnodes(i);
     for j=1:3
       nj=elnodes(j);
       K(3*ni-2:3*ni, 3*nj-2:3*nj)=K(3*ni-2:3*ni, 3*nj-2:3*nj) ...
                                  +k(i*3-2:i*3, j*3-2:j*3);
     end
   end
 end
 % set up the global  force  vector
 for i=1:size(forces,1)
   node = forces(i,1);
   F(3*node-2:3*node)= forces(i,2:4);
 end

 % apply the displacement boundary condition using the penalty method
 penalty=abs(max(max(K)))*1e7;
 for i=1:size(bcsdisp,1)
   node = bcsdisp(i,1);
   disp_type = bcsdisp(i,2);
   K(3*node + disp_type - 3, 3*node + disp_type - 3)= penalty;
   F(3*node + disp_type - 3)= penalty*bcsdisp(i,3);
 end

 d=K\F    %solve

 % plotting the plate deformation
 figure(1);
 clf;   hold on;
 % setup deformed nodes
 dnodes=nodes;
 for i=1:nNodes
   dnodes(i,4)=nodes(i,4)+ d(i*3-2);
 end

 % for-loop: plot the undeformed plates system
 for e=1:nElements
   for j=2:3
     n1=nodes(elements(e,j),2:4);
     n2=nodes(elements(e,j+1),2:4);
     plot3([n1(1) n2(1)],[n1(2) n2(2)],[n1(3) n2(3)],'k--');
   end
   n1=nodes(elements(e,2),2:4);
```

```
117    plot3([n1(1) n2(1)],[n1(2) n2(2)],[n1(3) n2(3) ],'k--');
118    end
119    %plot the deformed plate system
120    p=patch('Vertices',dnodes(:,2:4),'Faces',elements(:,2:4) );
121    set(p,'facecolor',[0.7 0.7 0.7],'edgecolor','black');
122    axis([-15 15 0 20 -1 0.2]);
123    xlabel('X'); ylabel('Y'); zlabel('Z');
124    view(110,20);
```

8.5 Summary

Upon completion of this chapter, you should be able to

- understand the fundamental theories of truss bars, beams, frames and plates, and know how to derive the governing equations for each type of structural component.
- understand the difference between bars in a truss and beams in a space frame structure.
- derive the weak form of beams by using the energy method.
- derive the weak form of Kirchhoff plates by using the principle of virtual work.
- derive the expressions of element matrices and vectors for truss bar, beam, and plate elements in their local coordinate systems.
- know how to perform coordinate transformation between the local and global coordinate systems.
- know how to assemble element matrices and vectors into the global matrix and vector for the three types of structural elements.
- know how to apply the essential boundary conditions for the three types of structures.
- implement MATLAB functions to perform the post-processing steps for visualization.
- implement a complete MATLAB program to perform FEA of the three types of structures.

8.6 Problems

8.1 A truss system is subjected to loading as shown in Fig. 8.23. The cross sectional area of the bars is 10 mm^2. Young's modulus of the material is 2×10^9 Pa. Manually calculate the element stiffness matrices and assemble the global systems (global stiffness matrix and force vector). Use MATLAB to solve for the displacements of the nodes.

8.2 Write a MATLAB program to solve Problem 8.1. Compare the numerical results with the results you obtained from Problem 8.1.

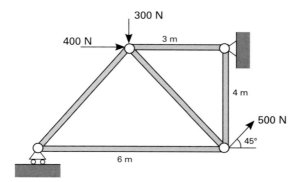

Figure 8.23 Truss system for Problem 8.1.

8.3 A truss system is subjected to loading as shown in the Fig. 8.24. The cross sectional area of the bars in the system is 10 mm². Young's modulus of the material is 2×10^9 Pa. Write a MATLAB program to compute the displacements of the nodes and stresses of the bars. Plot the deformed shape of the truss system.

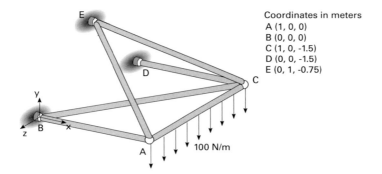

Figure 8.24 Truss system for Problem 8.3.

8.4 A Howe scissors roof truss with loading is shown in Fig. 8.25. The cross sectional area of the bars in the system is 10 cm². Young's modulus of the material is 2×10^9 Pa. Write a MATLAB program to compute the displacements of the nodes and stresses of the bars. Plot the deformed shape of the truss system.

8.5 A frame structure is fixed at the bottom and subjected to a moment and a point force as shown in Fig. 8.26. The geometry and dimensions are as shown. The cross section of the frame is a solid square of 10 cm × 10 cm. Young's modulus of the material is 1×10^9 Pa and Poisson's ratio is 0.3. Write an FEA code to compute the deformation of the frame by using

(a) 4-node plane stress quadrilateral elements
(b) beam elements.

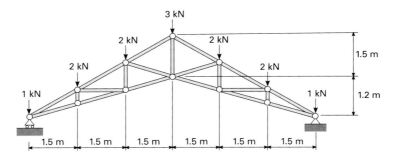

Figure 8.25 Roof truss for Problem 8.4.

Refine your mesh to obtain converged results for both cases. Compare the results obtained and discuss your FEA results.

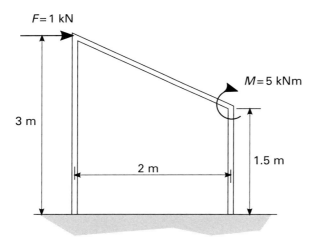

Figure 8.26 Frame structure for Problem 8.5.

8.6 A simply supported beam of length L is subjected to a point force P located at a distance a from the left end and b from the right end (i.e., $a + b = L$) as in Fig. 8.27. Classical mechanics using the Euler–Bernoulli beam theory predicts the deflection of the beam $v(x)$ to be

$$v(x) = \frac{-Pbx}{6EIL}\left(L^2 - b^2 - x^2\right) \qquad 0 \leq x \leq a$$

and the slopes of the deflected beam at the two ends are

$$\theta_1 = \frac{-Pab(L+b)}{6EIL} \qquad \theta_2 = \frac{Pab(L+a)}{6EIL}$$

where E is the Young's modulus of the beam material and I is the moment of inertia of the beam cross section.

(a) Assuming the beam is 1.0 m × 0.01 m × 0.01 m, E is 70 GPa, and a is 0.7 m, write FEA code with beam elements to compute the deflection of the beam and its slopes at the two ends.
(b) Refine your mesh to obtain a converged result.
(c) Compare the numerical results with the analytical results and discuss your observations.

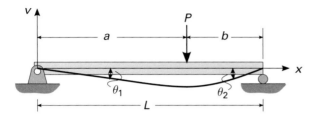

Figure 8.27 Beam for Problem 8.6.

8.7 A cantilever beam of length L is subjected to a uniform distributed load w as shown in Fig 8.28. Classical mechanics using the Euler–Bernoulli beam theory predicts the deflection of the beam $v(x)$ to be

$$v(x) = \frac{-wx^2}{24EI}\left(x^2 - 4Lx + 6L^2\right)$$

and the maximum deflection and slope of the deflected beam are

$$v_{max} = \frac{-wL^4}{8EI}, \qquad \theta_{max} = \frac{-wL^3}{6EI}$$

where E is the Young's modulus of the beam material and I is the moment of inertia of the beam cross section.
(a) Assuming the beam is 1.0 m × 0.01 m × 0.01 m, E is 70 GPa, and w is 5 kN/m, write FEA code with beam elements to compute the deflection of the beam and its slope along the length of the beam.
(b) Refine your mesh to obtain a converged result.
(c) Compare the numerical results with the analytical results and discuss your observations.

8.8 A steel frame structure is subjected to loading as shown in Fig 8.29. The square cross sectional area of the beams is 1.0 in². Young's modulus of the material is 2×10^9 Pa. Write a MATLAB program to compute the deflection of the frame. Refine the mesh until convergence is obtained. Plot the deformed shape of the frame.

8.9 A Howe scissors roof frame structure with loading is shown in Fig. 8.30. The square cross sectional area of the rods in the system is 10 cm². Young's modulus of the material is 2×10^9 Pa. Write a MATLAB program to compute the displacements of the frame. Refine the mesh until convergence is obtained. Plot the deformed shape of the frame.

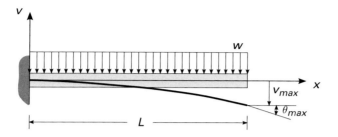

Figure 8.28 Cantilever beam for Problem 8.7.

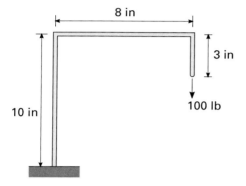

Figure 8.29 Steel frame for Problem 8.8.

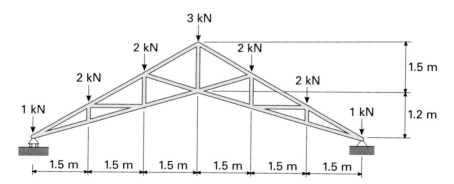

Figure 8.30 Roof frame for Problem 8.9.

8.10 A steel frame structure is subjected to loading as shown in Fig 8.31. The frame is fixed to the ground at the bottom. The force vector is $\mathbf{p} = \{-200N \ -200N \ 100N\}^T$ applied at the top of the frame. The cross sectional area of the beams is 10 cm^2. Young's modulus of the material is 2×10^9 Pa. Write a MATLAB program to compute the deflection of the frame. Refine the mesh until convergence is obtained. Plot the deformed shape of the frame.

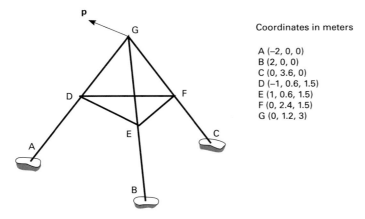

Figure 8.31 Frame for Problem 8.10.

8.11 For the Kirchhoff plate element (pure bending) discussed in Section 8.4.1, derive the force vector due to a uniform distributed transverse (normal) load, $q(x, y)$, on the surface of the element.

8.12 For the Kirchhoff plate element (pure bending) discussed in the text, derive the force vector due to a bending moment M_t acting on an edge of the element.

8.13 A 3 m × 4 m simply supported Kirchhoff plate is subjected to a uniform distributed load $q(x, y) = q_0 = 100$ Pa as shown in Fig. 8.32. The thickness of the plate is 5 mm. The material properties are: E=70 GPa, $\nu = 0.3$. From the classical plate theory, by using Fourier series expansion, an analytical solution of deflection of simply supported rectangular plates can be obtained as

$$w(x, y) = \sum_{n=1}^{\infty} \sum_{m=1}^{\infty} W_{mn} \sin \alpha x \sin \beta y$$

where $\alpha = m\pi/a$ and $\beta = n\pi/b$, a and b are the length and width of the plate, and

$$W_{mn} = \frac{Q_{mn}}{d_{mn}}$$

$$d_{mn} = \pi^4 \left(D_{11} \frac{m^4}{a^4} + 2(D_{12} + 2D_{33}) \frac{m^2 n^2}{a^2 b^2} + D_{22} \frac{n^4}{b^4} \right)$$

$$Q_{mn} = \frac{16 q_0}{\pi^2 mn} \qquad m, n = 1, 3, 5, \ldots$$

D_{ij} are the bending stiffness matrix elements.

(a) Use MATLAB to compute and plot the analytical solution of the deflection along the two (perpendicular) center lines of the plate. Experiment with m and n until a converged solution is obtained.

(b) Write a MATLAB FEA program to compute the deflection of the plate. Refine your mesh to obtain a converged result.

(c) Compare the numerical results with the analytical results and discuss your observations.

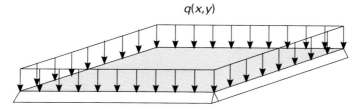

Figure 8.32 Kirchhoff plate for Problem 8.13.

8.14 An I-beam is modeled as three plates welded together. The I-beam is fixed at one end and loaded at the other end as shown in Fig. 8.33. The thickness of the plate is 5 mm. The material properties are: $E = 200$ GPa, $\nu = 0.3$. Write a MATLAB FEA program to compute the deflection of the I-beam. Refine your mesh to obtain a converged result. Obtain the maximum deflection of the I-beam and plot the deformed shape.

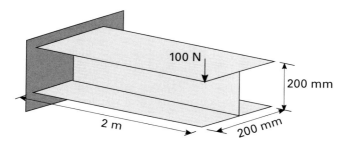

Figure 8.33 I-beam for Problem 8.14.

References

Rao, S. S. (2017), *The finite element method in engineering*, Butterworth-Heinemann.
Ventsel, E. & Krauthammer, T. (2001), *Thin plates and shells*, Mercel Dekker.

9 FEA for Linear Time-Dependent Analysis

9.1 Overview

In this chapter, we discuss the general finite element analysis procedure for time-dependent problems. Two types of problems are presented: transient heat transfer problems and structural dynamics (or elastodynamics) problems. These two types of problem are described by the transient heat equation and the equation of motion, respectively. As the physical quantities in time-dependent problems vary with time, these problems are also referred to as initial value problems. Compared to the static or steady state problems, the governing partial differential equations (PDEs) contain terms of time derivatives. Mathematically, the transient heat transfer equation is a parabolic type PDE and the equation of motion is a hyperbolic type PDE. Although both equations are of second order and have constant coefficients, their solutions possess very different properties. This is also reflected in the treatment of the time derivative terms in the finite element solutions. In finite element analysis, similar to the discretization of the spatial domain, the continuous time axis is discretized into time steps for time-dependent solutions. However, the time integration (or time marching) schemes for calculating the physical quantities at each time step are different for transient heat transfer and structural dynamics. The time integration schemes represent the major difference in the solution step for time-dependent problems.

This chapter contains two sections. The first section introduces the FEA procedure for transient heat transfer analysis by using a 2-D example problem. The second section discusses elastodynamic analysis by demonstrating the procedure for calculating the dynamic response of a 2-D cantilever beam as well as finding its natural frequencies and vibrational modes. MATLAB codes for solving these problems are presented.

9.2 2-D Transient Heat Transfer

In this section, by solving step by step an example problem of transient heat transfer in a 2-D square plate, the FEA procedure for transient heat transfer analysis is illustrated. We first derive the differential equation of transient heat transfer and define the boundary and initial conditions. The weak form, discretization, approximation, and element equations are similar to the steady state heat transfer case. The description of the steps in this section emphasizes the differences due to the added degree of freedom: time. The solution of the time-dependent discrete equations is the new content which is explained in detail. Finally, computer implementation and associated MATLAB codes are presented for the analysis.

9.2.1 Transient Heat Transfer in a Square Plate

Step 1: Physical Problem

Similar to the steady state heat transfer problem, here we consider a thin plate as shown in Fig. 9.1. The size of the plate is 0.2 m × 0.1 m. The temperature and the normal heat flux are specified on the four sides of the plate. Assuming the initial temperature of the plate is 0 °C, the objective is to compute the transient variation of the temperature distribution in the plate. The relevant material properties are given as thermal conductivity $\kappa = 150$ W/m · K, mass density $\rho = 1000$ kg/m^3 and the specific heat $c = 20$ J/kg · K.

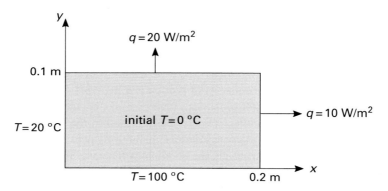

Figure 9.1 A 2-D transient heat transfer problem.

Step 2: Mathematical Model

In the steady state heat transfer problem, the right hand side of the energy conservation law

$$\text{heat in} - \text{heat out} = \text{heat causing temperature increase} \tag{9.1}$$

vanishes since the temperature does not vary as a function of time. In the transient analysis, however, the temperature is a function of time and the right hand side of the energy conservation law must be included. The governing PDE is then given by (compare with Eq. (5.14))

$$\frac{\partial}{\partial x}\left(k\frac{\partial T}{\partial x}\right) + \frac{\partial}{\partial y}\left(k\frac{\partial T}{\partial y}\right) - h_z(T - T_\infty) = c\rho\frac{\partial T}{\partial t} \tag{9.2}$$

where c is the specific heat and ρ is the mass density. In the transient analysis, other than the boundary conditions, one must specify the initial state of the temperature profile. This initial state of the temperature is referred to as an initial condition. For the

2-D heat transfer example shown in Fig. 9.1, we can specify the initial and boundary conditions as follows:

At $t = 0$

$$T = \overline{T}_0 \qquad 0 \leq x \leq 0.2; \ 0 \leq y \leq 0.1 \qquad (9.3)$$

At $t > 0$

$$T = \overline{T} \quad \text{at} \quad x = 0 \text{ and } y = 0 \qquad (9.4)$$

$$k\frac{\partial T}{\partial \mathbf{n}} = \begin{cases} \overline{q}_1 & y = 0.1 \\ \overline{q}_2 & x = 0.2. \end{cases} \qquad (9.5)$$

Equation (9.3) is referred to as the initial condition and Eqs. (9.4, 9.5) are the boundary conditions. Equations (9.2–9.5) represent the strong form of the 2-D transient heat transfer problem.

Step 3: Weak Form

We employ the Galerkin weighted residual method to derive the weak form. While the surface convection term in Eq. (9.2) can be treated as described previously, in this section, for the sake of simplicity, it is assumed that surface convection is negligible, i.e., $h_z = 0$. The residual can then be obtained as

$$R = c\rho \frac{\partial T}{\partial t} - \left[\frac{\partial}{\partial x}\left(k\frac{\partial T}{\partial x}\right) + \frac{\partial}{\partial y}\left(k\frac{\partial T}{\partial y}\right) \right]. \qquad (9.6)$$

Multiplying the temperature variation δT by the residual and integrating the product over the 2-D domain, we obtain

$$\int_\Omega R \delta T d\Omega = 0 \qquad (9.7)$$

$$\rightarrow \int_\Omega c\rho \frac{\partial T}{\partial t} \delta T d\Omega - \int_\Omega \left[\frac{\partial}{\partial x}\left(k\frac{\partial T}{\partial x}\right) + \frac{\partial}{\partial y}\left(k\frac{\partial T}{\partial y}\right) \right] \delta T d\Omega = 0. \qquad (9.8)$$

Note that the second integral is the same integral as obtained in the steady state analysis, Eq. (5.19). Therefore, we can directly use the result shown in Eq. (5.31) and obtain the weak from for the transient analysis as

$$\int_\Omega c\rho \frac{\partial T}{\partial t} \delta T d\Omega - \left[\int_{\Gamma_q} \delta T k \frac{\partial T}{\partial \mathbf{n}} d\Gamma - \int_\Omega k \nabla T \cdot \nabla \delta T d\Omega \right] = 0. \qquad (9.9)$$

Step 4: Discretization

The discretization of the 2-D domain remains the same for transient analysis. In fact, the meshing process depends only on geometry. While different physical problems may have different requirements on mesh properties, the meshing process itself is largely independent from the physics of the problem. Figure 9.2 shows a mesh generated for the example problem.

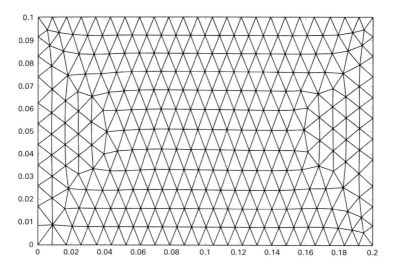

Figure 9.2 A mesh for the transient heat transfer problem.

Step 5: Approximation of Time-Dependent Variables

In the transient analysis, the temperature is not only a function of position (x, y) but also a function of time. The time-dependent temperature is typically approximated by

$$T(x, y, t) = \sum_{i=1}^{n} N_i(x, y) T_i(t). \tag{9.10}$$

In Eq. (9.10), the shape functions $N_i(x, y)$ are not functions of time and the nodal temperature is now a function of time. In the approximation, the shape functions are still for the spatial approximation and the temperature change in the time domain is represented by the time-dependent nodal temperature, as shown in Fig. 9.3. The derivatives of the temperature are written as

$$\frac{\partial T(x, y, t)}{\partial x} = \sum_{i=1}^{n} \frac{\partial N_i(x, y)}{\partial x} T_i(t) \tag{9.11}$$

$$\frac{\partial T(x, y, t)}{\partial y} = \sum_{i=1}^{n} \frac{\partial N_i(x, y)}{\partial y} T_i(t) \tag{9.12}$$

$$\frac{\partial T(x, y, t)}{\partial t} = \sum_{i=1}^{n} N_i(x, y) \frac{dT_i(t)}{dt}. \tag{9.13}$$

Similarly, the temperature variation and its derivatives are written as

$$\delta T(x, y, t) = \sum_{i=1}^{n} N_i(x, y) \delta T_i(t) \tag{9.14}$$

$$\frac{\partial \delta T(x, y, t)}{\partial x} = \sum_{i=1}^{n} \frac{\partial N_i(x, y)}{\partial x} \delta T_i(t) \tag{9.15}$$

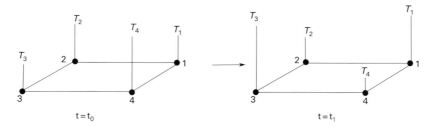

Figure 9.3 Time-dependent nodal temperature.

$$\frac{\partial \delta T(x,y,t)}{\partial y} = \sum_{i=1}^{n} \frac{\partial N_i(x,y)}{\partial y} \delta T_i(t). \quad (9.16)$$

Note that the time derivative of δT does not appear in the weak form.

Steps 6-7: Element and Global Matrices and Vectors

For an element e, one can write the element weak form as

$$\int_{\Omega}^{(e)} c\rho \frac{\partial T}{\partial t} \delta T d\Omega - \left[\int_{\Gamma_q}^{(e)} \delta T k \frac{\partial T}{\partial \mathbf{n}} d\Gamma - \int_{\Omega}^{(e)} k \nabla T \cdot \nabla \delta T d\Omega \right] = 0. \quad (9.17)$$

Since all the terms except for the first integral do not contain $\partial T/\partial t$, and all these terms are the same as those in the steady state case, the element matrices and vectors are the same for these terms. For the first term

$$\int_{\Omega}^{(e)} c\rho \frac{\partial T}{\partial t} \delta T d\Omega, \quad (9.18)$$

without loss of generality, assuming a linear quadrilateral element and substituting

$$\delta T = \sum_{i=1}^{4} N_i(x,y) \delta T_i \quad (9.19)$$

and

$$\frac{\partial T}{\partial t} = \sum_{i=1}^{4} N_i(x,y) \frac{dT_i}{dt} \quad (9.20)$$

we have

$$\begin{bmatrix} \delta T_1 & \delta T_2 & \delta T_3 & \delta T_4 \end{bmatrix} \left(\int_\Omega^{(e)} \begin{bmatrix} N_1 \\ N_2 \\ N_3 \\ N_4 \end{bmatrix} c\rho \begin{bmatrix} N_1 & N_2 & N_3 & N_4 \end{bmatrix} d\Omega \right) \begin{Bmatrix} \dfrac{dT_1}{dt} \\ \dfrac{dT_2}{dt} \\ \dfrac{dT_3}{dt} \\ \dfrac{dT_4}{dt} \end{Bmatrix}. \quad (9.21)$$

The middle part of the expression above is the element heat capacity matrix. Note that $[\delta T_1 \ \delta T_2 \ \delta T_3 \ \delta T_4]$ is a vector containing arbitrary nodal temperature variations, and will be canceled in the discretized element weak form. The final discretized element weak form can be obtained as

$$\left(\int_\Omega^{(e)} \begin{bmatrix} N_1 \\ N_2 \\ N_3 \\ N_4 \end{bmatrix} c\rho \begin{bmatrix} N_1 & N_2 & N_3 & N_4 \end{bmatrix} d\Omega \right) \begin{Bmatrix} \frac{dT_1}{dt} \\ \frac{dT_2}{dt} \\ \frac{dT_3}{dt} \\ \frac{dT_4}{dt} \end{Bmatrix}$$

$$+ \left(\int_\Omega^{(e)} \begin{bmatrix} \frac{\partial N_1}{\partial x} & \frac{\partial N_1}{\partial y} \\ \frac{\partial N_2}{\partial x} & \frac{\partial N_2}{\partial y} \\ \frac{\partial N_3}{\partial x} & \frac{\partial N_3}{\partial y} \\ \frac{\partial N_4}{\partial x} & \frac{\partial N_4}{\partial y} \end{bmatrix} k \begin{bmatrix} \frac{\partial N_1}{\partial x} & \frac{\partial N_2}{\partial x} & \frac{\partial N_3}{\partial x} & \frac{\partial N_4}{\partial x} \\ \frac{\partial N_1}{\partial y} & \frac{\partial N_2}{\partial y} & \frac{\partial N_3}{\partial y} & \frac{\partial N_4}{\partial y} \end{bmatrix} d\Omega \right) \begin{Bmatrix} T_1 \\ T_2 \\ T_3 \\ T_4 \end{Bmatrix}$$

$$= \left(\int_{\Gamma_q}^{(e)} \begin{Bmatrix} N_1 \\ N_2 \\ N_3 \\ N_4 \end{Bmatrix} k \frac{\partial T}{\partial \mathbf{n}} d\Gamma \right). \quad (9.22)$$

In short form,

$$\mathbf{c}^{(e)} \begin{Bmatrix} \frac{dT_1}{dt} \\ \frac{dT_2}{dt} \\ \frac{dT_3}{dt} \\ \frac{dT_4}{dt} \end{Bmatrix} + \mathbf{k}^{(e)} \begin{Bmatrix} T_1 \\ T_2 \\ T_3 \\ T_4 \end{Bmatrix} = \mathbf{f}^{(e)}. \qquad (9.23)$$

Equation (9.23) is then assembled to the global matrices as

$$\mathbf{C}_{N \times N} \begin{Bmatrix} \frac{dT_1}{dt} \\ \frac{dT_2}{dt} \\ \frac{dT_3}{dt} \\ \vdots \\ \frac{dT_N}{dt} \end{Bmatrix} + \mathbf{K}_{N \times N} \begin{Bmatrix} T_1 \\ T_2 \\ T_3 \\ \vdots \\ T_N \end{Bmatrix} = \mathbf{F}_{N \times 1} \qquad (9.24)$$

where $\mathbf{c}^{(e)}$ is assembled to $\mathbf{C}$, $\mathbf{k}^{(e)}$ is assembled to $\mathbf{K}$, and $\mathbf{f}^{(e)}$ is assembled to $\mathbf{F}$. $\mathbf{C}$ and $\mathbf{K}$ are referred to as the heat capacity matrix and heat conduction matrix, respectively. $\mathbf{F}$ is the heat source vector. The assembling procedure has been discussed in Section 5.2.1. It is shown in Eq. (9.24) that the total number of unknowns is $2N$ (N unknown nodal temperatures and N unknown time derivatives of the temperature) while the total number of algebra equations is N. Therefore, the vector of unknown time derivatives of the temperature must be treated.

Steps 8-9: Solution in the Time Domain
As shown in Fig. 9.4, the time-dependent nodal temperature varies with time. In most cases, along the time t axis, the nodal temperature T_i can be regarded as a smooth function of t. In the finite element analysis, discrete time steps, $t_0, t_1, t_2, \ldots$, are taken along the continuous time axis to calculate the "snapshots" of the temperature. The time derivative of the nodal temperature is then approximated by using the "snapshots." There are many methods available for time domain approximation. In this book, we introduce three methods, namely the backward difference method, central difference method (Crank–Nicholson method), and the forward difference method.

Backward Difference Method
For a given time step t_m, we repeat the global system of equations here

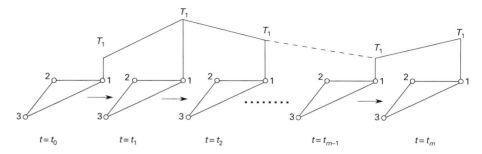

Figure 9.4 Time marching of nodal temperature.

$$\mathbf{C} \left\{ \begin{array}{c} \frac{dT_1}{dt} \\ \frac{dT_2}{dt} \\ \frac{dT_3}{dt} \\ \vdots \\ \frac{dT_N}{dt} \end{array} \right\}_{t=t_m} + \mathbf{K} \left\{ \begin{array}{c} T_1 \\ T_2 \\ T_3 \\ \vdots \\ T_N \end{array} \right\}_{t=t_m} = \mathbf{F}_{t=t_m}. \tag{9.25}$$

Denote

$$\left\{ \begin{array}{c} \frac{dT_1}{dt} \\ \frac{dT_2}{dt} \\ \frac{dT_3}{dt} \\ \vdots \\ \frac{dT_N}{dt} \end{array} \right\}_{t=t_m} \text{as } \frac{d\mathbf{T}_m}{dt} \quad \text{and} \quad \left\{ \begin{array}{c} T_1 \\ T_2 \\ T_3 \\ \vdots \\ T_N \end{array} \right\} \text{as } \mathbf{T}_m. \tag{9.26}$$

Assuming $\mathbf{T}_1, \mathbf{T}_2, \ldots, \mathbf{T}_{m-1}$ have been calculated, we would like to compute $\mathbf{T}_m$. Note that $\mathbf{T}_0$ is given as the initial condition. Since

$$\mathbf{C}\frac{d\mathbf{T}}{dt} + \mathbf{K}\mathbf{T} = \mathbf{F} \tag{9.27}$$

should be satisfied at any time step m, in the backward difference method we write

$$\mathbf{C}\frac{d\mathbf{T}_m}{dt} + \mathbf{K}\mathbf{T}_m = \mathbf{F}_m \tag{9.28}$$

and

$$\frac{d\mathbf{T}_m}{dt} = \frac{\mathbf{T}_m - \mathbf{T}_{m-1}}{t_m - t_{m-1}} = \frac{\mathbf{T}_m - \mathbf{T}_{m-1}}{\Delta t_m} \tag{9.29}$$

where $\Delta t_m = \mathbf{T}_m - \mathbf{T}_{m-1}$ is the time interval for time step m. Substituting Eq. (9.29) into Eq. (9.28),

$$\mathbf{C}\frac{\mathbf{T}_m - \mathbf{T}_{m-1}}{\Delta t_m} + \mathbf{K}\mathbf{T}_m = \mathbf{F}_m. \tag{9.30}$$

Rearranging the equation, we have

$$\left(\frac{1}{\Delta t_m}\mathbf{C} + \mathbf{K}\right)\mathbf{T}_m = \mathbf{F}_m + \frac{1}{\Delta t_m}\mathbf{C}\mathbf{T}_{m-1} \tag{9.31}$$

where the right hand side quantities are all known and the number of unknowns is reduced to N (nodal temperature).

Central Difference Method (Crank–Nicolson Method)
In this method, we write

$$\mathbf{C}\frac{d\mathbf{T}_{m-1/2}}{dt} + \mathbf{K}\mathbf{T}_{m-1/2} = \mathbf{F}_{m-1/2}. \tag{9.32}$$

The time derivative of T at $t = t_{m-1/2}$ is written as

$$\frac{d\mathbf{T}_{m-1/2}}{dt} = \frac{\mathbf{T}_m - \mathbf{T}_{m-1}}{\Delta t_m} \tag{9.33}$$

and $\mathbf{T}_{m-1/2}$ is taken as the average of $\mathbf{T}_m$ and $\mathbf{T}_{m-1}$,

$$\mathbf{T}_{m-1/2} = \frac{\mathbf{T}_m + \mathbf{T}_{m-1}}{2}. \tag{9.34}$$

Substituting the approximations into Eq. (9.32), we obtain

$$\left(\frac{1}{\Delta t_m}\mathbf{C} + \frac{1}{2}\mathbf{K}\right)\mathbf{T}_m = \mathbf{F}_{m-1/2} + \left(\frac{1}{\Delta t_m}\mathbf{C} - \frac{1}{2}\mathbf{K}\right)\mathbf{T}_{m-1}. \tag{9.35}$$

Note that $\mathbf{F}_{m-1/2}$ is evaluated at $t = t_{m-1/2}$.

Forward Difference Method
In this method, we write

$$\mathbf{C}\frac{d\mathbf{T}_{m-1}}{dt} + \mathbf{K}\mathbf{T}_{m-1} = \mathbf{F}_{m-1}. \tag{9.36}$$

The time derivative of T at $t = t_{m-1}$ is written as

$$\frac{d\mathbf{T}_{m-1}}{dt} = \frac{\mathbf{T}_m - \mathbf{T}_{m-1}}{\Delta t_m}. \tag{9.37}$$

Substituting the approximation into Eq. (9.36), we obtain

$$\frac{1}{\Delta t_m}\mathbf{C}\mathbf{T}_m = \mathbf{F}_{m-1} + \left(\frac{1}{\Delta t_m}\mathbf{C} - \mathbf{K}\right)\mathbf{T}_{m-1}. \tag{9.38}$$

Typically, in the forward difference method, $\mathbf{C}$ is lumped, i.e., $\mathbf{C}$ is replaced by an "effective" capacity matrix $\tilde{\mathbf{C}}$ where $\tilde{\mathbf{C}}$ is diagonal and $\tilde{C}_{ii} = \sum_{j=1}^{N} C_{ij}$. By using $\tilde{\mathbf{C}}$, the matrix equation, Eq. (9.38), becomes explicit, and $\mathbf{T}_m$ can be solved without solving a linear system. Therefore, the forward difference method is very efficient. However, the explicit method is only conditionally stable. The stability of the method depends on the time step size Δt. It has been shown that the explicit method is stable when

$$\Delta t < \frac{2}{\lambda_{max}}, \quad (9.39)$$

otherwise it is unstable. λ_{max} is the maximum eigenvalue of the problem

$$\mathbf{K}\boldsymbol{\phi} = \lambda \mathbf{C}\boldsymbol{\phi}. \quad (9.40)$$

Table 9.1 shows the performance comparison of the three methods.

Once a time approximation method is chosen to perform the transient analysis, essential boundary conditions are applied to the global system. It is straightforward to use the penalty method or the direct substitution method for this purpose. Note that the essential boundary conditions should be applied for each time step. After applying the essential boundary conditions, $\mathbf{T}_m$ can be obtained by solving the global system of equations.

Step 10: Post-processing

In this step, the temperature distribution as a function of time needs to be visualized. Such results can be displayed by using "snapshots" of the temperature profile at certain given time steps. For example, Fig. 9.5 shows the temperature profile of the plate at $t = 0.5$ s. With multiple temperature profiles displaying along the time axis, an animation can be created. Alternatively, we can examine the temperature variation at individual nodes. Figure 9.6 shows the temperature variation curves of three nodes.

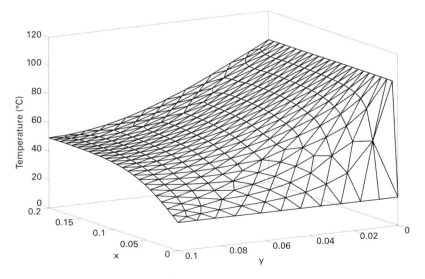

Figure 9.5 Plate temperature profile at $t = 0.5$ s.

Table 9.1 Performance comparison of the three time integration methods

	Backward difference	Central difference	Forward difference
Rate of convergence	$O(\Delta t)$	$O(\Delta t^2)$	$O(\Delta t)$
Matrix equation	Implicit	Implicit	Explicit with lumped **C**
Stability	Unconditionally stable	Unconditionally stable	Conditionally stable
Behavior			

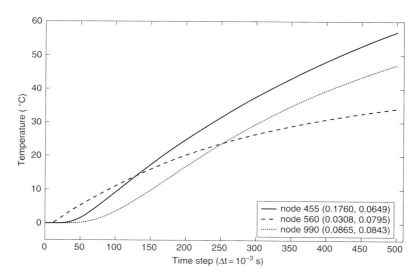

Figure 9.6 Nodal temperature variation as function of time.

9.2.2 Computer Implementation

In this section, we present the data files and the MATLAB code for the finite element analysis of the transient heat transfer example.

Data Files

The input data files are shown in Table 9.2. The nodes, elements, boundary conditions, material properties, and model options are stored as data files and loaded into MATLAB for calculation. Note that quadratic triangular elements are used in the code for solving the example problem.

MATLAB Code

The code of the transient heat transfer analysis is listed below:

```
1  clear all ;
2  % next 3 lines : read the input files
3  filenames = {'nodes.dat','elements.dat','materials.dat', ...
4               'options.dat','edgeFlux.dat', 'nodalTemp.dat'};
5  for i = 1:numel(filenames); load(filenames{i}); end;
6
7  % next 9 lines : set up global material properties and constants
8  kappa=materials(1,1);
9  cv=materials(2,1);
10 rho=materials(3,1);
11 thickness=options(2,1);
12 extTemp=options(4,1);
```

9.2 2-D Transient Heat Transfer

Table 9.2 Data files for analysis

nodes.dat

1	0.000000	0.0000
2	0.200000	0.0000
3	0.200000	0.1000
4	0.000000	0.1000
5	0.004243	0.0000
⋮	⋮	⋮

elements.dat

1	916	1	6	137	5	796
2	916	135	1	915	136	137
3	1151	955	48	138	837	151
4	1151	954	955	149	757	138
5	1152	946	69	139	855	147
⋮	⋮	⋮	⋮	⋮	⋮	⋮

nodalTemp.dat

1	20.000
2	100.000
4	20.000
5	100.000
6	100.000
⋮	⋮

edgeFlux.dat

11	1	4	114	−20
13	2	72	3	−20
14	1	3	69	−10
16	2	51	2	−10
504	1	53	51	−10
⋮	⋮	⋮	⋮	⋮

materials.dat

% Thermal conductivity (W/Km)
150
% Specific heat capacity (J/kgK)
20
% Mass density (kg/m^3)
1000

options.dat

% Dimensions
2
% Element type: no of edges, no of nodes
36
% Initial temperature
0
%time period (second)
0.5
%time step, dt, (second)
1e-3
% node for tracking temperature variation
990

```
13   initTemp=options(5,1);
14   timePeriod=options(6,1);
15   timeStep=options(7,1);
16   probeNode=options(8,1);
17   % next 9 lines : bookkeeping
18   n_nodes=size(nodes,1);
19   n_elements=size(elements,1);
20   n_edgeFlux=size(edgeFlux,1);
21   n_nodalTemp=size(nodalTemp,1);
22   n_timeSteps=timePeriod/timeStep+1;
23   prevTemp=ones(n_nodes,1)*initTemp;
24   currTemp=prevTemp;
25   tempHistory=zeros(n_nodes,n_timeSteps);
26   tempHistory(:,1)=currTemp;
27
28   K=CompK(nodes, elements, kappa);      % compute global K matrix
29   C=CompC(nodes, elements, cv, rho);    % global heat capacity matrix
30   F=CompF(nodes, elements, edgeFlux);   % compute global F vector
31
32   % next 10 lines : time integration
33   penalty=max(max(abs(K*0.5 + C*(1/timeStep))))*1e+6;
34   tempHistory(:,1)=currTemp;
35   for t=2:n_timeSteps
36     currTemp=CrankNicolson(timeStep,K,C,F,nodalTemp,prevTemp,currTemp,
37         penalty);
38     tempHistory(:,t)=currTemp; % record for this time step
39     prevTemp=currTemp;
40     if rem(t*100,(n_timeSteps-1)*5)==0
41       fprintf('%d %%\n', floor(t*100/(n_timeSteps-1)));
42     end
43   end
44
45   % next 3 lines : save the transient temperature results in file
46   Tout=tempHistory(probeNode,:);
47   save -ascii -double Tout.dat Tout
48   disp('Temperature results stored in Tout.dat');
```

```
1   function K=CompK(nodes, elements, kappa)
2   % next 5 lines : set up constants and empty matices
3   n_nodes = size(nodes,1);
4   n_elements = size(elements,1);
5   n_nodes_per_element = size(elements,2)-1;
6   K=zeros(n_nodes,n_nodes);
7   DN=zeros(2,6);
8   % get Gauss points, weights and compute shape functions
9   [gauss_points, gauss_weights]=GetTriGauss();
10  n_gauss_points=size(gauss_points,1);
11  [N,Nx,Ny]=CompNDNatPointsTri6(gauss_points(:,1), gauss_points(:,2));
12
```

9.2 2-D Transient Heat Transfer

```
13  % for-loop: compute K matrix
14  for e=1:n_elements     % loop over all the elements
15    ke=zeros(n_nodes_per_element, n_nodes_per_element);
16    [element_nodes, node_id_map]= SetElementNodes(e,nodes, elements);
17    for g=1:n_gauss_points
18      J= CompJacobian2DatPoint(element_nodes, Nx(:,g), Ny(:,g));
19      detJ=det(J);
20      Jinv=inv(J);
21      DN(1,:)=Nx(:,g);
22      DN(2,:)=Ny(:,g);
23      ke=ke+DN'*Jinv'*kappa*Jinv*DN*detJ*gauss_weights(g);
24    end
25    K= AssembleGlobalMatrix(K,ke,node_id_map,1); % assemble global K
26  end
```

```
1   % Compute heat capacity matrix
2   function C=CompC(nodes, elements, cv, rho)
3   % next 5 lines: set up constants and empty matices
4   n_nodes = size(nodes,1);
5   n_elements = size(elements,1);
6   n_nodes_per_element = size(elements,2)-1;
7   C=zeros(n_nodes,n_nodes);
8   Nv=zeros(1,6);
9   % get Gauss points, weights and compute shape functions
10  [gauss_points, gauss_weights]=GetTriGauss();
11  n_gauss_points=size(gauss_points,1);
12  [N,Nx,Ny]=CompNDNatPointsTri6(gauss_points(:,1), gauss_points(:,2));
13
14  % for-loop: compute C matrix
15  for e=1:n_elements     % loop over all the elements
16    ce=zeros(n_nodes_per_element,n_nodes_per_element);
17    [element_nodes, node_id_map]=SetElementNodes(e, nodes, elements);
18    for g=1:n_gauss_points
19      J= CompJacobian2DatPoint(element_nodes, Nx(:,g), Ny(:,g));
20      detJ=det(J);
21      Jinv=inv(J);
22      Nv(1,:)=N(:,g);
23      ce=ce+Nv'*Nv*detJ*cv*rho*gauss_weights(g);
24    end
25    C= AssembleGlobalMatrix(C,ce,node_id_map,1); % assemble global C
26  end
```

```
1  function F=CompF(nodes, elements, edgeFlux)
2  % next 5 lines: set up constants and empty matices
3  n_nodes = size(nodes,1);
4  n_elements = size(elements,1);
5  n_flux_edges = size(edgeFlux,1);
```

```
6   n_nodes_per_element = size(elements,2)-1;
7   F=zeros(n_nodes,1);
8   % next 3 lines : get Gauss points, weights and compute shape functions
9   [gauss_points, gauss_weights]=GetTriEdgeGauss();
10  n_edge_gauss_points=size(gauss_points,1)/3;
11  [N,Nx,Ny]=CompNDNatPointsTri6(gauss_points(:,1), gauss_points(:,2));
12
13  % for-loop: loop over the number of edges affected by heat flux
14  % to compute the element f vector and assemble the global F
15  for t=1:n_flux_edges
16      fe=zeros(n_nodes_per_element, 1);
17      eid=edgeFlux(t,1);
18      [element_nodes, node_id_map]=SetElementNodes(eid,nodes,elements);
19      edge=edgeFlux(t,2);
20      for g=1:2
21          gid=2*(edge-1)+g;
22          J= CompJacobian2DatPoint(element_nodes, Nx(:,gid), Ny(:,gid));
23          if   edge==1
24              lengthJ=sqrt(J(1,1)^2+J(1,2)^2);
25          elseif edge==3
26              lengthJ=sqrt(J(2,1)^2+J(2,2)^2);
27          else
28              lengthJ=norm(J(1,:) - J(2,:) )/sqrt(2.0);
29          end
30          Nv=N(:,gid);
31          fe =fe+ Nv*edgeFlux(t,5)*lengthJ*gauss_weights(gid);
32      end
33      F=AssembleGlobalVector(F,fe,node_id_map,1);    % Assemble F
34  end
```

```
1   % Time integration using the central difference (CrankNicolson) scheme
2   % Input: dt: time step
3   % Input: K, C, F: condution matrix, heat capacity matrix, heat source
4   %           vector, respectively
5   % Input: nodalTemp: temperature boundary condition vector
6   % Input: prevTemp, currTemp: temperature vector at previous and current
7   %           time step, respectively
8   % Input: penalty : penalty number for apply nodal temperature BC
9   % Output: currTemp: current temperature vector
10  function currTemp=CrankNicolson(dt,K,C,F,nodalTemp,prevTemp,currTemp,
        penalty)
11  Cp=C*(1.0/dt);
12  Kp=K*0.5;
13  LHS = Cp+ Kp; %left hand side matrix
14  RHS = F + (Cp - Kp)*prevTemp;  %right hand side vector
15  % for-loop: apply nodal temperature BC
16  for i=1:size(nodalTemp,1);
17     row=nodalTemp(i,1);
```

```
18    RHS(row,1) =nodalTemp(i,2)*penalty;
19    LHS(row, row)=penalty;
20  end
21  currTemp=LHS\RHS; % solve for the current temperature vector
```

```
1   % Get Gauss points and weights of triangle elements
2   function [ gauss_points , gauss_weights]=GetTriGauss()
3   %  currently only 6 point Gauss quadrature, can be expanded
4   gauss_points =zeros(6,2) ;
5   gauss_weights=zeros(6,1) ;
6   gauss_points (1,1) =0.091576213509661;
7   gauss_points (1,2) =0.091576213509661;
8   gauss_points (2,1) =0.816847572980459;
9   gauss_points (2,2) =0.091576213509661;
10  gauss_points (3,1) =0.091576213509661;
11  gauss_points (3,2) =0.816847572980459;
12  gauss_points (4,1) =0.445948490915965;
13  gauss_points (4,2) =0.10810301816807;
14  gauss_points (5,1) =0.445948490915965;
15  gauss_points (5,2) =0.445948490915965;
16  gauss_points (6,1) =0.10810301816807;
17  gauss_points (6,2) =0.445948490915965;
18  gauss_weights(1)=0.109951743655322/2.0;
19  gauss_weights(2)=0.109951743655322/2.0;
20  gauss_weights(3)=0.109951743655322/2.0;
21  gauss_weights(4)=0.223381589678011/2.0;
22  gauss_weights(5)=0.223381589678011/2.0;
23  gauss_weights(6)=0.223381589678011/2.0;
```

```
1   function [ gauss_points , gauss_weights]=GetTriEdgeGauss()
2   % current : 2 points per edge, can be expanded
3   gauss_points =zeros(6,2) ;
4   gauss_weights=zeros(6,1) ;
5   p1=0.5−1/sqrt(3)*0.5;
6   p2=0.5+1/sqrt(3) *0.5;
7   % 1st point
8   gauss_points (1,1) =p1;
9   gauss_points (1,2) =0.0;
10  gauss_weights(1) =0.5;
11  % 2nd point
12  gauss_points (2,1) =p2;
13  gauss_points (2,2) =0.0;
14  gauss_weights(2) =0.5;
15  % 3rd point
16  gauss_points (3,1) =p2;
17  gauss_points (3,2) =p1;
```

```
18    gauss_weights(3)=0.5*sqrt(2);
19    % 4th point
20    gauss_points (4,1) =p1;
21    gauss_points (4,2) =p2;
22    gauss_weights(4)=0.5*sqrt(2);
23    % 5th point
24    gauss_points (5,1) =0.0;
25    gauss_points (5,2) =p1;
26    gauss_weights(5) =0.5;
27    % 6th point
28    gauss_points (6,1) =0.0;
29    gauss_points (6,2) =p2;
30    gauss_weights(6) =0.5;
```

In the MATLAB programs listed above, the functions which have already been discussed previously are not repeated here. They are listed below for your reference:

function [N,Nx,Ny]=CompNDNatPointsTri6(xi_vec, eta_vec): Section 5.2.1.
function [element_nodes, node_id_map]= SetElementNodes(ele, nodes, elements): Section 5.2.1.
function J=CompJacobian2DatPoint(element_nodes, Nxi, Neta): Section 5.2.1.
function K=AssembleGlobalMatrix(K, e, node_id_map, ndDOF): Section 5.2.1.
function F=AssembleGlobalVector(F, fe, node_id_map, ndDOF): Section 5.2.1.

9.3 Elastodynamics

The FEA procedure for linear structural dynamic analysis is illustrated in this section by solving step by step an example problem of dynamic behavior of a 2-D cantilever beam subjected to external loading.

9.3.1 Vibration and Dynamic Response of Continuum Structures

In this section, finite element dynamic analysis is illustrated by demonstrating the step-by-step procedure of dynamic analysis of a 2-D cantilever beam subjected to a concentrated load as shown in Fig. 9.7.

Step 1: Physical Problem
Consider the cantilever beam shown in Fig. 9.7. The beam is initially at rest when the tip load P is suddenly applied. The width of the beam is 0.5 m. Young's modulus of the beam is 90 GPa, Poisson's ratio is 0.25 and the mass density is 2330 kg/m^3. The

proportional mass and stiffness damping coefficients are 2.0 and 10^{-4}, respectively. We perform a finite element analysis for the dynamic response of the beam.

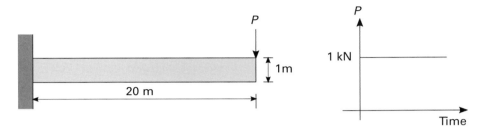

Figure 9.7 A cantilever beam subjected to a concentrated load.

Step 2: Mathematical Model

Generally speaking, three types of problems can be solved by using the same dynamic analysis procedure. They are:

1 Natural frequencies and modes
2 Transient response
3 Frequency response.

In this section, we discuss the first two types of problems. While the frequency response is not discussed in detail, the main idea is described briefly and the implementation is similar to the other two types of problems.

To understand the governing equation of a dynamic structure, we start with a single degree of freedom (SDOF) system as shown in Fig. 9.8. The basic equation of a single DOF spring–mass–dashpot system shown in the figure can be obtained by simply using Newton's second law.

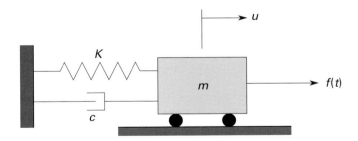

Figure 9.8 A spring–mass–dashpot system.

In Fig. 9.8, u is the displacement, m is the mass, k is the spring stiffness, c is the damping coefficient, and f is the force. From Newton's law of motion, we have

$$m\frac{d^2u}{dt^2} = f(t) - ku - c\frac{du}{dt} \qquad (9.41)$$

where d^2u/dt^2 is the acceleration, du/dt is the velocity. In the case of free vibration with no damping, i.e., $f(t) = 0$ and $c = 0$, we have

$$m\frac{d^2u}{dt^2} = -ku. \tag{9.42}$$

The solution u has the following form

$$u(t) = U\sin(\omega t) \tag{9.43}$$

where ω and U are the angular frequency and amplitude of oscillation, respectively. Substituting Eq. (9.43) into Eq. (9.42), we obtain

$$-mU\omega^2 \sin(\omega t) + kU\sin(\omega t) = 0$$
$$\Rightarrow (-\omega^2 m + k)U = 0$$
$$\Rightarrow -\omega^2 m + k = 0$$
$$\Rightarrow \omega = \sqrt{\frac{k}{m}}. \tag{9.44}$$

Therefore, the angular frequency of the object is a function of the spring stiffness and the mass of the object. The angular frequency obtained in Eq. (9.44) is also called the undamped natural frequency. One can also use the cyclic frequency $f = \omega/2\pi$ (Hz).

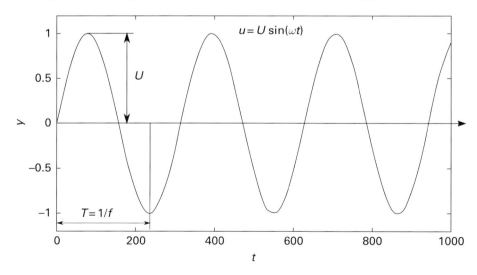

Figure 9.9 Single degree of freedom oscillation.

If the damping coefficient c is not zero, and c is less than a critical damping coefficient c_c, the damped natural frequency is given by

$$\omega_d = \omega\sqrt{1 - \xi^2} \tag{9.45}$$

where $\xi = c/c_c$ is the damping ratio. The critical damping c_c is given by

$$c_c = 2\sqrt{km} = 2\omega m. \tag{9.46}$$

In engineering structures, the damping ratio is typically small: $0 \leq \xi \leq 0.15$.

For continuum structures, the equations of motion are defined for an infinitesimal differential volume. The equations have been discussed in Section 7.2.1 as shown in Eqs. (7.8–7.10). They are repeated here as shown below

$$\frac{\partial \sigma_{xx}}{\partial x} + \frac{\partial \sigma_{xy}}{\partial y} + \frac{\partial \sigma_{xz}}{\partial z} + b_x = \rho \frac{\partial^2 u}{\partial t^2} \tag{9.47}$$

$$\frac{\partial \sigma_{xy}}{\partial x} + \frac{\partial \sigma_{yy}}{\partial y} + \frac{\partial \sigma_{yz}}{\partial z} + b_y = \rho \frac{\partial^2 v}{\partial t^2} \tag{9.48}$$

$$\frac{\partial \sigma_{xz}}{\partial x} + \frac{\partial \sigma_{yz}}{\partial y} + \frac{\partial \sigma_{zz}}{\partial z} + b_z = \rho \frac{\partial^2 w}{\partial t^2}. \tag{9.49}$$

In short form, they can be rewritten as

$$\nabla \cdot \boldsymbol{\sigma} + \mathbf{b} = \rho \frac{\partial^2 \mathbf{u}}{\partial t^2}. \tag{9.50}$$

Step 3: Weak Form

As discussed previously in the case of 2-D elasticity, there are multiple approaches to obtain the weak form of the governing equations. Once again, although it is possible to use the method of weighted residual (see Section 7.2.2), the process is relatively tedious. For the equations of motion, the principle of virtual work and variational calculus are more straightforward. In this section, we obtain the weak form by using the principle of virtual work. In the static analysis, the principle of virtual work is given in Eq. (7.102), stating that the internal virtual work done by the stresses is equal to the virtual work done by the external forces.

$$\int_\Omega \delta\boldsymbol{\epsilon} \cdot \mathbf{C}\boldsymbol{\epsilon} d\Omega = \int_{\Gamma_t} \delta\mathbf{u} \cdot \mathbf{t} d\Gamma + \delta\mathbf{u}(\mathbf{a}) \cdot \mathbf{p}/h + \int_\Omega \delta\mathbf{u} \cdot \mathbf{b} d\Omega. \tag{9.51}$$

In a dynamic system, the principle of virtual work is extended by adding two types of internal virtual work: the internal virtual work done by the inertial force and the internal virtual work done by the damping force, i.e.,

$$\int_\Omega \delta\mathbf{u} \cdot \rho \ddot{\mathbf{u}} d\Omega + \int_\Omega \delta\mathbf{u} \cdot \mathbf{C}\dot{\mathbf{u}} d\Omega + \int_\Omega \delta\boldsymbol{\epsilon} \cdot \mathbf{C}\boldsymbol{\epsilon} d\Omega$$
$$= \int_{\Gamma_t} \delta\mathbf{u} \cdot \mathbf{t} d\Gamma + \delta\mathbf{u}(\mathbf{a}) \cdot \mathbf{p}/h + \int_\Omega \delta\mathbf{u} \cdot \mathbf{b} d\Omega. \tag{9.52}$$

In Eq. (9.52), the first term on the left hand side is the internal virtual work done by the inertial force reacting to the acceleration. The second term is the virtual work done by the damping force (i.e., energy dissipation) where C is the damping coefficient.

Step 4: Discretization

For this example, we discretize the 2-D beam into 40×2 equal size 4-node linear quadrilateral elements. The uniform mesh shown in Fig. 9.10 shows a part of the mesh. There are a total of 80 elements and 123 nodes in the mesh. The nodes and elements data are given in Table 9.3.

Table 9.3 Nodes and elements

nodes.dat			elements.dat				
1	0.000	0.000	1	1	2	43	42
2	0.500	0.000	2	2	3	44	43
3	1.000	0.000	3	3	4	45	44
4	1.500	0.000	4	4	5	46	45
5	2.000	0.000	5	5	6	47	46
...	...	...	...	...	...	...	...

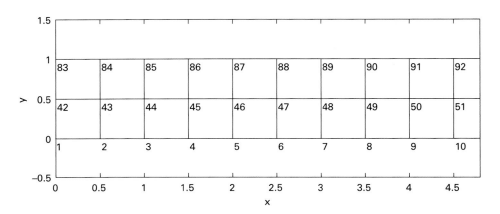

Figure 9.10 A part of the uniform mesh of the beam. Node numbers are also shown.

Steps 5-7: Approximation, Element and Global Matrices and Vectors

In the dynamic analysis, the weak form shown in Eq. (9.52) is the same as that of the static analysis except for the two additional terms. Therefore, in this section we only derive the element matrices for the inertia and damping terms. In the first two terms of Eq. (9.52), the finite element approximations of the virtual displacement, velocity, and acceleration are written as

$$\delta \mathbf{u} = \begin{Bmatrix} \delta u \\ \delta v \end{Bmatrix} = \begin{Bmatrix} \sum_i N_i(x,y) \delta u_i \\ \sum_i N_i(x,y) \delta v_i \end{Bmatrix} \tag{9.53}$$

$$\dot{\mathbf{u}} = \begin{Bmatrix} \dot{u} \\ \dot{v} \end{Bmatrix} = \begin{Bmatrix} \sum_i N_i(x,y) \dot{u}_i \\ \sum_i N_i(x,y) \dot{v}_i \end{Bmatrix} \tag{9.54}$$

$$\ddot{\mathbf{u}} = \begin{Bmatrix} \ddot{u} \\ \ddot{v} \end{Bmatrix} = \begin{Bmatrix} \sum_i N_i(x,y) \ddot{u}_i \\ \sum_i N_i(x,y) \ddot{v}_i \end{Bmatrix}. \tag{9.55}$$

Equations (9.54, 9.55) can be rewritten in matrix form. For example, for 4-node quadrilateral elements, we have

$$\delta \mathbf{u} = \begin{bmatrix} N_1 & 0 & N_2 & 0 & N_3 & 0 & N_4 & 0 \\ 0 & N_1 & 0 & N_2 & 0 & N_3 & 0 & N_4 \end{bmatrix} \begin{Bmatrix} \delta u_1 \\ \delta v_1 \\ \delta u_2 \\ \delta v_2 \\ \delta u_3 \\ \delta v_3 \\ \delta u_4 \\ \delta v_4 \end{Bmatrix} = \mathbf{N} \delta \mathbf{d} \quad (9.56)$$

where the vector of the displacement variations (or virtual displacements) is denoted in short by $\delta \mathbf{d}$, and $\mathbf{N}$ is used to denote the transpose of the 8×2 matrix containing the shape functions. Similarly, we have

$$\dot{\mathbf{u}} = \begin{bmatrix} N_1 & 0 & N_2 & 0 & N_3 & 0 & N_4 & 0 \\ 0 & N_1 & 0 & N_2 & 0 & N_3 & 0 & N_4 \end{bmatrix} \begin{Bmatrix} \dot{u}_1 \\ \dot{v}_1 \\ \dot{u}_2 \\ \dot{v}_2 \\ \dot{u}_3 \\ \dot{v}_3 \\ \dot{u}_4 \\ \dot{v}_4 \end{Bmatrix} = \mathbf{N} \dot{\mathbf{d}} \quad (9.57)$$

$$\ddot{\mathbf{u}} = \begin{bmatrix} N_1 & 0 & N_2 & 0 & N_3 & 0 & N_4 & 0 \\ 0 & N_1 & 0 & N_2 & 0 & N_3 & 0 & N_4 \end{bmatrix} \begin{Bmatrix} \ddot{u}_1 \\ \ddot{v}_1 \\ \ddot{u}_2 \\ \ddot{v}_2 \\ \ddot{u}_3 \\ \ddot{v}_3 \\ \ddot{u}_4 \\ \ddot{v}_4 \end{Bmatrix} = \mathbf{N} \ddot{\mathbf{d}}. \quad (9.58)$$

Substituting Eqs. (9.56–9.58) into the first two terms in Eq. (9.52), for an element e, we obtain

$$\int_\Omega^{(e)} \delta\mathbf{u} \cdot \rho\ddot{\mathbf{u}} d\Omega = \int_\Omega^{(e)} \rho\delta\mathbf{u}^T \ddot{\mathbf{u}} d\Omega = \delta\mathbf{d}^T \left(\int_\Omega^{(e)} \rho\mathbf{N}^T\mathbf{N} d\Omega \right) \ddot{\mathbf{d}} = \delta\mathbf{d}^T \mathbf{m}^{(e)} \ddot{\mathbf{d}} \quad (9.59)$$

and

$$\int_\Omega^{(e)} \delta\mathbf{u} \cdot C\dot{\mathbf{u}} d\Omega = \int_\Omega^{(e)} C\delta\mathbf{u}^T \dot{\mathbf{u}} d\Omega = \delta\mathbf{d}^T \left(\int_\Omega^{(e)} C\mathbf{N}^T\mathbf{N} d\Omega \right) \dot{\mathbf{d}} = \delta\mathbf{d}^T \mathbf{c}^{(e)} \dot{\mathbf{d}} \quad (9.60)$$

where $\mathbf{m}^{(e)}$ and $\mathbf{c}^{(e)}$ are called element mass matrix and element damping matrix, respectively. By following the assembling procedure for the stiffness matrix, the global mass and damping matrices can be assembled. The global finite element equation of motion is then

$$\mathbf{M}\ddot{\mathbf{d}} + \mathbf{C}\dot{\mathbf{d}} + \mathbf{K}\mathbf{d} = \mathbf{f} \quad (9.61)$$

with initial conditions

$$\mathbf{d}_{t=0} = \mathbf{d}_0 \quad \text{(initial displacement)} \quad (9.62)$$
$$\dot{\mathbf{d}}_{t=0} = \dot{\mathbf{d}}_0 \quad \text{(initial velocity)} \quad (9.63)$$

and boundary conditions

$$d_i = \bar{d}_i \quad \text{on } \Gamma_u \quad (9.64)$$
$$\mathbf{t} = \bar{\mathbf{t}} \quad \text{on } \Gamma_t \quad (9.65)$$

where d_i can be either u or v.

Steps 8-9: Solution
Solution for Free Vibration

We can calculate the natural frequencies and corresponding modes by solving the free vibration problem of a structure. In a free vibration problem, there is no external forces $\mathbf{f}$ acting on the structure. In addition, no damping is included in the free vibration problem, i.e.,

$$\mathbf{f} = 0 \quad \text{and} \quad \mathbf{C} = 0. \quad (9.66)$$

The equation of motion is rewritten as

$$\mathbf{M}\ddot{\mathbf{d}} + \mathbf{K}\mathbf{d} = \mathbf{0}. \quad (9.67)$$

It can be shown that the solution of the displacement has the form of

$$\mathbf{d} = \mathbf{D}\sin(\omega t). \quad (9.68)$$

That is, without external force and damping, all the nodes move in phase at the same frequency ω. The acceleration is then

$$\ddot{\mathbf{d}} = -\omega^2 \mathbf{D}\sin(\omega t). \quad (9.69)$$

Table 9.4 First 10 natural frequencies of the cantilever beam

Mode	Frequency (Hz)
1	2.649598
2	16.449825
3	45.415768
4	77.746582
5	87.297663
6	141.001837
7	205.186378
8	233.341219
9	278.580302
10	360.044817

Substituting Eqs. (9.68, 9.69) into Eq. (9.67), we have

$$-\omega^2 \mathbf{MD}\sin(\omega t) + \mathbf{KD}\sin(\omega t) = 0 \tag{9.70}$$
$$\Rightarrow \mathbf{KD} = \omega^2 \mathbf{MD}. \tag{9.71}$$

Equation (9.71) is the equation of motion of the undamped free vibration. It is a generalized eigenvalue problem. The eigenvalue, ω^2, determines the natural frequencies of the structure. **D** is the eigenvector containing the normalized nodal displacements of the structure vibrating at ω. **D** is also called a vibrational mode corresponding to the vibrational frequency ω. For a given finite element discretization, the total number of frequencies and modes is equal to the total degrees of freedom. For the cantilever beam in the example problem of this section, the first four modes obtained are shown in Fig. 9.11. Table 9.4 lists the first 10 natural frequencies of the beam. Note that the lowest natural frequency is called the fundamental frequency of the system.

Solution for Time History Dynamic Analysis

The global equation of motion can be solved by using direct time integration methods. Newmark family of schemes are a set of such methods. The popular set of Newmark schemes are listed in Table 9.5. For any time step $n + 1$, we have

$$\mathbf{M}\ddot{\mathbf{d}}_{n+1} + \mathbf{C}\dot{\mathbf{d}}_{n+1} + \mathbf{K}\mathbf{d}_{n+1} = \mathbf{f}_{n+1} \tag{9.72}$$

$$\mathbf{d}_{n+1} = \mathbf{d}_n + \Delta t \dot{\mathbf{d}}_n + \frac{\Delta t^2}{2}(1 - 2\beta)\ddot{\mathbf{d}}_n + \frac{\Delta t^2}{2}(2\beta)\ddot{\mathbf{d}}_{n+1} \tag{9.73}$$

$$\dot{\mathbf{d}}_{n+1} = \dot{\mathbf{d}}_n + \Delta t (1 - \gamma)\ddot{\mathbf{d}}_n + \Delta t \gamma \ddot{\mathbf{d}}_{n+1}. \tag{9.74}$$

Implicit Schemes

As shown in Table 9.5, based on the values of β and γ selected, the schemes are categorized as implicit and explicit schemes. As will be demonstrated below, an implicit scheme is a set of solution steps to compute the displacement, velocity, and acceleration of the nodes at each time step (referred to as time marching), in which at

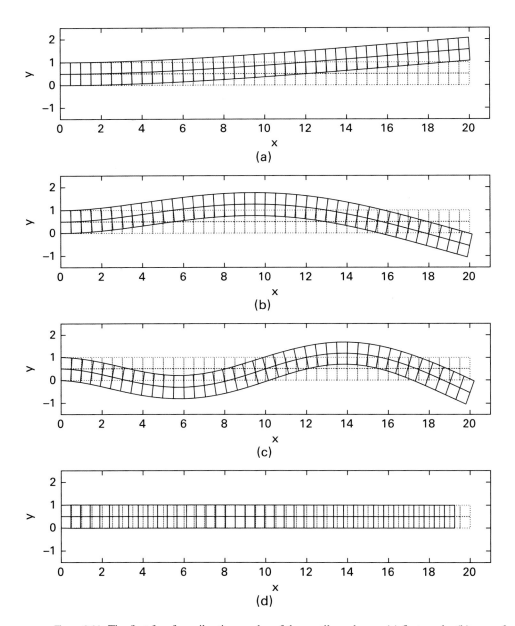

Figure 9.11 The first few free vibration modes of the cantilever beam: (a) first mode; (b) second mode; (c) third mode; and (d) fourth mode.

some point solving a linear system of simultaneous equations is required. On the contrary, explicit schemes calculate the relevant physical quantities through the time steps without solving a linear system.

In the following, the steps of time marching in an implicit scheme are illustrated.

9.3 Elastodynamics

Table 9.5 Newmark schemes

Scheme	Type	β	γ	Stability	Order of accuracy
Average Acceleration (Trapezoid Rule)	Implicit	$\frac{1}{4}$	$\frac{1}{2}$	Unconditional	2
Linear Acceleration	Implicit	$\frac{1}{6}$	$\frac{1}{2}$	Conditional	2
Fox–Goodwin	Implicit	$\frac{1}{12}$	$\frac{1}{2}$	Conditional	2
Central Difference	Explicit	0	$\frac{1}{2}$	Conditional	2

Step 1

Starting from $t = 0$, $\dot{\mathbf{d}}_0$ and $\mathbf{d}_0$ are known from the initial conditions, calculate $\ddot{\mathbf{d}}_0$ by using

$$\mathbf{M}\ddot{\mathbf{d}}_0 + \mathbf{C}\dot{\mathbf{d}}_0 + \mathbf{K}\mathbf{d}_0 = \mathbf{f}_0 \tag{9.75}$$

i.e.,

$$\mathbf{M}\ddot{\mathbf{d}}_0 = \mathbf{f}_0 - \mathbf{C}\dot{\mathbf{d}}_0 - \mathbf{K}\mathbf{d}_0 \tag{9.76}$$

$\ddot{\mathbf{d}}_0$ can be obtained.

Step 2

For time step = 1, substitute Eqs. (9.73, 9.74) into Eq. (9.72). That is, substituting

$$\mathbf{d}_1 = \mathbf{d}_0 + \Delta t \dot{\mathbf{d}}_0 + \frac{\Delta t^2}{2}(1 - 2\beta)\ddot{\mathbf{d}}_0 + \frac{\Delta t^2}{2}(2\beta)\ddot{\mathbf{d}}_1 \tag{9.77}$$

and

$$\dot{\mathbf{d}}_1 = \dot{\mathbf{d}}_0 + \Delta t(1 - \gamma)\ddot{\mathbf{d}}_0 + \Delta t \gamma \ddot{\mathbf{d}}_1 \tag{9.78}$$

into the equation of motion for time step 1

$$\mathbf{M}\ddot{\mathbf{d}}_1 + \mathbf{C}\dot{\mathbf{d}}_1 + \mathbf{K}\mathbf{d}_1 = \mathbf{f}_1, \tag{9.79}$$

we obtain

$$\mathbf{M}\ddot{\mathbf{d}}_1 + \mathbf{C}\left[\dot{\mathbf{d}}_0 + \Delta t(1-\gamma)\ddot{\mathbf{d}}_0 + \Delta t \gamma \ddot{\mathbf{d}}_1\right]$$
$$+ \mathbf{K}\left[\mathbf{d}_0 + \Delta t \dot{\mathbf{d}}_0 + \frac{\Delta t^2}{2}(1 - 2\beta)\ddot{\mathbf{d}}_0 + \frac{\Delta t^2}{2}(2\beta)\ddot{\mathbf{d}}_1\right] = \mathbf{f}_1. \tag{9.80}$$

Moving all the known quantities to the right hand side, we obtain

$$\left(\mathbf{M} + \gamma \mathbf{C} \Delta t + \beta \Delta t^2 \mathbf{K}\right) \ddot{\mathbf{d}}_1 =$$
$$\mathbf{f}_1 - (\mathbf{C} + \Delta t \mathbf{K})\dot{\mathbf{d}}_0 - \left(\Delta t(1-\gamma)\mathbf{C} + \frac{\Delta t^2}{2}(1-2\beta)\mathbf{K}\right)\ddot{\mathbf{d}}_0 - \mathbf{K}\mathbf{d}_0. \tag{9.81}$$

We can solve this equation system to obtain $\ddot{\mathbf{d}}_1$.

Step 3
Substituting $\mathbf{d}_0, \dot{\mathbf{d}}_0, \ddot{\mathbf{d}}_0$, and $\ddot{\mathbf{d}}_1$ into Eqs. (9.73, 9.74), $\mathbf{d}_1$ and $\dot{\mathbf{d}}_1$ can be calculated.

Repeat steps 2 and 3 for time steps 2, 3, ...

Since the acceleration is the unknown to be solved in the linear system shown in Eq. (9.81), the scheme is also called the "a-form." One can also formulate the scheme so that the displacement is the unknown of the linear system as follows. We rewrite Eq. (9.73) as

$$\ddot{\mathbf{d}}_{n+1} = \frac{1}{\beta \Delta t^2} \left(\mathbf{d}_{n+1} - \mathbf{d}_n - \Delta t \dot{\mathbf{d}}_n \right) - \frac{1-2\beta}{2\beta} \ddot{\mathbf{d}}_n. \quad (9.82)$$

Substituting Eq. (9.82) into Eq. (9.74), we obtain

$$\dot{\mathbf{d}}_{n+1} = \dot{\mathbf{d}}_n + \Delta t \left(\frac{2\beta - \gamma}{2\beta} \ddot{\mathbf{d}}_n + \frac{\gamma}{\beta \Delta t^2} \left(\mathbf{d}_{n+1} - \mathbf{d}_n - \Delta t \dot{\mathbf{d}}_n \right) \right). \quad (9.83)$$

Substituting Eqs. (9.82, 9.83) into the equation of motion, (9.72), we have

$$\left(\frac{1}{\beta \Delta t^2} \mathbf{M} + \frac{\gamma}{\beta \Delta t} \mathbf{C} + \mathbf{K} \right) \mathbf{d}_{n+1} =$$

$$\mathbf{f}_{n+1} + \left(\frac{1}{\beta \Delta t^2} \mathbf{M} + \frac{\gamma}{\beta \Delta t} \mathbf{C} \right) \mathbf{d}_n + \left(\frac{1}{\beta \Delta t} \mathbf{M} + \frac{\gamma - \beta}{\beta} \mathbf{C} \right) \dot{\mathbf{d}}_n +$$

$$\left(\frac{1-2\beta}{2\beta} \mathbf{M} + \Delta t \frac{\gamma - 2\beta}{2\beta} \mathbf{C} \right) \ddot{\mathbf{d}}_n. \quad (9.84)$$

Equation (9.84) can be applied and can replace Eq. (9.81) to solve for the new displacement instead of acceleration. The scheme is then called the "d-form." Note that "a-form" and "d-form" are mathematically equivalent, i.e., the numerical results from "a-form" and "d-form" are identical. As shown in Table 9.5, several β, γ parameter sets can be used in the implicit schemes. For example, the "d-form" scheme with Trapezoid rule ($\beta = 1/4, \gamma = 1/2$) can be obtained as

$$\left(\frac{4}{\Delta t^2} \mathbf{M} + \frac{2}{\Delta t} \mathbf{C} + \mathbf{K} \right) \mathbf{d}_{n+1}$$

$$= \mathbf{f}_{n+1} + \left(\frac{4}{\Delta t^2} \mathbf{M} + \frac{2}{\Delta t} \mathbf{C} \right) \mathbf{d}_n + \left(\frac{4}{\Delta t} \mathbf{M} + \mathbf{C} \right) \dot{\mathbf{d}}_n + \mathbf{M} \ddot{\mathbf{d}}_n. \quad (9.85)$$

Explicit Scheme (Central Difference)
The central difference scheme shown in Table 9.5 is an explicit method with $\beta = 0$ and $\gamma = 1/2$. Note that, the central difference scheme is explicit only when $\mathbf{M}$ and $\mathbf{C}$ are diagonal matrices. Given that $\beta = 0$ and $\gamma = 1/2$, we obtain from Eqs. (9.73, 9.74),

9.3 Elastodynamics

$$\mathbf{d}_{n+1} = \mathbf{d}_n + \Delta t \dot{\mathbf{d}}_n + \frac{\Delta t^2}{2}\ddot{\mathbf{d}}_n \quad (9.86)$$

$$\dot{\mathbf{d}}_{n+1} = \dot{\mathbf{d}}_n + \frac{\Delta t}{2}\ddot{\mathbf{d}}_n + \frac{\Delta t}{2}\ddot{\mathbf{d}}_{n+1}. \quad (9.87)$$

Substituting Eqs. (9.86, 9.87) into Eq. (9.72), we have

$$\mathbf{M}\ddot{\mathbf{d}}_{n+1} + \mathbf{C}\left(\dot{\mathbf{d}}_n + \frac{\Delta t}{2}\ddot{\mathbf{d}}_n + \frac{\Delta t}{2}\ddot{\mathbf{d}}_{n+1}\right) + \mathbf{K}\left(\mathbf{d}_n + \Delta t \dot{\mathbf{d}}_n + \frac{\Delta t^2}{2}\ddot{\mathbf{d}}_n\right) = \mathbf{f}_{n+1}. \quad (9.88)$$

Rearranging the equation

$$\left(\mathbf{M} + \frac{\Delta t}{2}\mathbf{C}\right)\ddot{\mathbf{d}}_{n+1} = \mathbf{f}_{n+1} - \mathbf{C}\left(\dot{\mathbf{d}}_n + \frac{\Delta t}{2}\ddot{\mathbf{d}}_n\right) - \mathbf{K}\left(\mathbf{d}_n + \Delta t \dot{\mathbf{d}}_n + \frac{\Delta t^2}{2}\ddot{\mathbf{d}}_n\right). \quad (9.89)$$

If $\mathbf{M}$ and $\mathbf{C}$ are diagonal, $\ddot{\mathbf{d}}_{n+1}$ can be computed from Eq. (9.89) without solving a linear system.

Mass and Damping Matrices
The element mass matrix formulation has been obtained in Section 9.3.1, which is given by

$$\mathbf{m}^{(e)} = \int_\Omega^{(e)} \rho \mathbf{N}^T \mathbf{N} d\Omega \quad (9.90)$$

where $\mathbf{N}$ is the matrix of shape functions. For example, for a 2-D quadrilateral element in elasticity problems,

$$\mathbf{N} = \begin{bmatrix} N_1 & 0 & N_2 & 0 & N_3 & 0 & N_4 & 0 \\ 0 & N_1 & 0 & N_2 & 0 & N_3 & 0 & N_4 \end{bmatrix} \quad (9.91)$$

and

$$\mathbf{m}^{(e)} = \int_\Omega^{(e)} \rho \begin{bmatrix} N_1 & 0 \\ 0 & N_1 \\ N_2 & 0 \\ 0 & N_2 \\ N_3 & 0 \\ 0 & N_3 \\ N_4 & 0 \\ 0 & N_4 \end{bmatrix} \begin{bmatrix} N_1 & 0 & N_2 & 0 & N_3 & 0 & N_4 & 0 \\ 0 & N_1 & 0 & N_2 & 0 & N_3 & 0 & N_4 \end{bmatrix} d\Omega \quad (9.92)$$

$$= \int_\Omega^{(e)} \rho \begin{bmatrix} N_1^2 & 0 & N_1N_2 & 0 & N_1N_3 & 0 & N_1N_4 & 0 \\ 0 & N_1^2 & 0 & N_1N_2 & 0 & N_1N_3 & 0 & N_1N_4 \\ N_1N_2 & 0 & N_2^2 & 0 & N_2N_3 & 0 & N_2N_4 & 0 \\ 0 & N_1N_2 & 0 & N_2^2 & 0 & N_2N_3 & 0 & N_2N_4 \\ N_1N_3 & 0 & N_2N_3 & 0 & N_3^2 & 0 & N_3N_4 & 0 \\ 0 & N_1N_3 & 0 & N_2N_3 & 0 & N_3^2 & 0 & N_3N_4 \\ N_1N_4 & 0 & N_2N_4 & 0 & N_3N_4 & 0 & N_4^2 & 0 \\ 0 & N_1N_4 & 0 & N_2N_4 & 0 & N_3N_4 & 0 & N_4^2 \end{bmatrix} d\Omega. \tag{9.93}$$

This matrix is called a consistent mass matrix. It can be shown that for any rectangular element, the consistent mass matrix can be pre-calculated and written as

$$\mathbf{m}^{(e)} = \frac{m}{36} \begin{bmatrix} 4 & 0 & 2 & 0 & 1 & 0 & 2 & 0 \\ 0 & 4 & 0 & 2 & 0 & 1 & 0 & 2 \\ 2 & 0 & 4 & 0 & 2 & 0 & 1 & 0 \\ 0 & 2 & 0 & 4 & 0 & 2 & 0 & 1 \\ 1 & 0 & 2 & 0 & 4 & 0 & 2 & 0 \\ 0 & 1 & 0 & 2 & 0 & 4 & 0 & 2 \\ 2 & 0 & 1 & 0 & 2 & 0 & 4 & 0 \\ 0 & 2 & 0 & 1 & 0 & 2 & 0 & 4 \end{bmatrix} \tag{9.94}$$

where m is the mass of the element (because the elements are 2-D, the unit of mass is kg per unit thickness). Note that m can be easily calculated as $m = \rho A$ where ρ is the material mass density and A is the area of the element. For any 3-node triangular element,

$$\mathbf{m}^{(e)} = \frac{m}{12} \begin{bmatrix} 2 & 0 & 1 & 0 & 1 & 0 \\ 0 & 2 & 0 & 1 & 0 & 1 \\ 1 & 0 & 2 & 0 & 1 & 0 \\ 0 & 1 & 0 & 2 & 0 & 1 \\ 1 & 0 & 1 & 0 & 2 & 0 \\ 0 & 1 & 0 & 1 & 0 & 2 \end{bmatrix}. \tag{9.95}$$

If, for each row, we add all the entries in the row and put the result on the diagonal, we can rewrite Eq. (9.93) as

$$\mathbf{m}^{(e)} = \int_{\Omega}^{(e)} \rho \begin{bmatrix} N_1 & 0 & 0 & 0 & 0 & 0 & 0 & 0 \\ 0 & N_1 & 0 & 0 & 0 & 0 & 0 & 0 \\ 0 & 0 & N_2 & 0 & 0 & 0 & 0 & 0 \\ 0 & 0 & 0 & N_2 & 0 & 0 & 0 & 0 \\ 0 & 0 & 0 & 0 & N_3 & 0 & 0 & 0 \\ 0 & 0 & 0 & 0 & 0 & N_3 & 0 & 0 \\ 0 & 0 & 0 & 0 & 0 & 0 & N_4 & 0 \\ 0 & 0 & 0 & 0 & 0 & 0 & 0 & N_4 \end{bmatrix} d\Omega. \qquad (9.96)$$

Equation (9.96) is called a lumped mass matrix. The lumped mass matrix is a diagonal matrix containing element integrals of the shape functions. Physically, the mass lumping process distributes the total mass of an element to its nodes. The lumped mass matrix of Eq. (9.94) is

$$\mathbf{m}^{(e)} = \frac{m}{4} \begin{bmatrix} 1 & 0 & 0 & 0 & 0 & 0 & 0 & 0 \\ 0 & 1 & 0 & 0 & 0 & 0 & 0 & 0 \\ 0 & 0 & 1 & 0 & 0 & 0 & 0 & 0 \\ 0 & 0 & 0 & 1 & 0 & 0 & 0 & 0 \\ 0 & 0 & 0 & 0 & 1 & 0 & 0 & 0 \\ 0 & 0 & 0 & 0 & 0 & 1 & 0 & 0 \\ 0 & 0 & 0 & 0 & 0 & 0 & 1 & 0 \\ 0 & 0 & 0 & 0 & 0 & 0 & 0 & 1 \end{bmatrix}. \qquad (9.97)$$

The lumped mass matrix of Eq. (9.95) is

$$\mathbf{m}^{(e)} = \frac{m}{3} \begin{bmatrix} 1 & 0 & 0 & 0 & 0 & 0 \\ 0 & 1 & 0 & 0 & 0 & 0 \\ 0 & 0 & 1 & 0 & 0 & 0 \\ 0 & 0 & 0 & 1 & 0 & 0 \\ 0 & 0 & 0 & 0 & 1 & 0 \\ 0 & 0 & 0 & 0 & 0 & 1 \end{bmatrix}. \qquad (9.98)$$

It is shown that the diagonal entries represent 1/4 and 1/3 of the element mass, respectively. Mass lumping has several advantages: (1) the diagonal matrix requires less storage; (2) it is necessary in the explicit scheme; and (3) in free vibration analysis, it can reduce the generalized eigenvalue problem to a standard eigenvalue problem.

The damping matrix $\mathbf{C}$ in the global equation of motion represents an approximation of the overall energy dissipation during the motion of the structure. In general, the determination of the damping matrix $\mathbf{C}$ requires the solution of the energy equations for the system. A simplified construction of the damping matrix can be obtained by assuming a proportional damping, i.e.,

$$\mathbf{C} = \alpha \mathbf{M} + \beta \mathbf{K} \tag{9.99}$$

where α and β are constants that can be determined from two given damping ratios corresponding to two different frequencies of vibration. In many cases, the constants α and β are given as properties of a system. For single DOF systems, the proportional damping coefficient is then

$$c = \alpha m + \beta k \tag{9.100}$$

where m is the mass and k is the stiffness. The damping ratio is then

$$\xi = \frac{c}{c_c} = \frac{\alpha m + \beta k}{2\omega m} = \frac{1}{2}\left(\frac{\alpha}{\omega} + \beta\omega\right). \tag{9.101}$$

Equation (9.101) shows that, for small ωs, α is the critical constant for the damping ratio while β become a more significant factor for ξ for large ωs. In other words, α and β determine the damping of the low and high frequency modes, respectively. In the explicit scheme shown in Eq. (9.89), the damping matrix $\mathbf{C}$ is required to be diagonal. We have shown that the mass matrix can be lumped to be diagonal. However, the stiffness matrix can not be made diagonal since that would break the connections between the nodes. Therefore, for the proportional damping shown in Eq. (9.99), β must be set to zero. Since β determines the damping of the high frequency modes, setting $\beta = 0$ could impair the accuracy of the analysis of high frequency problems.

Step 10: Post-processing

In the solution (time marching) step, the displacement, velocity, and acceleration of the nodes are calculated at each time step. With these solutions, we can visualize the deformation of the structure as a function of time. Animations can be produced for that purpose. In addition, the trajectory of motion of certain points in the structure can be visualized. For example, the trajectory of motion of point (20,0.5) with and without damping is shown in Fig. 9.12.

Other than the motion, the stresses and strains in the structure can be calculated as described in the elastostatics case. The only difference is that in the dynamic case they are all time-dependent quantities, in the same fashion as displacement, velocity, and acceleration.

9.3.2 Computer Implementation

In this section, by using the beam vibration problem as an example, we present the data files and the complete MATLAB code for the finite element analysis.

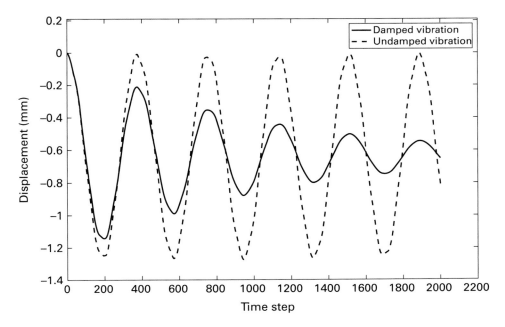

Figure 9.12 Motion trajectory of point (20,0.5) as a function of time.

Data Files

The input files "nodes.dat" and "elements.dat" are already shown in Table 9.3. The material properties, loads, boundary conditions, and other model options are shown in the tables below.

bcsdisp.dat	bcsforce.dat	materials.dat
1 1 0	123 0 −1000	% Young's modulus (Pa)
1 2 0		9e10
42 1 0		
42 2 0		% Poisson's ratio
83 1 0		0.25
83 2 0		
		% Mass density (kg/m^3)
		2330.000000
		% Damping coefficients for K and M
		1e-4
		2.0

options.dat

% Dimensions
2

% Thickness of the structure (m, only 2-D analysis)
0.5

% Element type: No. of edges or faces, No. of nodes
44

% time period (second)
2

% time step, dt, (second)
1e-3

% node for tracking temperature variation
82

MATLAB Code
The code of the elastodynamics analysis is listed below:

```matlab
clear all ;
% next 3 lines : read the input files
filenames = {'nodes.dat', 'elements.dat', 'materials.dat', ...
             'options.dat', 'bcsforce.dat', 'bcsdisp.dat'};
for i = 1:numel(filenames); load(filenames{i}); end;

% next 9 lines : set up global material properties and constants
E=materials(1,1) ;
nu=materials(2,1) ;
rho=materials(3,1) ;
dampK=materials(4,1);
dampM=materials(5,1);
dimension=options(1,1) ;
thickness =options(2,1) ;
timePeriod=options(4,1) ;
dt=options(5,1) ;
probeNode=options(6,1);
% next 5 lines : bookkeeping
n_nodes=size(nodes,1);
n_elements=size(elements,1);
```

```
21  n_bcsforce=size(bcsforce,1);
22  n_bcsdisp=size(bcsdisp,1);
23  n_timeSteps=timePeriod/dt+1;
24  % next 3 lines : set up empty matrices
25  U=zeros(n_nodes*dimension,n_timeSteps);
26  V=zeros(n_nodes*dimension,n_timeSteps);
27  A=zeros(n_nodes*dimension,n_timeSteps);
28
29  K=CompK(nodes, elements, materials);  % compute global K matrix
30  M=CompM(nodes, elements, rho);         % compute global M matrix
31  C = dampM*M + dampK*K;                 % compute global C matrix
32  F=CompF(nodes, elements, thickness, bcsforce);  % compute global F
         vector
33
34  % next 12 lines : time marching using Newmark Scheme
35  beta = 0.25;
36  gamma = 0.5;
37  LHS = M*(1.0/(beta*dt*dt))+ C*(gamma/(beta*dt)) + K;
38  penalty=max(max(abs(LHS)))*1e+6;
39  A(:,1)=M\F;  % initilization of accelecration
40  for t=2:n_timeSteps
41      [U,V,A]=Newmark(t,dt,beta, gamma, dimension,...
42          K,M,C,F,bcsdisp, penalty ,U,V,A);
43      if rem(t*100,(n_timeSteps−1)*5)==0
44          fprintf('%d %%\n', floor ( t *100/( n_timeSteps −1)));
45      end
46  end
47
48  % next 9 lines : save the displacement results in file , plot
49  Uout=U(probeNode*2,:);
50  save −ascii −double Uout2.dat Uout
51  disp('Vertical displacement results stored in Uout.dat');
52  Uout(200:205)*1000
53  plot(Uout, 'LineWidth',2);
54  axis([0  2100 −1.2e−3 0]);
55  xlabel('Time step');
56  ylabel('Displacement (m)');
57  set(gca,'fontsize',16);
```

```
1  % Compute mass matrix
2  function M=CompM(nodes, elements, rho)
3  n_nodes = size(nodes,1);
4  n_elements = size(elements,1);
5  n_nodes_per_element = size(elements,2)−1;
6  M=zeros(n_nodes*2,n_nodes*2);
7  Nv=zeros(8,2);
8  [gauss_points , gauss_weights]=GetQuadGauss(2,2);
9  [N,Nx,Ny]=CompNDNatPointsQuad4(gauss_points(:,1),gauss_points(:,2));
```

```
10
11  % for-loop: compute M matrix: loop over all the elements
12  for e=1:n_elements
13    me=zeros(n_nodes_per_element*2, n_nodes_per_element*2);
14    [element_nodes, node_id_map]= SetElementNodes(e, nodes, elements);
15    % for-loop: compute element mass matrix me
16    for g=1:size(gauss_points,1)
17      J=CompJacobian2DatPoint(element_nodes, Nx(:,g), Ny(:,g));
18      detJ=det(J);
19      for p=1:4
20        Nv(2*p-1,1)=N(p,g);
21        Nv(2*p,2)=N(p,g);
22      end
23      me=me+Nv*Nv'*detJ*rho*gauss_weights(g);
24    end
25    M= AssembleGlobalMatrix(M,me,node_id_map,2); % assemble global M
26  end
```

```
1  % Compute global force vector
2  function F=CompF(nodes, elements, thickness, bcsforce)
3  F=zeros(size(nodes,1)*(size(nodes,2)-1),1);
4  n_force_nodes = size(bcsforce,1);
5  % for-loop: apply point forces
6  for i=1:n_force_nodes
7    row=2*bcsforce(i,1)-1;
8    F(row,1) = F(row,1)+ bcsforce(i,2) / thickness;
9    F(row+1,1) = F(row+1,1)+ bcsforce(i,3) / thickness;
10 end
```

```
1  % Time integration using Newmark scheme Trapezoid rule
2  function [U,V,A]=Newmark(t,dt,beta,gamma,dim,K,M,C,F,bcsdisp,penalty,
       U,V,A)
3  n_nodes=size(F,1)/dim;
4  % get u,v,a vectors
5  u1= U(:,t-1);
6  vel1 = V(:,t-1);
7  accel1 = A(:,t-1);
8
9  % next 3 lines: compute the LHS matrix and RHS vector
10 LHS = M*(1.0/(beta*dt*dt))+ C*(gamma/(beta*dt)) + K;
11 rhsvec = u1 + vel1*dt + accel1*((.5-beta)*dt*dt);
12 RHS = F + M*(rhsvec*(1.0/(beta*dt*dt))) + C*(rhsvec *...
13      (gamma/(beta*dt)) - vel1 - accel1*(dt*(1.0-gamma)));
14
15 % for-loop: apply displacement BC using the penalty method
```

```
16    for j=1:size(bcsdisp,1);
17        nid=bcsdisp(j,1);
18        k=bcsdisp(j,2);
19        RHS(dim*(nid−1)+k,1) =bcsdisp(j,3)*penalty;
20        LHS(dim*(nid−1)+k,dim*(nid−1)+k)=penalty;
21    end
22
23    U(:,t)=LHS\RHS;  % solve for the displacement
24    % next two lines : calculated acceleration and velocity
25    A(:,t) = (U(:,t)− U(:,t−1))/(beta*dt*dt)...
26            −V(:,t−1)/(beta*dt) −A(:,t−1)*(0.5−beta)/beta;
27    V(:,t) =V(:,t−1) + A(:,t−1)*(1.0−gamma)*dt + A(:,t)*dt*gamma;
```

In the MATLAB programs listed above, the functions which have already been discussed previously are not repeated here. They are listed below for your reference:

function K=CompK(nodes, elements, materials): Section 7.2.3.
function B= CompB4x8Quad4atPoint(Nxi_vec, Neta_vec): Section 7.2.3.
function H=CompH(): Section 7.2.3.
function C= CompCPlaneStress(materials): Section 7.2.3.
function [gauss_points, gauss_weights]=GetQuadGauss(rows, cols): Section 5.2.1.
function [N,Nx,Ny]=CompNDNatPointsQuad4(xi_vector, eta_vector): Section 5.2.1.
function [element_nodes, node_id_map]= SetElementNodes(ele, nodes, elements): Section 5.2.1.
function J=CompJacobian2DatPoint(element_nodes, Nxi, Neta): Section 5.2.1.
function K=AssembleGlobalMatrix(K, ke, node_id_map, ndDOF): Section 5.2.1.
function F=AssembleGlobalVector(F, fe, node_id_map, ndDOF): Section 5.2.1.

9.4 Summary

In this chapter, FEA procedures for transient heat transfer and structural dynamic analysis are presented by solving the example problems step by step. It is demonstrated that, compared to the steady state and static problems, the major new development for initial value problems lies in the time integration schemes. For transient heat transfer analysis, forward, backward, and central difference schemes are discussed. For elastodynamics analysis, the Newmark family time integration schemes are introduced. The characteristics of the time integration schemes are described and compared. MATLAB codes for solving these problems are presented.

Upon completion of this chapter, you should be able to:

- understand the fundamental theory of transient heat transfer and derive the differential governing equations
- derive the weak form of transient heat transfer by using the Galerkin weighted residual method
- derive the expressions of element matrices and vectors from the weak form for transient heat transfer problems
- understand the forward, backward, and central difference schemes for time integration in transient heat transfer analysis
- implement a complete MATLAB program to perform FEA of transient heat transfer problems. At this point, you should be able to implement MATLAB programs for solving 1-D, 2-D, and 3-D transient heat transfer problems
- understand the fundamental theory of structural dynamics and derive the differential governing equation of motion
- derive the weak form for elastodynamics by using the principle of virtual work
- derive the expressions of element matrices and vectors from the elastodynamics weak form
- understand the Newmark family time integration schemes for time integration in elastodynamics analysis
- implement a complete MATLAB program to perform FEA of elastodynamics problems. Similar to the transient heat transfer case, at this point, you should be able to implement MATLAB programs for solving 1-D, 2-D, and 3-D elastodynamic problems
- analyze the validity and accuracy of the numerical results.

9.5 Problems

9.1 A solid cylinder of radius R and temperature $T_0 = 100\,°C$ is placed in a medium of temperature $T_\infty = 0\,°C$. The governing equation of transient heat transfer in a solid cylinder is given by

$$\rho c \frac{\partial T}{\partial t} - \frac{1}{r}\frac{\partial}{\partial r}\left(r\kappa \frac{\partial T}{\partial r}\right) = 0.$$

The boundary conditions for this problem are

$$\frac{\partial T}{\partial r}(0,t) = 0, \qquad \left(r\kappa \frac{\partial T}{\partial r} + \beta T\right)\bigg|_{r=R} = 0$$

where κ, c, β is the thermal conductivity, specific heat, and convection heat transfer coefficient, respectively. Assuming $R = 2$ cm, $\kappa = 150$ W/m·°C, $c = 1.0$ kJ/kg·°C, $\beta = 300$ W/m·°C, $\rho = 2700$ kg/m^2, determine the temperature distribution along the radial direction as a function of time.

(a) Derive the weak form.
(b) Derive the expressions of element matrix and vector.
(c) Describe the time marching process using central difference scheme.

(d) Implement a MATLAB program using linear elements.
(e) Compute the temperature distributions on $r = [0, R]$ at $t = 1, 5, 20$ s, and plot the distributions on a single figure for comparison.

9.2 A series of heating cables have been placed in a conducting medium, as shown in Fig. 9.13. The medium has a conductivity of $k = 10$ W/cm·°C, a mass density of 5 g/cm^3, and a specific heat of 1 J/g °C . Each cable has a radius of 0.3 cm and generates heat at a rate of 2000 W/cm^3. The lower surface is bounded by an insulating medium. The upper surface of the conducting medium is exposed to the environment of a temperature of -5 °C and the initial temperature of the conducting medium is -5 °C too. Take the convection coefficient between the medium and the upper surface to be 5 W/(cm^2· C). Write a MATLAB code to compute the temperature variation as a function of time in the conducting medium. Use any symmetry available in the problem. Note: (1) the insulation boundary condition implies the heat flux= 0; (2) the temperature field is symmetric about the vertical lines in the middle of any two heating cables, which implies heat flux =0 on these vertical lines.

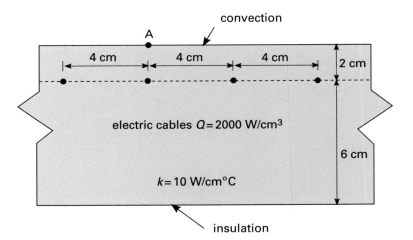

Figure 9.13 Conducting medium for Problem 9.2.

Solve the transient problem for the time period of 0–100 s using a proper mesh. Note that the cables are thin and can be treated as points in the 2-D domain. Use a uniform mesh of 0.25 cm × 0.25 cm linear square elements. Generate a time history plot of the temperature at point A on the top surface of the conducting medium. Point A is located right above a heating cable.

9.3 Compute the consistent and lumped mass matrices of the 2-D elements shown in Fig. 9.14. The nodes and their coordinates (unit: cm) are indicated in the figure. Assume the mass density $\rho = 2330$ kg/m^3.

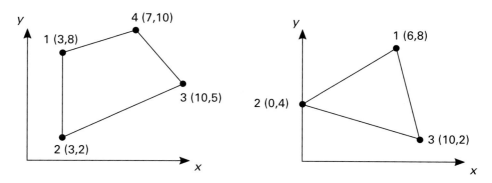

Figure 9.14 2-D elements for Problem 9.3.

9.4 Compute the consistent and lumped mass matrices of the two 3-D elements shown in Fig. 9.15. The nodes and their coordinates (unit: cm) are indicated in the figure. Assume the mass density $\rho = 2330$ kg/m^3.

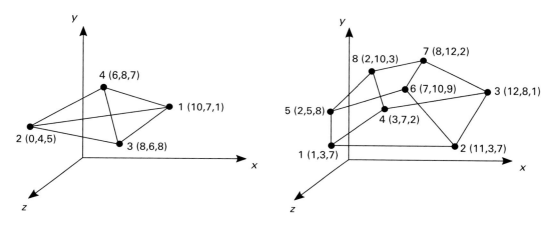

Figure 9.15 3-D elements for Problem 9.4.

9.5 Modify the MATLAB code listed in Section 9.3.2 and obtain the solution of the example problem of Section 9.3 by using lumped mass matrices. Keep the time marching scheme and all the other parameters the same as were used in the section.

9.6 Implement the explicit central differencing scheme and obtain the solution of the example problem of Section 9.3. Experiment with the time steps until a convergence solution is obtained.

9.7 The governing partial differential equation of the axial vibration of a 1-D rod is given by

$$-EA\frac{\partial^2 u}{\partial x^2} + \rho A\frac{\partial^2 u}{\partial t^2} = 0. \qquad 0 \leq x \leq L$$

The rod has a length of 1 m and is fixed at one end. The constants are: Young's modulus $E = 200$ GPa, Poisson's ratio, $\nu = 0.3$, mass density $\rho = 7800$ kg/m^3, cross sectional area $A = 0.01$ m^2.

(a) Write a MATLAB code and calculate the first three natural frequencies of the rod and their associated vibrational modes.

(b) Assuming there is a concentrated mass (volume of the mass is negligible) of 200 kg attached to the free end of the rod, calculate the first three natural frequencies of the rod and their associated vibrational modes by using a lumped mass matrix.

9.8 An "F" shaped structure is fixed at the bottom as shown in Fig. 9.16. The dimensions of the structure are shown in the figure. Young's modulus of the structure is 1.2×10^{10} Pa, Poisson's ratio is 0.25, and the mass density of the structure is 2330 kg/m^3. Write an FEA code to calculate the lowest four frequencies and corresponding mode shapes of the structure, refine your mesh and compare the results obtained from different meshes, and discuss the physical meaning of your FEA results.

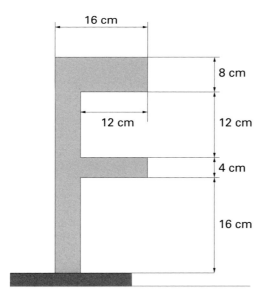

Figure 9.16 "F" shaped structure for Problem 9.8.

9.9 Using beam elements, calculate the first three natural frequencies and associated vibrational modes of the Howe scissors roof frame structure shown in Fig. 9.17. The cross sectional area of the rods is 10 mm^2. Young's modulus of the material is 2×10^9 Pa. Mass density of the beam is 2330 kg/m^3.

9.10 Repeat Problem 9.9 with the two leg tips of the frame fixed to a rigid support. Compare the results obtained using the two types of support (roller support and fixed) and explain the difference.

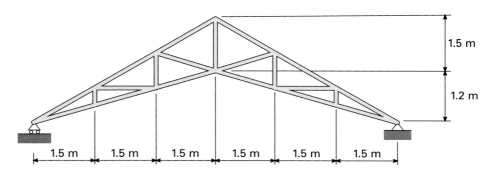

Figure 9.17 Roof frame for Problem 9.9.

9.11 Calculate the first three natural frequencies and associated vibrational modes of the truss system shown in Fig. 8.1.

9.12 Calculate the first three natural frequencies and associated vibrational modes of the space frame structure shown in Fig. 8.7.

9.13 A uniform bar is fixed at one end and subjected to a force $f(t) = f_0$ at the other, as in Fig. 9.18. The axial displacement is governed by the wave equation

$$\frac{\partial^2 u}{\partial^2 t} = \frac{E}{m} \frac{\partial^2 u}{\partial x^2}$$

with the boundary condition

$$u(x,t)|_{x=0} = 0$$

and initial condition

$$\sigma(x,t)|_{(t=0,\ x=L)} = \frac{f_0}{A}$$

where E, m, A, L are the Young's modulus, mass density and cross sectional area, and the length of the bar, respectively. An approximate solution of the problem is

$$u(x,t) = \frac{f_0 x}{AE}\left(1 - \cos\left(\sqrt{\frac{3E}{m}}\frac{t}{L}\right)\right).$$

(a) Assuming $f_0 = 100$ N, $A = 0.01 \text{m}^2$, $L = 1$ m, E is 70 GPa and total mass is 50 kg, write an FEA code with beam elements to compute the axial displacement and stress of the bar.
(b) Refine your mesh to obtain a converged result.
(c) Compare the numerical results with the analytical results and discuss your observations.

9.14 A vehicle is moving over a steel beam at a speed of 50 km/hr as in Fig. 9.19. Model the system as a 2-D problem. Write a computer code and compute the deformation and the normal stress in the longitudinal direction of the beam as a function

Figure 9.18 Problem 9.13.

of time. The material properties of the beam are: Young's modulus $E = 200$ GPa, Poisson's ratio, $\nu = 0.3$, mass density $\rho = 7800$ kg/m^3.

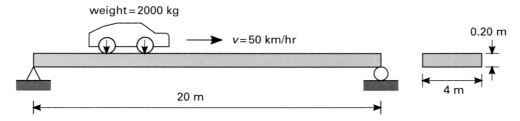

Figure 9.19 Vehicle moving on steel beam for Problem 9.14.

Index

C^n continuity, 95

abscissas, 69
admissible function, 37
advection, 139
advection–diffusion equation, 140
approximation, 52
area coordinates, 170
assembly process, 104, 206, 354, 385

basis functions
   Lagrange, 57
   monomial, 53, 168
beam, 419
bilinear functional, 41
bilinear shape function, 163
Boolean operation, 261
boundary
   integration, 349
boundary conditions, 88
   direct substitution method, 107
   essential, 107
   penalty method, 109
boundary representation (B-rep), 261
bulk modulus, 325

Cauchy stress, 321
circumcenter, 269
circumcircle, 268, 276
circumradius, 276
coefficient of convective heat transfer, 127
compatibility, 240
completeness, 240
conductivity matrix, 193
connectivity, 104
conservation
   linear momentum, 379
   mass, 140, 380
constitutive law, 324
constrained Delaunay triangulation, 275
convection, 126
convection matrix, 193

convergence, 240, 242
   monotonic, 240
   rate, 240
convex quadrilateral, 316
copy-on-write, 213
CSG tree, 261

damping
   critical, 482
   damping coefficient, 483, 494
   damping matrix, 494
   ratio, 482
data structure, 284
Delaunay refinement, 275
Delaunay triangle, 269
Delaunay triangulation, 269
diametral circle, 277
differential equation of equilibrium, 322
differential equation of motion, 323
differential operator, 42
diffusion, 140
diffusivity, 142
Dirac delta function, 332
direct method, 410
divergence theorem, 36
divide-and-conquer, 270
dual graph, 269

elastic constants, 324
elasticity, 320
elastodynamics, 323, 480
elastostatics, 322, 330, 364
element
   1-D, 92
   1-D bar, 407
   1-D beam, 424
   1-D linear, 93
   1-D master, 118
   1-D quadratic, 93
   1-D spring, 407
   2-D, 162, 163
   2-D Lagrange type, 163

2-D Serendipity type, 165
2-D bar, 408
2-D master, 175
2-D plate, 446
2-D rectangular, 162
2-D transition, 166
2-D triangular, 167
3-D, 220
3-D bar, 409
3-D beam, 426
3-D hexahedral, 224
3-D prismatic, 229
3-D pyramidal, 230
3-D tetrahedral, 221
3-D master, 222
non-conforming, 446
quality, 244
element radii ratio, 245
element weak form, 187
elliptic partial differential equation, 246
encroached edge, 275
energy
　conservation, 126, 156, 464
　external, 89
　internal, 89
　potential, 89
equation of motion, 323, 486
Euler equation, 39
Euler–Bernoulli beam equation, 420

Fick's first law, 141
flow
　cavity driven, 378
　incompressible viscous, 378, 380
Fourier series, 461
Fourier's law, 127
free body diagram, 87
frequency
　fundamental frequency, 487
　natural frequency, 487
functional, 38

Galerkin, 90
Gauss points, 73
Gauss' theorem, 36
general procedure, 85
generalized Hooke's law, 324
global
　force vector, 109
　node index, 104
　stiffness matrix, 109
Green's first identity, 160
Green's formula, 36

half-space, 261
heat flux, 127

heat source, 127
heat transfer, 126
Hooke's law, 88
Howe scissors roof, 457, 459, 503

in-circle test, 295
in-place modification, 213
incompressible flow, 380
infinitesimal element, 127
initial
　conditions, 464, 486
　displacement, 486
　velocity, 486
initial value problems, 463
integration by parts, 34
interpolation, 52
　Hermite, 61
　Lagrange, 53
　piecewise polynomial, 64, 95
isoparametric mapping, 118

Jacobian, 121, 182, 233, 368
joint, 405

Kronecker delta, 172

Lagrange multipliers, 356
Lagrange type elements, 163
Laplace equation, 217
linear system of equations, 78
　direct methods, 78
　homogeneous, 77
　iterative methods, 78
　LU factorization, 78
　non-homogeneous, 77
　over-determined, 77
　UMFPACK, 78
　under-determined, 77
　unsymmetric multifrontal method, 78
local
　node index, 104

mass matrix
　consistent, 492
　lumped, 493
mass transport, 140
material coordinates, 379
matrix, 18
　band, 21
　condition number, 29
　determinant, 23
　diagonal, 19
　eigenvalues, 24
　eigenvectors, 24
　identity, 19
　integration, 27

inversion, 23
norms, 28
rotation, 408
singular, 77
skew, 20
sparse, 212
symmetric, 20
transpose, 19
triangular, 20
Vandermonde, 56
memory management, 212
mesh
  1-D, 92
  2-D, 160
  advancing front, 265
  Delaunay refinement, 275
  Delaunay triangulation, 269
  structured, 263
  uniform, 262
  unstructured, 264
mesh convergence, 240
modulus of elasticity, 88
monomial basis, 168
monotonic convergence, 240

Navier–Stokes equations, 139
Newmark schemes, 489
  a-form, 490
  central difference, 490
  d-form, 490
  explicit, 490
  implicit, 487
Newtonian fluid, 139
nodal displacement, 92
nodal temperature, 129
nodes, 92
non-uniform isoparametric mapping, 121
normal strain, 324
numerical differentiation
  finite difference, 66
numerical integration, 69
  Gauss quadrature, 73, 196, 223, 226, 231
  Newton–Cotes rules, 72
  Simpson's rule, 72
  Trapezoid rule, 71

order of error, 240
orthotropic material, 325

Pascal's triangle, 242
patch test, 243
penalty number, 109
penalty parameter, 110
planar straight line graph, 266
plane strain, 326
plane stress, 328

plate, 438
  Kirchhoff–Love, 439
point force, 331
Poisson's equation, 247
post-processing, 85

quadrature points, 73
queue, 284
quiver plot, 387

rate of convergence, 240
reference coordinate system, 118
rotation matrix, 408
Runge phenomenon, 61

scalar field
  gradient, 30
  Laplacian, 32
serendipity type elements, 165
shape function
  1-D, 94
  2-D quadrilateral master, 180
  2-D rectangular, 162
  2-D triangular, 167
  2-D triangular master, 181
  3-D hexahedral, 224
  3-D prismatic, 229
  3-D pyramidal, 230
  3-D tetrahedral, 221
shear modulus, 325
shear strain, 324
solid modeling, 261
space frame, 419
spatial coordinates, 379
specific heat, 127
spring, 406
spring–mass–dashpot, 481
stable
  conditional, 473, 489
  unconditional, 473, 489
stationary condition, 38
steady state heat transfer, 126
Stokes flow, 381
strain energy, 339
strain–displacement matrix, 343, 367
strains, 323
strong form, 40, 88, 158
support reactions, 414
surface traction, 321
surface traction vector, 370

Taylor series expansion, 33, 66
test function, 91
thermal conductivity, 128
time derivative, 467
time derivative discretization

    backward difference, 469
    central difference, 471
    Crank–Nicolson, 471
    forward difference, 471
transient heat transfer, 463
transition elements, 166
trial function, 91
truss, 405
two force members, 405

uniform isoparametric mapping, 121

variational problem, 36
vector, 16
    norms, 27
    scalar product, 16
    vector product, 17
vector field, 31
    divergence, 31
    gradient, 31
vertex angle, 245
vibration modes, 487
virtual displacement, 90
virtual strain, 90
virtual work, 90, 338
Voronoi diagram, 270

wave equation, 504
weak form, 40, 88
weighted residual, 90

Young's modulus, 88, 325